ROCK FRAGMENTATION BY BLASTING

PROCEEDINGS OF THE FIFTH INTERNATIONAL SYMPOSIUM
ON ROCK FRAGMENTATION BY BLASTING – FRAGBLAST-5
MONTREAL/QUEBEC/CANADA/25-29 AUGUST 1996

Rock Fragmentation by Blasting

Edited by
B. MOHANTY
ICI Explosives Canada, McMasterville, Québec, Canada
McGill University, Montréal, Québec, Canada

Taylor & Francis
Taylor & Francis Group
LONDON AND NEW YORK

Photo cover: Blasting in Open Pit Gold Mine (© ICI Explosives)

The texts of the various papers in this volume were set individually by typists under the supervision of each of the authors concerned.

Published by Taylor & Francis
2 Park Square, Milton Park, Abingdon, Oxon, OX14 4RN
270 Madison Ave, New York NY 10016

Transferred to Digital Printing 2007

ISBN 90 5410 824 X
© 1996 Taylor & Francis

Publisher's Note
The publisher has gone to great lengths to ensure the quality of this
reprint but points out that some imperfections in the original
may be apparent

Rock Fragmentation by Blasting, Mohanty (ed.) © 1996 Taylor & Francis. ISBN 90 5410 824 X

Table of contents

Blast induced fractures and rock damage

Fracture control blasts

Rock Fragmentation by Blasting, Mohanty (ed.) © 1996 Taylor & Francis. ISBN 90 5410 824 X

Preface

This series of symposia has grown in scope and immediacy since the first one held in Luleå, Sweden in 1983. This has been in keeping with the advances made in the science and technology of rock fragmentation over that period. The mandate of this 5th Symposium, like its predecessors, is to bring together the leading researchers and the practitioners in rock fragmentation around the world. It will provide a forum for re-assessment, discussion and dissemination of the latest theory and practice on the subject.

As the papers (and the abstracts of the Poster Session) contained in this volume show, the symposium has fulfilled its mandate admirably. It covers a very wide range of topics on rock fragmentation, from carefully documented case studies to attempts, for example, at fractal representation of the fracture process itself. The blast modelling efforts, significantly shorn of some of the earlier empiricism and arbitrariness, are increasingly aimed at tackling realistic blasting variables. This is being supported by an improved understanding of the detonation behaviour of commercial explosives, and advanced measurement techniques in the field, especially in assessing size distribution in large volumes of fragmented rock mass. The reliability and reproducibility of these techniques in actual mine and quarry environments continue to pose significant challenges.

Specific blasting techniques such as wall-control blasts are being examined in much greater detail due to their increasing use in both open-pit and underground operations. Limiting blast-induced damage and its measurement and control by means of near-field explosion seismology and measurement of transmitted explosion gas pressures are becoming central issues in maintaining efficient blasting operations. The fragmentation process is being viewed from both the simpler energy balance (comminution) as well as the more complex fracture micromechanics points of view. Understanding the creation and localization of micro-cracks and their coalescence and propagation to form macro-cracks and ultimately, the stockpile, constitutes very active areas of research. However, despite the need to characterize rock mass with its inherent lithological and structural discontinuities, in terms of their response to blasting action, work in this area seriously lags those in the other. All these represent formidable scientific and technical challenges, and will no doubt, provide rich sustenance for future Fragblast symposia for years to come.

The editor gratefully acknowledges the dedication and commitment of his colleagues in the Canadian Organizing Committee for FRAGBLAST'5, and the support of the International Organizing Committee for making the symposium a success.

B. Mohanty
Montréal, August 1996

Rock Fragmentation by Blasting, Mohanty (ed.)© 1996 Taylor & Francis. ISBN 90 5410 824 X

Organization

INTERNATIONAL COMMITTEE

S. Bhandari, J. N. V. University, Jodhpur, India
C. V. B. Cunningham, AECI Explosives, South Africa
W. L. Fourney, University of Maryland, USA
R. Holmberg, Gytorp, Dyno-Nobel, Sweden
C. McKenzie, Australian Blasting Consultants, Australia
B. Mohanty, ICI Explosives, McMasterville, Québec, Canada
P. A. Persson, New Mexico Tech. Socorro, USA
H. P. Rossmanith, Technical University of Vienna, Austria
A. Rustan, Luleå University of Technology, Sweden

CANADIAN ORGANIZING COMMITTEE

P. P. Andrieux, Noranda Technology Centre, Montréal
W. Comeau, McGill University, Montréal
R. Elliot, Dyno-Nobel, Toronto
F. P. Hassani, McGill University, Montréal
P. D. Katsabanis, Queen's University, Kingston
W. Lidkea, INCO, Sudbury
Y. C. Lizotte, CANMET, Val D'Or
B. Mohanty, ICI Explosives, McMasterville
N. Procyshyn, McGill University, Montréal
M. J. Scoble, McGill University, Montréal
A. Cameron, Golder Associates, Sudbury
C. G. Watson, ICI Explosives, Toronto

Rock Fragmentation by Blasting, Mohanty (ed.) © 1996 Taylor & Francis. ISBN 90 5410 824 X

Sponsors

The Canadian Organizing Committee gratefully acknowledges the generous support provided by the following organizations:

ICI EXPLOSIVES CANADA

BARRICK

CAMBIOR

McGill
175th anniversary

INCO

Fracture dynamics

Rock Fragmentation by Blasting, Mohanty (ed.) © 1996 Taylor & Francis. ISBN 90 5410 824 X

Fracture characteristics and micromechanical theory of rock as a quasibrittle material: Apperçu of recent advances

Zdeněk P. Bažant
Department of Civil Engineering, Northwestern University, Evanston, Ill., USA

Pere C. Prat
Department of Geotechnical Engineering, Technical University of Catalunya, Barcelona, Spain

ABSTRACT: Realistic characterization of rock fracture properties is essential for successful predictions of rock fracture under any situation, including the fragmentation in blasting. The paper presents a review of some interesting recent results on the characterization of rock fracture by point-wise, line-wise and diffuse models for the fracture process zone. The effect of the rate of loading and of crack growth on the fracture characteristics is also reviewed and the micromechanical aspects are discussed. Attention is then focused on the problem of the effect of microcracks on the global stiffness tensor of a microcracked material such as rock, and some new results which take into account the growth of cracks retaining their criticality are described. The lecture is documented by numerical results.

1 INTRODUCTION

Although no-tension models for rock failure are useful for some geotechnical engineering problems, analysis of rock failure requires fracture mechanics (Bažant, 1996). Proper characterization of the fracture properties of rock is a necessary tool for dependable predictions of rock failure, including rock fragmentation by blasting. The fracture behavior of rock is complicated by the fact that rock is a quasibrittle material, that is, a material in which the fracture tip is normally not sharp but is surrounded by a sizeable fracture process zone. This is best manifested by the fact that the size effect on the nominal strength of geometrically similar rock specimens does not follow the size effect of linear elastic fracture mechanics (LEFM) but exhibits a transitional size effect between plasticity and LEFM.

Various new mathematical formulations to describe the quasibrittle behavior of rock, as well as other materials such as concrete, ice, toughened ceramics and various composites, have recently been developed, and factors that significantly affect the quasibrittle response, such as the rate of loading and of fracture growth, have been intensely studied. To gain understanding of the fracture mechanism in a diffuse fracture process zone of a quasibrittle material, some interesting micromechanical models have been formulated.

The purpose of the present lecture is to provide an apperçu of recent advances in this subject, with a focus on the results recently obtained at Northwestern University. In addition, one new result, namely the extension of the theory of elastic constants of a randomly and uniformly microcracked elastic material to tangential stiffness calculation of a material with growing microcracks, will be presented in detail. It must be warned that the intent of the review that follows is not a broad coverage of all the results in the literature but an exposition selectively highlighting only some recent contributions.

2 REVIEW OF RECENT RESULTS

The fracture properties of a quasibrittle material can be described in three ways: (1) in a point-wise fashion, in which the energy dissipation in the fracture process zone is lumped into a point, the crack tip, (2) line-wise fashion, in which this energy dissipation is represented by stress-displacement relations for a cohesive crack, and (3) in a multi-dimensional diffuse fashion, representing a continuous smearing of the discrete microcracking and other inelastic phenomena such a frictional slip and fragment pull out in the fracture process zone.

2.1 *Point-wise characterization of fracture*

2.1.1 Size effect method and determination of R-curve

The size effect of quasibrittle fracture (Bažant, 1984b; Bažant, 1984a) can be effectively exploited to identify quasibrittle nonlinear fracture characteristic solely from measured maximum loads of geometrically similar specimens of sufficiently different sizes (Bažant and Pfeiffer, 1987; Bažant and Kazemi, 1990a; Bažant and Kazemi, 1990b; Bažant et al., 1991). The size effect is due to difference in scaling of energy release and energy dissipation of large fractures (and not to a possible fractal character of fracture (Bažant and Li, 1995)). The size effect law represents an asymptotic matching between the large size and small size asymptotic expansions of the size effect in quasibrittle fracture (Bažant and Li, 1995). By fitting the test results for specimens of different sizes, one can obtain size-independent as well as shape-independent values of the effective fracture energy, the effective length of the process zone, and the effective critical crack-tip opening displacement. Using the size effect law, one can also easily calculate the R-curve (resistance curve), which turns out to be geometry and shape dependent.

The main advantage of the size effect method is its simplicity and unambiguity. Assuming that specimens of positive geometry are used, measurements of the crack-tip locations, which are notoriously difficult in the case of a microfracturing material such as rock, are not necessary. Only the maximum loads of specimens of different sizes need to be measured, which means that a stiff machine and fast servo-control are not needed.

To eliminate the need for producing specimens of different sizes, the size effect method has recently been modified in a way which allows using specimens of only one size but with notches of different lengths, so as to provide a sufficient range of brittleness number (Bažant and Li, 1996b). This method uses also the value of the flexural strength of unnotched specimens. It is based on a universal size effect law — a generalization of size effect law that provides the transition from failures at large cracks to failures at no crack (Bažant, 1995), as recently derived by asymptotic analysis.

Extensive studies have demonstrated that the size effect method provides a simple yet realistic representation of the results on fracture properties of various types of rock (Takahashi, 1988; Schmidt and Lutz, 1979; Carpinteri, 1980; Schmidt, 1977; Hoagland et al., 1973; Bažant et al., 1991; Labuz et al., 1985).

2.1.2 Rate influence

The R-curve model, characterizing the variation of the critical energy release rate with the crack propagation length, has been generalized to describe both the rate effect and size effect observed in rock or other quasibrittle materials (Bažant and Jirásek, 1993). In this generalization, it is assumed that the crack propagation velocity depends on the ratio of the stress intensity factor to its critical value based on the R-curve, and that this dependence has the form of a power function with an exponent much larger than 1. The shape of the R-curve is determined, taking into account the specimen geometry, as the envelope of the fracture equilibrium curves corresponding to the maximum load values for geometrically similar specimens of different sizes. The formulation also allows taking into account the creep in the bulk of the specimen, however, this is not important for rock in most cases. Good representations of test data have been demonstrated for rock as well as concrete (in the case of concrete, the creep in the bulk is also very important, and causes a change of brittleness).

The tests which were used to calibrate the rate-dependent R-curve model used limestone specimens subjected to loading rates ranging over four orders of magnitude of the loading rate (Bažant et al., 1993). The loading rates were all in the static regime of response of the specimen, which simplified evaluation (the times to the peak load of specimens ranged from 2 seconds to 83,000 seconds). It is likely that the results can be extrapolated to dynamic loading rates in which the inertia and wave propagation effects are important. These tests utilized the size effect method with three specimen sizes.

A phenomenon of considerable interest for dynamic fracture analysis as recent tests have shown (Bažant and Gettu, 1992) is that a sudden increase of the loading rate causes a reversal of softening response to hardening response followed by a second peak, while a sudden decrease of the loading rate causes a decrease of the softening slope. These characteristics are described by the rate generalization of the R-curve model quite well (Bažant and Jirásek, 1993).

2.2 Line-wise characterization of fracture

A more realistic model for fracture is the cohesive crack model, in which the fracture properties are characterized by the dependence of the cohesive (crack-bridging) stress on the crack opening displacement. For concrete this model is also known as the fictitious crack model, pioneered by Hillerborg et al. (Hillerborg et al., 1976), and is based on the original ideas of Barenblatt (Barenblatt, 1962) and Dugdale (Dugdale, 1960).

This model has recently been formulated on the basis of energy principles and variational equations useful for numerical solutions have been presented (Bažant and Li, 1995). Using the energy formulations, the conditions of stability loss of a specimen or structure with a growing cohesive crack have been obtained from the condition of vanishing of the second variation of the complementary energy or the potential energy. They were found to have the form of a homogeneous Fredholm integral equation for the derivatives of the cohesive stress or crack opening displacement with respect to the crack length. Based on this equation, the criterion of stability limit of geometrically similar specimens or structures of different sizes was transformed to an eigenvalue problem, with the size of the specimen or structure as the eigenvalue. This formulation makes it possible to solve the size effect curve directly, without actually calculating the load-deflection curves of the specimens. One solves directly for the structure size for which a given relative crack length corresponds to stability loss, i.e., the maximum load. This formulation greatly simplifies studies of the size effect.

The cohesive (fictitious) crack model serves as the basis of the work-of-fracture method (Nakayama, 1965; Tatersall and Tappin, 1966) for measuring the fracture energy of rock or concrete. This method has recently been subjected to critical analysis which clarified its limitations (Bažant, 1987). It was shown that measuring the unloading compliance at a sufficient number of states on the post peak descending load-deflection curve, one can calculate the so-called pure fracture energy, representing the energy dissipated by the fracture process alone, excluding the energy dissipated by plastic frictional slips and fragment pull outs. However, this value of fracture energy is pertinent only if the material model, consisting of a fracture law and a constitutive law for structural analysis, takes into account separately the fracture-damage deformations and the plastic-frictional deformations. Otherwise, one must use the conventional fracture energy, which includes all the plastic-frictional energy dissipation in the fracture process zone. Either type of fracture energy should properly be determined by extrapolating the results of the work of fracture method to infinite specimen size, or else unambiguous, size-independent results cannot be obtained.

It was also shown that the work of fracture method can be improved by averaging the work done by fracture over only a central portion of the ligament. Other valuable advances were recently reported by Elices and Planas (1989), Elices et al. (1992), Hu and Wittmann (1991), and Planas et al. (1992).

For numerical purposes, the crack band model developed by Bažant (1982,1984a) and Bažant and Oh (1983, 1984) is normally more convenient in the finite element context. This model is essentially equivalent to the cohesive (fictitious) crack model and yields nearly the same results. The idea of the model is to replace a line crack with a crack band of a fixed width considered as a material property and assume a uniform distribution of cracking strain across the crack band such that the accumulated cracking strain be equal to the opening of the cohesive crack model.

2.2.1 Rate influence on cohesive crack model

Application of the cohesive crack model to dynamic fracture requires incorporating the time dependence, particularly the effect of loading rate and the rate of crack growth. It was shown (Bažant and Li, 1996a) that the rate effect can be realistically described by introducing a rate dependent softening law between the cohesive stress and the crack opening displacement (Bažant and Li, 1996a). The proper form of this softening law has been derived from the activation energy theory of the rate process of bond ruptures on the atomic level (Bažant, 1993).

It was shown that the phenomenon that causes the sudden reversal of softening to hardening when the loading rate is suddenly increased is the rate dependence of the softening stress-displacement law. The effect of viscoelasticity in the specimen or structure has also been incorporated into the cohesive crack model, but this is of little importance for rock.

2.3 *Multidimensional diffuse characterization of fracture and micromechanics*

Several new results will now be described, which are, however, limited in their present form to static failure of rock. Further generalization will be required to extend them to dynamic crack growth.

2.3.1 Effect of microcracks and their interactions on macroscopic stiffness

Ideally, for the purpose of structural analysis, fracture must be characterized on the continuum level. In the case of a quasibrittle material with a diffuse fracture process zone, this means that the model for the fracture process zone should be a multidimensional continuum which incorporates microcracking (as well as plastic-frictional phenomena) in a smeared manner. This involves two problems: (1) determination of the stiffness properties, including softening, under the assumption that the microcracking is distributed uniformly throughout a material element, which can be done by a constitutive stress-strain tensorial relation, and (2) interactions between microcracks which control the localization of damage due to microcracking. The second problem is at present the most challenging one and is the object of intense debates. Some new results on the first problem will be presented in detail in section 3 of the present paper.

Recently, a nonlocal continuum model for strain-softening was derived by micromechanical analysis of a macroscopically non-homogeneous (nonuniform) system of interacting and growing microcracks. Kachanov's version of the superposition method (Kachanov, 1985; Kachanov, 1987a) was used as the point of departure. Homogenization (or smearing) of the microcrack system was achieved by seeking a continuum field equation whose possible discrete approximation coincides with a matrix equation governing a system of interacting microcracks.

The result of this homogenization was a Fredholm integral equation for the unknown non-local inelastic stress increments, which involved two spatial integrals. One integral, which ensues from the fact that crack interactions are governed by the average stress over the crack length rather than the crack center stress, represents short-range averaging of inelastic macrostresses. The kernel of the second integral is the long-range crack influence function, which is a second-rank tensor and varies with directional angle (i.e., is anisotropic), exhibiting sectors of crack shielding and crack amplification. For long distances r, the weight function decays as r^{-2} in two dimensions and as r^{-3} in three dimensions.

Application of the Gauss-Seidel iteration method which can conveniently be combined with iteration in each loading step of a nonlinear finite element code, simplifies the handling of the nonlocality by allowing the nonlocal inelastic stress increments to be calculated from the local ones explicitly. This involves evaluation of an integral involving the crack influence function for which closed-form expressions are derived (for three dimensions they are based on the recent results of Fabrikant, 1990). Because the constitutive law for the microcracked material is strictly local, no difficulties arise with the unloading criterion or the continuity condition of plasticity. This micromechanical theory puts the previously proposed phenomenological nonlocal models for strain softening damage on a solid footing.

2.3.2 Micromechanical model for compression crushing with microbuckling

Compression failure, also called crushing, requires lateral expansion of the damaged material in the directions normal to the maximum compressive stress. Fracture mechanics and micromechanics of this behavior is a rather difficult subject which has recently received considerable attention. One recently proposed model which appears to give a good description of compression failure including the size effect is based on the idea of transverse propagation of a band of axial splitting cracks and microbuckling of the microslabs between the axial splitting cracks. This model has first been used for explaining the size effect observed experimentally in the breakout of boreholes in rock (Nesetova and Lajtai, 1973; Haimson and Herrick, 1986; Haimson and Herrick, 1989; Carter et al., 1992; Carter, 1992; Martin et al., 1994; Dzik and Lajtai, 1994). The analysis involved calculations of the energy release from the surrounding rock mass by approximating the failure zone as elliptical and using the Eshelby theorem of elasticity. This analysis, as well as the subsequent analysis of lateral propagation of a band of axial splitting cracks indicated that the asymptotic effect of size on the nominal strength can be described by a power law in which the nominal strength is approximately proportional to the -2/5 power of the characteristic size (Bažant and Xiang, 1996).

2.3.3 Spacing of cracks at their initiation

The spacing of parallel cracks in rock or other materials is a matter of stability of crack system and bifurcation in its evolution. Under static loading, parallel cracks such as cooling cracks tend to evolve in such a manner that every other crack closes and the intermediate ones open more widely while advancing further (Bažant and Ohtsubo, 1978; Bažant et al., 1979; Bažant and Cedolin, 1991). However, the initial spacing of cracks when they initiate is a matter of fracture characteristics of the material.

The initiation of parallel cracks in a half space under static loading has recently been studied by Li, Hong and Bažant (1995). It was shown that the following three conditions govern the initial spacing and the initial equilibrium length of parallel cracks: (1) the energy release rate of the initial cracks must be equal to its critical value, (2) the total energy dissipated by the formation of the initial cracks must be equal to the total energy release as a result of the finite initial cracks, and (3) the stress before the appearance of the initial cracks must be equal to the tensile strength of the material. From these three conditions, one can calculate the initial length of the cracks, their initial spacing, and the load at which the cracks form.

3 NEW RESULTS ON THE STIFFNESS TENSOR OF A MATERIAL WITH GROWING MICROCRACKS

Microcracks in a brittle material affect its stiffness, strength and toughness. Their evolution is a mechanism of failure. Prediction of the response of structures made of damaging materials such as rock (as well as concrete, ice, ceramics or composites), requires modeling of the effect of microcracks on the macroscopic constitutive law. The basic problem is the effect that a crack system statistically uniform in space has on the overall elastic constants of the material.

In rock mechanics, careful attention must be paid to the families of cracks, preexisting or man-made. They can have a strong influence on the response of the rock during blasting operations in tunneling, excavation and mining. The distribution of these cracks within the rock mass is usually characterized by aligned patterns with some preferred crack orientations that render the body macroscopically anisotropic. The crack density can vary from the case of dilute (non-interacting) cracks to highly concentrated (heavily interacting) cracks depending on the geological history and loading. On crack systems in rock, one may distinguish cracks of at least two typical scales: (1) microcracks of the order of the grain size, and (2) intersecting families of rock joints. For the former, the material can be treated as a continuum on the scale over 1 ft, and for the latter on the scale over 1 km. The continuum treatment means that the effect of a crack system statistically uniform in space can be captured by the macroscopic stiffness tensor of a continuum.

The problem of calculating the macroscopic stiffness tensor of elastic materials intersected by various types of random or periodic crack systems has been systematically explored during the last two decades and effective methods such as the self-consistent scheme (Budiansky and O'Connell, 1976; Hoenig, 1978), the differential scheme (Roscoe, 1952; Hashin, 1988), or the Mori-Tanaka method (Mori and Tanaka, 1973; Benveniste, 1987) have been developed. A serious limitation of the current knowledge is that all the studies have dealt with cracks that neither propagate nor shorten (Fig. 1a). This means that, in the context of response of a material with growing damage illustrated by the curve in Fig. 1a, the existing formulations predict only the secant elastic moduli (such as E_s in Fig. 1a). Such information does not suffice for calculating the response of a body with progressing damage due to cracking.

To calculate the response of a material with cracks that can grow or shorten, it is also necessary to determine the tangential moduli, exemplified by E_t in Fig. 1b. Knowledge of such moduli makes it possible, for a given strain increment, to determine the inelastic stress drop $\Delta\sigma_{cr}$ (Fig. 1b). This problem will be addressed in this paper. For a more detailed derivation see (Prat and Bažant, 1995).

Knowledge of the secant and tangential moduli is of course still insufficient to predict the response of a structure with growing cracks. It is now well known that softening damage caused by cracking tends to localize into cracking bands or other regions. The localization of cracking is caused and governed mainly by interactions among propagating cracks. The interactions cause that the average behavior of a representative volume of the material with cracks does not follow the local stress-strain curve for growing cracks but follows a slope that is either smaller or larger, as shown in Fig. 1c. This problem has recently been analyzed and an integral equation in space governing the nonlocal behavior of such material has been formulated (Bažant, 1994; Bažant and Jirásek, 1994; Jirásek and

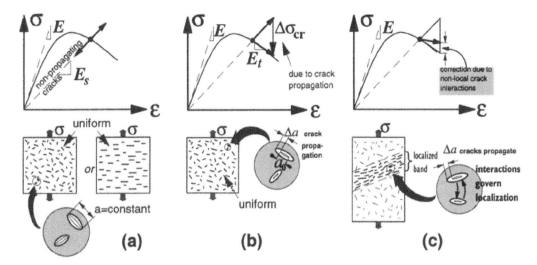

Figure 1. (a) Effective Secant Moduli; (b) Tangential moduli and stress increment due to crack growth; (c) Tangential stiffness of material with localizing cracks.

Bažant, 1994) on the basis of continuum smoothing of a matrix equation for crack interactions. However, a complete solution is beyond the scope of this paper.

3.1 Equivalent elastic modulus of a material with uniformly distributed cracks

We consider a representative volume V of an elastic material containing on the microscale many cracks (microcraks). On the macroscale, we imagine the cracks to be smeared and the material to be represented by an approximately equivalent homogeneous continuum whose local deformation within the representative volume can be considered approximately homogeneous over the distance of several average crack sizes. Let ϵ and σ be the average strain tensor and average stress tensor within this representative volume. To obtain a simple formulation, we consider only circular cracks of average radius $\bar{a}$.

To derive the tangential compliance tensor of the material on the macroscale, the cracks must be allowed to grow during the prescribed strain increment $\Delta\epsilon$. This means that the energy release rate per unit length of the front edge of one crack, $\mathcal{G}_1$, must be equal to its critical value, i.e., to the fracture energy G_f of the material. For the sake of simplicity, we will enforce the condition of criticality of cracks only in an overall (weak) sense, by assuming that the average overall energy release rate of all the cracks within the representative volume equals their combined energy dissipation rate.

Let n be the number of cracks per unit volume of the material with cracks, characterizing the representative volume. The surface area of one circular crack of radius a is $A = \pi a^2$ and its change when the crack radius increases by δa is $\delta A = 2\pi a\delta a$. We assume we can replace the actual

crack radii a by their average radius $\bar{a}$. The critical state of crack growth is obtained when the energy release rate per unit volume of the material, $\mathcal{G}_1$, equals the rate of energy dissipation by all the cracks in the unit volume, i.e.,

$$\frac{\partial\Pi^*}{\partial\bar{a}} = 2\pi\bar{a}nG_f \qquad (1)$$

where $\Pi^* = \Pi^*(\sigma, \bar{a}) = \frac{1}{2}\sigma : \mathbf{C}(\bar{a}) : \sigma$ is the complementary energy per unit volume and $\mathbf{C}(\bar{a})$ is the macroscopic secant compliance tensor of the material with the cracks.

Calling $G^* = 2\pi nG_f$, the following equation for $\Delta\bar{a}$ is obtained:

$$\sigma : \mathbf{C} : \Delta\sigma + \left[\frac{1}{2}\sigma : \frac{\partial\mathbf{C}}{\partial\bar{a}} : \sigma - G^*\bar{a}\right]\Delta\bar{a} = 0 \qquad (2)$$

in which $\Delta\bar{a} = \bar{a} - \bar{a}_0 =$ is the average crack radius increment. Eqn. (2) can be easily solved numerically, and the current average crack size is obtained. The strains are then computed from $\epsilon = \mathbf{C}(\bar{a}) : \sigma$, which is explicit. Thus no iterations within the constitutive subroutines for structural analysis by incremental loading are required.

3.2 Equivalent elastic modulus of a material with arbitrarily oriented cracks

In the preceding section we have considered all the orientation of cracks to be equally probable, which must obviously lead to isotropic effective elastic properties. In reality, when the cracks are produced by a non-isotropic strain tensor such as is the case in rock mechanics problems, a preferential crack orientation must exist. So we will now assume that the crack orientations are not uniformly distributed but that there exist several families of cracks, each with a different orientation, different average crack size and different crack density.

Consider now that the elastic body is intersected by N families of random cracks, labeled by subscripts $\mu = 1, 2, \ldots, N$. Each crack family may be characterized by its spatial orientation $\boldsymbol{\nu}_\mu$, its average crack radius $\bar{a}_\mu$, and the number n_μ of cracks in family μ per unit volume of the material. Thus, the compliance tensor may be considered as the function $\mathbf{C} = \mathbf{C}\left(\bar{a}_1, \bar{a}_2, \ldots, \bar{a}_N; n_1, n_2, \ldots, n_N\right)$. Approximate estimation of this function has been reviewed by Kachanov and co-workers (Kachanov, 1992; Kachanov, 1993; Sayers and Kachanov, 1991; Kachanov et al., 1994).

The incremental constitutive relation can be obtained by differentiation of Hooke's law, which yields:

$$\Delta\boldsymbol{\epsilon} = \mathbf{C} : \Delta\boldsymbol{\sigma} + \sum_{\mu=1}^{N} \frac{\partial\mathbf{C}}{\partial\bar{a}_\mu} : \boldsymbol{\sigma}\Delta\bar{a}_\mu + \sum_{\mu=1}^{N} \frac{\partial\mathbf{C}}{\partial n_\mu} : \boldsymbol{\sigma}\Delta n_\mu \quad (3)$$

where Δ denotes small increments over a loading step.

Our analysis will be restricted to the case when the number of cracks in each family is not allowed to change ($\Delta n_\mu = 0$, i.e., no new cracks are created and no existing cracks are allowed to close). With the number of cracks constant, the third term on the right-hand side of Eqn. 3 vanishes, yielding:

$$\Delta\boldsymbol{\epsilon} = \mathbf{C} : \Delta\boldsymbol{\sigma} + \sum_{\mu=1}^{N} \frac{\partial\mathbf{C}}{\partial\bar{a}_\mu} : \boldsymbol{\sigma}\Delta\bar{a}_\mu \quad (4)$$

The crack radius increments $\Delta\bar{a}_\mu$ must be determined in conformity to the laws of fracture mechanics. Let us assume that the cracks (actually microcracks) follow linear elastic fracture mechanics (LEFM). This assumption means that the energy release rates must be equal to the fracture energy of the material, G_f. To make the problem tractable, we impose the energy balance condition only in the overall, weak sense, which leads to the following N_g conditions for $\Delta\bar{a}_\mu > 0$ ($1 \leq \mu \leq N_g$):

$$\frac{\partial\Pi^*}{\partial\bar{a}_\mu} = 2\pi\bar{a}_\mu n_\mu G_f \quad (5)$$

in which N_g is the number of families of growing cracks. (Repetition of subscript μ in this and subsequent equations does not imply summation.)

When the cracks are shortening, their faces are coming in contact, which requires no energy. Therefore, for shortening cracks ($\Delta\bar{a}_\mu < 0$, $N_g < \mu \leq N_s$),

$$\frac{\partial\Pi^*}{\partial\bar{a}_\mu} = 0 \quad (6)$$

where N_s is the number of all the families of growing and shortening cracks. For the remaining crack families ($N_s < \mu \leq N$) no equations are necessary since for them $\Delta\bar{a}_\mu \equiv 0$.

Substitution of Eqns. (5) or (6) into

$$\frac{\partial\Pi^*}{\partial\bar{a}_\mu} = \boldsymbol{\sigma} : \mathbf{C} : \frac{\partial\boldsymbol{\sigma}}{\partial\bar{a}_\mu} + \frac{1}{2}\boldsymbol{\sigma} : \frac{\partial\mathbf{C}}{\partial\bar{a}_\mu} : \boldsymbol{\sigma} \quad (7)$$

(where $1 \leq \mu \leq N$) leads to the following incremental relations which must be satisfied by the crack radius increments (positive or negative, or 0):

$$\boldsymbol{\sigma} : \mathbf{C} : \Delta\boldsymbol{\sigma} + \left(\frac{1}{2}\boldsymbol{\sigma} : \frac{\partial\mathbf{C}}{\partial\bar{a}_\mu} : \boldsymbol{\sigma} - 2\pi\bar{a}_\mu n_\mu G_f\right)\Delta\bar{a}_\mu = 0$$

$$\boldsymbol{\sigma} : \mathbf{C} : \Delta\boldsymbol{\sigma} + \frac{1}{2}\boldsymbol{\sigma} : \frac{\partial\mathbf{C}}{\partial\bar{a}_\mu} : \boldsymbol{\sigma}\Delta\bar{a}_\mu = 0 \quad (8)$$

$$\Delta\bar{a}_\mu = 0$$

The first equation applies to the families of growing cracks ($1 \leq \mu \leq N_g$), the second to the families of shortening cracks ($N_g < \mu \leq N_s$), and the third to the families of stationary cracks ($N_s < \mu \leq N$).

The constitutive law of Eqn. (4) and the energy equilibrium conditions of Eqn. (8) together represent a system of $N + 6$ equations for the increments of crack radii in N crack families of different orientations and 6 increments of stress tensor components, $\Delta\boldsymbol{\sigma}$. If the strain tensor increment $\Delta\boldsymbol{\epsilon}$ is prescribed, these $N + 6$ unknowns can be solved from this system of equations. The tangential stiffness tensor can be obtained by solving the stress increments for all the cases in which a unit value is assigned to each component of $\Delta\boldsymbol{\sigma}$, with all the other components being 0.

To solve the problem, we must have the means to evaluate the effective secant stiffness $\mathbf{C}$ as a function of the vector crack radii. For its simplicity, we will use a technique developed by Sayers and Kachanov (1991) using the symmetric second-order crack density tensor $\boldsymbol{\alpha} = \sum_{\mu=1}^{N} n_\mu\bar{a}_\mu^3 \boldsymbol{\nu}_\mu\otimes\boldsymbol{\nu}_\mu$ (Vakulenko and Kachanov, 1971; Kachanov, 1980; Kachanov, 1987b).

The effective secant compliance $\mathbf{C}$ can be derived from an elastic potential F which may be considered as a function of the crack density tensor $\boldsymbol{\alpha}$ in addition of the stress tensor $\boldsymbol{\sigma}$:

$$\boldsymbol{\epsilon} = \frac{\partial F(\boldsymbol{\sigma}, \boldsymbol{\alpha})}{\partial\boldsymbol{\sigma}} = \mathbf{C} : \boldsymbol{\sigma} \quad (9)$$

Also, $\mathbf{C} = \mathbf{C}^\circ + \mathbf{C}^{cr}$ in which $\mathbf{C}^\circ$ is the elastic compliance and $\mathbf{C}^{cr}$ is the additional compliance due to the crack system. The elastic potential $F(\boldsymbol{\sigma}, \boldsymbol{\alpha})$ can be expanded into a tensorial power series. According to the Cayley-Hamilton theorem, this expansion can always be reduced to a cubic tensor polynomial. Sayers and Kachanov (1991) proposed to approximate potential $\mathbf{F}$ by a tensor polynomial that is quadratic in $\boldsymbol{\sigma}$ and linear in $\boldsymbol{\alpha}$:

$$F(\boldsymbol{\sigma}, \boldsymbol{\alpha}) =$$
$$\frac{1}{2}\boldsymbol{\sigma} : \mathbf{C}^\circ : \boldsymbol{\sigma} + \eta_1(\boldsymbol{\sigma} : \boldsymbol{\alpha})\mathrm{tr}\,\boldsymbol{\sigma} + \eta_2(\boldsymbol{\sigma}\cdot\boldsymbol{\sigma}) : \boldsymbol{\alpha} \quad (10)$$

in which η_1 and η_2 are assumed to depend only on the first invariant of $\boldsymbol{\alpha}$ (Sayers and Kachanov, 1991).

The additional secant compliance $\mathbf{C}^{cr}$ due to the cracks is written in terms of η_1, η_2 and $\boldsymbol{\alpha}$. The functions $\eta_1(\rho)$ and

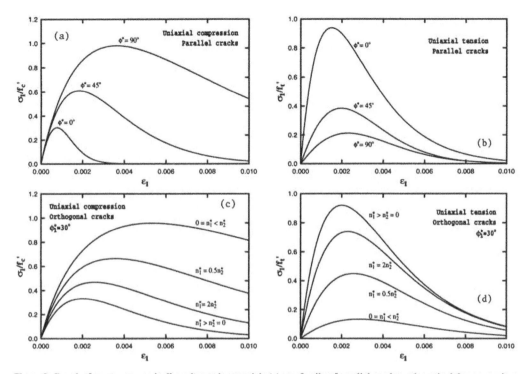

Figure 2. Results for a macroscopically orthotropic material: (a) one family of parallel cracks under uniaxial compression; (b) one family of parallel cracks under uniaxial tension; (c) two families of orthogonal cracks under uniaxial compression; (d) two families of orthogonal cracks under tension.

$\eta_2(\rho)$ can be obtained by taking the particular form of the preceding formulation for the case of random cracks, and equating the results to those obtained using, e.g., the differential scheme (Hashin, 1988):

$$\eta_1 = \frac{3}{2\rho} \frac{3(E_{eff} - E) + 4(\nu E_{eff} - E \nu_{eff})}{E_{eff} E}$$
$$\eta_2 = \frac{6}{\rho} \left(\frac{1 + \nu_{eff}}{E_{eff}} - \frac{1 + \nu}{E} \right) \qquad (11)$$

where E_{eff} and ν_{eff} are the effective Young's modulus and Poisson's ratio for random cracks obtained using the differential scheme. Substituting η_1 and η_2 into $\mathbf{C}^{cr}$, the effective compliance tensor is obtained. Then $\Delta \bar{a}_\mu$ is calculated from Eqns. 4 and 8 for a given $\Delta \epsilon$

3.3 Examples

Fig. 2a shows the results obtained when uniaxial compression is applied to an elastic material with a single family of parallel cracks. The angle of crack normals with the direction of compression is ϕ^*. The solution has been obtained for three different values of the angle: $0°$, $45°$ and $90°$. As expected, the calculated peak stress (strength) is the lowest when the cracks are parallel to the loading direction ($\phi^* = 0°$), and the material response is the most brittle as

indicated by the steepest postpeak decline of stress. Fig. 2b shows the results obtained when uniaxial tension is applied. In that case, the minimum strength occurs when the cracks are perpendicular to the tension direction, as known from experimental evidence.

The analysis has been repeated for an elastic material with two families of orthogonal cracks, which in general are not aligned with the tension direction ($\phi_1^* = 30°$, $\phi_2^* = 120°$). The analysis has been performed for several combinations of the number of cracks n_1^* and n_2^* corresponding to the two crack families. These combinations are indicated in Figs. 2c and 2d. Fig. 2c depicts the results obtained for uniaxial compression and Fig. 2d the results for uniaxial tension. The analytical results obtained for both cases agree with the expected trends.

The model has also been used to simulate the response to biaxial stress applied to an initially isotropic material. Fig. 3a shows the stress-strain curves obtained for biaxial tests with different stress ratios σ_1/σ_2. These curves again agree with the observed behavior of brittle materials such as rock (or concrete). Furthermore, combining the peaks of the curves for different stress ratios, the biaxial failure envelope (Fig. 3b) is obtained. The envelope agrees well with experimental results. In particular, the ratios of the uniaxial tensile strength and of the biaxial compression

9

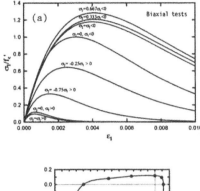

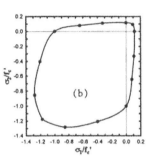

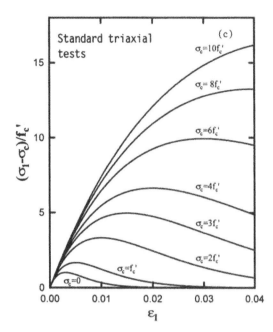

Figure 3. Multiaxial tests: (a) biaxial tests, stress-strain curves; (b) biaxial failure envelope; (c) triaxial tests, stress-strain curves.

strength to the uniaxial compression strength have values that are normal for some rocks or concrete.

Finally, the model has been verified for the case of standard triaxial tests, in which a certain hydrostatic pressure of maximum value σ_c is applied first and then uniaxial compression is superimposed. The analytical results are presented in Fig. 3c, for several values of σ_c ranging from 0 to $10f_c'$ where f_c' is the uniaxial compression strength. So we see that, at least qualitatively, the new model is able to capture the decreasing brittleness of the material for increasing hydrostatic pressure. Eventually this approach is likely to lead to an improvement of the microplane model (Bažant et al., 1996a; Bažant et al., 1996b), a very effective and versatile model for fracturing materials.

4 CONCLUDING REMARKS

As is clear from the preceding exposition, fracture of rock is a complex phenomenon with many facets, which is only now beginning to be reasonably well understood. A number of effective models for rock fracture, varying from simple or sophisticated, have been outlined. Attention has then been focused on the micromechanical aspects, whose understanding is essential for the development of realistic models. The micromechanical modeling of fracture of rock or other quasibrittle materials is still in its infancy, and should represent the focus of further research. One difficult problem which deserves thorough further study

if the problem of localization of microcracking into large macroscopic fractures, and its continuum description.

ACKNOWLEDGEMENTS

Partial financial support under NSF Grant MSS-9114426 to Northwestern University is gratefully acknowledged. Part of the work on crack systems was also supported by the ACBM Center at Northwestern University. The second author wishes to acknowledge the financial support received from DGICYT Project Grant PB94-1204 (Ministerio de Educación y Ciencia, Madrid, Spain), and from CIRIT (Generalitat de Catalunya, Barcelona, Spain) AIRE Grant 1995BEAI300213 which provided funds for the visit to Northwestern University where this work was completed.

REFERENCES

Barenblatt, G.I. (1962). The mathematical theory of equilibrium cracks in brittle fracture. *Advances in Applied Mechanics*, 7:55-129.

Bažant, Z.P. (1982). Crack band model for fracture of geomaterials. In *Proceedings of the Fourth International Conference on Numerical Methods in Geomechanics*, Edmonton, Alberta, Canada. Vol. 3.

Bažant, Z.P. (1984a). Mathematical modeling of progressive cracking and fracture of rock. In Dowding, C.H. and Singh,

M., editors, *Proceedings, 25th U.S. National Symposium on Rock Mechanics*, pp. 29–37, Northwestern University, Evanston, IL. Soc. of Mining Engrs.—Am. Institute of Mining, Met. & Petr. Engrs., New York.

Bažant, Z.P. (1984b). Size effect in blunt fracture: Concrete, Rock, Metal. *ASCE J. Engng. Mech.*, 110:518–535.

Bažant, Z.P. (1987). Fracture energy of heterogenenous material and similitude. In Shah, S.P. and Swartz, S.E., editors, *SEM-RILEM Int. Conf. on Fracture of Concrete and Rock*, pp. 390–402, Bethel. SEM.

Bažant, Z.P. (1993). Current status and advances in the theory of creep and interaction with fracture. In Bažant, Z.P. and Carol, I., editors, *Creep and Shrinkage of Concrete (ConCreep 5)*. Chapman & Hall.

Bažant, Z.P. (1994). Nonlocal damage theory based on micromechanics of crack interactions. *ASCE J. Engng. Mech.*, 120(3):593–617. Addendum and Errata 1401–1402.

Bažant, Z.P. (1995). Scaling of quasibrittle fracture and the fractal question. *ASME J. of Materials and Technology*, 117(Oct.):361–367.

Bažant, Z.P. (1996). Is no-tension design of concrete or rock structures always safe? *ASCE J. Engng. Mech.*, 122(1):2–10.

Bažant, Z.P., Bai, S.-P., and Gettu, R. (1993). Fracture of rock: Effect of loading rate. *Engineering Fracture Mechanics*, 45:393–398.

Bažant, Z.P. and Cedolin, L. (1991). *Stability of Structures: Elastic, Inelastic, Fracture and Damage Theories*. Oxford University Press. Sec. 12.6.

Bažant, Z.P. and Gettu, R. (1992). Rate effects and relaxation in static fracture of concrete. *ACI Materials Journal*, 89:456–468.

Bažant, Z.P., Gettu, R., and Kazemi, M.T. (1991). Identification of non-linear fracture properties from size effect tests and structural analysis based on geometry-dependent R-curves. *Int. J. Rock Mech. Min. Sci.*, 28:43–51.

Bažant, Z.P. and Jirásek, M. (1993). R-curve modeling of rate and size effects in quasibrittle fracture. *Int. J. Fracture*, 62:355–373.

Bažant, Z.P. and Jirásek, M. (1994). Damage nonlocality due to microcrack interactions: statistical determination of crack influence function. In Bažant, Z.P., Bittnar, Z., Jirásek, M., and Mazars, J., editors, *Fracture and damage in quasi-brittle structures*, pp. 3–17. E & FN SPON, London.

Bažant, Z.P. and Kazemi, M.T. (1990a). Determination of fracture energy, process zone length and brittleness number from size effect, with application to rock and concrete. *Int. J. Fracture*, 44:111–131.

Bažant, Z.P. and Kazemi, M.T. (1990b). Size effect in fracture of ceramics and its use to determine fracture energy and effective process zone length. *J. Am. Ceram. Soc.*, 73(7):1841–1853.

Bažant, Z.P. and Li, Y.N. (1995). Stability of cohesive crack model. *J. Appl. Mech.*, 62(Dec.):959–969.

Bažant, Z.P. and Li, Y.N. (1996a). Cohesive crack model with rate dependent softening and viscoelasticity. *Int. J. Fracture*, (submitted to).

Bažant, Z.P., Li, Y.N., and Hong, A.P. (1995). Initiation of parallel cracks from surface of elastic half-plane. *Int. J. Fracture*, 69:357–369.

Bažant, Z.P. and Li, Z. (1996b). Zero-brittleness size-effect method for one-size fracture test of concrete. *ASCE J. Engng. Mech.*, 122(5). In press.

Bažant, Z.P. and Oh, B.H. (1983). Crack band theory for fracture of concrete. *Materials & Structures RILEM*, 16:155–177.

Bažant, Z.P. and Oh, B.H. (1984). Rock fracture via strain-softening finite elements. *ASCE J. Engng. Mech.*, 110(7):1015–1035.

Bažant, Z.P. and Ohtsubo, R. (1978). Geothermal heat extraction by water circulation through a large crack in dry hot rock mass. *Int. J. Num. Anal. Methods in Geomechanics*, 2:317–327.

Bažant, Z.P., Ohtsubo, R., and Aoh, K. (1979). Stability and post-critical growth of a system of cooling and shrinkage cracks. *Int. J. Fracture*, 15:443–456.

Bažant, Z.P. and Pfeiffer, P.A. (1987). Determination of fracture energy from size effect and brittleness number. *ACI Materials Journal*, 6:463–480.

Bažant, Z.P. and Xiang, Y. (1996). Size effect in compression fracture: Splitting crack band propagation. *ASCE J. Engng. Mech.*, 122. In press.

Bažant, Z.P., Xiang, Y., Adley, M.D., Prat, P.C., and Akers, S.A. (1996a). Microplane for concrete: II. Data delocalization and verification. *ASCE J. Engng. Mech.*, 122(3):263–268.

Bažant, Z.P., Xiang, Y., and Prat, P.C. (1996b). Microplane for concrete: I. Stress-strain boundaries and finite strain. *ASCE J. Engng. Mech.*, 122(3):245–262.

Benveniste, Y. (1987). A new approach to the application of Mori-Tanaka's theory in composite materials. *Mechanics of Materials*, 6:147–157.

Bediansky, B. and O'Connell, R.J. (1976). Elastic moduli of a cracked solid. *Int. J. Solids and Structures*, 12:81–97.

Carpinteri, A. (1980). Static and energetic fracture parameters for rock and concretes. Technical report, Istituto di Scienza delle Costruzioni–Ingegneria, University of Bologna, Italy.

Carter, B.J. (1992). Size and stress gradient effects on fracture around cavities. *Rock Mech. and Rock Eng.*, 25(3):167–186.

Carter, B., Lajtai, E.Z., and Yuan, Y. (1992). Tensile fracture from circular cavities loaded in compression. *Int. J. Fracture*, 57:221–236. Figs. 5,6,9.

Dugdale, D.S. (1960). Yielding of steel sheets containing slits. *J. Mech. Phys. Solids*, 8:100–104.

Dzik, E.J. and Lajtai, Z. (1994). Primary propagations from circular voids loaded in compression. Report, Dept. of Civil and Geological Engineering, University of Manitoba, Winipeg.

Elices, M., Guinea, G., and Planas, J. (1992). Measurements of the fracture energy using three-point bend tests: Part 3 – Influence of cutting the p-δ tail. *Materials & Structures RILEM*, pp. 212–218.

Elices, M. and Planas, J. (1989). Material models (chapter 3). In Elfgren, L., editor, *Fracture Mechanics of Concrete Structures*, pp. 16–66, London, U.K. Chapman & Hall.

Fabrikant, V.I. (1990). Complete solutions to some mixed boundary value problems in elasticity. In Hutchinson, J. and Wu, S., editors, *Advances in Applied Mechanics*, Vol. 27, pp. 153–223. Academic Press, New York.

Haimson, B. and Herrick, C. (1986). Borehole breakouts — a new tool for estimating the in situ stress? In *Proceedings, Int. Symp. On Rock Stress and Rock Measurements*, Stockholm.

Haimson, B. and Herrick, C (1989). Borehole breakouts and in situ tests. *Proceedings Drilling Symposium at ETCE, ASME*, 22:17–22.

Hashin, Z. (1988). The differential scheme and its application to cracked materials. *J. Mech. Phys. Solids*, 36(6):719–734.

Hillerborg, A., Modéer, M., and Petersson, P.E. (1976). Analysis of crack formation and crack growth in concrete by means of Fracture Mechanics and Finite Elements. *Cement and Concrete Research*, 6:773–781.

Hoagland, R.G., Hahn, G.T., and Rosenfeld, A.R. (1973). Influence of microstructure on fracture propagation in rock. *Rock Mechanics*, 5:77–106.

Hoenig, A. (1978). The behavior of a flat elliptical crack in an anisotropic elastic body. *Int. J. Solids and Structures*, 14:925–934.

Hu, X.Z. and Wittmann, F.H. (1991). An analytical method to determine the bridging stress transferred within the fracture process zone: I. General theory. *Cement and Concrete Research*, 21:1118–1128.

Jirásek, M. and Bažant, Z.P. (1994). Localization analysis of nonlocal model based on crack interactions. *ASCE J. Engng. Mech.*, 120(3):1521–1542.

Kachanov, M. (1980). Continuum model of medium with cracks. *ASCE J. Engng. Mech. Div.*, 106(EM5):1039–1051.

Kachanov, M. (1985). A simple technique of stress analysis in elastic solids with many cracks. *Int. J. Fracture*, 28:11–19.

Kachanov, M. (1987a). Elastic solids with many cracks: A simple method for analysis. *Int. J. Solids and Structures*, 23:23–43.

Kachanov, M. (1987b). On modelling of anisotropic damage in elastic-brittle materials — a brief review. In Wang, A. and Haritos, G., editors, *Damage Mechanics in Composites*, ASME Winter Annual Meeting, pp. 99–105, Boston. ASME.

Kachanov, M. (1992). Effective elastic properties of cracked solids: critical review of some basic concepts. *Applied Mechanics Review*, 45(8):304–335.

Kachanov, M. (1993). Elastic solids with many cracks and related problems. In Hutchinson, J. and Wu, T., editors, *Advances in Applied Mechanics*, Vol. 30, pp. 259–445. Academic Press.

Kachanov, M., Tsukrov, I., and Shafiro, B. (1994). Effective moduli of solids with cavities of various shapes. *Applied Mechanics Review*, 47(1):S151–S174.

Labuz, J., Shah, S., and Dowding, C.H. (1985). Experimental analysis of crack propagation in granite. *Int. J. Rock Mech. Min. Sci.*, 22:85–98.

Martin, C.D., Martino, J., and Dzik, E. (1994). Comparison of borehole breakouts from laboratory and field tests. In *EUROCK '94*, Delft, The Netherlands.

Mori, T. and Tanaka, K. (1973). Average stress in matrix and average elastic energy of materials with misfitting inclusions. *Acta Metallurgica*, 21:571–574.

Nakayama, T. (1965). Direct measurements of fracture energies of brittle heterogeneous materials. *J. Am. Ceram. Soc.*, 48(11).

Nesetova, V. and Lajtai, Z. (1973). Fracture from compressive stress concentrations around elastic flaws. *Int. J. Rock Mech. Min. Sci.*, 10:265–284.

Planas, J., Elices, M., and Guinea, G. (1992). Measurements of the fracture energy using three-point bend tests: Part 2 – Influence of bulk energy dissipation. *Materials & Structures RILEM*, pp. 305–312.

Prat, P.C. and Bažant, Z.P. (1995). Tangential stiffness of elastic materials with systems of growing and closing cracks. Technical Report 95-12/C402t, Dept. of Civil Engng., Northwestern University, Evanston, IL 60208, USA.

Roscoe, R.A. (1952). The viscosity of suspensions of rigid spheres. *Brit. J. Appl. Phys.*, 3:267–269.

Sayers, C.M. and Kachanov, M. (1991). A simple technique for finding effective elastic constants of cracked solids for arbitrary crack orientation statistics. *Int. J. Solids and Structures*, 27(6):671–680.

Schmidt, R.A. (1977). Fracture mechanics of oil shale unconfined fracture toughness, stress corrosion cracking, and tension test results. In *Proceedings 18th U.S. Symposium on Rock Mechanics*, Colorado School of Mines, Golden, Colorado. Paper 2A2.

Schmidt, R.A. and Lutz, T.J. (1979). K_{IC} and J_{IC} of Westerly granite — effect of thickness and in-plane dimensions, Vol. ASTM STP 678. American Society for Testing of Materials.

Takahashi, H. (1988). Application of rock fracture mechanics to HDR geothermal reservoir design. In *Proceedings, Int. Workshop on Fracture Toughness and Fracture Energy Test Methods for Concrete and Rock*, pp. 453–472, Sendai, Japan.

Tatersall, H.G. and Tappin, G. (1966). The work of fracture and its measurement in metals, ceramics and other materials. *J. Mat. Sci.*, 1(3):296–301.

Vakulenko, A.A. and Kachanov, M. (1971). Continuum theory of medium with cracks. *Mechanics of Solids*, 6:145–151.

Rock Fragmentation by Blasting, Mohanty (ed.)© 1996 Taylor & Francis. ISBN 90 5410 824 X

Blast-induced dynamic fracture propagation

A. Daehnke
CSIR Mining Technology, Johannesburg, South Africa

H. P. Rossmanith & R. E. Knasmillner
Institute of Mechanics, Technical University of Vienna, Austria

ABSTRACT: A coupled solid, fluid and fracture mechanics numerical model is used to analyse the gas driven fracture propagation phase, and predict propagation rates and give insights into pressure profiles and gas velocities within fractures. The model incorporates dynamic material properties and quasi-dynamic fracture mechanics which are shown to significantly affect the numerical predictions.

Three-dimensional cube-type laboratory models fabricated from PMMA are dynamically loaded with explosives and the resulting contained fracture networks are studied. The dynamic evolution of the spatial crack system is monitored by taking a sequence of high-speed recordings with a Cranz-Schardin type camera. The chosen model geometries are designed to indicate the effect of stemming in borehole blasting, and it is shown that stemming has a decisive influence on fracture profiles and orientation. It is found that stress waves rapidly out-pace the slower dynamic fractures, and the majority of fracturing occurs due to pressurisation by detonation gases.

1 INTRODUCTION

Rockmass discontinuities in the stope vicinity of deep-level mines can be classified into (i) pre-existing fractures and planes of weakness due to the structural geology, and (ii) mining-induced fractures. Fractures and planes of weakness due to the structural geology typically form between the host rock and intrusions such as dykes, on fault interfaces and between rock strata, e.g. bedding planes. Mining-induced fractures are formed as stoping progresses through highly stressed rock, and they can be classified according to their quasi-static or dynamic formation. Quasi-static fractures are characterised by their comparatively slow growth rate (in the order of mm per day), are ubiquitous in the stoping area and are formed during the slow disintegration of highly stressed rock mass. Fractures of dynamic origin (propagation rates up to 1000 m/s) are propagated by (i) stress waves due to seismic events such as the sudden brittle failure of rock and slip on fault interfaces, and (ii) stress waves and gas pressure due to detonation of explosive media in boreholes drilled into, for example, stope faces and development ends.

Due to their immediate stope vicinity blast-induced fractures have a direct bearing on the hangingwall stability and support requirements in deep-level mines. Hence, in improving the safety underground and streamline efficiency of the mining operation, knowledge of blast-induced fracturing is essential.

This investigation deals with blast-induced fracture propagation and attempts to contribute towards gaining insights into the underlying mechanisms of propagating fractures from blast holes. Kutter and Fairhurst (1971) were among the first to consider the complex coupling of solid, fracture and fluid mechanics in blast-induced fracturing. They divided the blast problem into two phases: (i) local fracturing around the borehole driven by stress waves, which are initiated by the extremely rapid expansion of the gaseous detonation products by the reaction of borehole explosives, and (ii) subsequent global fracture extension due to detonation gases penetrating into the fractures. The authors of further studies (Fourney et al. 1979, Worsey et al. 1981 and McHugh 1983) concluded that in many cases the second stage, i.e. gas driven fracture propagation, is the dominant mechanism, and it is feasible that the gas pressurisation increases the fracture length by a factor of 10-100, as compared with crack extension due to stress wave loading alone (McHugh 1983).

The objective of this work is to conduct controlled three-dimensional laboratory experiments in transparent PMMA (Polymethylmethacrylate), and to record, by means of high speed photographic techniques, the evolution of blast fractures. This yields a qualitative (fracture structure), as well as a quantitative (fracture surface area and propagation rates versus time after detonation) description of a blast model with known input parameters. As the detonation gases play a vital role during the fracture evolution, two-dimensional plate models, which would vent the gases as the fractures propagate through the plate, are not suitable for investigating blast-induced fracturing. Thus, three-dimensional cube type PMMA models are used, and the amount of explosive is restricted such that the fractures do not extend to the cube surface, and the detonation products within the fractures are contained.

A numerical method proposed by Nilson (1986) is used to back-analyse the gas driven stage during fracture extension.

2 EXPERIMENTAL PROCEDURE

High speed photography was utilised to record the evolution of blast induced fractures. A Cranz-Schardin type multiple spark gap camera provides 24 frames photographed at discrete time intervals at a framing rate of approximately 50 000 frames per second. The experiments were conducted in transparent PMMA cubes (250 x 250 x 240 mm³), thus allowing full field visualisation of the fracture evolution. The use of circularly polarised monochromatic light, in conjunction with the bi-refringent material property of PMMA, permits the recording of isochromatic fringe patterns of the propagating stress waves (an isochromatic contour is a line of equal maximum shear stress). Thus the exact fracture front and stress wave position in time is recorded for further assessment of the growth of fracture.

In laboratory experiments Kutter and Fairhurst (1971) compared blast induced fracture patterns in laboratory models fabricated from PMMA with patterns in equivalent models constructed from homogeneous monolithic blocks of rock. They found that the fracture patterns in PMMA appeared to be practically identical with those in rock, and only the scale, i.e. the length of the cracks, to differ. As the objective of this investigation is to understand the underlying mechanisms governing dynamic fracture, PMMA was chosen as a suitable material to study fracture mechanisms which also apply to brittle rock encountered in deep-level mines.

Two blast model configurations are compared:

- Fracture propagation from a stemmed charge, where part of the borehole is loaded with explosives and part is filled with a dummy material effectively preventing the immediate venting of the gaseous combustion products.
- Fracture propagation from a non-stemmed, totally sealed charge; here the charge is totally isolated from the outside.

In mining operations the explosive cannot be totally sealed and a stemming effect is always present.

However the two models are considered to represent a valuable comparison to illustrate the effect of borehole stemming.

For the stemmed charge experiment (Figure 1 a) the explosive is placed at the end of a hole drilled into the PMMA cube, and the hole is then filled with a comparatively weak epoxy glue representing the rock fines commonly used as stemming material. Upon detonation the stem is deformed and partially pushed out of the borehole.

In order to investigate the hypothetical case of a totally sealed borehole (Figure 1 b), two PMMA slabs are bonded together with the explosive encased in the centre. The explosive is triggered and ignited by thin copper wires bonded between the slabs which conduct the 2 kV triggering pulse. The slabs are glued with liquid PMMA, resulting in a homogeneous PMMA cube.

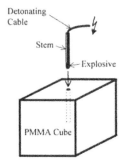

a): Stemmed Charge Experiment

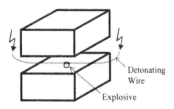

b): Sealed Charge Experiment

Figure 1: Schematic exploded view illustrating components of borehole breakdown experiments: a) stemmed charge experiment and b) sealed charge experiment.

In both experiments a micro-charge of 160 mg Lead Azide (Pb N_6) is used as explosive; the amount of explosive is restricted such that (i) the fractures do not propagate to the cube sides and vent their gases, and (ii) that the effect of the stress waves reflected by the cube boundaries and interacting with the propagating gas-driven fractures is negligible.

3 EXPERIMENTAL RESULTS

3.1 Stemmed Charge Experiment

Photographs depicting fractures which have propagated from a stemmed borehole are shown in Figure 2 a) and b). As shown by the view in-line with the borehole axis, Figure 2 a), three prominent fractures have extended from the borehole at angles close to 120° relative to each other (actual angles: 121°, 127° and 112°) with lengths of approximately 65 mm, 60 mm and 45 mm for fractures A, B and C respectively. It is interesting to note that this quasi-symmetry with respect to the fracture angles is in accord with theoretical predictions based on static considerations (Ouchterlony 1983).

At the borehole bottom a dish-shaped fracture has developed and propagated to form a cone type crack.

The view normal to the borehole axis (Figure 2 b) shows that the three fractures are roughly semi-circular in shape and extend back from the explosive source along the stem. The epoxy bond between the stem and the bulk material is broken, and the stem is pushed back and out of the borehole by a distance of approximately 33 mm.

3.2 Sealed Charge Experiment

The photographs shown in Figures 3 a) and b) reveal the final structure of the fractures due to the detonation of a short cylindrical charge, where the charge was sealed such that no gaseous detonation products could escape.

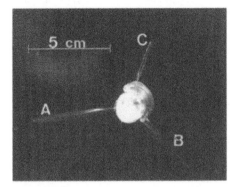

Figure 2 a)

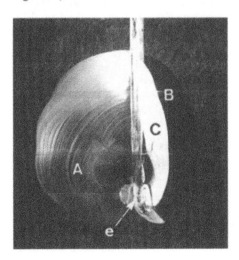

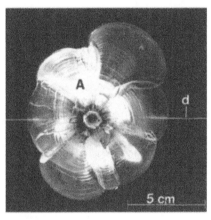

Figure 3 a)

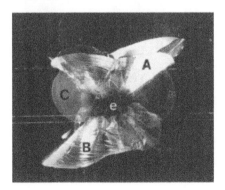

Figure 2: Final fracture structure of stemmed charge experiment shown: a) in-line with the borehole axis and b) normal to the borehole axis. *e* refers to the position of the explosive charge.

Figure 3: Final fracture structure of sealed charge experiment shown: a) in-line with the borehole axis and b) normal to the borehole axis. *e* refers to the position of the explosive charge and *d* to the detonation wire.

It is obvious that the fracture pattern propagating from a sealed borehole is fairly complex and very different compared with the stemmed borehole. Instead of fractures propagating along the borehole axis, two dish-shaped conical fractures (Fractures A and B in Figure 3) have propagated from the circular ends of the cylindrical charge cavity which act as discontinuities. These dish-shaped fractures represent most of the fracture area, and the fractures parallel to the borehole axis linking the two main dish-shaped fractures (e.g. Fracture C) are much smaller.

To the authors' knowledge this is the first time the effect of stemming is illustrated in a transparent material and the effect on the fracture profiles is described. In previous studies (Simha et al. 1987) experiments were conducted to investigated the usefulness of stemming in crater blasting, but did not study the final fracture profiles of a sealed versus stemmed charge.

Figure 4 pertains to an instant in time 65 µs after charge detonation of the dynamic fracture propagation and stress waves for the sealed charge experiment. Isochromatic contours corresponding to the position of the reflected longitudinal wave, which has been reflected by the cube boundaries and is now propagating at C_P = 2500 m/s back towards the model centre, are visible (Label P). The detonation of the explosive and sudden expansion of the gaseous detonation products should only generate a longitudinal wave, however borehole breakdown and the ensuing fracture initiation and propagation give rise to shear waves. The isochromatic contours associated with the outgoing shear waves propagating at C_S = 1400 m/s are labelled in Figure 4 by S.

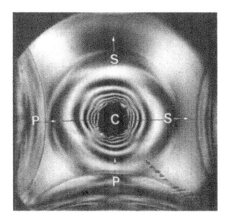

Figure 4: Photograph showing fracture and stress wave positions 65 µs after charge detonation. P and S refer to the position of the longitudinal and shear wave respectively, whilst C labels the silhouette of the dish-shaped fracture.

By comparing the fracture positions visible in the frame taken before and after the frame shown in Figure 4, the average expansion velocity of the dish shaped fracture is calculated as 200 m/s. The maximum fracture extension rate recorded in the experiments occurred immediately after charge detonation and was approximately 500 m/s, which is in agreement with maximum fracture velocities in PMMA measured in experiments described in the literature, e.g. Hernandez Gomez and Ruiz (1993).

The most important feature of Figure 4 is that the longitudinal stress wave rapidly out-paces the fractures propagating at a much slower rate of maximum 500 m/s; hence fracture extension occurs mainly due to pressurisation by combustion gases. Considering a maximum pulse length of the outgoing longitudinal wave of 6 cm (measured in the experiments, and including the tensile pulse tail) propagating at a rate of 2500 m/s, fractures propagating at their maximum rate of 500 m/s would be outpaced within 12 mm of fracture growth. Thus for the dish-shaped fractures with an average final radius of 63 mm, a maximum of 8 % of the fracture surface area is due to the stress waves and the remaining 92 % due to the gas pressurisation.

4 NUMERICAL ANALYSIS

A numerical method for predicting fracture propagation driven by gases as proposed by Nilson (1986) is used to simulate the gas flow in the fractures observed in the sealed charge experiment. The two conical fractures (labelled A and B in Figure 3) are approximated as two independent penny-shaped cracks. The analysis includes convective and conductive heat transfer from the hot combustion gases to the surrounding host material, and friction losses due to turbulent flow. In this work, Nilson's method is extended to incorporate fracture velocity dependant toughness of PMMA according to experimental data by Green and Pratt (1974), and velocity dependant dynamic stress intensity factors according to Freund (1972) for cracks propagating at constant rates. It is found that by incorporating dynamic material properties and velocity dependent stress intensity factors the initial fracture extension rate is reduced by a factor of 0.6 compared with a static analysis, and it is recognised that dynamic PMMA material properties significantly influence propagation rates and need to be incorporated in the numerical analyses.

The input to the numerical analysis consists of the detonation temperature and detonation pressure. The detonation temperature can be estimated by the Rankine-Hugoniot equations (Persson et al. 1994) and for Lead Azide at the density used in the experiments it is approximately 2700° Kelvin. The borehole pressurisation is assumed to occur instantaneously throughout the borehole, as the detonation wave propagates through the explosive in less than 1 μs (detonation wave velocity in Lead Azide: 5200 m/s), and thus, relative to the time period the fractures extend, the effect of the borehole pressure rise as the explosive detonates is considered negligible. The initial borehole pressure is taken as 160 MPa. This pressure level is chosen to yield an initial fracture propagation rate and total fracture extent of similar magnitude as observed in the experiments. To date in this study no experiments have been conducted were the borehole pressure is measured directly, however future 3D PMMA cube experiments are planned to incorporate direct pressure measurements in the borehole and along the fracture.

During the initial phase the fractures are assumed to be driven by stress waves, and as these outpace the fractures, combustion gas induced cracking takes over as the propagating mechanism. In the calculation the cracks are assumed to have extended 10 mm due to stress wave loading, and only the subsequent crack extension due to gas pressurisation is modelled numerically.

Figure 5 compares the experimental with the numerical fracture length and speed versus time after the detonation gases take over as the fracture driving mechanism.

Experimental and numerical data indicate that the initially high fracture propagation speed of about 500 m/s (38 % of the Rayleigh wave speed in PMMA) rapidly decrease to below 100 m/s for the main duration of the fracture extension. The crack speed is governed by the rate hot combustion gases penetrate the crack. As the crack propagates and is pressurised by combustion gases, the borehole pressure decreases; due to increased friction and the divergent geometry of the penny-shaped crack, the driving gas speed is reduced and the speed of fracture propagation decreases.

Figure 6 shows the pressure profiles in the propagating fracture at various time instances after the detonation gases become the dominant driving mechanism. At t = 0 μs, i.e. at the moment the blast induced stress waves outpace the propagating fractures and propagation occurs due to pressurisation by the combustion gases, the high pressure gas penetrates only up to 90% of the fracture length. At high fracture propagation rates PMMA becomes significantly tougher compared with the static fracture toughness. To illustrate the effect of dynamic material properties, fracture growth was simulated using static material properties, and in this case the high pressure gas front penetrated the fracture up to 62 % of its initial length.

After 400 μs the high pressure gases penetrate up to the fracture tip. At 800 μs the fracture is about to be uniformly pressurised along the total length, and at this stage the fracture slows to a halt.

Figure 7 gives the gas Mach number profiles at various times after gas flow commenced. Although absolute gas speeds approach 750 m/s, the gas speeds are subsonic due to the high gas temperatures.

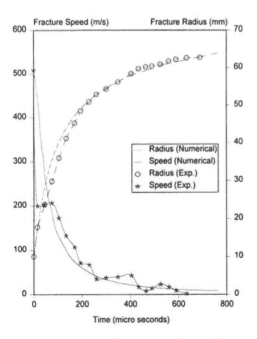

Figure 5: Numerical and experimental fracture lengths and speeds versus time after the gas driving mechanism becomes dominant.

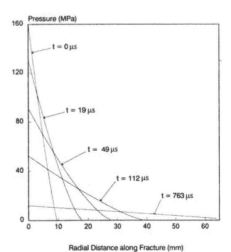

Figure 6: Fracture pressure profiles at various time instances after the detonation gases are the dominant driving mechanism.

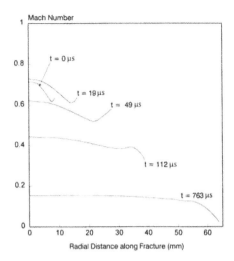

Mach Number

Radial Distance along Fracture (mm)

Figure 7: Mach number profiles along the fracture at various times after detonation gases take over as the main driving mechanism.

5 CONCLUSIONS

This paper has given a qualitative as well as quantitative description of three-dimensional fracture propagation due to detonation of explosives in two models. The chosen model geometries are designed to indicate the effect of stemming in borehole blasting. It has been demonstrated that stemming has a significant influence on fracture profiles and fracture orientation. As a result of stemming, fractures are aligned along the borehole axis. This feature is important to recognise in the stoping environment where unfavourably orientated blast-induced fractures could compromise the hangingwall stability.

Descriptions of blast-induced fractures obtained under controlled conditions in the laboratory are valuable for developers of dynamic fracture codes to validify and check their results. Here, upon recognising that stress waves rapidly outpace the slower dynamic fractures and the majority of fracturing occurs due to pressurisation by detonation gases, a coupled solid, fluid and fracture mechanics numerical model is used to analyse the gas driven fracture propagation phase. The model can predict propagation rates and the final fracture extent, and gives insights into pressure profiles and gas velocities within fractures. The model incorporates dynamic material properties and quasi-dynamic fracture mechanics which are shown to significantly affect the numerical predictions.

ACKNOWLEDGEMENTS

The authors would like to express their thanks to the South African Department of Mineral and Energy Affairs for permission to publish. The use of the facilities of the Photomechanics Laboratory of the Technical University of Vienna during the course of this work is kindly acknowledged. H.P. Rossmanith would like to acknowledge the support granted by the Austrian National Science Foundation FWF under project number P10326-GEO.

REFERENCES

Kutter, H.K. and C. Fairhurst 1971. On the fracture process in blasting. *Int. J. Rock Mech. Min. Sci.* 8: 181-202.

Fourney, W.L., D.C. Holloway and D.B. Barker 1979. Fracture initiation from the packer area. *University of Maryland Research Report prepared for the National Science Foundation.*

Worsey, P.N., I.W. Farmer and G.D. Matheson 1981. The mechanics of pre-splitting in discontinuous rock. *Proc. 22nd U. S. Symposium on Rock Mechanics:* 218-223. Massachusetts Institute of Technology.

McHugh, S. 1983. Crack extension caused by internal gas pressure compared with extension caused by tensile stress. *Int. J. of Fracture* 21: 163-176.

Nilson, R.H. 1986. An integral method for predicting hydraulic fracture propagation driven by gases or liquids. *Int. J. for Num. and Analyt. Methods in Geomech.* 10: 191-211.

Ouchterlony, F. 1983. *Analysis of cracks related to rock fragmentation.* Rock Fracture Mechanics, CSIM course 275 (Ed. H.P. Rossmanith), Springer Verlag.

Simha, K.R.Y., W.L. Fourney and R.D. Dick 1987. An investigation of the usefulness of stemming in crater blasting. *Proc. of the 2nd Int. Symp. on Rock Frag. by Blasting*, Keystone, Colorado: 591-599.

Hernandez Gomez, L.H. and C. Ruiz 1993. Experimental evaluation of crack propagation velocity in PMMA under dynamic pressure loading. *Int. J. of Fracture* 61: 21-28.

Green, A.K. and P.L. Pratt 1974. Measurement of the dynamic fracture toughness of Polymethylmethacrylate by high-speed photography. *Eng. Frac. Mech.* 6: 71-80.

Freund, L.B. 1972. Crack propagation in an elastic solid subjected to general loading - I. Constant rate of extension. *J. Mech. Phys. Solids* 20: 129-140.

Persson, P., R. Holmberg and J. Lee 1994. *Rock Blasting and Explosives Engineering.*

Rock Fragmentation by Blasting, Mohanty (ed.) © 1996 Taylor & Francis. ISBN 90 5410 824 X

Research on the fragment-size model for blasting in jointed rock mass

Jichun Zhang
Sichuan Union University, Chengdu, People's Republic of China

ABSTRACT: Rock masses have been cut by various kinds of weak planes such as joints and cracks which play an important role in blasted fragment sizes. Based on the spacing composition of field joints and the equivalent spacing regularity of the same class-order joints, a *Cantor set* distribution of joint spacings is advanced. The method of calculating natural fragment sizes is given by *Monte-Carlo* simulation theorem. The process of blasted fragment formation is abstractly described by the stochastic fractal theory. In accordance with the treatment of continuum damage mechanics, the distribution of joints and cracks is introduced to the physical process of rock masses blasting and the fragmentation mechanism is introduced to the physical process of rock masses blasting and the fragmentation mechanism is emphatically analysed. The damage model of fragment-size calculation of jointed rock masses blasting is established by statistical analyses.

1 INTRODUCTION

The fragment-size distribution of ore and rock masses blasting is an important index to evaluate the blasting quality quantitatively. It affects the efficiency of each follow-up producing process and the total cost in mining. In building dam engineering with blasting, the blasted fragment-size distribution has a principal influence on the dam quality. Therefore, it is of momentous theoretical and practical significance to thoroughly analyse the mechanism of fragment-size formation in rock masses blasting and put forward a calculating model of fragment sizes which reflects the essence of jointed rock masses blasting.

Since 1960s, many researchers in various countries have successively put forward more than ten theoretical and empirical models of fragment-size calculation of rock masses blasting using the stress waves theory, energy theory and distribution function of fragment sizes (Zhang et al. 1992), in which rock masses are mostly considered as homogeneous continuum media. Rock masses, however, have been cut by various kinds of weak planes such as joints and cracks. These discontinuities not only play a major role in controlling the behaviour of a rock mass, but also determine its failure forms (Sun 1988). The blasted fragments are produced by rock masses cracking along the trace directions of weak planes. To completely solve the problem of fragment-size calculation, we must introduce the distribution of joints and cracks to the physical process of rock masses blasting, analyse the fragmentation mechanism and the law of rock fragment formation and, moreover, develop a calculating model of blasted fragment sizes.

2 DISTRIBUTION OF JOINT SPACINGS AND SIMULATION OF NATURAL FRAGMENT SIZES

2.1 Cantor set distribution of joint spacings

The distribution of joints is one of the principal factors controlling the fragment sizes of rock masses blasting and the composition of spacings reflects the characteristic of rock masses structure. As long as the distribution law of joints is mathematically described, the joints can be introduced to the model of blasted fragment-size calculation.

The relations between the joint spacing of three regions in Hua Ziyu magnesium mine z and the number of joints $N(z)$, whose spacing values are greater than z, are shown in Fig. 1. Practically, $N(z)$ is plotted against z on double logarithmic scale, the graph is almost linear between $z = 0.1$ m and $z = 0.8$ m. This indicates that the distribution of joint spacings has a self-similar structure over this range of scales.

The fractal structure of spacings can reappear through the construction of a uniform *Cantor set*. The construction process of a one-third *Cantor set* is illustrated in Fig. 2. The solid line of unit length is divided into three parts and the center third is removed. The process is repeated and, finally, numbers of periodic points distributed on a straight line are gained. If taking finite periodic points to observe, the combination of these points is quite similar to the distribution of points, of which the discontinuities intersect a scanline. Thus, based on the construction process shown in Fig. 2, the *Cantor set* distribution of spacings can be deduced (Zhang et al. 1993):

$$y(\geqslant x) = \frac{(x/x_0)^{-D} - (x_m/x_0)^{-D}}{1 - (x_m/x_0)^{-D}} \qquad (1)$$

where $(y \geqslant x)$ = distribution probability of joints with a spacing value equal to or greater than x; x_0 = minimum value of spacing; x_m = maximum value of spacing; D = fractal dimension of spacing.

The calculating results, based on the measured data, showed that the *Cantor set* distribution contains much more information of spacing values than the negative exponential distribution and can describe the joint composition of each spacing class accurately (Zhang et al. 1993). Moreover, the new distribution may restrict the range of joint spacings, which embodies the dimension effect of engineering rock masses.

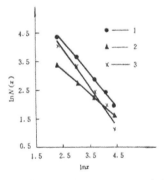

Fig. 1 The relation between $N(x)$ and x in Hua Ziyu mine

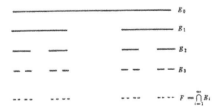

Fig. 2 The construction process of joint spacing composition

2.2 Monte-Carlo simulation of natural fragment sizes

The mutual intersection of more than two sets of joints cuts a rock mass to natural fragments with a certain size distribution. The field investigation discovered that even if the traces of a set of joints are longer, in the case of them intersecting other joint sets, the size of a natural fragment equals the maximum value of spacings of two adjacent joints among all the joint sets. This indicates that the sizes of natural fragments are controlled by the distribution of spacings.

Consider three mutually orthogonal scanlines through a rock mass intersecting joints (more exactly discontinuities) at a variety of points and measure the spacing values of joints intersected by each scanline x_i, $(i = 1, 2, 3)$. The *Cantor set* distributions of joint spacings in three orthogonal directions can be given by:

$$y_i(x) = \frac{(x/x_0^i)^{-D_i} - (x_m^i/x_0^i)^{-D_i}}{1 - (x_m^i/x_0^i)^{-D_i}} \qquad (2)$$

where x_0^i, x_m^i, D_i = minimum spacing value, maximum spacing value and spacing fractal dimension on the ith scanline, respectively.

Considering that the value of $y_i(x)$ is given over the interval $[0, 1]$, according to the *Monte-Carlo* simulation, three different random variables following the distribution of equation (2) may be produced by using inverse transformation theorem. Thereby, the formula of inverse transformation can be obtained from equation (2):

$$x = x_0^i \{R[1 - (x_m^i/x_0^i)^{-D_i}] + (x_m^i/x_0^i)^{-D_i}\}^{-1/D}. \qquad (3)$$

where R = random number uniformly distributed over the range of dimensions $[0,1]$.

The distribution of natural fragment sizes is actually to calculate the volume distribution of which a fragment volume equals the product of three random variables created by the above sub-distributions. In calculation, first of all, determine the number of simulated rock fragments N, on the basis of the accuracy of spacing simulation. In general, when N is 5000 to 10000, the relative accuracy of simulation is smaller than 2%. Secondly, get three random numbers from the three sub-distributions respectively, multiply one number with other two numbers and obtain the volume of a rock fragment. This calculation is repeated until all the volumes of N fragments are obtained. Finally, regard the maximum value among the three random numbers of each fragment as the size of this fragment and screen all the fragments with computer calculation in accordance with various size classes. The natural fragment-size distribution can be given in this way.

In the light of the simulation method mentioned above, authors have developed a programme called NSS (Natural Size Simulation) for calculating the natural fragment sizes. Based on the measured data from six experiments of blasting at Hua Ziyu mine, the average relative error of simulations of natural fragment sizes does not exceed 8.7%.

3 LAW OF BLASTED FRAGMENT FORMATION

3.1 Fractal characteristic of blasted fragments

Taking any blasted rock fragment to observe, we can not exactly describe its shape. But the geometric shapes of a number of rock fragments with various sizes seem to have a common characteristic.

Take more than ten samples at random from the pictures of magnesite fragments of Hua Ziyu mine and count up the projection shapes of magnesite fragments. Among the one hundred and sixty-eight counted fragments, the projection quadrilaterals account for 80. 36% of the total number of fragments. The others are projection triangles and pentagons. Measure the maximum and minimum interior angles of these projection quadrilaterals and draw the envelope curves of distributions of interior angles as shown in Fig. 3. The maximum and minimum interior angles vary on the interval 98°~ 131° and 48°~ 78° respectively, which shows that the magnesite fragments produced in blasting have a similarity of geometric shapes.

From the angle of statistics, it can be considered that for a rock type the shapes of blasted fragments are approximately similar to one another. Small-scale fragments can be regarded as the reduction of large-scale ones, i.e. the shapes of blasted fragments have a self-similar structure.

Through counting up the fragment sizes of three production blasts with a random sampling at Hua Ziyu magnesium mine, it was found that the number of blasted fragments $N(x)$, with a characteristic linear dimension greater than x, is plotted against x on double logarithmic scale, the graph is almost linear within the range of $x = 0.05 \sim 0.60$m. The relation curves between $N(x)$ and x for No. 1 and No. 3 blasts are illustrated in Fig. 4, which indicates that the

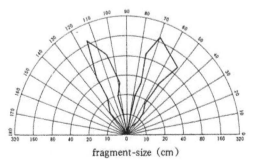

Fig. 3 The envelope curves of interior angles of projection quadrilaterals

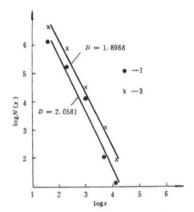

Fig. 4 The relation between $N(x)$ and x of No. 1 and No. 3 blasting

distribution of blasted fragment sizes is also a fractal structure and the fractal dimensions of three fragment-size distributions are 2.0581, 1.5936 and 1.8968 respectively. The process of blasted fragment formation may be described using the fractal construction of rock blocks.

3.2 Fractal description of blasted fragment formation

From the angle of fractal geometry, the process of rock masses blasting could be described as: rock masses are initially fragmented into finite blocks whose shapes are similar to one another, among of which part of the fragmented blocks, under the action of blasting loads, are individually fragmented into first-order sub-blocks with the similar shape to original ones; part of these sub-blocks are again broken into second-order sub-blocks with the same shapes. The process is repeated at successively higher orders and, as a result, a mass of sub-blocks with various sizes are produced as illustrated in Fig. 5.

Suppose the total volume of rock fragments in the process of fractal construction is a constant, the probability that a fragment will be broken into sub-fragments is f and the similar ratio of fragmentation is r. In accordance with the fractal construction model of blasted fragments shown in Fig. 5, the fragment-size distribution of blasted fragments can be deduced (Zhang et al. 1994) as:

$$y = (x/x_m)^{3-D} \qquad (4)$$

where y = cumulative volume probability of blasted fragments with a linear dimension smaller than or equal to x; x_m = maximum value of fragment sizes; D = fractal dimension of fragment-size distribution.

Equation (4) is essentially the same as the classical Gates-Gaudin-Schumann (G-G-S) distribution

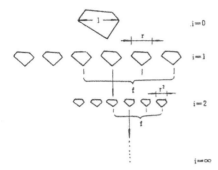

Fig. 5 The fractal construction model of blasted fragments

function: $y = (x/x_m)^n$. Comparing the function with equation(4), we have

$$D = 3 - n \qquad (5)$$

where n =distribution modulus of fragment sizes.

According to the fractal definition of blasted fragments (Mandelbrot 1983), the fractal dimension D can be expressed as $D = 3 - (\log f / \log r)$. Thus, from equation(5) we can obtain:

$$f = r^n \qquad (6)$$

Therefore, the fractal model of blasted fragment formation not only demonstrates that the fractal structure of fragments is the mathematical fundament of G-G-S distribution function, but points out the fact that the fragment-size distribution depends upon both the breakage probability and the fragmentation similar ratio.

If the fragmentation similar ratio of all the rock elements in the same blasting region is considered as a constant, the primary problem of determining the fragment-size distribution of the whole blasting region is to calculate the breakage probability of each rock mass element.

4 DAMAGE MODEL OF FRAGMENT-SIZE CALCULATION IN BLASTING

4.1 Damage view point and damage variable

A lot of investigations into blasted rock fragments discovered that most of the fragment surfaces are original weak planes and many discontinuities still exist in these fragments. This shows that under the action of explosion loads, part of the joints in rock masses simultaneously burst and propagate, which results in rock masses fragmentation. The fragmentation process in blasting is just an object of study and description of

continuum damage mechanics. Seeing that the size of joints and cracks in a blasting region is much smaller than that of rock masses and they occupy only a little volume, joints and cracks may be regarded as defects of rock masses and the problem of joint cracking in blasting process can be solved by means of treatment of comtinuum damage mechanics. In this way, we can put forward a damage mechanics model of fragment-size calculation of jointed rock masses (BDM) in this paper.

If the fragmentation of a rock mass element in a blasting region is considered as rock cracking and joint cracking, the original cracks in a rock mass can be treated as initial damage and the joints as a latent source of damage developing. In accordance with the damage definition (Chaboche 1988), the damage variable of a rock mass element can be given by:

$$\omega = \omega_0 + \left[\sum_{i=1}^{m} N_i' \bar{l}_i^2 / V \sum_{i=1}^{m} \lambda_i^{3/2} \bar{l}_i^2 \right] \qquad (7)$$

where ω_0 = initial value of damage, $\omega_0 = \sum_{i=1}^{m} N_i' \bar{l}_i^2 / [V \sum_{i=1}^{m} \lambda_i^{3/2} \bar{l}_i^2]$; $\lambda_i, \bar{l}_i, N_i^0, N_i'$ = area density, mean trace length, original crack number and cracked joint number of the ith set of discontinuities, respectively; m = number of discontinuity sets; V = volume of a rock mass element.

Under the action of explosion gases, part of the joints simultaneously burst and propagate until cracking planes occur along the trace directions of joints, which causes the increase of damage variable. When all the joints are completely broken, the damage variable $\omega = 1$. Equation(7) embodies the internal relation of which the damage value of a rock mass increases with the growth of the number of cracked joints.

4.2 Fundamentals of fragment-size distribution model

The existent models of blasting fragment-size calculation consider stress waves to be the principal power of breaking rock masses. For the complete and homogeneous rock masses with non-growing joints and cracks, we can approximately calculate the energy of stress waves created in explosion or analyse the attenuation of stress waves energy resulted from a single joint. In actual engineering, however, the blasting operations are usually carried out in rock masses containing amounts of joints and cracks. Even though several joints and cracks exist in rock masses, they will cause the energy of stress waves to attenuate rapidly and their propagation directions to change. For this reason, it is quite difficult to accurately and reliably calculate the stress waves energy distributed in a blasting region. In addition, the propagation velocity of stress waves is much greater than that of cracking (in general, over 3~4 times), therefore, the stress waves only fragment a small part of rock masses near the charges and cause original cracks to radialize (Gao et al. 1991). The BDM model will consider the quasistatic pressure of explosion gases as the primary power of fragmenting rock masses.

There are two possibilities in the blasting process that a rock mass cracks along the trace directions of discontinuities and rock cracks itself accompanying with producing new fracture planes. According to the fractal construction model of blasted fragments, it is obvious that the blasted frogment sizes depend upon both the breakage probability and the fragmentation similar ratio. Now that the explosion stress is uniformly distributed in a rock mass element by and large, we can assume the fragmentation similar ratio of every element ⟩ be a constant. Thus, the number of cracks produced ⟩y explosion must determine the fragment-size composition of a muckpile. we put forward a new concept — rock mass fracture ratio which includes two factors:

$$K = K_j + K_r, \tag{8}$$

where K = rock mass fracture ratio; K_j = ratio of cracked joints and original cracks, $K_j = N_j/N$; N_j = number of cracked joints and original cracks; N = total number of joints and cracks; K_r = ratio of rock cracks created in blasting, $K_r = N_r/N$; N_r = number of rock cracks.

Rock mass fracture ratio varies with the positions of elements in a blasting region. To calculate fragment sizes accurately, it is necessary that rock masses should be divided into finite rock mass elements by the non-homogeneous treatment as shown in Fig. 6. Thus, the

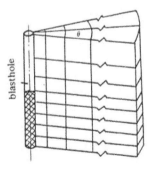

Fig. 6 The illustration of dividing rock masses into elements

stress decrement of a element in the direction of loading is much small and the rock mass fracture ratio of each element can be assumed to be a constant.

According to the damage mechanism of rock masses blasting (Zhang 1993), the rock mass fracture ratio is given by:

$$
\begin{cases}
K_j = \dfrac{F(\sigma) - \omega_0}{1 - \omega_0} - \dfrac{1 - F(\sigma)}{1 - \omega_0} \cdot \dfrac{\sum\limits_{i=1}^{m} N_i^0}{\sum\limits_{i=1}^{m} N_i} \\[4mm]
K_r = \dfrac{\{\mathrm{INT}[e/e_t]/M_\theta + \mathrm{INT}[\sigma_v/\sigma_c]h/H_c\}}{\sum\limits_{i=1}^{m} N_i}
\end{cases}
\tag{9}
$$

where $F(\sigma)$ = function of rock mass strength, in general, $F(\sigma)$ follows $Weibull$ distribution; N_i = total number of the ith set of joints and cracks; e = circumferential tensile strain; e_t = strain value of rock dynamic tensile strength; σ_c = stress value of rock dynamic tensile strength; σ_v = vertical stress component of a rock mass element; h = vertical height of a rock mass element; H_c = charging length of a blasthole; M_θ = element number of uniformly dividing rock masses in the circumferential direction of a blasthole.

4.3 Calculation method of B D M model

From analysing the fragment-size distributions before and after blasting, it was found that the fragment-size class, in which fragment content is higher before blasting, still contains more volume of fragments after blasting (Wang 1986), which shows that the fragment sizes are determined by natural fragment-size composition and rock mass fracture ratio. If a cube rock is exerted a external load and broken into pieces, usually, its fragment sizes follow a certain distribution function. In this way, it may be considered that the blasted fragments of a rock mass element have a certain fragment-size composition. But the ratios of all the elements are different from one another, resulting in the various fragment-size compositon. Supposing the fragment sizes after blasting are divided into l classes, i. e. $x_1 < x_2 < \cdots \cdots < x_l$, the cumulative distribution of volumes of blasted fragments can be expressed as:

$$Y_j = \sum_{i=1}^{M} [V(i)y_j(i)] \Big/ \sum_{i=1}^{M} V(i) \tag{10}$$

where Y_j = volume distribution of fragments with a size smaller than x_j, $j = 1 \sim l$; $V(i)$ = volume of the ith rock mass element, $i = 1 \sim M$, M is the number of elements of a blasting region; $y_j(i)$ = volume ratio of blasted fragments with a size smaller than x_j in the ith rock mass element after blasting.

The mutual intersection of each cracked joint and each rock crack contributes to forming blasted fragments. In accordance with the fractal construction model of blasted fragments, the rock mass fracture ratio of each fragment-size class in a rock mass element can be assumed to be constant. For this reason, the volume contents of all the fragment-size classes in a rock mass element will keep constant respectively before and after blasting. In this way, the fragment-size composition of a muckpile is calculated through the fracture ratios of rock mass elements.

Owing to the equivalent spacing regularity of the same class-order joints (Sun 1988), we can assume various kinds of cracks created in blasting process to follow this regularity. In the light of statistical analyses, it is considered that the sizes of blasted fragments are the natural sizes divided by K. Suppose the distribution function of natural fragment sizes is F_0 which is given by the above-mentioned *Monte-Carlo* simulation. Thus, the volume ratio of blasted fragments of the ith rock mass element can be obtained:

$$y_j(i) = F_0 \big[K(i) x_j(i) \big] \tag{11}$$

where $K(i)$, $x_j(i) =$ rock mass fracture ratio and fragment size of the ith element, respectively.

The BDM model consists of equations (10) and (11). This model embodies the fractal characteristic of blasted fragment-size composition and reflects the essence of joint cracking and rock cracking in rock masses blasting. It becomes possible to relate the fragment-size calculation with the fragmentation mechanism of rock masses blasting through rock mass fracture ratios.

5 CONCLUSION

From the above studies, we can obtain some important results as follows:

1. The uniform *Cantor set* can reappear the fractal structure of joint spacing composition and the *Cantor set* distribution embodies the characteristic of dimension effects of engineering rock masses. The composition of natural fragment sizes may be calculated using *Monte-Carlo* simulation.

2. The blasted fragments of a rock type have a property of statistical self-similarity and the distribution of blasted fragment sizes is also a fractal structure. The fractal dimension of fragment sizes depends upon breakage probability and fragmentation similar ratio.

3. The mechanism of jointed rock masses blasting can be analysed by means of treatment of continuum damage mechanics. The BDM model bridges the relation between the process of rock masses blasting and its fragment-size composition and makes up the deficiency of existent models which can not reflect joint cracking.

ACKNOWLEDGEMENTS

The author is particularly indebted to Professor Xu Xiaohe and Niu Qiang (Mining Engineering Department of Northeastern University) for their guidance of this work.

REFERENCES

Chaboche, J. L. 1988. Continuum damage mechanics: part I — general concepts. *J. Applied Mechanics.* 55:55—59.

Gao, J. S. & Zhang, J. C. 1991. Dynamic analysis of the mechanism of rock fragmentation by explosicon. *Proc. Int. Conf. Eng. Blasting Tech:* 288—294. China: Beijing.

Mandelbrot, B. B. 1983. *The fractal geometry of nature.* San Francisco: Freeman.

Sun, G. Z. 1988. *Rock masses structural mechanics.* Beijing: Science press.

Wang, W. X. 1986. Influence of discontinuity distribution in rock masses on blasted fragment-size distribution and its computer simulation. *Metal Mine.* 3:15—19.

Zhang, J. C., Niu, Q. & Xu, X. H. 1992. Summary of fragment-size predicting model in rock masses blasting. *Blasting.* 9(4):63—69.

Zhang, J. C. 1993. *Study on fragment-size resulting from jointed rock masses blasting.* Doctorate Thesis of Northeastern University. 56—59.

Zhang, J. C. & Xu, X. H. 1993. Fractal study on fracture geometry of rock masses. *Trans. of Nonferrous Metals Soc. of China.* 3(4):11—15.

Zhang, J. C., Niu, Q. & Xu, X. H. 1994. Fractal study on forming law of fragmentation in process of rock mass blasting. *Metal Mine.* 11:9—12.

Modelling of blasting process

Rock Fragmentation by Blasting, Mohanty (ed.) © 1996 Taylor & Francis. ISBN 90 5410 824 X

'Blastability' and blast design

Andrew Scott

Julius Kruttschnitt Mineral Research Centre, The University of Queensland, Qld, Australia

ABSTRACT: Rock mass properties largely control blasting performance. However, despite many years of fundamental work aimed at describing the role of rock mass properties in blasting, these properties are not routinely included in blast design procedures. This paper reviews the conventional methods used to account for rock mass properties in the design of blasts and discusses the concept of 'blastability'. In the absence of suitable methods for the incorporation of rock mass properties in blast design, an approach is suggested where blasting engineers can develop a 'local blastability scheme' in order to derive basic designs for different rock mass conditions. These basic designs can then be optimised through operating experience assisted by monitoring, analysis and modelling.

1. INTRODUCTION

There are four critical aspects to a rock blasting system: the rock mass, the explosive, its distribution and the detailed detonation timing. The latter two aspects have received quite rigorous analysis in recent years with the development of a number of useful computer based blast design editors and models. Explosive performance can be described by complex detonic codes but these have largely remained the domain of a few experts and have yet to be linked to blasting performance in a manner suited to routine blast design. Rock mass properties have a controlling influence on blast performance but remain too complex to be explicitly incorporated into current blast design procedures.

Geotechnical engineers have approached this problem from a different perspective. The design of most rock excavations is guided by rock mass classification schemes. These schemes rate the suitability of a rock mass for a particular duty based on simple measures of rock mass properties. The properties used are those that can be readily measured and that affect the critical engineering requirements for that duty. Each property is weighted in the overall analysis to reflect its influence. A similar approach is being used by the JKMRC to describe the behaviour of a rock mass when blasted. The challenge is to define a 'blastability' index which incorporates the most relevant parameters according to their influence over blasting performance. The resulting description of 'blastability' can then be used to derive basic blast designs for new blasting situations.

2. BLASTING MECHANISMS

The properties of the rock mass are of fundamental importance to the design of blasts. The use of standard designs without regard to any variation in rock mass properties will lead to either over blasting in some areas and under blasting in others or, more typically, consistent over-blasting. Fines generation and damage to adjacent structural rock are also common results. The real cost of these departures from targeted performance can be several times the cost of the initial drilling and blasting operation (Scott 1992). If the desired blasting results are to be achieved over a range of rock mass conditions, it is necessary to incorporate the most relevant rock mass properties into the blast design process.

Blasting performance is determined by the interaction of the detonation products of an explosive and the confining rock mass. Rock mass properties dominate this process. The blasting engineer is therefore faced with the challenge of determining which rock mass properties most influence blasting performance in each situation and of deciding how designs should be changed to suit different geological conditions.

Sarma (1994) provides a useful description of the blasting process as a background to modelling explosive-rock interaction. According to Sarma's model, when an explosive detonates, the ingredients of the explosive are rapidly converted into gaseous products at very high temperature and pressure. The high pressure gases impact the blast hole wall and transmit a shock wave into the surrounding rock as shown on Figure 1. The stresses resulting from the shock wave compress and crush the rock in the vicinity of the blast hole in response to the blast hole pressure and the strength and stiffness of the rock. As the rock is compressed and crushed, the volume of the blast hole increases until it reaches a quasi-static equilibrium state where the explosion gas pressure is matched by the stress of the blast hole wall.

Beyond the crushed zone, the shock wave compresses the material at the wave front and induces a tangential tensile stress. If the intensity of this stress is greater than the dynamic tensile strength of the rock, radial fractures will develop.

During and after the stress wave propagation, high pressure explosion gases penetrate the available fractures extending and dilating them as shown on Figure 2. This leads to an increase in the volume and permeability of the rock mass and a subsequent reduction in the pressure of the explosion gases. The lack of restraint available at a free face (including the surface) creates an imbalance of forces and causes the burden rock to move. Eventually the gases find their way through the fracture network or stemming to the atmosphere. The confining pressure is then rapidly reduced and rock movement continues as a result of the momentum imparted up until this point in the process.

This model of the blasting process identifies several rock mass properties important to blasting performance. These include:

- *rock mass stiffness,* which controls the distortion of the blast hole wall and hence the pressure developed in the blast hole and the partition of the explosion energy into shock and heave
- *dynamic compressive strength,* which controls the crushing that occurs at the blast hole wall

- *the attenuation parameters of the rock mass,* which control how far the stress wave travels before its energy falls below the levels that cause primary breakage
- *the dynamic tensile strength of the rock,* which influences the extent of new fracture generation in both the shock and gas phases of breakage
- *the in-situ fracture frequency, orientation and character* which together define the in-situ block size distribution and influence the attenuation of the shock wave and the migration of the explosion gases
- *the density of the rock mass,* which affects its inertial characteristics and hence how the rock mass moves in response to the forces applied during blasting.

Any significant change in these properties requires some aspect of the blast design to be modified to suit the different blasting conditions. The blast geometry and initiation timing control how the explosion energy is utilised to achieve the outcome sought by the blasting engineer. All of these design parameters must be juggled to produce the final desired result - a muckpile of the required shape, looseness and fragment size distribution.

3. TRADITIONAL APPROACHES TO BLASTABILITY

3.1 The Rock Constant "c"

For many practicing engineers, Langefors and Kihlström (1978) mark the beginning of the modern

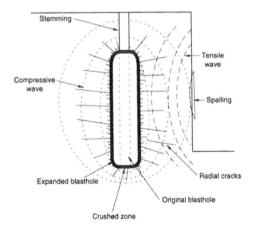

Figure 1   Shock Wave Effects

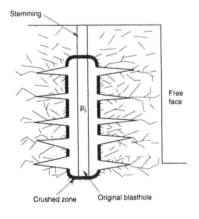

Figure 2   Gas Expansion

era of blast engineering. In their book "The Modern Technique of Rock Blasting", which was first published in 1963, the influence of the rock is represented by a 'rock constant', designated c, representing the base charge concentration required for satisfactory blasting performance. Formulae are provided to describe how the powder factor and other blast design parameters should be varied for particular blasting geometries. Langefors and Kihlström worked mainly in the brittle, stiff rocks of Scandinavia and considered a nominal value of $c = 0.4$ kg/m^3 to be an adequate starting point for most designs. Specific field trials were recommended to optimise designs for rock that varied either in strength or structural characteristics from the standard Scandinavian granites.

This concept is further developed by Persson et al (1993) who describe a test blast arrangement to determine the specific charge just required to loosen and displace a cubic metre of rock. Values for c of around 0.4 kg/m^3 are reported for most rock materials, although values as low as 0.2 kg/m^3 and as high as 1.0 kg/m^3 are noted depending on the extent and orientation of fissures or strata. Once determined, the rock constant can be used to estimate the required burden and other blast design parameters.

3.2 Fracture Mechanics Approach

A great deal of effort has gone into model blasting where small rock samples are blasted under controlled conditions (e.g. Rustan et al 1983). Many of these studies have been reported in previous Fragblast proceedings and are too numerous to list here. Parameters such as critical burden, critical charge, break out angle and resulting fragmentation size distribution have been related through these empirical experiments. While each experimental data set produces relationships between the various parameters, no unified theoretical description of blasting performance based on material properties and blast design parameters has been devised. The scatter in the data invariably leads to the possible selection of a wide range for each parameter and the designer is left with broad design recommendations such as those summarised by Hagan (1983) or Hoek and Bray (1981).

Most of the theoretical and small scale blasting investigations deal with essentially homogeneous intact rock, whereas in reality, the rock to be blasted will be affected to some extent by jointing, bedding, foliation and small scale fractures. Many of these features are not significant in terms of the rock mass static behaviour or the stability of an excavation and are often overlooked in conventional geotechnical logging. They are, however, critical in terms of the blasting behaviour of the rock mass.

3.3 Common Practice

Most blast optimisation involves the modification of designs based on observed performance. Hagan (1983) provides extensive commentary as to how varying particular blast parameters will affect blasting performance. Advice is provided as to how particular aspects of the design should be altered when the rock is found to be more structured, more massive, stronger, weaker etc. General guidelines for blast design are readily found in the commercial literature (DuPont Blaster's Handbook 1977, Atlas Powder Company 1987, AECI 1986, ICI 1991) but little information is available on which new blast designs can be based for particular rock types. Common practice in most mines is to adopt standard designs that have been found by trial and error to provide satisfactory results in similar blasting environments.

However, rock mass conditions continually change, and it is necessary to modify blast designs to suit different rock mass conditions. Many operations maintain a conservative basic pattern which has been proven in the most difficult material on site. This results in many of the mine's blasts providing excessive energy to the weaker or more structured materials. A better approach would be to develop a number of basic blast designs each suited to the different rock mass conditions.

4. ASSESSMENT OF ROCK MASS PROPERTIES

4.1 Controlling Parameters

There have been many attempts to predict blasting performance from the physical and mechanical properties of rock specimens measured in the laboratory. These attempts have generally not been successful and Hagan and Harries (1977) attribute this to the effects of rock mass structure on the detailed mechanisms involved in blasting. However, certain relationships appear valid:

- strong rock requires greater shock energy to create new primary fractures
- the absence of fractures or discontinuities increases the blasting effort required to achieve a given degree of fragmentation
- soft or plastic rock tends to absorb shock energy and requires more heave energy to create a loose muckpile
- high density rock requires more energy to loosen and displace it than low density rock.

Such relationships can be used as a basis for a description of the blasting characteristics (or 'blastability') of a rock mass. To be effective, rules based on these sort of relationships must be based on an adequate measure of the property concerned and must be applied in the absence of 'rogue' factors that have not been included in the analysis but have a significant influence over the results.

A rock mass can comprise several different rock types and be affected by different degrees of fracturing and varying stress conditions. A number of different rock mass rating schemes have been developed for geotechnical purposes (RQD - Deere et al 1967; NGI Q - Barton et al 1974; RMR - Bieniawski et al 1974). Each of these schemes has been developed for particular design applications and they have been modified by subsequent workers to suit their particular requirements. These traditional rock mass ratings have not been widely applied to blast design although further work in this area is warranted.

Experience at the JKMRC in blasting a wide range of rock types suggests that the controlling rock mass parameters fall into the following categories:

- strength parameters
- mechanical parameters
- absorption parameters
- structure parameters
- comminution parameters.

In order to incorporate these properties into blast designs, they must be quantified in a consistent and representative manner.

4.2 Strength

Static compressive, tensile and shear strength are generally determined by testing specially prepared samples in the laboratory. Large numbers of tests are required for the results to have statistical relevance as the results can vary considerably. Several excellent texts have been published describing the recommended procedures for intact rock testing and the reader is referred to Hoek and Brown (1980) and Hoek and Bray (1981) for descriptions of the tests used. Standards have also been prepared (e.g. ISRM 1972, 1977) covering the performance of these tests.

Rock, in common with many other materials, is strain rate sensitive. The strength of rock increases as the rate of loading increases. This means that the strengths that are appropriate for the analysis of blasting are quite different from the classic static unconfined tests. Mokhnachev and Gromova (1970) found that weak rocks exhibit a stronger dependence upon strain rate than strong rocks indicating that a universal adjustment where dynamic strength is taken to be x times larger than static strength is inappropriate.

The JKMRC is modifying a laboratory method to determine the minimum energy required to break a rock sample of known competence. The technique utilises the Hopkinson Split Bar (Figure 3) to simulate both the magnitude and rate of loading experienced by rock under blasting conditions. The resulting breakage energies will be used to represent the breakage aspects of the blasting task in a future 'blastability' analysis. These energies can also be input into engineering breakage models such as extended Kuz-Ram or comminution based models to predict fragmentation.

4.3 Mechanical Properties

The mechanical properties of Young's Modulus and Poisson's Ratio control how the rock behaves under loading and the partition of explosive energy. Young's Modulus relates stress and strain under linear behaviour and Poisson's ratio is the ratio of lateral strain to longitudinal strain for uniaxial stress in the longitudinal direction.

If a critical strain is taken as a criterion of failure, then as the value of Young's modulus increases, the stress and strain energy which must be provided by the explosive must also increase. Young's modulus is also dependent upon the state of stress in the rock, the

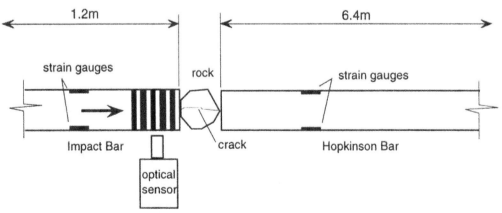

Figure 3   Hopkinson's Bar (JKMRC 1996)

loading rate and heterogeneous features such as bedding, foliation and microstructure.

4.4 Absorption Parameters

The ability of the rock mass to transmit or absorb blast energy influences the choice of explosive, its distribution and timing. The amplitude of the seismic pulse diminishes as it propagates through the rock mass. This occurs as result of two mechanisms:

- Geometric expansion of the wave which results in a lowering of the energy contained per unit of rock volume, but does not cause any overall loss of energy
- Energy dissipative mechanisms effectively remove energy from the pulse. This energy loss is a result of the internal friction within the material. The energy may either be stored locally in the material or dissipated as heat.

In rock masses which dissipate a high proportion of the blast energy through internal friction there will be a high attenuation of the shock wave. Consequently, the fracturing processes will be severely impeded. The seismic characteristics can be measured during detailed blast monitoring (JKMRC 1991, 1994).

4.5 Structure

The blasting process may be visualised as a process of size reduction brought about by the addition of explosive energy, as shown in Figure 4. The properties of both the intact rock and the discontinuities play a role in determining the amount of explosive energy which must be applied to the rock mass to achieve the required breakage.

The distribution of the size and shape of the natural blocks and intact bridges that make up a structured rock mass have a profound effect on fragmentation and blasting performance. In many blasting situations the muckpile is dominated by natural rock fragments that have simply been loosened and freed by the blast. If the natural block size is substantially larger than the fragment size needed from the blast, then considerable explosive energy will have to be provided to fragment the natural blocks down to the desired size range.

Conventional geotechnical mapping and logging of rock mass fractures is generally conducted for the purpose of assessing the stability of a slope or bench. Many of the rock mass discontinuities of importance to blasting are of smaller scale and not considered relevant to stability. It is thus possible to find that a rock mass described as massive or sparsely jointed may in fact contain close spaced bedding or foliation that has an important influence on fragmentation. Specialised mapping and structural analysis procedures have therefore been developed to record and model the rock mass structures important to blasting (Villaescusa 1991, Hudson and Priest 1979).

4.6 Comminution Characteristics

The blasting mechanisms described earlier refer to crushing of the blast hole wall close to the explosive and the dynamic fracturing of rock within the blast burden by both shock and gas related mechanisms. Comminution is the reduction of the size of a particle through the application of energy. The term is usually applied to the crushing and grinding processes commonly carried out downstream of the mining operation. However, the parameters used to describe the breakage behaviour of rock when it is crushed or ground has the potential to help describe the breakage during blasting as well.

Characterisation of comminution behaviour can be conveniently divided into two basic types of tests: those which specify a hardness or strength parameter, and those that describe the extent of breakage achieved after a known breakage event (JKMRC 1996a). The former include the conventional rock mechanics tests referred to in Section 4.2 and the latter refer to single particle breakage tests. Although normally applied to crushing and grinding behaviour, the JKMRC has recently had considerable success linking these descriptors of breakage behaviour to the modelling of the linked production processes of blasting, handling and crushing (Kojovic et al, 1995).

5. BLASTABILITY

5.1 Traditional Approaches

There is no proven theoretical approach to the definition of blasting requirements based on simple rock mass properties. This is testimony to the complexity of rock as a material and blasting as a process. However, there have been many attempts to define the

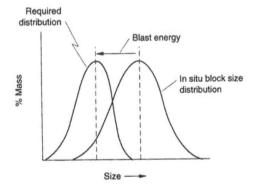

Figure 4   The Fragmentation Process

'blastability' of rock. Many of these approaches are based on observations in particular rock types and most treat the rock as a homogeneous, brittle material. Examples of these schemes have been reviewed by Kropp, 1996. They include an empirical formula proposed by Fraekel in 1944 for the height and diameter of the required charge based on blast geometry and "blastability" (Langefors and Kihlström 1978).

Hino (1959) argues that the upper limit of rock mass strength is the compressive strength and the lower limit is the tensile strength. He defined a 'Blasting Coefficient' to be the ratio of the two strengths. Heinen and Dimock (1976) proposed a method for describing rock mass blastability based on their experience at Kennecott Copper's Ely mine in Nevada, USA. They related the seismic velocity of the rock mass to the average powder factor required.

Borquez (1981) defined a critical burden based on the tensile strength of the rock, explosive detonation pressure, diameter of the charge and a blastability factor which was linked to the rock mass RQD, modified to account for the strength of the joints. Borquez's blastability factor could be determined directly from drill core making it easy to apply to 'greenfield' sites. Rakishev (1982) defined a critical fracture velocity to describe the blastability of rock based on a combination of compressive and tensile strengths and a coefficient related to the structural properties of the rock mass.

There have been many other relationships proposed and used to guide blast design in particular blasting situations. The approaches outlined above are typical of the forms used.

5.2 Empirical Approaches

5.2.1 Ashby

Many blasters develop their own approach to define how to blast different materials. The problem is that most of these approaches have never been formalised or structured in any systematic manner and they depend on casual observation of local conditions and resulting performance. Nevertheless, the innate skill of experienced practitioners has served the industry well for many years. An impressive example of an empirical blastability scheme was developed by Ashby for the Bougainville Copper Mine (Hoek and Bray 1977).

Ashby used the graphical relationships in Figure 5 to describe the powder factor required to adequately blast material at Bougainville based on the fracture frequency and effective friction angle. Ashby had apparently determined that the density of fracturing and a measure of the strength of the structured rock mass (represented by the friction angle) had the greatest impact on blasting performance. These rock mass descriptors were determined for various areas in the pit. It was then a simple matter to consult the chart to derive a target powder factor for design.

5.2.2 Lilly

Lilly (1986) developed a blasting index based on rock mass, joint density and orientation, specific gravity and hardness. The index was closely correlated with powder factor for blasting in open cut iron ore mines.

$$BI = 0.5(RMD + JPS + JPO + SGI + H)$$

Lilly's characterisation parameters are described in Table 1.

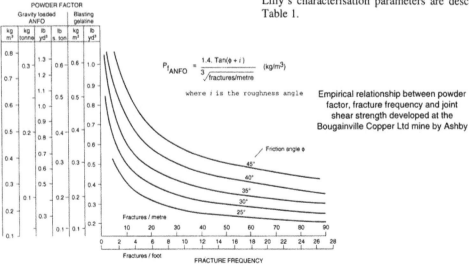

Figure 5   Ashby's Blastability Scheme for Bougainville (Hoek and Bray 1977)

32

Table 1: Ratings for Blastability Index Parameters (Lilly 1986)

	Parameter	Rating
1.	Rock mass description	
	Powdery/Friable	10
	Blocky	20
	Totally Massive	50
2.	Joint Plane Spacing (JPS)	
	Close (<0.1m)	10
	Intermediate (0.1 to 1 m)	20
	Wide (>1 m)	50
3.	Joint Plane Orientation (JPO)	
	Horizontal	10
	Dip out of face	20
	Strike normal to face	30
	Dip into face	40
4.	Specific Gravity Influence (SGI)	
	SGI = 25 x SG - 50	
	(where SG is in t/m³)	
5.	Hardness (H)	1 to 10

To use Lilly's blastability model as a blast design tool, it is necessary to develop a relationship between the blastability index and the powder factor required to break the rock mass. This relationship is site specific and must be developed from either historical data or a series of trial blasts under various rock conditions.

5.3 Incorporating Blasting Outcomes

The JKMRC has had some success using an approach that classifies the rock mass according to the properties known to affect blasting performance and taking into account the objectives of the blast. A blastability analysis has been developed for coal measure strata incorporating the following parameters:

Rock Mass	Strength
	Density
	Young's Modulus
Structure	Average In Situ Block Size
	Influence of Structure
Design	Target Fragment Size
	Heave Desired
	Confinement Provided
	Scale of Operations
Environment	*Water*

These parameters can be measured or estimated fairly easily in the field. The parameters in *italics* are regarded as design modifiers - factors which make blasting easier or more difficult depending on whether they favour the desired result or not. A certain amount of judgement or site experience is inevitably required to interpret the effect of these parameters on blasting performance.

The required powder factor using ANFO was estimated for a range of materials and blasting situations for an open cut coal mine in the Hunter Valley of New South Wales. An example of the analysis is shown in Table 2. Current blasting practices were used to derive

Table 2: Blastability Analysis for Coal Operations

		Dragline Operatn.	Dragline Operatn. Cast Blast	Shovel Operatn.	Shovel Operatn. Wet	Parting F.E.L
Rock Mass						
Strength	MPa	60	60	50	50	40
Density	gm/cc	2.51	2.51	2.47	2.47	2.42
E	GPa	12	12	10	10	10
Structure						
Block Size	m	2	2	2	2	0.5
Structure	1 to 9	5	5	5	5	3
Design						
Target Fragment						
Size	m	0.5	0.5	0.3	0.3	0.15
Heave	1 to 9	5	10	5	5	7
Confine	1 to 9	5	5	5	5	7
Scale	1 to 9	3	3	5	5	7
Environment						
Water	1 to 9	1	1	1	5	1
Indices						
Strength		0.30	0.30	0.25	0.25	0.20
Breakage		0.08	0.08	0.13	0.13	0.06
Heave		0.25	0.51	0.26	0.26	0.36
Modifiers		-0.02	0.03	0.00	0.08	0.02
Powder Factor						
	kg/T	0.18	0.24	0.17	0.21	0.16
	kg/m³	0.44	0.61	0.42	0.52	0.39

the blastability relationships. These were then used to estimate future blasting requirements in near-by, but quite different strata and mining conditions.

The unconfined compressive strength, density and Young's Modulus are input to describe the basic strength and stiffness characteristics of the rock material. The average in-situ block size is estimated from observation of rock mass exposures or from the analysis of structural mapping.

The impact of structure is the first of the design modifiers. The appropriate input can only be derived from experience in the same or similar materials. The structure of some rock masses makes blasting very difficult (e.g. 4 metre massive blocks of strong rock separated by weak, open joints), and some structure makes blasting quite straightforward (intensely fractured, but otherwise brittle ground). The structure parameter in this part of the analysis allows this aspect of the ground behaviour to be taken into account. A value of 5 is neutral, 1 indicates strongly favourable structure and 9 indicates difficult, blocky ground. It should be noted that massive ground is not necessarily difficult to blast and should be given a value of 5 in this analysis.

The ratio of the target fragment size and the in-situ block size provides a measure of the new or fundamental breakage required from the design. The Heave,

Confinement and Scale parameters are further design modifiers intended to help the transfer of experience gained from one blasting environment to a new or different environment. Smaller loading equipment tends to have a lower breakout force than larger machines. A loose, more swollen muckpile is therefore required to maximise the efficiency of this equipment. This looseness comes from achieving greater heave from the blast, and the heave parameter can be used to account for this in the blastability analysis.

Confinement is sometimes inevitable in blasting because of awkward bench or pit geometry. If a blast, or portion of a blast, is destined to be over-confined or 'tight', then extra blast energy will be required. This is achieved by setting this parameter >5. By comparison, a fully exposed, open full face might attract a value of <5 for this parameter. The parameter accounting for scale is treated in the same fashion and is useful when comparing blasts in the same material, but quite different geometry.

The internal relationships utilised in this scheme are very simple, but they do recognise the impact of rock mass properties on explosive performance. The relationships used in the example shown are based on the assumptions described below. Four indices are calculated from the blastability parameters shown on Table 2.

Strength Index

The powder factor required to blast a given rock is assumed to be directly proportional to the unconfined compressive strength of the rock. In this instance, the component of powder factor attributed to overcoming the material strength is simply calculated from the UCS in MPa divided by 200.

Breakage Index

The energy required to achieve a given degree of breakage is assumed to increase with the stiffness of the rock as represented by Young's modulus. The degree of breakage required is represented by the ratio of the in-situ block size to the target block size. The higher this ratio, the greater is the breakage required. Thirty percent of the explosive energy is assumed to contribute to this breakage. This is in line with the partition of explosion energy modelled for typical coal measure strata.

Heave Index

The heave modifier is combined with Young's modulus to estimate the energy required. A rock with low Young's modulus will require more heave energy to achieve a given swell or heave. Seventy percent of the explosive energy is assumed to contribute to heave.

Modifiers

The modifiers simply increase or reduce the powder factor according to whether they assist or hamper the blasting process. For the structure, scale and confinement modifiers a value of 5 is neutral. Each unit above or below 5 increases or reduces the powder factor by 1%.

Dry blasting conditions are indicated by a value of 1 for the water parameter. Wet conditions can be accounted for by increasing the value of this parameter. Each unit above 1 adds 2% to the required powder factor.

The required powder factor is estimated from a combination of the four calculated indices. In this example, the strength and design modifiers are given greater weight than the breakage and heave indices. Provided the allocation of the subjective parameters is consistent, the scheme provides a simple and reliable basis for the estimation of powder factor.

5.4 Recommended Approach to Blastability

The blastability scheme developed by Lilly places importance on the material hardness, joint frequency and joint orientation. Ashby's 'blastability scheme' for the Bougainville mine is strongly dependent on fracture frequency and joint shear strength, and the blastability analysis described above depends on simple measures of strength, the degree of breakage sought, aspects of the blasting environment and the performance objectives of the blast. Each of these schemes is different, but each has served its creators and users effectively.

Given the broad range of rock mass conditions and the different types of mining operations and hence outcomes required, it is unlikely that a single universal 'blastability' scheme can be very effective. The creation of such a general scheme inevitably leads to equally general predictions or guidelines throughout the range of possible applications. Based on this philosophy, the following section outlines an approach to the definition and development of blastability analyses that can be easily created and used at any mine site.

6. DEVELOPING A BLASTABILITY SCHEME

6.1 Framework

There is no formal structure recommended for each blastability scheme, but the general approach described in Section 5.3 has been found to be quite serviceable in practice and is used as an example for the purposes of this discussion. The essence of the proposed approach is that different rock mass properties (domains) and different blasting operations (objectives) are identified so that the detailed analyses used for optimising the basic design can be conducted in similar materials and similar mining methods. This avoids the need to cater for radically different conditions within a single scheme.

6.2 Domains

The suggested starting point is to identify regions within the mining operation where different rock mass properties control blasting performance. These areas can be characterised as belonging to different 'domains'. A domain is defined here to be an area of the mine where the lithology and structure are such that all areas within that domain behave similarly with regard to blasting. It is likely that most operations will contain several domains, but the approach is likely to become unwieldy if more than half a dozen or so distinctive blasting domains are defined.

6.3 Blasting Objectives

Not all blasts in a mining operation have the same objective. Overburden blasts for a dragline which produce fragments of a cubic metre in size may be regarded as satisfactory, whereas a blast in a 5 metre gold bench or partings horizon in a coal mine that produces similar boulders would be regarded as having failed. The extent of movement sought in the muckpile is another important blasting target. In the blastability example in Section 5.3, the blasting objectives and environment were allowed for through the adjustment of parameters involving the target fragment size, heave desired (i.e. low heave or paddock blast, mid heave or conventional bench blasts, or high heave such as required for a dragline cast blast), confinement and the scale of the operation. On Table 2, a clear distinction is made between dragline, shovel and front end loader operations.

6.4 Critical Mechanisms

In each blasting domain and type of blast, certain mechanisms can be identified as being critical. Experience might indicate that oversize is the most critical factor in a bench blast. In this case the variation of in-situ rock structure within the bench might be the most obvious factor driving the need for changes in the blast design. In this instance, in-situ block size would have a high weighting in the resulting blastability scheme. Fines generation is often a critical issue in coal blasting. The proportion of bright, friable coal in the face might therefore control the detailed blast design required. The proportion of weathered material in the face might be the most critical factor when blasting upper mine benches.

In each case, the blastability scheme must be structured to react adequately to the most influential parameters affecting blast performance. Other factors may also need to be represented, but to a lesser degree. In the example being used in this section, rock mass strength and the degree of breakage required are the critical factors, with heave, scale and confinement parameters being of less importance.

6.5 Basic Designs

Basic designs must be derived for each type of blast in each domain. The term 'basic' is used here rather than 'standard' as many operations adopt standard designs for their operations and show great reluctance to vary these regardless of blasting conditions. The application of the same standard design across different domains must lead to inferior blasting results and is not recommended. As a minimum, different standard designs may be applied within each blasting domain. However, to optimise blasting performance, these domain specific standard designs should be regarded as the 'basic designs' for each domain and subject to tuning or optimisation as conditions vary within each domain.

6.6 Performance Assessment and Record Keeping

No ordered optimisation of blasting performance can be undertaken without a systematic process of assessing the performance of current operations and defining the conditions in which each blast takes place, so that the rules and guidelines used to up-date designs are themselves continually up-dated. Some useful software database systems are now available (JKMRC 1996b) to store measurements and observations with rock mass properties and blast design details for future reference and comparison.

6.7 Optimisation and Tuning

The final stage in the development and use of a blastability scheme is the optimisation or 'tuning' of designs through further monitoring, analysis and experience. This is the routine blast engineering process that has been practiced either formally or informally by blasting engineers for generations. The purpose of the blastability scheme described in this paper is to give this process a head start by adequately representing the rock mass properties and blasting objectives in the design process, and to help to formalise the process so that it is less dependent on the skills and interest of just one person in an organisation.

7. CONCLUSIONS

The foregoing does not represent the result of extensive fundamental research into blast design and the impact of blasting operations on mining performance. However, the approach outlined for the development of a suitable blastability scheme for any blasting situation appears to the author to provide a useful link between field practices and what is currently known and understood about the impact of rock mass properties on blasting performance.

8. REFERENCES

AECI Explosives Today, Series 2, No. 41 - March 1986.

Atlas Powder Company, 1987. Explosives and rock blasting, R.C. Morhord (ed.) ISBN 0-961-6284-0-5.

Barton, N.R., Lien, R. and Lunde, J., 1974. Engineering classification of rock masses for the design of tunnel support, Rock Mech. 6(4), 189-239.

Bieniawski, Z.T., 1974. Geomechanics classification of rock masses and its application in tunnelling, Advances in Rock Mechanics, 2 (A), 27-32, Washington, D.C. Nat. Acad. Sci.

Borquez, G.V., 1981. Estimating drilling and blasting costs - An analysis and prediction model. *Eng and Mining J.*, pp 83-89.

Deere, D.U., Hendron, A.J., Patton, F.D. & Cording, E.J. 1967 Design of surface and near surface construction in rock. *Proc. 8th US Symp. Rock Mech.*, AIME, New York, pp237-302.

DuPont Blasting Handbook, 1977. E.I. DuPont de Nemours & Co. Wilmington, Delaware, USA.

Hagan, T.N. & Harries, G., 1977. The effects of rock properties on blasting results, *A M F Inc. Drilling and Blasting Tech.*, Adelaide, May, 4/1-4.31.

Hagan, T.N., 1983. The influence of controllable blast parameters on fragmentation and mining costs, *Proc. First Int. Symp. on Rock Frag. by Blasting*, Lulea, Sweden, August 1, 31-51.

Heinen, R.H. and Dimock, R.R., 1976. The use of seismic measurements to determine the blastability of rock. *Proc 2nd Conf. on Explosives and Blasting Techniques*, (Ed. C.J.Konya), Society of Explosives Engineers, Louisville, Kentucky, pp 234-248.

Hino, K., 1959. Theory and practice of blasting, Nippon Kayaku Co. Ltd., Japan.

Hoek, E. and Bray, J.W., 1981. *Rock Slope Engineering*, 3rd Ed., The Institute of Mining and Metallurgy, London.

Hoek, E. and Brown, E.T., 1980. *Underground Excavations in Rock*, London. Instn Min. Metall.

Hudson, J.A. and Priest, S.D., 1979. Discontinuities and rock mass geometry, *Int J. Rock Mech. Min. Sci. & Geomech. Abstr.*, **16**: 339-362.

ICI 1991 Safe and efficient blasting in open cut mines and quarries, ISBN 0 646 05887 8

ISRM 1972, 1977 Report by the Commission on Standardisation of Laboratory and Field Tests. Documents Nos 4, 5 and 8, March 1977 and Document No. 1, October 1972.

JKMRC, 1991. Advanced Blasting Technology, Final Report on AMIRA project No. P93D

JKMRC, 1994. Advanced Blasting Technology, Final Report on AMIRA project No. P93E

JKMRC, 1996a. Mineral Comminution Circuits - Their Operation and Optimisation. 25th Anniversary Monograph, JKMRC, Brisbane, Australia.

JKMRC, 1996b. Open Pit Blast Design - Analysis and Optimisation. 25th Anniversary Monograph, JKMRC, Brisbane, Australia.

Kojovic, T., Michaux, S. and McKenzie, C., 1995. Impact of blast fragmentation on crushing and screening operations in quarrying, EXPLO 95 Conference, AusIMM, Brisbane, September, pp 427-436.

Kropp 1996 MEng Sci Thesis, University of Queensland - under preparation.

Langefors and Kihlström, 1978. The Modern Technology of Rock Blasting, John Wiley & Sons, Inc.: New York, 438pp.

Lilly, P., 1986. An empirical method of assessing rock mass blastability, *Large Open Pit Mining Conf.*, Newman (The AusIMM) pp 89-92.

Mokhnachev, M.P. and Gromova, N.V., 1970. Laws of variation of tensile strength. Indices and deformation properties of rocks with rate and duration of loading. Sov. Min. Sci., No. 6 p 609.

Persson, P.A., Holmberg, R. and Lee, J., 1993. Rock blasting and explosives engineering, CRC press, Boca Raton, Florida.

Rakishev, B.R., 1982. A new characteristic of the blastability of rock in quarries. *Soviet Mining Science*, Vol. 17, pp 248-251.

Rustan, A., Vutukuri, V.S. and Naarttijarvi, T., 1983. The influence from specific charge, geometric scale and physical properties of homogenous rock on fragmentation. Trans. 1st Int. Symp. Rock Fragmentation by Blasting, Lulea, 1, 115-42.

Sarma, K.S., 1994. Models for assessing the blasting performance of explosives, PhD Thesis, The University of Queensland.

Scott, A., 1992. A technical and operational approach to the optimisation of blasting operations, Proceedings MASSMIN 92, South African Institute of Mining and Metallurgy, Johannesburg, 1992.

Villaescusa, E.C., 1991. A three dimensional model of rock jointing, PhD Thesis, The University of Queensland, Brisbane.

Rock Fragmentation by Blasting, Mohanty (ed.) © 1996 Taylor & Francis. ISBN 90 5410 824 X

The use of ideal detonation computer codes in blast modelling

M. Braithwaite
ICI Research and Technical Centre, Wilton, Cleveland, UK

W. Byers Brown
University of Manchester, UK

Alan Minchinton
ICI Explosives, Kurri Kurri, N.S.W., Australia

ABSTRACT: The ideal detonation characteristics of an explosive can be predicted using procedures based on equilibrium thermodynamics and high pressure–high density equations of state (EoS). Important decisions on the allocation of blasting resources are often based on energies predicted by ideal detonation computer codes employing these EoS. A number of such codes are in common use though there is frequently a large disparity in their predictions generally accounted for in terms of the weaknesses inherent in the empirical equations of state employed. These results are compared with those from more accurate codes such as IDeX which employ sophisticated intermolecular potential based EoS.

The use of simplified representations of the thermodynamics of detonation products in procedures employing finite chemical rate models which no longer have the limitations of the currently preferred equations (JWL or polytropic EoS) is also briefly discussed.

1 INTRODUCTION

The ideal detonation characteristics of an explosive provide an estimate of the maximum performance that can be achieved based on a particular explosive formulation and initial density. Few laboratories have the capability of providing a complete set of experimental data, covering energies, pressures and velocities of interest, to accurately describe the performance of an explosive. It is therefore necessary to resort to a computer model in order to determine ideal detonation properties and use these results as a basis for comparing different explosive formulations. The predicted performance is largely governed by the equations of state (EoS) used to describe the different products and phases produced.

Estimates of performance, including velocity of detonation (VoD), energy release, detonation pressures and particle velocities will be at variance with most experimental data for commercial explosives due to lateral losses and finite rate chemistry in these systems. These calculational procedures can, however, be validated against military (HE) explosives where these losses are small: in many circumstances the EoS of individual product species can also be compared with experimental compression data or, for small systems, with ab initio calculations.

Results from the predictions of an ideal detonation code can provide the following information for use in rock blasting computer codes:

- an estimate of the maximum obtainable velocity of detonation;

- a first estimate of the energy split and the total energy release;

- a database for simpler EoS.

There are a large number of different ideal detonation codes in existence. The differences in the predictions of these codes are almost entirely due to the different fluid and solid product EoS used.

2 IDEAL DETONATION

An ideal detonation is illustrated in Figure 1. It consists of a one–dimensional process where mass, momentum and energy are conserved whilst thermal, mechanical and chemical equilibrium are attained in the end products. The ideal (CJ or Chapman–Jouguet) detonation criterion further requires that the VoD is a minimum subject to the above constraints.

In addition to the product EoS, an ideal detonation calculation requires only a knowledge of the explosive

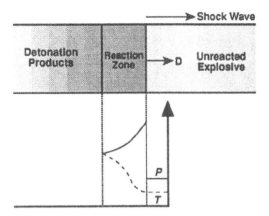

Fig. 1: Ideal detonation (ZND) process

composition (in terms of elemental makeup), its initial density and heat of formation.

An ideal detonation prediction therefore ignores the following known physics:

- CJ assumption: most observed detonations probably lie on the *weak* branch;

- 1–D planar structure: cell structure, flow divergence and shock front curvature are known to occur, albeit to only a limited extent in HE systems;

- attainment of chemical equilibrium: it is likely that there are both diffusional and chemical kinetic constraints to achieving full conversion to chemical equilibrium.

The above assumptions limit ideal detonation codes in their application in commercial explosives to:

1. ideal (infinite diameter) VoDs: CJ state conditions;

2. equilibrium expansions (isentropes, isotherms, isobars) from a starting state at equilibrium and energy releases during expansions.

This excludes calculations where there is a chemical kinetic component or loss term (VoD vs diameter effects, explosion fume, shock and heave energy partitioning): where these calculations have been attempted with ideal detonation codes it has only been accomplished by resort to empirical fixes which cannot be generally extrapolated.

3 EQUATIONS OF STATE

The thermodynamics of all the product species are defined in terms of their EoS for each phase in which

they may be present. For the conditions that prevail in a condensed phase detonation it is normally sufficient to calculate the intra–molecular properties of a single molecule and inter–molecular forces between molecules separately, describing the former as an ideal (low pressure) system. It is therefore assumed that there is no change in molecular vibration or rotation due to high pressure.

The thermodynamic parameters such as energies, entropies and pressure can be determined from a sum of the ideal and non–ideal fluid contributions. The former quantities can be calculated using fitted representations of standard ideal gas tables for the temperature range of interest.

The non–ideality of the products (due to inter–molecular forces) is described by various EoS, one for each phase present. These EoS have been chosen on the basis of their ability to represent the repulsive intermolecular forces which will dominate at high pressures. Over the past 50 years a wide range of fluid and solid EoS have been used for detonation products. Explosives performance in commercial products is largely governed by the fluid phase EoS and discussion will here be restricted to this and to three typical example of EoS in common use:

1. Empirical: BKW (Becker–Kistiakowsky–Wilson) (Fickett & Davis 1979).
   Codes that have this simple EoS as options include Fortran BKW (Fickett & Davis 1979), Tiger (Cowperthwaite & Zwisler 1976) and Cheetah (Fried 1993);

2. Semi-Empirical: JCZ3 (Jacobs–Cowperthwaite–Zwisler) (Cowperthwaite & Zwisler 1976)) The Tiger, Cheetah and Quartet (Heuzé. 1991) codes are among several with this EoS as an option;

3. Fundamental: Intermolecular Potential Based – WCA (Weeks–Chandler–Anderson) (Chirat & Pittion–Rossillon 1981)
   A number of ideal detonation codes are based on fundamental EoS eg. WCA (specific relation for hard sphere diameter) – CARTE (Turkel & Charlet 1995) , IDeX (Freeman et al. 1990): an alternative approach using variational criterion for hard sphere diameter, has also been used but requires considerably more computation eg. CHEQ (Ree 1984) , PANDA (Kerley 1990).

The preferred EoS for any detonation calculation must be that which provides the most realistic representation of the thermodynamics of the product species. This can be established based on the following:

- agreement with shock Hugoniot data for pure unreactive species – pressure (and temperature, where available);

- self consistency in EoS parameters used – in accord with expectations based on critical properties;

- incorporation of appropriate chemistry – polar terms for polar molecules etc.;

- well behaved EoS and its derivatives eg. adiabatic, Gruneisen gammas etc.;

- correct high and low pressure asymptotic behaviour;

- reasonable agreement between measured and predicted VoD's for HE's.

The major features of the individual EoS are discussed below.

3.1 BKW equation of state

The BKW EoS (Fickett & Davis 1979) in its normal form is given by

$$\frac{PV}{RT} = 1 + xe^{\beta x} \tag{3.1}$$

where,

$$x = \frac{\kappa \sum X_i k_i}{V(T + \theta)^\alpha} \tag{3.2}$$

and P, V, T, R, X denote pressure, volume, temperature, gas constant and species mole fractions: α, β, κ and θ are constants and k, individual species covolumes. The constants can be chosen to give reasonable fits to Hugoniot PV data but agreement with published shock temperatures and general PVT behaviour is poor – see later comparison. This EoS has been found to require parameterization for different explosive media: it should also be noted that there is no polar term included.

In a recent review of different high pressure fluid EoS (Byers Brown & Amaee 1991) it is shown that the excess heat capacity at constant volume for this EoS becomes negative at very high density: this might lead to thermal stability problems at very high pressure as a result of the functional form of this equation.

3.2 JCZ3 EoS

The JCZ3 EoS (Cowperthwaite & Zwisler 1976) is a semi–empirical hybrid EoS for mixtures based on an exponential 13.5:6 potential. In its basic form for one component, the Helmholtz Free Energy, A is expressed as

$$A = A_i + E_0(v) + RT \ln f(v, t) \tag{3.3}$$

where A_i denotes an ideal gas contribution to this energy, $E_0(v)$ is a volume potential for a solid lattice (Lennard–Jones Devonshire cell theory) and the f factor is the sum of a gaseous and solid contribution based on fits to Monte Carlo simulations. In order to apply this EoS to mixtures of detonation products, specific mixture rules and the standard combination rules were used.

Its behaviour in the vicinity of the critical region is poor (Byers Brown & Amaee 1991) but it has met with some success in predicting the detonation velocities of a range of C, H, N, O high explosives. The intermolecular potential parameters used for the product molecules have been adjusted on a number of occasions in order to achieve improved prediction: care must be exercised in deciding on which set of constants or allowed product species to adopt in any particular calculation. A number of computer codes have this EoS as an option though the absence of a polar contribution for H_2O in particular does limit its reliability for commercial water–in–oil emulsion explosives.

3.3 WCA EoS (KLRR extension)

Advances in theoretical EoS for fluids based on both perturbational and variational approaches have resulted in a number of EoS available for high pressure fluids based on the fundamental equations of statistical mechanics (Kang et al. 1985). The predictions of these different fundamental EoS are similar since the equations themselves can be validated against Monte–Carlo simulations with the intermolecular potential of interest.

The major drawback of ideal detonation calculations using theoretical EoS has been the large amount of computer time required to calculate the excess thermodynamic contributions. In general terms the Helmholtz Free Energy, A, for a perturbational approach, is expressed by

$$A = A_i + A_{excess} \tag{3.4}$$

where A_{excess}, the intermolecular contribution to the energy is given by

$$A_{excess} = A_{HS} + A_1 + higher\ order\ terms \ldots \quad (3.5)$$

where the hard sphere (HS) contribution is determined by judicious choice of hard sphere diameter minimizing the higher order term contributions.

For a mixture of detonation products, standard mixture and combination rules are applied. Corrections to the normally repulsive forces can be readily made to allow for the attractive force contribution from polar molecules such as H_2O, NH_3 etc. Intermolecular parameters used can be obtained from molecular beam scattering studies, ab initio calculation, by fitting to high pressure shock Hugoniot or static data or even by resort to Corresponding States criteria based on known critical properties.

Calculation of ideal detonation characteristics, formally solving all the statistical mechanical equations, requires substantial computational resources and is beyond the means of small desktop and portable computers. This difficulty can be circumvented by using an analytic representation of the fluid EoS (Byers Brown 1987). This is achieved in two stages:

1. choice of *canonical* variables
   - the adoption of variables such that the thermodynamic excess properties vary more smoothly in the P, V, T region of interest;

2. appropriate polynomial fit
   - provision of optimum interpolation procedure.

It is now possible to perform complete ideal detonation calculations in a matter of a minute on a 486 series PC or equivalent.

3.4 Comparison of equation of state predictions and hugoniot data

Shock Hugoniot data are available for a number of pure, non-reactive materials. For a high pressure EoS to have any credence there must be a reasonable correspondence between predicted and experimental shock Hugoniot data. In some instances, shock temperature information is available for some species.

For the case of the major common explosive detonation products, these data (both velocities and temperatures) have been published for both H_2O and N_2 (Byers Brown & Braithwaite 1990). These data are compared with Hugoniot predictions for the three EoS described here (Figures 2 to 5).

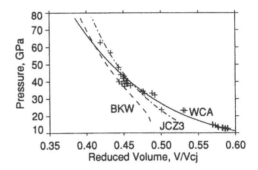

Fig. 2: Water Hugoniot: shock pressure

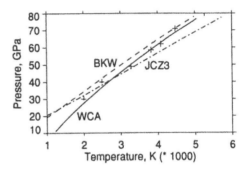

Fig. 3: Water Hugoniot: shock temperature

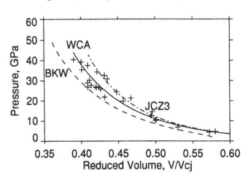

Fig. 4: Nitrogen Hugoniot: shock pressure

For the case of the Nitrogen Hugoniot, the WCA EoS clearly fits the experimental shock Hugoniot well: the other EoS lie on the periphery of the P,V data. Comparison with measured shock temperatures indicate both the WCA and JCZ3 are in reasonable agreement over a wide range of conditions. For the case of the polar water system it is clear that only the WCA EoS both adequately fits the water Hugoniot and is in reasonable agreement with the published shock temperature data: as the WCA EoS in IDeX contains a polar term this was only to be expected.

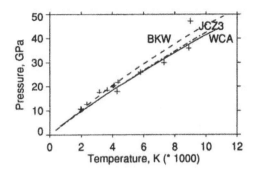

Fig. 5: Nitrogen Hugoniot: shock temperature

4 IDEAL DETONATION CODES

All ideal detonation computer codes essentially comprise a means of calculating chemical equilibrium in a multi–phase multi–component non–ideal mixture subject to a variety of constraints or conditions. An outline flow diagram for ICI's IDeX code is given in Figure 6.

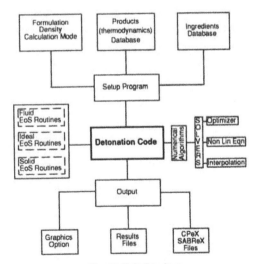

Fig. 6: Code flowsheet

The code consists of:

- user interface (via Windows or similar);

- databases – explosives ingredients, product species thermodynamic parameters, results;

- chemical equilibrium solver: constrained minimization of Helmholtz Free energy;

- Equation of state subroutines (each phase);

- output files and graphics interface;

- Help facility.

Other codes have similar facilities but, in addition to EoS used and product species and phases included, will differ in the method of calculating chemical equilibrium: this can have implications for code robustness. With the advent of faster small computers and more efficient compilers the majority of ideal detonation codes can reside on a portable machine.

4.1 Comparison of code predictions

In addition to the validation of the EoS discussed earlier there are two further criteria for discriminating between the predictions of different ideal detonation codes:

1. Comparison of predicted detonation velocities (monomolecular explosives - experimental diameters much larger than critical) with those measured experimentally. Detonation velocities can be measured to accuracies less than 1% and for the case of military explosives, it is generally accepted that measured velocities should correspond to Chapman Jouguet estimates (Mader 1979). A comparison of predictions for a range of these explosives with the three fluid equations of state is given in Table 1 using standard BKW and JCZ3 parameter sets (Cowperthwaite & Zwisler 1976; Byers Brown & Amaee 1991) and ICI's IDeX.

2. Behaviour of isentrope slopes - adiabatic gamma.
   For detonation product media at modest densities and in the absence of phase transitions, the slope of the isentrope, γ where,

$$\gamma = -\frac{V}{P}\left(\frac{\partial P}{\partial V}\right)_S \tag{4.1}$$

would be expected to be monotonic i.e. with no turning points or inflections.

A comparison of the three EoS predictions for PETN, detonating from its maximum crystal density of 1.77 g/cc, is given in Figure 7 alongside a JWL EoS (see later) fit. It is clear that both the JWL and JCZ3 EoS give characteristic turning points, well below CJ densities. The WCA EoS at all but the highest density is monotonic: the decrease in γ at the high density can be attributed to the formation of solid Carbon.

Table 1: Detonation velocity comparison

Explosive	Initial Density g/cc	VOD, m/sec			
		Expt.	WCA	JCZ3	BKW
BTF	1.76	8262	8317	8246	7984
HMX	1.189	6713	6741	6651	6381
HN	1.626	8691	8692	8797	8082
NG	1.60	7700	7650	7531	7712
NM	1.128	6290	6283	6110	6351
NO	1.294	5620	5614	5732	5665
NQ	1.00	5460	5534	5285	5486
PETN	1.770	8295	8247	8213	8446
PETN	0.95	5330	5360	5531	5701
RDX	1.00	6100	5981	6094	6143
TATB	1.895	7860	8133	8071	8118
TETRYL	1.710	7850	7751	7602	7651
TNT(s)	1.64	6950	6896	6279	7137
TNT(l)	1.45	6590	6489	6319	6570

5 SIMPLIFIED REPRESENTATIONS

Ideal detonation estimates of the detonation velocities of commercial explosives can only predict a maximum attainable velocity: the predictions take no account of finite chemical rates or energy dissipation terms. Where a more complete study of the fluid dynamic characteristics of a non-ideal detonation process is undertaken it is not practical to incorporate a full EoS and simplified representations, chemistry implicit, such as the JWL and Williamsburg EoS are used. In the selection of a simplified EoS it is desirable to include properties such as:

1. simple analytic form;

2. statistical mechanical basis;

3. correct asymptotic behaviour;

4. complete EoS (include Pressure, Volume, Temperature, Energy);

5. readily fitted to a principal isentrope.

5.1 JWL EoS

The JWL EoS (Fickett & Davis 1979) has been extensively used by explosives engineers to describe isentropic expansion of detonation products. It is an entirely empirical (and incomplete) EoS which, for an isentrope, has the form:

$$P = a_1 e^{-R_1 V} + a_2 e^{-R_2 V} + \frac{a_3}{V^{1+\omega}} \quad (5.1)$$

where a_1, a_2, a_3, R_1, R_2, and ω are constants and $V = v/v_0$: P, v denote pressure and volume. An expression for energy is readily obtained by integrating this expression. The parameter ω can be shown to be equal to the Gruneisen gamma coefficient, $1/V(\partial E/\partial P)_V$: the equation is therefore of a constant Gruneisen Gamma form and this is a poor approximation for a quantity that can vary by more than factor of two. The arbitrary form of this equation also gives anomalies in the adiabatic gamma plot. It therefore satisfies only (1, 3 and 5) of the above criteria.

5.2 Williamsburg EoS

The Williamsburg EoS (Byers Brown & Braithwaite 1991) for an isentrope is given by

$$E = \frac{PV}{g - 1} \quad (5.2)$$

where g is given by

$$g = \gamma_0 + \sum_{k=1}^{N} \frac{\gamma_k}{1 + \beta_k v} \quad (5.3)$$

where γ_k and β_k are constants: the order of approximation, N, is less than or equal to 4. Were g a constant this would reduce to a polytropic form. This equation, described more fully elsewhere (Byers Brown & Braithwaite 1991) can be shown to satisfy all the prerequisites listed earlier with little penalty in terms of complexity.

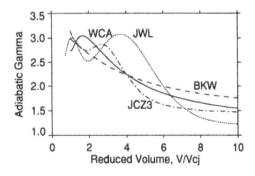

Fig. 7: Adiabatic gamma plots

5.3   Application to an IDeX Isentrope

Table 2 lists the properties of a typical commercial explosive as predicted by IDeX.

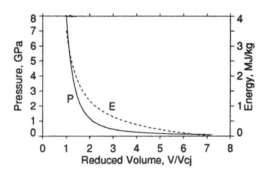

Fig. 8: Isentrope of standard explosive — pressure and relative energy

Table 2: Explosive properties for an AN–based emulsion

Properties

    Initial Density: 1.00 g/cc

    Heat of Formation: -4.88 MJ/kg

    Elemental composition(/kg)

        C (G–atom) 4.350

        H (G–atom) 57.84

        N (G–atom) 22.40

        O (G–atom) 35.99

CJ Detonation

    VOD: 5625 m/s

    γ_{CJ}: 2.93

    Density: 1.34 g/cc

    Pressure: 8.05 GPa

    Energy: 3.51 MJ/kg

    Temperature: 2748 K

    Product: 44 moles/kg

Energies

    Heat of reaction: 3.54 MJ/kg

    Effect. energy (100 MPa): 2.48 MJ/kg

    Energy loss CJ (100 MPa): 3.51 MJ/kg

    RWS, RBS: 96, 120

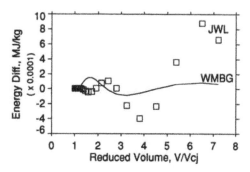

Fig. 9: Errors in energy prediction along an IDeX isentrope

The calculated CJ isentrope for this explosive (Figure 8) has been fitted by both the Williamsburg (WMBG) and JWL EoS. The isentropic expansion in terms of the pressure and energy (referenced to 100 MPa) are shown as a function of reduced volume: differences between the calculated energies and those derived from the WMBG and JWL EoS are illustrated in (Figure 9). Predictions at larger volumes with the JWL EoS are in error: this can lead to significant errors in computations with commercial explosives at volumes appreciably larger than the initial explosive volume. This uncertainty is further compounded by the lack of a reliable method of obtaining off isentrope points with this EoS. The WMBG EoS, whilst constrained to match CJ conditions and isentrope gradient (at CJ state) has smaller errors than JWL – typically two orders of magnitude based on sums of squares of residuals.

6   CONCLUSION

Over the past 25 years there have been considerable advances in the field of high pressure–high temperature EoS for condensed phase systems. A number of intermolecular potential based EoS are in use in a variety of ideal detonation computer codes. Computations using intermolecular potential based EoS have been much facilitated by using various polynomial fits as representations over Pressure, Volume, Temperature zones of interest. Thus modern ideal detonation codes offer the following advantages at little extra computational cost:

- no further EoS parameterization required;

- where a new element is added to an explosive's formulation, intermolecular potential data for the corresponding new products can be estimated using critical properties and Corresponding

States principles;

- predictions are without anomalies in the thermodynamics and associated derivatives of the detonation product mixture.

Ideal detonation codes provide reasonable estimates of the maximum performance attainable from an explosive: in the case of a commercial explosive this would correspond to a completely reacting medium of large diameter and well confined where there are neither diffusional constraints or lateral losses. In addition they provide insight into the thermodynamic behaviour of the detonation products: this information can be represented in terms of a simpler EoS for later modelling studies. Without resort to somewhat arbitrary fixes, ideal detonation codes cannot be used to assess behaviour where there is a degree of chemical kinetic/diffusion control or where there are lateral losses.

For a wide variety of detonation modelling where a chemical equilibrium calculation is too computationally intensive, a chemistry implicit EoS is usually adopted: this is standard practice in non–ideal detonation, computational fluid and hydro–codes where much use has been made of the JWL EoS or simple variants on polytropic descriptions. In modelling detonations of commercial explosives it has been shown that the use of JWL EoS can lead to erroneous results: an alternative, WMBG EoS, is proposed that is both a complete EoS and more reliable over volumetric expansions of interest.

REFERENCES

Byers Brown W. 1987. Analytical representation of the excess thermodynamic equation of state for classical fluid mixtures of molecules interacting with α-exponential-six pair potentials up to high densities. *J. Chem. Phys.* 87(1):566

Byers Brown W. & M. Braithwaite 1990. Sensitivities of adiabatic and gruneisen gammas to errors in molecular properties of detonation products. *9th Detonation Symp. (Intnl.)* OCNR 113291-7:513

Byers Brown W. & B. Amaee 1991. Review of equations of state of fluids valid to high density. *HSE Contract Research Report*, No 39 (1992)

Byers Brown W. & M. Braithwaite 1991. Analytical representation of the adiabatic equation for detonation products based on statistical mechanics and intermolecular forces. *Shock Compression of Condensed Matter*, Elsevier, P. 325

Chirat R. & G. Pittion–Rossillon 1981. Detonation properties of condensed explosives calculated with an equation of state based on intermolecular potentials. *6th Detonation Symp. (Intnl.)*, NSWC MP 82-334

Cowperthwaite M. & W.H. Zwisler 1976. The JCZ equations of state for detonation products and their incorporation into the TIGER code. *6th Detonation Symp. (Intnl.)*, 162, ACR-221.

Freeman T.L., I. Gladwell, M. Braithwaite, W. Byers Brown, P.M. Lynch & I.B. Parker 1990. Modular software for modelling of detonation of explosives. *Math. Engng. Ind.* 3:97

Fickett W. & W.C. Davis 1979. *Detonation* Univ. California Press.

Fried L.E. 1993. CHEETAH: a fast thermochemical code for detonation. *Lawrence Livermore National Laboratory Report* UCRL-ID-115752

Heuzé O. 1991. *Quartet: Systéme de Calcul Thermochimique.* Softworld, Paris

Kang H.S., C.S. Lee, T. Ree & F.H. Ree 1985. A perturbation theory of classical equilibrium fluids. *J. Chem. Phys.*, 82(1):414

Kerley G.I. 1990. Theoretical model of explosive detonation products: tests and sensitivity studies. *9th Detonation Symp. (Intnl.)* OCNR 113291-7:443

Mader C.L. 1979. *Numerical Modeling of Detonations,* Univ. California Press

Ree F.H. 1984. A statistical mechanical theory of a chemically reacting multiphase mixture: application to the detonation properties of PETN. *J. Chem. Phys.* 81(3):1251

Turkel M.L. & F. Charlet 1995. Carbon in Detonation Products. *J. de Physique IV*, 5:C4–407

Rock Fragmentation by Blasting, Mohanty (ed.) © 1996 Taylor & Francis. ISBN 90 5410 824 X

An optimal design of blasting

Q. Zeng, P. Navidi & J. Zarka
Laboratoire de Mécanique des Solides, Ecole Polytechnique, Palaiseau, France

Abstract : Rock blasting is often considered to be still an art performed by only a few good experts. Numerical simulations are one of the means considered to help them. However, many aspects of such numerical simulations need to be improved such as the explosive model, the rock's response and of course, their coupling. This paper provides a review of blasting simulations. Some developement are presented, especially in the rock damage model. It is also shown that an optimal design of blasting can be reached thanks to automatic learning techniques coupled to optimization tools which are discussed.

1 Introduction

Blasting is one of the basic operations in mining and quarrying industry. An efficient blasting for given geology conditions, rock volume to break and security conditions, results from a correct choice of explosive types, explosive quantity and the blasting planning. One can think of diameter, and length of boreholes, drilling pattern, initiation modes, firing sequence, delay time... Efficiency is then determined by the quality of rock fragmentation and a minimum cost. In this sense it is an optimization problem. At present, almost all of the blast design is based on experiences, at first because the behavior of the explosive and the surrounding fractured rocks are still not well known, secondly because of the complex coupling effects. However, the computer aided design techniques, numerical blasting models and expert systems can considerably reduce both the cost and duration of blast design while increasing the quality of prediction. This is the motivation of our research.

The key to a good blast design is a precise prediction. The first part of our work focus on the prediction by numerical simulations, in which the modelling is the discussion center.

The second part is on to the optimization of blast design. After introducing the general methodology of our approach we show an application example for blast design.

2 Simulation of blasting

Many computational models of rock blasting are based on data obtained from experiments or/and empirical relations derived from experimental data and analytical results for very simple cavities problems [27],[7],[13].... They are either too much simplified or very specific. Recently, more fundamental models based on the material constitutive laws, usually implemented in a finite element or finite difference code, have been attempted [16],[10],[12]....

From a mechanical point of view, the blasting problem couples two difficult fields : the detonation of explosives and the rock mechanics. At present, the two fields are still in course of development. This section begins by a brief introduction of an existing 3D computer code we used for our simulations. An explicit scheme algorithm enable us to handle the non-linear aspects and the fluid-structure coupling. Then a review of the models for the explosive and the rock is given. After, we introduce a new damaged rock constitutive relation. Finally some numerical results based on this new model are presented to show its tractability.

2.1 Computer code description

The program used can simulate most of the nonlinear dynamic transient phenomena [23]. This code can handle solid and fluid mechanics problem and the coupling between them. The main features used for our specific problems are :

- explicit central-difference time-integration algorithm

- Arbitrary Lagrangian-Eulerian (ALE) formulation [3] for fluid/structure interaction

- artificial viscosity in order to smear shock fronts[29]

- Incremental form of constitutive relations

2.2 Detonation model

Computation of the detonation propagation in explosives is a complex problem. It results from strong chemico-physics interaction during the detonation process[14],[6].... When a coupling with a structure is considered, the numerical modeling of the detonation is often simplified. This is because the response of the structure is the main interest. Such models are often of continuum type and admit the following hypotheses :

1. reduction of the reactive fluid in a fluid with two components, an initial substance and a final substance (explosive reaction products);

2. each of the two substances is separately in thermo-chemical equilibrium;

3. quasi Chapman-Jouguet detonation (a quasi-sonic detonation velocity).

These hypotheses reduce the complex computation to a simple fluid flow one with discontinuities which are then often smeared as follows :

- an artificial viscosity is used for the thermophysical variables of discontinuity (shock wave);

- a decomposition law is used for the thermochemical variables of discontinuity.

Therefore, the behavior of the explosive is represented by an equation of state and a decomposition law.

Often, in case of the detonation/structure simulation, the equation of state has a pseudo-potential form $p(v, E)$ (pressure-volume-energy) to avoid the thermo-chemical iterative calculation. The decomposition law is of «bulk burn» type due to the homogeneity consideration. A variety of explosive behavior ranging from the simple to complex models are available, we use the following models[30]:

- The Jones-Wilkins-Lee (JWL) equation of state for the explosive products pressure :

$p = A(1 - \frac{\omega}{R_1 V})e^{-R_1 V} + B(1 - \frac{\omega}{R_2 V})e^{-R_2 V} + \frac{\omega E}{V}$

where

$V = \frac{v}{v_0}$	:	relative volume
p	:	explosive products pressure
E	:	detonation energy per unit volume
A, B, ω	:	
$, R_1, R_2$		constant parameters

- if $A = B = 0$, we refind the simple «gamma-law» with

$\omega = \gamma - 1$ $(\gamma = \Gamma = \frac{\partial \ln p}{\partial \ln v})$

- the «C.-J volume burn» chemical decomposition law :

$m = \frac{1 - V}{1 - V_{CJ}}$

where the burn fraction m is the indicator of decomposition progress.

- the current pressure is calculated by :

$$P = m \cdot p$$

The detonation velocity, the pressure at Chapman-Jouguet point P_{CJ} and the initial energy E_0 should also be known.

After having done several simulations (explosion in water, explosion in an elastic medium), we noticed some interesting results [31] :

- computation with the equation of state «gamma law» has a similar result compared to the computation with the equation of state JWL.

- the explosive energy partition (shock energy/bubble energy) can be characterized by the coefficient γ of the «gamma law», the greater γ is, the greater is the shock energy.

- the modeling of blast loading by using defined pressure-time history is only a crude approximation specially for the case of the point initiation.

2.3 Rock mechanical model

Another essential point for the blasting simulation is a proper representation of the rock behavior. The approach which involves the formulation of appropriate material models (constitutive relation) is preferred. The main phenomena that must be accounted for is crack initiation and propagation. Two alternatives are then possible. One is the discrete model which takes into account the physical presence of the cracks in a direct approach[11],[8].... For the computation, this one is often very costly and suitable for the post-period of the blasting processus. Another way is to take a continuum damage model. We have chosen the last method which is more consistent with a finite element approach. A continuum model [31] is therefore developed and implemented into the explicit code. This is an elastic anisotropically damaging behavior based on the damage mechanics theory introduced by Kachanov [20].

The constitutive relation is :

$$\sigma = \bar{C}(D) : \epsilon$$

where $\bar{C}$ is the elastic effective stiffness tensor and D is the damage tensor.

In an incremental form :

$$d\sigma = \bar{C}(D) : d\epsilon + d\bar{C}(D) : \epsilon \qquad (1)$$

The main hypotheses are the following:

1. the dominant mode of failure is mode I

   as proven by numerous experiments [21],[22],[15]....

2. the material non linearity is totally governed by damage and thus plastic deformations are disregarded

3. The expanding plan directions of the penny-shaped cracks are perpendicular to the tensile principal stress directions.

   this concept is deduced from the impact experiment results [9].

2.3.1 Damage description

The nature of the rock flaws and the possible complex loadings require an anisotropic damage description. Generally, an eight order tensor should be introduced to describe the anisotropic damage, but the difficulty to identify all the parameters is obvious. It is thus desirable to reduce the order of that tensor. Based on the third hypothesis above, a symmetric second-order damage tensor D is introduced, reducing the model to an orthotropic one.

We can consider each of the eigenvalues D_i as a group of parallel cracks perpendicular to the associated eigenvector. As if :

$$D_i = \frac{4}{3}\pi \cdot f \qquad (2)$$

where $f = N_\alpha \cdot c_\alpha^3$ is the cracks (N_α cracks of radius c_α) group density .

This definition is the same as the one for a homogenization method [25], [19] so that the results of homogenization can be used to determine the effective elastic matrix which is going to be shown lately.

D_i can be also defined with the main parameters as introduced by Grady and Kipp[16] :

$D_i = Nv$; with $v = \frac{4}{3}\pi \cdot c^3$ (c and N are mean values)

2.3.2 Damage evolution law

Generally, the second thermodynamic principle is used to determine the damage evolution law as in an elastoplastic problem. However the dissipative energy is not easily established. Grady and Kipp have developed an evolution law based on the activation and growth of an initial Weibull distribution of fracture-producing flaws, for an isotropic behavior [16]. It can be extended to an orthotropic material :

$$\dot{D}_i = \begin{cases} \left[\frac{8\pi(m+3)^2}{(m+1)(m+2)}\right]^{1/3} c_g n(\frac{\sigma_i^p - \sigma_u}{E})^{\frac{1}{3}} D_i^{\frac{2}{3}} & \sigma_i^p > \sigma_u \\ 0 & else \end{cases}$$

$$(3)$$

with $i = 1, 3$ $n(x) = kx^m$
where

c_g	:	constant fracture growth velocity
σ_i^p, σ_u	:	principal stress and a constant depending on the material cohesion
E	:	Young's modulus
k, m	:	constant Weibull parameters

2.3.3 Damage criterion

The damage criterion is usually derived from the Griffith theory [17], [4], with the damage evolution law (3) shown above, we just take the following simple form :

$$\sigma_i^p > \sigma_u$$

2.3.4 Effective elastic matrix

Based on the results of the homogenization methods [25], [19], the symmetric effective elastic matrix takes the form :

$$[\bar{C}] = \begin{bmatrix} [\bar{C}^{(1)}] & 0 \\ 0 & [\bar{C}^{(2)}] \end{bmatrix}$$

$\bar{C}_{11}^{(1)} = \frac{E}{\alpha}(1 - \nu^2 d_2 d_3)d_1$ $\bar{C}_{22}^{(1)} = \frac{E}{\alpha}(1 - \nu^2 d_3 d_1)d_2$
$\bar{C}_{33}^{(1)} = \frac{E}{\alpha}(1 - \nu^2 d_1 d_2)d_3$ $\bar{C}_{12}^{(1)} = \frac{E}{\alpha}\nu(1 + \nu d_3)d_1 d_2$
$\bar{C}_{13}^{(1)} = \frac{E}{\alpha}\nu(1 + \nu d_2)d_1 d_3$ $\bar{C}_{23}^{(1)} = \frac{E}{\alpha}\nu(1 + \nu d_1)d_3 d_2$
$\bar{C}_{11}^{(2)} = 2\mu\frac{(1-\beta D_2)(1-\beta D_3)}{2-\beta D_2 - \beta D_3}$
$\bar{C}_{22}^{(2)} = 2\mu\frac{(1-\beta D_1)(1-\beta D_3)}{2-\beta D_1 - \beta D_3}$
$\bar{C}_{33}^{(2)} = 2\mu\frac{(1-\beta D_2)(1-\beta D_1)}{2-\beta D_2 - \beta D_1}$
$\bar{C}_{ij}^{(2)} = 0$ $(i \neq j)$
with
$d_i = 1 - D_i$ $(i = 1, 3)$
$\alpha = 1 - \nu^2(d_1 d_2 + d_2 d_3 + d_1 d_3 + 2\nu d_2 d_3 d_1)$
$\beta = (1 - \nu)(2 - \nu)$ ν : poisson's ratio

The result is in $0(f^2)$ order to the constant strain or constant stress method and the formulae are easy to calculate.

2.3.5 Incremental algorithm

During each time step the following algorithm is used for the determination of the stress and the damage tensor values : given $\sigma(t)$, $\varepsilon(t)$, $D(t)$, $\dot{\varepsilon}(t+dt)$:

$$D(t+dt) = D(t) + \dot{D}(\sigma(t))$$
$$d\tilde{C} = \tilde{C}(D(t+dt)) - \tilde{C}(D(t))$$
$$\sigma(t+dt) = \sigma(t) + \tilde{C}(t) : \dot{\varepsilon}(t+dt)dt + d\tilde{C} : \varepsilon(t)$$

2.3.6 Dominant fragment size

The calculation of the dominant fragment size (fragment size corresponding to the largest volume fraction of material) is also derived from the result given by Grady and Kipp which is strain-rate dependant. For sake of simplicity the formula is now fracture stress dependant :

$$L_M = \eta(k,m)\left(\frac{\sigma_c}{E(m+3)}\right)^{-\frac{m}{3}}$$

where η is a constant and σ_c the tensile fracture stress.

2.4 Simulation results

The damage evolution law based on the Weibull distribution leads to predictions which are in good agreement with the strain-rate effects discussed by different authors[16],[2][10]....). In order to show the capability of the model proposed to take into account anisotropic effects, the following numerical experiment is done : we consider several cases of constant strain rate loading on a single element. The first loading is carried out up to a value of $D_1 = 0.4$ (largest eigenvalue of the damage tensor). After changing the loading direction, a second loading is applied to the same specimen, the curves in dot lines correspond the results for various directions of the second loading respectively $0°, 30°, 45°, 90°$ Fig(1).

2.4.1 Calculations in plane strain with an imposed pressure function

We use the model presented above to carry out the analysis on various geometries. One of them is a circular cylindrical cavity problem. Due to symmetry a quarter of the geometry is considered. The configuration and the loading, a sine function pressure imposed inside the cavity, are given in Fig 2 and 3 :

The surrounding rock is an oil shale which has the properties :

$\rho(\frac{g}{mm^3})$	$E(MPa)$	ν	m	$k(\frac{1}{mm^3})$	$c_g(\frac{m}{s})$
0.002	$1.065 \cdot 10^4$	0.4	8	1.7×10^{18}	1300

(4)

In Fig 4 and 5, the distributions of the largest principal value of the damage tensor are respectively

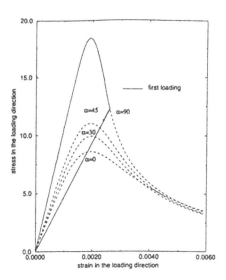

Figure 1: Anistropic effect

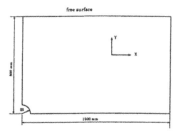

Figure 2: Configuration

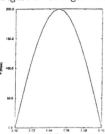

Figure 3: Pressure function

shown at time $t = 0.25ms$ et $t = 0.5ms$. This result underlines that :

- the damage begins around the cavity and progresses with decreasing value.

- a second damage area appears when the stress wave reaches the free surface and is reflected.

- the eigenvector directions of D tells us that the

Figure 4: First principal value of **D** at $t = 0.25ms$

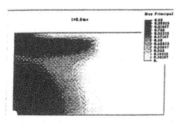

Figure 5: First principal value of **D** at $t = 0.5ms$

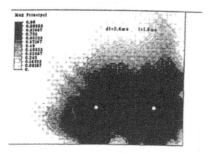

Figure 7: simultaneous detonation

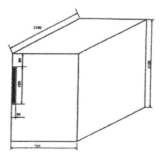

Figure 8: with a delay time $\tau = 0.6ms$

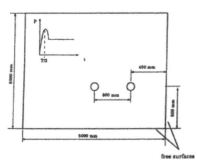

Figure 6: Configuration of two blasts

cracks in the first area are radial and those in the second are parallel at the free surface.

To see the influence of blasting with delay time, two calculations are carried, one with simultaneous detonations and the second with a delay $\tau = 0.6ms$. The configuration is shown in Fig 6 with $T = 0.05(ms)$.

The results in Fig 7) and (8 are given for the largest eigenvalues of the damage tensor at $t = 1.5ms$ where the damage results are convergent.

We observe that the detonation with delay modifies the damage distribution and increases the damage volume (10% for this example).

2.4.2 3-D calculation coupled with detonation

The configuration is shown in Fig (9) :

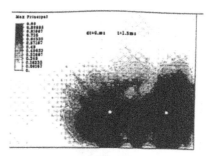

Figure 9: Configuration in 3-D

The explosive is idealized. Its brisance is near the ANFO. The equation of state used is the ≪gamma law≫ where the parameters are as follows :

γ	$\rho_0(\frac{g}{cm^3})$	$E_0(MPa \cdot \frac{cm^3}{cm^3})$	$p_{cj}(MPa)$	$D_{cj}(\frac{cm}{\mu s})$
2.8	0.9	770	2.765×10^3	0.4

The space between the explosive and the surrounding rock is occupied by air. the decoupling ratio (borehole diameter to explosive diameter) is 1.2, the confinement is perfect (the stemming is also occupied by rock). A point detonator is put at the bottom of the borehole. The rock is the same as in the first example (4).

In Fig (10), the damage result (largest eigenvalue of the damage tensor) is shown at $t = 0.375\,ms$,

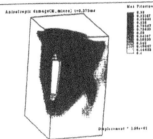

Figure 10: damage result in 3-D

3   Blast optimization

The blasting simulation based on the advanced modeling allows us to predict more generally and more precisely the blasting results. It can be used also to better understand fragmentation mechanisms, the roles played by the different design parameters [31]. However, It can not be yet a perfect tool for the blast design mainly because:

- it is difficult to simulate the entire blasting process with a single model often much simplified.

- the strong non linearity and the local/global problems in blasting make the computation too expensive. Even if it is technically possible to simulate the entire blasting process by more sophisticated modeling (e.g.: a combined finite element/discrete element model [8]), it is practically impossible to simulate the real sequence blasting problem.

In this context, the use of the automatic learning technics for searching rules (or models) seems to be a rational solution, and opens a realistic way to the blasting optimization.

In this section we will first give a brief introduction to the automatic learning technics. The methodology of blasting optimization is then discussed. Finally an example of automatic learning and optimization are given.

3.1   Automatic learning

In many technical domains, experts are aware that, very often, there is no a complete solution for a real world problem. But there is a possibility to build

a data base of examples. Each sample of this base is obtained experimentally or numerically with sometimes some fuzzy or missing informations. The automatic $\mathcal{L}$earning $\mathcal{E}$xpert $\mathcal{S}$ystems generator ($\mathcal{LES}$), developed at LMS during 1986-1990, can build a set of rules based on the description of such a examples base. Several applications have been shown [26], [24], [28], [18], [5].... The main difficulty is to provide a **good description** of the examples base. The initial descriptors (let us call descriptor each single field of the data base) appear then as variables for the final model.

Suppose that the expert defines a set of these initial descriptors describing the observed and/or measurable phenomena, (x_i) generally in a limited number (e.g. the radius of the borehole, explosive energy etc...). Indeed only a subset of this initial set is sufficient to make the describtion of one example but it is not known a priori. These descriptors are then integrated within our actual limited knowledge to determine a set of more complex ones which are the intelligent descriptors, (X_i) (e.g. results of a numerical simulation or the ftime-history of the radius of the bubble in the water blast test,etc...). These descriptors are function of the initial descriptors and some of them may be considered among the conclusions (C_k) and then also be learned (e.g. the damaged rock volume).

In order to build the full knowledge of the problem or the model, the expert must prepare a file containing for each example :

The input descriptors:

$x_1\ |\ ....\ |\ x_6\ |\ X_1\ |\ .....\ |\ X_{40}\ |\ ......\ |$

The output descriptors:

$C_1\ |\ C_2\ |\ ....$

The symbolical option of $\mathcal{LES}$ will give a set of rules while its numerical option will generates a polynomial based model which is often easier to handle.

3.1.1   Generating rules

For a simple problem, the data base, with the initial descriptors (x_i) and the conclusions C_I, can be presented to the system. The system generates then a mathematical expression based on a polynomial expression :

$C_I = POL\,(x_1, x_2, \cdots, x_n)$

For more complex problems, it is better to use the data base with the intelligent descriptors (X_i) in order to help the system to find a more reliable model for the conclusion C_I :

$C_I = POL\,(X_1, X_2, \cdots, X_k)$.

Once the model is created, new unknown examples may be introduced for evaluating the conclusions.

The building of the set of (X_i) requires making them intrinsic in order to be able to reproduce them for various types of problems independently of the nature or type of problem considered.

3.2 Optimization

Many optimization technics are available. But we have the following points in mind :

- Design variables can be discrete and also take their value in a finite set.

- From a technological point of view many solutions can lead to the same type of results. This means that if C_I is the cost function then in the design variable space POL can have flat domains (with no gradient). Many optima can exist simultaneously that is POL does not define a convex .

- Constraints generally varies depending on the domain in which the design set is looked for.

Blasting optimization by the traditional methods based on the mathematical programming theory such as quasi-Newton, conjugate gradient... seems to be impossible. On the other hand, the evolution methods such as the genetic algorithms are more suitable for a general optimization problem.

Contrary to the traditional methods based on the continuum concept, the genetic algorithms make evolve a population of solutions with a manner suggested by the genetic. Their advantages compared with the traditional methods are as following:

- the probabilistic characteristics can handle local optima.

- there is no other restriction for the functions (cost functions, constraint function), excepted to be calculable.

- the algorithm is extremely general.

- design variables can be of any type.

The main disadvantage is costly computation time. *Optimization by coupling the genetic algorithms with automatic learning* is then a good strategy because the generation of the specific rules (for the cost function, the constraint functions) by automatic learning allows to compensate the handicap of the genetic algorithms method.

It is interesting to observe that the fact of expressing all the relations in a simple polynomial way enable us to carry out genetic optimization very effectively.

The other intereseting points of this approach can be summarized as follows:

- It gives us a way of dealing with inverse problems, that cannot be handled by classical schemes.

- It lets us understand the essential descriptors which drives the desired conclusion. This can be a starting point for experts to build new PHYSICAL or MATHEMATICAL MODELS.

- By a clever choice of the intelligent descriptors, it is possible to extrapolate models resulting from a reduced set of simple samples to more complex situations.

3.3 Generation a damaged volume rule by $\mathcal{LES}$

In this section we use the above methodolgy in a simple blasting design problem. Suppose that the explosive type and the rock site are given. We search a damaged volume v_d (broken rock) expression in function of the drilling pattern parameters in order to carry out next the optimization calculation.

The database is built by numerically simulating blasting in plane strain. The choosen descriptors are :

borehole diameter d ; burden B;decoupling ratio R_d and two ratios d/R_d , d/B.

The result in Tab(1) is given by eleven simulations

$d(m)$	$B(m)$	R_d	v_d	d/R_d	d/B
0.050	0.600	1.0	1.952774	0.05	0.0833
0.050	0.5066	1.677	0.79336	0.0298	0.0987
0.050	0.600	1.172	1.467875	0.0427	0.0833
0.050	0.600	1.667	0.675639	0.0298	0.0833
0.040	0.500	1.3333	0.94768	0.03	0.08
0.040	0.600	1.333	0.871909	0.03	0.0667
0.100	0.750	1.257	1.520870	0.0796	0.133
0.100	0.750	1.474	1.265204	0.0678	0.133
0.200	1.000	1.2067	1.570654	0.1657	0.2
0.200	0.750	1.2067	2.957848	0.1657	0.267
0.200	1.000	1.843	1.182392	0.1085	0.2

Table 1: damaged volume database

First we ask the $\mathcal{LES}$ to build a polynomial model. Limiting the system to terms of order 2 with cross terms. the $\mathcal{LES}$ output is :

$$v_d =$$

$$0.85B{\cdot}d - 2.44B\left(\frac{d}{R_d}\right)^2 + (24.134 - 45.6R_d + 23.17R_d^2)\frac{d}{B} \tag{5}$$

(and $e_m = 0.00477$.)

3.4 Optimization problem

The problem can be set as following :

For a given explosive type and rock in-site, we search the design parameters values (d, R_d, B) which maximize the specific charge

$$E_s = \frac{v_d}{v_e \, (\text{explosive volume})}.$$

Taking back the result (equation (5)) divided by $\left(\frac{d}{R_d}\right)^2$ which is in proportion to the explosive volume, the problem becomes :

$\max f =$
$4.256 B \frac{R_d}{d} - 88.95 B + (90.39 - 92.644 R_d + 25.54 R_d^2) \frac{R_d^2}{B d}$

with the constraints :

$1.0 \leq R_d \leq 2.0$
$0.4 \leq B \leq 1.0$
$0.5 \leq d \leq 0.2$
$d/R_d \geq 0.03$ (the explosive cartridge diameter is supposed superior to $0.03m$)

This "simple" problem is firstly resolved by one of the traditional methods, then by the genetic algorithm.

- Conjugate Gradient

  With a starting point : $\{d = 0.04, B = 0.8, Rd = 1.0\}$, a solution has be found such that :

  $\{d = 0.04, B = 0.5, Rd = 1.09358\}$

  $\Rightarrow f = 1187.1658$

  With another starting point :$\{d = 0.1, B = 0.8, Rd = 2.0\}$,a different solution has be found such that :

  $\{d = 0.06, B = 0.5, Rd = 1.09358\} \Rightarrow f = 995.79$

  This means that their are at least two local solution. In fact the solution found depends highly on the starting point.

- Genetic Algorithm

  With several different starting populations, the following solution is found :

  $\{d = 0.04, B = 0.5, Rd = 1.094118\}$

  $\Rightarrow f = 1187.1654$

Now, suppose that borehole diameter d and the explosive cartridge diameter can only take some fixed discrete values :

$d \quad \in \quad \{0.048, 0.05, 0.066, 0.084, 0.1, 0.15, 0.2\}$
$\frac{d}{R_d} \quad \in \quad \{0.03, 0.036, 0.04, 0.05, 0.64\}$

The solution found by genetic algorithm and changing R_d by d/R_d is:

$\{d = 0.048, B = 0.5, d/R_d = 0.045 \rightarrow Rd = 1.0667\}$

$\Rightarrow f = 980.96$

One can notice that these points where not in the initial database. These values can then be used for a new numerical simulation (or experiment). The damaged volume obtained is then compared with the one predicted by the model. If they agree, the model generated by the $\mathcal{LES}$ is consistent. Otherwise we can add a new example to our data base and build a new model. This work can be done iteratively until the model is consistent.

4 Conclusion

In this paper, it has been shown that the modelling of micro-craks by a continuum damage model may improve the quality of the prediction. For the continum rock model considered, the mixed phenomenological - microphysical approach seems to be tractable; the microphysical approach allows to build a consistent model thus limiting the number of parameters, it also allows quantitative results for the fragmentation. The phenomenological approach simplify the model and decrease the computing time.

However, even if the blast prediction by simulation appears to be a powerfull prediction tool, it can only be an element of the industrial blasting design. Indeed, many other relevant parameters and technical aspects cannot directly enter the mathematical model. The use of the automatic learning lets us overcome this step efficiently, as we showed it on a simple example. The coupling between genetic algorithms and automatic learning seems to be a powerful potential approach to blasting optimization.

References

[1] First International Symposium on Rock Fragmentation By Blasting, Luleå, Sweden, 1983.

[2] L.G Margolin & T.F Adams. Numerical simulation of fracture. In First International Symposium on Rock Fragmentation By Blasting [1].

[3] C.W. Hirt A.A. Amsden and J.L. Cook. An arbitrary lagrangian-eulerian computing method for all flow speeds. J. of Computational Physics, 14:227–253, 1974.

[4] H.D. Bui. Mécanique de la Rupture Fragile, volume 1. MASSON, 1978.

[5] Michale Bulik. Sur l'Optimisation de la Protection Parasismique. PhD thesis, Université Paris-VI, 1994.

[6] R. Chéret. La Détonation Des Explosifs Condensés, volume 1 of Scientifique. MASSON, 1988.

[7] W.I. Duvall. Vibration associated with a spherical cavity in an elastic medium. *U.S. Bureau of Mines*, RI(4692), 1950.

[8] A. Munjiza et al. On a rational approche to rock blasting. *Computer Methods and Advances in Geotechnics*, pages 857–862, 1994.

[9] D.A. Shockey et al. Fragmentation of rock under dynamic loads. *Int. J. Rock Mech. Min. Sci. & Geomech. Abstr.*, 11:303–317, 1974.

[10] P.J. Digby et al. Computer simulation of blasking-induced vibration, frature and fragmentation processes in brittle rock. In *First International Symposium on Rock Fragmentation By Blasting* [1].

[11] R. Hamajima et al. Analysis for discontinuous medium considering elemental deformation. *Computer Methods and Advances in Geotechnics*, pages 877–880, 1994.

[12] T.F Adams et al. Simulation of rock blasking with the shale code. In *First International Symposium on Rock Fragmentation By Blasting* [1].

[13] R.F. Favreau. Generation of strain waves in rock by an explosion in a spherical cavity. *Journal of Geophysical Research*, 74(17):4267–4280, 1969.

[14] W. Fickett and W.C. Davis. *Detonation*. University of California Press, 1979.

[15] D. E. Grady and M. E. Kipp. *Dynamique Rock Fragmentation*, volume 1 of *Acadmic Press Geology*, section 10. Academic Press Inc, second edition, 1987.

[16] D.E. Grady and M.E. Kipp. Continuum modelling of explosive fracture in oil shale. *Int. J. Rock Mech. Min. Sci. & Geomech. Abstr.*, 17:147–159, 1980.

[17] A. A. Griffith. The phenomenon of rupture dans flow in solids. *Philosophical Transactions of the R. Soc. A*, 221:163–170, 1920.

[18] J.M. Hablot. *Construction de solutions exactes en élastro-plasticité. Application à l'estimation d'erreur par apprentissage*. PhD thesis, Ecole Nationale des Ponts et Chaussées, 1990.

[19] A Hoenig. Elastic mduli of non-randomly cracked body. *Int. J. Solids Structures*, 15:137–154, 1979.

[20] Mark L. Kachanov. Time of the rupture process under creep condition. *Izvestiya Akademii Nauk;SSSR, Otd Tekhn Nauk*, 8:26–31, 1958.

[21] U. Langefors and B. Kilhlström. *The Modern Technique of Rock Blasting*, volume 1. JOHN WILEY and SON, 1963.

[22] S. McHugh. Computational simulation of dynamically induced fracture and fracgmentation. In *First International Symposium on Rock Fragmentation By Blasting* [1].

[23] MECALOG. *RADIOSS*. FRANCE, 1986.

[24] P. Navidi and J. Zarka. Clever optimal design of materials and structures. In *Second French-Korean Conference on Machine Learning*, 1993.

[25] S. Nemat-Nasser and M. Hori. *Micromechanics overall properties of heterogeneous materials*, volume 37 of *Applied Mathematics And Mechanics*. North-Holland, 1993.

[26] M. Schoenauer and M. Sebag. Incremental learning of rules and meta-rules. In *7th Int. Conference on Machine Learing*, 1990.

[27] J.A. Sharpe. The production of elastic wave by explosive pressure, part i - theory and empirical field observations. *Geophysics*, 7(2):144–154, 1942.

[28] Michel Terrien. Systèmes experts par apprentissage en contrôle non-destructif. In *Les systèmes experts & leurs applications*, volume 3, pages 439–447. EC2, 1991. Onzièmes Journées Internationales "Les systèmes experts & leurs applications", Avignon, 27–31 Mai 1991.

[29] J. Richtmayer Von Neumann. A method for the numerical calculation of hydrodynamical shock. *J. of applied physics*, 1950.

[30] Mark L. Wilkins. Calcul de détonations mono et bidimensionnelles. In *Symposium High Dynamic Pressures*, Paris, 1967. I.U.T.A.M., Dunod.

[31] Q. Zeng. *Optimisation et l'utilisation des explosifs en Génie civil*. PhD thesis, Ecole Nationale des Ponts et Chaussées, 1995.

Rock Fragmentation by Blasting, Mohanty (ed.)© 1996 Taylor & Francis. ISBN 90 5410 824 X

Modeling of shock- and gas-driven fractures induced by a blast using bonded assemblies of spherical particles

D.O. Potyondy & P.A. Cundall
Itasca Consulting Group, Inc., Minneapolis, Minn., USA

R.S. Sarracino
AECI Explosives Ltd, Modderfontein, South Africa

ABSTRACT: A computational methodology is described for simulating explosive rock breakage that incorporates the separate and combined effects of shock- and gas-induced damage within a single, three-dimensional model, using the spherical discrete-element code, PFC^{3D} (Itasca, 1995). The rock mass is modeled as an assembly of spherical particles bonded together into a dense packing. Use of the distinct-element method allows dynamic stress waves to propagate through the assembly and allows the conglomerate to slip or separate, with unlimited displacement, under the action of applied forces. The development of a discrete, gas-filled fracture network of inter-connected, penny-shaped crack reservoirs distributed throughout the PFC^{3D} model at "crack" locations (i.e., where bonds have broken) is modeled explicitly, with fracture, flow and pressure distributions evolving dynamically during the simulation. The reservoir pressures load the rock mass, making this a fully coupled model that is dynamic in both gas flow and structural response.

1 INTRODUCTION

Because the mechanisms of explosive rock breakage are not well understood, effort has been directed toward empirical methods of designing blasts to optimize various criteria — e.g., fragment size distribution, damage localization, material throw — (Persson et al., 1994). Significant improvements in these methods, and in the efficiency of the engineering operations that they support, stand to be made through accurate computational modeling of the rock-breakage process. Detonation of the explosive produces both a shock wave and a large volume of gas. Energy is released from these two sources nearly simultaneously over an extremely small time period, making them difficult to differentiate experimentally. Fourney (1993) reviews theories describing the mechanisms of explosive rock breakage and argues that stress waves and gas pressures both play important roles in the rock-breakage process. Similar conclusions derive from the blasthole-liner experiments of Brinkmann (1987, 1990) in which a metal liner allows shock transfer but prevents gas penetration. Brinkmann concludes that the shock energy is most effective at fracturing the rock mass near the blasthole, where it produces the initial fracture network for gas penetration; the subsequent gas expansion extends the fractures and produces final fragmentation and heave of the burden rock.

Sarracino and Brinkmann (1994) suggest the following qualitative model of the damage process:

> The compressive wave, close to the blasthole, causes hydrodynamic deformation in the rock. In the wake of this wave the rock resolidifies into a highly damaged — crushed — state. In a region beyond the hydrodynamic zone the rock is presumably deformed plastically .... Beyond these two zones of compressive deformation we would expect there to be a zone of tensile fracture .... Without gas penetration, damage will be slight in this region. However, this tensile fracture creates a sparse network into which gas can penetrate. The gas extends the network and causes the familiar damage and fragmentation observed far beyond the zone of compressive failure. A revised damage model should treat shock and gas, and these two very different regimes, separately.

Most computational models of explosive rock breakage are continuum descriptions of the phenomena, based on the activation, growth and coalescence of inherent distributions of fracture-producing flaws and implemented as generalizations of strain-rate-dependent elasto-plasticity laws within wave-propagation codes (Grady and Kipp, 1987). These codes numerically integrate the conservation equations of mass, momentum, and energy, along with the constitutive equations for the material, and provide estimates of damage that result from passage (and possible reflection) of the shock wave. They cannot compute the throw of material; however, the throw can be computed by discrete-element codes (Butkovich et al., 1988). An example of coupling these two techniques (i.e., prediction of fragmentation caused by the shock wave, followed by rock motion caused by the gas expansion) is described by Preece et al. (1994).

Here, we describe a methodology for modeling the explosive rock-breakage process that incorporates the separate and combined effects of shock- and gas-induced damage within a single, three-dimensional model, using the spherical discrete-element code, *PFC3D* (Itasca, 1995). The development of a discrete, gas-filled fracture network of inter-connected, penny-shaped crack reservoirs is modeled explicitly, with fracture, flow and pressure distributions evolving dynamically during the simulation. The gas-flow model and crack-tracking logic are integrated with the mechanical computations, making this a fully coupled model that is dynamic in both gas flow and structural response. It is hoped that this model will contribute to a better understanding of the mechanisms of explosive rock breakage by providing a tool with which the separate and combined effects of the shock wave and gas expansion can be examined.

2 PREVIOUS MODELING OF ROCK BLASTING

The following review is not exhaustive; rather, it focuses upon explicit models of rock interaction with blast gases. All of the reviewed work concerns two-dimensional models only. Continuum damage approaches are not described.

Preece et al. (1994) have developed a numerical model that couples the gas flow and rock motion resulting from a blast. The model uses the discrete-element method to simulate circular particles, but it does not treat the fragmentation of the material by the explosive, since the rock is assumed to be already fragmented at the start of the calculations. The flow of explosive gases outward from the borehole is modeled by assuming that the rock is a porous medium. The loads on the circular particles are calculated from the gas-flow characteristics. Particle rearrangement modifies the porosity used in the gas-flow model, which is related to the specific internal energy and the pressure of the explosive gases via an equation of state.

Work similar to that of Preece has been done by Munjiza (1992) and Munjiza et al. (1994). Unlike the work of Preece, however, fragmentation is modeled using a strain-softening-based "fracture" model in which each discrete element is broken into two separate discrete elements whenever its material strength reaches zero. (The fracture model requires that the

solid material comprising each discrete element be modeled using a collection of finite elements.) Munjiza provides example blast-modeling results in which a constant gas pressure is assumed within a fixed distance from the borehole and a fragmented rock mass is produced; the flow of gas within the fractured rock mass is not modeled.

Schatz et al. (1987) describe a methodology for predicting the formation of multiple discrete radial fractures emanating from a borehole as the result of a propellant/explosive loading. Rock fragmentation is not modeled; rather, the trajectories of an initial number of discrete fractures are computed. Flaw rupture is viewed as a quasi-static fracture-mechanics problem with known stresses, material properties, and fracture-entry gas pressures. The pressure distribution along each crack is obtained by solving a one-dimensional fluid-flow problem. This pressure distribution causes a stress intensity at each crack tip, and fracture arrest occurs when the stress intensity becomes less than the fracture toughness of the rock. Similar work has been done by Nilson et al. (1985). Roberts and Swenson (1988) have extended this approach to include dynamic material deformation and elasto-dynamic fracture mechanics, thereby producing a fully coupled model of gas-driven fracture that is dynamic in both gas flow and structural response. In their work, the rock is modeled as a continuum using the finite-element method with automatic remeshing to accommodate crack growth.

3 MODELING METHODOLOGY

The computational methodology for simulating explosive rock breakage couples the *PFC3D* model of rock with an evolving gas-filled fracture network of penny-shaped crack reservoirs. A description of the methodology is presented here; a more detailed formulation can be found in Potyondy and Cundall (1995). The aim is to describe, and then to demonstrate, the feasibility of this approach — not to present a definitive model (which will require further development).

3.1 *Modeling the Rock*

A *PFC3D* model simulates the mechanical behavior of a collection of distinct, arbitrarily sized spherical particles that displace independently and interact only at contacts. The particles are assumed to be rigid, and the behavior at the contacts is characterized using a soft-contact approach, in which finite normal and shear stiffnesses are taken to represent the measurable contact stiffnesses. The behavior of a solid can be simulated by bonding groups of spheres together at their contact points using "contact bonds". These bonds exist between pairs of particles and reproduce the effect of an adhesion acting over the vanishingly small area of the contact point, thus transmitting a force (but no moment) between the two bonded particles. The existence of a contact bond precludes sliding and limits the allowable magnitudes of normal tension and shear force acting at the contact. If either the normal or shear limit is reached, then the bond breaks, and the contact cannot subsequently take tension, although compressive forces are allowed, together with appropriate shear forces that are limited by the friction coefficient.

The *PFC³D* model of rock consists of an assembly of spherical particles (similar to grains) bonded together in a dense packing. The resulting assembly can be regarded as a solid that exhibits deformability via contact stiffnesses, strength via bond strengths and particle friction coefficients, and damage via progressive bond breakage. The assembly is capable of fracturing into separate autonomous assemblies, or blocks, that interact with one another. Use of the distinct-element method (described below) allows dynamic stress waves to propagate through the assembly and allows the particles and blocks to slip or separate (with unlimited displacement) under the action of applied forces. The *PFC³D* model of rock produces complex macroscopic behaviors such as strain softening and hardening, dilation, and fracture by simulating the physical micro-mechanisms directly without resorting to complex constitutive laws (Potyondy et al., 1996; Lorig et al., 1995). The *PFC³D* model of rock has been used to predict strains in the near zone of a blasthole by Sarracino and Guest (1995).

The distinct-element method (DEM) is used to model the movement and interaction of the *PFC³D* particle assembly. A thorough description of the method is given in the two-part paper of Cundall (1988) and Hart et al. (1988). *PFC³D* is classified as a "discrete-element" code based on the definition in Cundall and Hart (1992) because it allows finite displacements and rotations of discrete bodies (including complete detachment) and recognizes new contacts automatically as the calculations progress.

In the DEM, the interaction of the particles is treated as a dynamic process in which movements result from the propagation through the particle system of disturbances caused by specified wall and particle motion and/or body forces. The speed of propagation depends on the physical properties of the discrete system. The dynamic behavior is represented numerically by a time-stepping algorithm identical to that used in the explicit finite-difference method for continuum analysis. The use of an explicit, as opposed to an implicit, numerical scheme makes it possible to simulate the non-linear interaction of a large number of particles without excessive memory requirements or the need for an iterative procedure.

3.2 *Modeling the Explosion*

In a rock mass that is unfractured prior to detonation,

the pressure of the explosive gases acts only on the borehole walls, and the resulting shock wave may trigger formation of a network of cracks. Once any part of this network intersects the borehole, it provides a pathway into which the gas will flow. Here, the crushed zone immediately adjacent to the borehole wall is not simulated; instead, a time-varying pressure [see Eq. (1)] is applied at the outer surface of a cylinder three times the diameter of the actual borehole. This model approximates the zone of tensile fracture referred to in Section 1. More quantitatively accurate simulations would require an estimate of pressure and temperature loss within this inner zone by using, for instance, a continuum code, or *PFC³D* itself.

3.3 *Modeling Gas-Driven Fracture Growth*

The overall model behavior is characterized by a force distribution within the rock mass (comprised of a bonded collection of densely packed spheres) and a flow and pressure distribution within the fracture network (comprised of a pipe network in which pipes connect penny-shaped reservoirs that are distributed throughout the *PFC³D* model at "crack" locations — i.e., where bonds have broken). Bond breakage is controlled by the bond strengths and force distribution. Gas flow occurs only within the pipe network. New pipes are added to the network as cracks form and their corresponding reservoirs physically overlap. The pressure within each reservoir produces two equal but opposite forces that act upon the two parent particles from which the reservoir originated.

3.3.1 *Reservoir Geometry*

Whenever a bond between two particles breaks, a penny-shaped reservoir is formed. The geometry and location of each reservoir are determined by the sizes and current locations of the two parent particles from which the reservoir originated. A typical reservoir is shown in Figure 1. The reservoir geometry is described by thickness, radius and unit-normal: the thickness equals the gap between the two parent particles; the radius is given by the intersection of the cylinder bisection plane with a membrane stretched tightly between the two parent particles; and the unit-normal is directed along the line joining the centers of the two parent particles. Because of the geometrical decoupling between particles and reservoirs, the relation between reservoir geometry and parent particle sizes could be modified (e.g., reservoir thickness could be made independent of particle size).

3.3.2 *Pipe Geometry*

Whenever two reservoirs overlap physically, they are connected by a straight pipe with two rectangular cross-sections (one for each reservoir). A typical pipe is shown in Figure 2. The geometry of each of the two pipe sections is described by aperture, length and depth: the aperture is equal to the thickness of the corresponding reservoir; the length is equal to the diameter of the corresponding reservoir; and the depth is chosen such that the total pipe volume equals the total volume of the two reservoirs. A minimum aperture is used here to represent the fact that once a crack forms in rock, a minimum aperture remains, even in the presence of significant compressive stress trying to close the crack. Also, the total pipe length used in the pipe-flow model is taken as the average of the lengths of the two pipe sections.

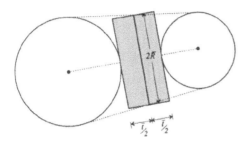

Fig. 1 Typical penny-shaped reservoir and its two parent particles

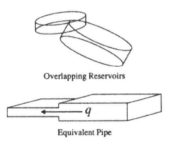

Fig. 2 Typical pipe formed by two overlapping reservoirs

3.3.3 Pipe-Flow Model

The flow within each pipe in the network is modeled by assuming mixed laminar-and-turbulent one-dimensional fluid flow of a compressible gas within a rock joint (Schatz et al., 1987). Mixed laminar-and-turbulent flow is described by separate flow terms that are derived under the assumption of flow between two parallel plates. The mixed-flow equation is quadratic in flow q (with coefficients that depend upon fluid viscosity, as well as roughness, length, aperture, and pressure drop across the pipe) and provides a smooth transition from laminar to turbulent flow. Under laminar-flow conditions (small q), the linear term dominates; under turbulent flow conditions (large q), the quadratic term dominates.

The pressure within each reservoir in the network can be related to reservoir volume and total mass of gas flowing into (or out of) the reservoir. Here, a simpler relation (based on the gas bulk modulus) between the fractional change in volume of a reservoir and the resulting change in reservoir pressure has been employed. The volume change arises from gas flow into the reservoir and mechanical deformation of the enclosing solid material.

3.3.4 Pipe-Flow Boundary Conditions

The boundary conditions for the network consist of specified pressures for a subset of the reservoirs. The borehole is treated as a reservoir of infinite volume whose pressure-time history is given as model input. At any time during a simulation, each reservoir is classified into one of three categories depending on its proximity to the borehole or the free surface of the model. The set of reservoirs "near" the borehole are designated as category-1 reservoirs, and the set of reservoirs "near" the free surface of the model are designated as category-2 reservoirs. All remaining reservoirs are designated as category-3 reservoirs. Category-1 reservoirs are assigned the current borehole pressure, and category-2 reservoirs are assigned atmospheric pressure. The pressures of category-3 reservoirs are computed based on the network conditions.

3.3.5 Calculation Cycle

Both the gas and the mechanical computations are performed using a finite-difference formulation; therefore, the calculation cycle for the coupled fluid-mechanical system involves a time-stepping algorithm. The cycle begins with an estimation of the mechanical timestep Δt^m. If T_1 is the global elapsed time at the start of the cycle, then $T_2 = T_1 + \Delta t^m$ will be the elapsed time at the end of the cycle. The applied particle forces arising from both the reservoir and borehole pressures at time T_2 are determined. Then, the law of motion is applied to each particle to update its velocity and position based on the resultant force and moment arising from the contact forces, body forces (if present), and the applied particle forces. Next, the set of contacts is updated from the known particle positions. The force-displacement law is then applied to each contact to update the contact forces — based on the relative motion between the two entities at the contact and the contact constitutive model. During this step, cracks will form if the contact force exceeds the contact bond strength. Finally, the pressures and flow rates in the pipe network are computed by the pipe-flow model, and any newly formed cracks are incorporated into the pipe network. (For all problems considered, it has been found that the estimated fluid timestep is less than the estimated mechanical timestep; thus, a set of fluid sub-cycles is performed within the pipe-flow model.)

4 DEMONSTRATION PROBLEM

Implementation of the methodology described above within *PFC³D* did not require modification of the source code; instead, the algorithms were written as a set of *FISH* functions. *FISH* is a programming language embedded within *PFC³D* that enables the user to define new variables and functions to monitor and control program execution. A demonstration problem of a borehole blast is presented to show that *PFC³D*, along with the set of *FISH* functions, correctly implements the algorithms comprising the methodology.

4.1 *Boulder Geometry*

The model simulates a blast that occurs in a thin slice of material surrounding a borehole in a rectangular-shaped granite boulder. The *PFC³D* model is shown in Figure 3, in which the particles adjacent to free surfaces are drawn in darker shades of gray. The *PFC³D* model contains 1391 uniform-radius (2.35×10^{-2} m) particles arranged in a regular, face-centered cubic packing. The borehole extends completely through the model, and the two surfaces at the top and bottom of the borehole axis are given boundary conditions of no motion in the normal direction and no flow. The other four surfaces are considered to be free surfaces; thus, any cracks intersecting these surfaces will vent to the atmosphere.

Fig. 3 Perspective view of boulder model

4.2 *Borehole Pressure Loading*

A "pressure zone" of one particle diameter surrounds the borehole. Any particle whose centroid lies within the pressure zone receives an external force proportional to its assigned portion of borehole-wall area multiplied by the borehole pressure, and any reservoir with at least one parent particle in the pressure zone receives the borehole pressure.

The user-defined borehole pressure-time history is given by

$$P = P_\mathrm{m} \begin{cases} t/t_r, & t < t_r \\ 1.0, & t_r \leq t < t_f \quad (1) \\ (0.76 + 2000t)^{-2}, & t \geq t_f \end{cases}$$

where P_m is the maximum pressure, t_r is the rise time, and t_f is the fall-off time. These values were set at 1.3×10^8 Pa, 0.6×10^{-4} s and 1.2×10^{-4} s, respectively, for the demonstration problem.

4.3 *Calibrating the Model to Represent Granite*

The micro-properties of the *PFC³D* model are chosen to simulate the behavior of granite. The constant normal and shear particle stiffnesses, constant friction coefficient, and normally distributed normal and shear contact bond strengths are set such that the value of both Young's modulus (69.9 GPa) and the unconfined compressive strength (200 MPa) representative of granite are obtained when an unconfined compression test is performed upon the *PFC³D* sample. For all simulations performed here, the default damping value of 0.7 was used. This corresponds to approximately 22 percent of critical damping, which may be too high for granite. Determination of an appropriate damping coefficient demands further study that is beyond the scope of the present investigation.

4.4 *Test Case A: Shock Wave Only*

The effect of the shock wave alone on fragmentation of the *PFC³D* model is demonstrated by two tests in which the gas-flow mechanism is disabled such that the only loading is from the pressure applied to the borehole walls. Two tests were performed for Case A, in which the maximum pressures [P_m of Eq. (1)] were set to 300 MPa and 130 MPa, respectively.

When the maximum borehole pressure is set to 300 MPa, damage begins in the region near the borehole, then localizes along four planes running from the borehole to near each of the four corners of the model. By the time crack formation ceases, the specimen has broken into four separate pieces that are moving away from the borehole as individual units (Figure 4). In Figures 4 and 5, each crack is drawn as a spherical icon. Icon radius equals 30 percent of crack radius, and icon shading represents time of crack formation with light shades representing early events and dark shades representing late events.

When the maximum borehole pressure is set to 130 MPa, damage remains contained within the region near the borehole. By the time crack formation ceases, a total of 10 cracks, distributed both around the borehole circumference and through the specimen thickness, have formed (Figure 5). In this Figure, each contact bond is drawn as a line between the two bonded particle centers. The contact force distribution at the time of 0.237 ms is shown in Figure 6, in

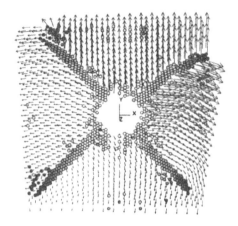

Fig. 4 Perspective view of crack icons and particle velocities for case A (maximum pressure of 300 MPa) at time of 1.88 ms

which the contact forces are drawn as lines — light gray for tension and black for compression — through each contact location, oriented in the direction of the contact force, with a thickness proportional to force magnitude. Note the presence of tensile forces directed parallel to the borehole walls that accompany the radial expansion; such forces are absent in the first case-A test, having been relieved by the extreme cracking.

4.5 Test Case B: Shock Wave and Gas Pressure

The combined effect of the shock wave and the gas pressure on fragmentation of the PFC^{3D} model is demonstrated by the case-B test, in which the gas-

flow mechanism is enabled. The maximum borehole pressure is set to 130 MPa in order to facilitate comparison with the second test of case A.

The case-B test specimen experiences greater damage than the corresponding case-A test specimen (Figure 5). At a time of 0.237 ms for the case-B test, 699 cracks and 2471 pipes, distributed both around the borehole circumference and through the specimen thickness, have formed. The contact force distribution at the time of 0.237 ms is shown in Figure 6. Note the absence of tensile hoop forces and the presence of large compressive forces in the region near the borehole for the case-B test. The maximum compressive force (37.6×10^4 N) for the case-B test is larger than the maximum compressive force (7.05×10^4 N) for the corresponding case-A test. These increased compressive forces result from the pressure acting within the cracks.

4.6 Discussion of Test Cases

The greater extent of both damage and compressive forces for the case-B test, as compared to the case-A test, confirms that the gas pressure is propagating beyond the near-borehole region. The case-B test also demonstrates that the crack-tracking, pipe-formation, and pipe-flow logic is implemented robustly, since it is possible to track the dynamic formation and interaction of 699 cracks and 2471 pipes.

The results from an additional test — in which the pipe-flow mechanism is disabled, but the cracks (reservoirs) in the borehole pressure zone receive the full borehole pressure immediately upon formation — suggest that the present approach may be triggering excessive cracking, as a miniature shock wave travels outward from each such newly formed crack. It may be more realistic to ramp-up the pressure in these cracks to the borehole pressure over a finite lag-time.

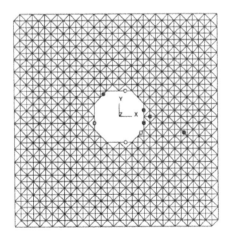 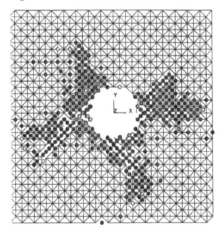

Fig. 5 Non-perspective view of crack icons and contact bond network for case A (left) and case B (right) at time of 0.237 ms

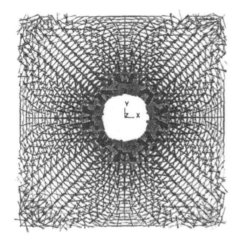 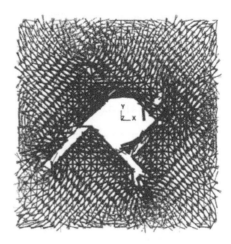

Fig. 6 Contact force distribution for case A (left) and case B (right) at time of 0.237 ms

5 CONCLUSIONS

The feasibility of using *PFC³ᴰ* to model explosive rock fragmentation has been demonstrated by simulating gas flow through a fracture network that is formed dynamically during *PFC³ᴰ* execution. The pipe-flow model and crack-tracking logic are integrated with the mechanical computations, making this a fully coupled model that is dynamic in both gas flow and structural response. The model provides a tool with which the separate and combined effects of the shock wave and gas expansion can be examined. The approach holds promise for this class of rock fracture problems, where the complex constitutive behavior (arising from extensive micro- and macro-cracking) is difficult to characterize accurately in terms of a continuum formulation.

Qualitative and quantitative correlations of model predictions with experience and with experiment are being pursued, as well as refinement of the methodology to incorporate more of the mechanisms underlying the physical processes. Refinements which would add greatly to the accuracy and predictive capability of the model are:

(1) implementation of a burn model incorporating an equation of state for the post-detonation gases;

(2) adjustment of reservoir thicknesses to correspond more accurately with physical crack widths; and

(3) addition of inertial terms to the fluid equations — i.e., solution of the one-dimensional, Navier-Stokes equations in each pipe. (This would introduce a time lag into the pressure history applied to the crack walls and could increase the critical fluid timestep.)

6 REFERENCES

Brinkmann, J. R. 1987. Separating shock wave and gas expansion breakage mechanisms. *Proc. 2nd Int. Symp. on Rock Fragmentation by Blasting*: 6–15. W. L. Fourney & R. D. Dick, Eds. Bethel, Connecticut: Society for Experimental Mechanics.

Brinkmann, J. R. 1990. An experimental study of the effects of shock and gas penetration in blasting. *Proc. 3rd Int. Symp. on Rock Fragmentation by Blasting*.

Butkovich, T. R., O. R. Walton & F. E. Heuze 1988. Insights in cratering phenomenology provided by discrete element modeling. *Key Questions in Rock Mechanics*: 359–368. P. A. Cundall, R. L. Sterling & A. M. Starfield, Eds. Rotterdam: Balkema.

Cundall, P. A. 1988. Formulation of a three-dimensional distinct element model — part I: a scheme to detect and represent contacts in a system composed of many polyhedral blocks. *Int. J. Rock Mech. Min. Sci. & Geomech. Abstr.* 25(3):107–116.

Cundall, P. A., & R. Hart 1992. Numerical modelling of discontinua. *Int. J. Engr. Comp.* 9:101–113.

Fourney, W. L. 1993. Mechanisms of rock fragmentation by blasting. *Comprehensive Rock Engineering: Principles, Practice and Projects*:39–69. J. A. Hudson, Ed. Oxford: Pergamon Press.

Grady, D. E., & M. E. Kipp 1987. Dynamic rock fragmentation. *Fracture Mechanics of Rock*:429–475. B. K. Atkinson, Ed. London: Academic Press.

Hart, R., P. A. Cundall & J. Lemos 1988. Formulation of a three-dimensional distinct element model — part II: mechanical calculations for motion and interaction of a system composed of many polyhedral blocks. *Int. J. Rock Mech. Min. Sci. & Geomech. Abstr.* 25(3):117–125.

Itasca Consulting Group, Inc. 1995. *PFC³ᴰ (Particle Flow Code in 3 Dimensions)*, Version 1.1. Minneapolis, Minnesota: ICG.

Lorig, L. J., M. P. Board, D. O. Potyondy & M. J. Coetzee 1995. Numerical modeling of caving using continuum and micro-mechanical models. *Proc. 3rd Canadian Conf. on Computer Applications in the Mineral Industry*: 416–425. Montréal: McGill University Press.

Munjiza, A. 1992. Discrete elements in transient dynamics of fractured media. University of Wales, University College of Swansea, England, Ph.D. Thesis.

Munjiza, A., D.R.J. Owen, N. Bicanic & J. R. Owen 1994. On a rational approach to rock blasting. *Computer Methods & Advances in Geomechanics*:857–862. H. J. Siriwardane & M. M. Zaman, Eds. Rotterdam: Balkema.

Nilson, R. H., W. J. Proffer & R. E. Duff 1985. Modelling of gas-driven fractures induced by propellant combustion within a borehole. *Int. J. Rock Mech. Min. Sci. & Geomech. Abstr.* 22(1):3–19.

Persson, P.-A., R. Holmberg & J. Lee 1994. Rock blasting and explosives engineering. Boca Raton, Florida: CRC Press, Inc.

Potyondy, D., & P. Cundall 1995. Addition of gas pressurization to fractures in the three-dimensional particle flow code. Itasca Consulting Group, Inc., report to AECI Explosives Limited (AEL).

Potyondy, D. O., P. A. Cundall & C. Lee 1996. Modeling rock using bonded assemblies of circular particles. To appear in *Proc. 2nd North American Rock Mechanics Symp.* Rotterdam: Balkema.

Preece, D. S., B. J. Thorne, M. R. Baer & J. W. Swegle 1994. Computer simulation of rock blasting: a summary of work from 1987 through 1993. Sandia National Laboratories, Report No. SAND92-1027.

Roberts, J. M., & D. V. Swenson 1988. A numerical model of gas-driven dynamic fractures with an application to borehole stimulation. *Computers in Engineering 1988*:1–6, Vol. 3. V. A. Tipnis & E. M. Patton, Eds. New York: American Society of Mechanical Engineers.

Sarracino, R. S., & J. R. Brinkmann 1994. Modeling of blasthole liner experiments. *Computer Methods and Advances in Geomechanics*:871–876. H. J. Siriwardane & M. M. Zaman, Eds. Rotterdam: Balkema.

Sarracino, R. S., & A. Guest 1995. Measured and predicted strains in the near zone of a blasthole. *Proc. 10th Int. Shock Compression Symp.*

Schatz, J. F., B. J. Zeigler, R. A. Bellman, J. M. Hanson, M. Christianson & R. D. Hart 1987. Prediction and interpretation of multiple radial fracture stimulations. Science Applications International Corporation (SAIC), report to Gas Research Inst., Contract No. 5084-213-1149, SAIC report No. SAIC-87/1056, GRI report No. GRI-87/0199.

Rock Fragmentation by Blasting, Mohanty (ed.) © 1996 Taylor & Francis. ISBN 90 5410 824 X

Numerical simulation of initial shock parameters of detonating charge on borehole wall

Zhifang Zi & Yumin Li
Blasting Institute, Shandong Institute of Mining and Technology, Taian, People's Republic of China

ABSTRACT: A physical model for calculating dynamic initial shock parameters of coupling charge on borehole wall has been established in this paper with regards to many factors such as interaction between detonation wave and borehole wall, deformation of borehole wall and the properties of both explosives and rocks. Based on this model, the calculating mothod has been set up by using the laws of the conservation of mass, momentum and energy, the detonation theory and Mach Reflection theory. The initial shock parameters such as pressure P_m, shock wave velocity D_m, etc. of borehole wall in Granite, Limestone and Marble have been calculated separately under the action of medium strength explosives. The varing characteristic of every parameter on borehore wall has been obtained from the ignited dot to the end of coupling charge borehole along the axial direction of borehole.

1 INTRODUCTION

When a coupled charge in borehoie is initiated, a complex process of reflection and refraction is formed because of interaction between the detonation wave of explosive and the borehole wall. An initial shock wave generated by the detonation wave impacting the walls of borehole will travels outwards into the rock mass. Then, a dynamic stress field in rock around the borehole will occur. It is unquestionable that the calculation and measurement of dynamic stress be the base for research on rock fragmentation by blasting. However, the more basic research should be how to calculate and measure the initial shock parameters because it is the propagation of them that result in formation of dynamic stress in rock mass.

Because of the difficulties encountered in measuring the initial shock parameters on the walls of borehole, numerical simulation becomes necessary to determine the parameters. Blasting researchers have studied this problem for many years and have generated models and formulas. But the models are mostly built up based on the cases: (1) the direction of propagation of detonation wave and the rock wall is parallel and (2) the borehole wall is rigid. In fact, the former is only true at the end of the borehole, or when spherical charges are used, or cylindrical charges are initiated by detonation

cord continously. For most blasting operations, the radius of the borehole is much smaller than the depth of borehole. Generally cylindrical charges are used. When the charges are ignited from one initial point or several points by detonator(s), the detonation wave travels from the initial point in spherical shape. As a result, the direction of propagation of the detonation wave and the rock wall is neither perpendicular nor parallel. Meanwhile, the deformation of the wall is inevitable because of the shock action of high detonation pressures. So, when establishing models or equations, we should consider not only the properties of both rock and explosives, but also the interaction between the detonation wave and rock wall and the deformation of rock wall. These important factors will be taken into account in this paper.

2 THE ESTABLISHMENT OF PHYSICAL MODEL

For a cylindrical charge, when the explosive is initiated at a point, the detonation wave will travel in spherical shape (see Fig. 1(a)). The oblique incidence will begin while the detonation wave runs into the borehole wall. The plane of incidence detonation wave of explosive, Φ_0 is the incidence angle of detonation and OR is the reflection wave. A oblique shock wave or rafraction shock wave (OB) is genarated in rock mass around borehole wall under the action of high detonation pressures. Furthermore, the wall of borehole will deform. The line OC representes the deformed wall (shown in Fig. 1 (b), (c), (d)). The deformation angle ε is determined by both detonation pressures and the comressibility of rock. According to hydrodynamic theory (pan, 1982), the flowing direction of detonation products will deflect to the deformed wall after they pass through the plane of incidence detonation wave OA. The deflection angle θ depends on both Φ_0 and the strength of detonation wave. In the primary process of reflection, the reflection of detonation wave is regular oblique reflection (see Fig. 1(b)) because of $\theta > \varepsilon$ and $\Phi_0 < \Phi_\infty$ (Φ_∞ is a critical angle). The incidence angle Φ_0 is getting bigger with the propagation of detonation wave. The irregular reflection i. e. Mach Reflection (see Fig. 1(c)) will occur on the wall while $\Phi_0 > \Phi_\infty$. When Φ_0 is big enough and makes $\theta < \varepsilon$, the reflection wave is expansion wave but not shock wave, just then, the Prantl-Meyer Eapansion forms (see Fig1 (d)).

To sum up the above, we know that the initial shock parameters of coupling charge borehole wall from the initial dot to the end of borehole along the axial can be calculated by three sections: (I)regular oblique reflection, (I)Mach Reflection and (Ⅲ) Prantl-Meyer Expansion. The parameters of refraction shock wave of every section is just the initial shock parameters of borehole wall. The length of every section is decided by the properties both explosive and rock.

3 A METHODOLOGY FOR CALCULATING THE INITIAL SHOCK PARAMETERS

Based on the physical model above, the calculating methods can be obtained by use of theories of hydrodynamics, detonation and Mach (Ni, 1994. 5).

3. 1 The parameters of incidence detonation wave

According to the detonation wave theory (Zhang, et al. 1979), we can easily get the calculating formulas of incidence detonation wave:

$$
\left.\begin{array}{l}
P_H = P_o D^2/(\gamma+1); \ \rho_H = \rho_0(\gamma+1)/\gamma; \\
u_H = D/(\gamma+1); \ C_H = \gamma D/(\gamma+1); \\
tg\theta = tg\Phi_0/[\gamma tg^2\Phi_0+(\gamma+1)]; \\
q_1 = C_H \sqrt{[(\gamma+1)/\gamma]^2 ctg^2\Phi_0+1}
\end{array}\right\} \quad (1)
$$

where P, ρ, C, u, q are respectively the pressure, density, sonic velocity, partical velosity and flowing velosity of detonation products in parallel to the wall. The subscript o is for the original state of explosives, H for detonation parametes behind the plane of incidence detonation wave. The specific heat ratio γ is a constant, let γ equal 3 when calculating in this paper.

The incidence angle Φ_0 is expressed by the following equation (Li, 1990. 12):

$$
\Phi_0 = arctg(L/R) \quad (2)
$$

where R is the radius of the borehole, L is the distance from the initial dot to the calculating point on the wall along the axial (shown in Fig. 2).

3. 2 The general equations of initial shock parameters of borehole wall

$$
\left.\begin{array}{l}
P_m = \rho_{mo}\dfrac{D^2}{Sin^2\Phi_0}Sin^2\Phi_3(1-\dfrac{\rho_{mo}}{\rho_m}) \\[2mm]
tg\varepsilon = \dfrac{(1-\dfrac{\rho_{mo}}{\rho_m})tg\Phi_3}{1+\dfrac{\rho_{mo}}{\rho_m}tg^2\Phi_3} \\[2mm]
q_m = \dfrac{\rho_{mo}}{\rho_m}\dfrac{D}{Sin\Phi_0}\dfrac{Sin\Phi_3}{Sin(\Phi_3-\varepsilon)} \\[2mm]
\dfrac{\rho_{mo}}{\rho_m} = \dfrac{b-1}{b}+\dfrac{a}{b}\dfrac{Sin\Phi_0}{DSin\Phi_3}
\end{array}\right\} \quad (3)
$$

Where the subscript m indicates the parameters for initial shock wave in rock around borehole wall and mo for the original parameters of rock mass. a, b are Hugoniot parameters of rock (shown in Table 1), Φ_3 is the included angle between the original borehole wall and the initial shock wave in rock mass around borehole wall.

Table 1   The main physical properties of rocks (Niu, 1990. 8)

	Granite	Limestone	Marble
Density (g/cc)	2. 6—2. 8	2. 3—3. 0	2. 7—2. 9
C.S (kb)	1. 0—2. 5	0. 9—1. 6	4. 4—5. 9
a (km/s)	2. 10	3. 40	4. 00
b	1. 63	1. 27	1. 32

C. S: Compressive Strength

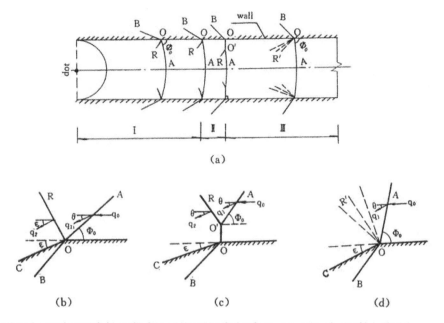

(a)

(b)   (c)   (d)

Fig. 1   The physical model for calculating the initial shock parameters of coupling charge
   (a) The process of action between the detonation wave and borehole wall for a cylindrical
       coupling charge borehole
   (b) The regular oblique reflection diagram representing section I of Fig. (a)
   (c) Mach Reflection diagram representing section I of Fig. (a)
   (d) Prantl—Meyer Expansion representing section II of Fig. (a)

3. 3 The equations of regular oblique reflection region

$$\left.\begin{aligned}
&\frac{P_2}{P_H} = \frac{2\gamma}{\gamma+1} M_1^2 Sin^2 \upsilon - \frac{\gamma-1}{\gamma+1} \\
&\frac{\rho_2}{\rho_H} = \frac{(\gamma+1)M_1^2 Sin^2 \upsilon}{(\gamma-1)M_1^2 Sin^2 \upsilon + 2} \\
&tg\varepsilon = \frac{(1+\frac{\rho_H}{\rho_2}tg^2\upsilon)tg\theta - (1-\frac{\rho_H}{\rho_2})tg\upsilon}{(1-\frac{\rho_H}{\rho_2})tg\upsilon tg\theta + (1+\frac{\rho_H}{\rho_2}tg^2\upsilon)} \\
&q_2 = \frac{\rho_H}{\rho_2} \frac{q_1 Sin\upsilon}{Sin[\upsilon-(\theta-\varepsilon)]} \\
&P_2 = P_m
\end{aligned}\right\} \quad (4)$$

Where: M is Mach Number, υ is the included angle between OR and q_1, the subscript 2 means the parameters behind OR.

Combinating the equations (3) and (4) and by use of computer we can calculating the nine parameters of regular reflection, such as υ, ρ_2, P_2, q_m, P_m, ρ_m, $u_m(D_m)$, Φ_3, ε, the latter five parameters are just wanted to be obtained for initial shock parameters of regular reflection.

3. 4 The equations of Mach Reflection region

$$\left.\begin{aligned}
&\frac{P_4}{P_H} = (\gamma+1)(1-\frac{\rho_0}{\rho_4})(\frac{Sin\beta}{Sin\Phi_0})^2 \\
&tg\varepsilon = \frac{(1-\frac{\rho_0}{\rho_4})tg\beta}{1+\frac{\rho_0}{\rho_4}tg^2\beta} \\
&q_4 = \frac{\rho_0}{\rho_4} \frac{D}{Sin\Phi_0} \frac{Sin\beta}{Sin(\beta-\varepsilon)} \\
&\frac{\rho_0}{\rho_4} = \frac{\gamma-1}{\gamma+1} + \frac{1}{\gamma+1}\frac{P_H}{P_4} \\
&P_4 = P_m
\end{aligned}\right\} \quad (5)$$

Where β is the included angle befween Mach bar oo´ and q_0, $\Phi_0 \leqslant \beta < \pi/2$, the subscript 4 indicates the parameters behind the Mach bar.

The parameters for example ρ_4, P_4, q_4, q_m, p_m, $u_m(D_m)$, ρ_m, ε, Φ_3 can be calculated by combining equations (3) and (5), the latter five parameters are the initial parameters for Mach reflection region.

The following equations is a critial condition for generating Mach reflection (Zhang, 1979)

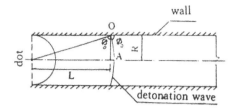

Fig. 2 The relationship between Φ_0 and L

$$\text{arctg} \frac{\text{tg}\Phi_\infty}{\gamma\text{tg}^2\Phi_\infty+(\gamma+1)}=\varepsilon+$$
$$\text{arctg}\frac{M_1^2\text{Sin}_v^2-1}{M_1^2((\gamma+1)/2-\text{Sin}^2v)+1}\text{ctg}v \qquad (6)$$

when $\Phi_0 \leqslant \Phi_\infty$, the Mach Reflection will occur.

3. 5   The equations for Prantl-Meyer Expansion

$$\left.\begin{array}{l}\dfrac{P_1}{P_5}=\{\dfrac{\frac{\gamma-1}{2}M_5^2+1}{\frac{\gamma-1}{2}M_1^2+1}\}^{\frac{\gamma}{\gamma-1}}\\[6pt]\dfrac{\rho_1}{\rho_5}=(\dfrac{P_1}{P_5})^{\frac{1}{\gamma}}\\[6pt]\dfrac{C_1}{C_5}=(\dfrac{\rho_1}{\rho_5})^{\frac{2}{\gamma-1}}\\[6pt]\varepsilon\text{-}\theta=\sqrt{\dfrac{\gamma+1}{\gamma-1}}\text{arctg}\sqrt{\dfrac{\gamma-1}{\gamma+1}(M_5^2-1)}\\[6pt]\quad-\text{arctg}\sqrt{M_5^2-1}-\zeta\sqrt{\dfrac{\gamma+1}{\gamma-1}}\\[6pt]\quad\text{arctg}\sqrt{\dfrac{\gamma-1}{\gamma+1}(M_1^2-1)}-\text{arctg}\sqrt{M_1^2-1}\rbrack\\[6pt]\dfrac{q_5}{C_5}=M_5\\[6pt]P_5=P_m\end{array}\right\} \qquad (7)$$

The subscript 5 indicates the parameters for the flowing products of Prantl-Meyer Expansion, C is sonic velosity.

We also can calculate the parameters: P_5, ρ_5, C_5, M_5, q_m, p_m, ρ_m, $D_m(u_m)$, ε, Φ_3 by combining the equations (3) and (7). The latter five parameters are the initial shock parameters for Prantl-Meyer Expansion.

4. THE ANALYSIS OF RESULTS

Based on the equations above, the iterative calculation for the initial shock parameters from initiation point to the end of borehole can be carried out easily.

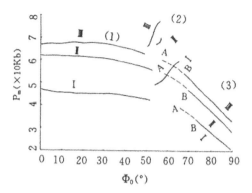

Fig. 3 The variance of initial shock pressure P_m with Φ_0 changing

The calculating results have given satisfactory accurancy. The following calculation results will show the variation of every parameter.

When the medium strength explosives such as №2AN, Emulsion and water Gel (shown in Thabe 2) (wang, 1984. 11) acts on the rocks for instance Granite, Limetone and Marble respectively, The regular oblique reflection, Mach Reflection and Prantl-Meyer Expansion occurs on the borehole wall successively from the initial dot to the end of borehole along axial direction. The regular oblique reflection usually generates in the region from $\Phi_0=0°$ to $\Phi_0=50°$, i. e. L=0 to L= $R\text{tg}50°=1.19R$. The Mach Reflection begin in $\Phi_0=50°$, it last to $\Phi_0=65°$, i. e. L=2. 14R. The regular oblique reflection again from about $\Phi_0=60°-75°$. When $\Phi_0>75°$, i. e. L>3. 73R the process belong to Prantl-Meyer Explansion. Because the calculating results are very similar when medium explosives detonate in the three rocks, the following disccuss is only to the cases that three kinds of explosives act on Granite separately.

Table 2   Detonation parameters of three types of explosives

Explosive	Density (g/cc)	VOD (m/s)	Pressure (kb)
№2AN	1. 00	3600	32. 4
Emulsion	1. 20	4000	48. 3
Water Gel	1. 24	4150	53. 3

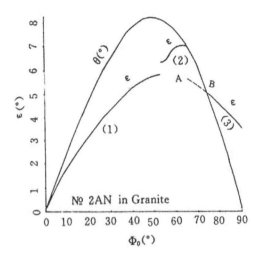

Fig. 4 The relation between ε and θ with Φ_0 changing

Table 3   The variance of initial shock pressure P_m with Φ_0 varing

In Granite	№2AN	Emulsion	Water Gel
$\Phi_0(°)$	0—52	0—53	0—49
P_m(Kb).	45—41	82—56	87—62
$\Phi_0(°)$	53—64	54—56	50—58
P_m(Kb)..	48—59	67—70	65—78
$\Phi_0(°)$	65—74	57—87	57—88
P_m(Kb).	38—32	54—48	59—54
$\Phi_0(°)$	75—90	88—90	87—80
P_m(Kb)...	32—19	47—29	54—31

∗ in regular oblique reflection process
∗ ∗ in Mach Reflection process
∗ ∗ ∗ in Prantl-Meyer Expansion process

4. 1 The dynamic initial shock pressure P_m

From Figure 3 we can see that the shock pressure P_m is higher than detonation pressure P_H and attenuates slowly in the regular oblique reflection region with Φ_0 increasing. The pressure P_m rises suddenly as soon as Mach Reflection begins. In all Mach Reflection rgeion the pressure P_m is almost uptumning. However, the pressure P_m goes down abruptly just as Mach Reflection ends. Then, it retains to regular oblique reflection (the dotted line AB shown in Fig. 3 and Fig. 4). The pressure P_m decreases in parabolical shape and the angle θ is getting smaller with Φ_0 increasing in the section of AB. On the point B, $P_m=P_H$, just then θ=ε, after then θ<ε and Prantl-Meyer Expansion occurs. In this process, the initial shock pressure P_m is smaller than detonation pressure P_H, and P_m attenuates in oblique straight line shape with Φ_0 increasing. The variation of P_m with Φ_0 changing is shown in Figure3 and Table 3. In all following figures, the symbol I、II、III represents №2 AN、Emulsion and water Gel in Granite respectively and (1)、(2)、(3) means oblique reflefction, Mach Reflection and pranfl-Meyer Expansion seqarately.

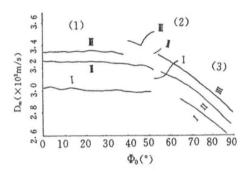

Fig. 5 The variance of initial shock wave velósity D_m with Φ_0 changing

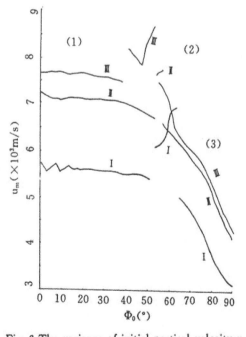

Fig. 6 The variance of initial partical velositv u_m with Φ_0 changing

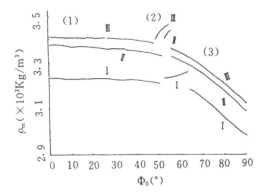

Fig. 7 The variance of dynamic initial density of rock around borehole wall with Φ_0 changing

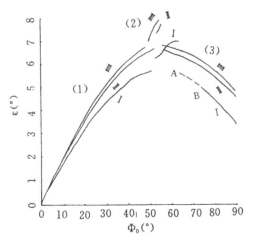

Fig. 8 The variance of deformation angle ε of borehole wall with Φ_0 changing

In addition, the P_m are affected by detonation velosity D (VOD) of explosive and the Hugoniot parameter a of rock. For same calculating point, if same explosive detonates in different rocks, the bigger the Hugoniot parameter a, the higher the pressure P_m, if different explosives in same rock separately, the higher the detonation velosity D of explosive, the bigger the pressure P_m.

4.2 The initial shock velocity D_m and particle velocity u_m

The variations of initial shock velocity D_m and particle velocity U_m with Φ_0 varying are shown in Figure 5 and Figure 6.

The velocity D_m and u_m are influenced directly by the detonation velocity D. For the same rock, the higher the detonation velocity D of explosive, the bigger is the initial velocity D_m and u_m.

Hugoniot parameter a affect the velocity D_m and u_m just as it affects pressure P_m.

4.3 Dynamic initial density ρ_m of rock

The density of rock has increased greatly due to the compressive deformation of borehole wall. it is noted that the higher the detonation velocity D, the bigger the density ρ_m (see in Fig. 7) when different explosives act on same rock respectively.

For same explosive the variance of density ρ_m is influenced by Hugoniot parameter b. The higher the Hugoniot parameter b, the bigger the magnitude of density ρ_m.

4.4 Deformative angle ε of borehole wall

The variation of deformation angle ε is shown in Fig. 8. It illustrates that the borehole walls deform due to the striking compressive actions of detonation wave. The maximum deformative angle ε can reach almost 8°. Even near the end borehole the deformation angle ε is bigger than 3°. So, we can say that the borehole wall is in deformed state from beginning to end when detonation wave impinges upon it.

4.5 The included angle Φ_3 between the plane of initial shock wave and the original wall of borehole.

When Granite is acted by three exlosives respectively, the included angle Φ_3 is getting bigger with Φ_0 increasing. The angle Φ_3 is going down only when $\Phi_0 > 80°$ (shown in Fig. 9). When $\Phi_0 = 50 - 90°$, whether which explosive works on Granite, Φ_3 is mostly between 40° to 60°. If $\Phi_0 = 50°$, it is easy to get $L = R tg\Phi_0 = R tg50° = 1.19R \approx 1.2R$, therefore, from 1.2 times radius of borehole along axial direction the wall of borehole is shocked by the initial shock wave which travels outwards into rock mass in 40—60° angle to the original wall of borehole.

5 CONCLUSIONS

A physical model and a methodology have been set up to calculate dynamic initial shock parameters for

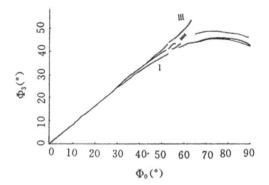

Fig. 9 The variance of included angle Φ_3 with Φ_0 changing

coupled charge on borehole wall under the actions of medium explosives.

The calculations show that each initial shock parameter varies from initiation point to the end of borehole wall along the axis. The initial shock pressure P_m is larger than detonation pressure P_H in regular oblique reflection and Mach Reflection regions and smaller than P_H when the process is in Prantl-Meyer Expansion. P_m can reach its maximum magnitude before Mach Reflection concludes, usually $\Phi_0 = 56 - 64°$. The maximum P_m is almost (1.5 - 1.8) P_H (see Table 3). From this result we can understand why some blasting operations can obtain good fragmentation around the bottom of borehole when utilizing reverse initiation technique in hard rock. Because the wall of borehole is not rigid, it deforms when detonation wave shocks on it. The deformation angle ε is about 0 - 12.5 . 5°. The initial shock wave travels outwards into rock mass with included angle 0 - 60° to the original borehole wall. The density of rock around the borehole wall increases greatly because of the compressive deformation of borehole wall.

This method and the results would be helpful in quantitative analysis of near-field explosion around a coupled charge and calculation of stress wave parameters.

REFERENCES

Li Yumin & Ni Zhifang, 1990. 12. Reseach on calculating methods for initial dynamic shock parameters. J. SIMT. Taian, China (in Chinese)

Ni Zhifang, 1994. 5. Numerical simulation on initial shock parameters of coupling charge on borehole wall and experimental study. Master Degree thesis, Taian, China (in Chinese)

Niu Qiang, 1990. 8, Rock blasting mechanism. Dong Bei Industrial College Press, Shen Yang, China (in Chinese)

Pan wen quan, 1982, The Base of hydrodynamics. Mechanical Industry Press, Beijing, China (in Chinese).

Wang Wenlong, 1994. 11, Drilling and Blasting. Coal Mine Industry Press, Beijing China (in Chinese)

Zhang Baoping, et al. 1979. Explosions and Their Applications. Publishing House of Defence Industry, Beijing, China (in Chinese)

Rock Fragmentation by Blasting, Mohanty (ed.) © 1996 Taylor & Francis. ISBN 90 5410 824 X

Fragmentation and heave modelling using a coupled discrete element gas flow code

Alan Minchinton
ICI Explosives, Kurri Kurri, N.S.W., Australia

Peter M. Lynch
ICI Explosives, Ardeer Site, Stevenston, UK

ABSTRACT: Experimental evidence indicates that a large percentage of the crack development and subsequent fragmentation and heave in the blasting process is due to the effects of the high pressure gas generated by the explosive in the blasthole. An explicit finite/discrete element code has been developed to dynamically model stress field development, crack generation and crack growth as well as the motion and stacking of the rock fragments. Cracking in the current model is limited to failure under tension (Rankine). This failure is due to both the stress and the gas flow which is treated by coupling the local discrete element porosity to a volume element gas flow model. Several examples are discussed which show striking qualitative differences between blasting with stemming to contain the gases in the blastholes, and without stemming.

1 INTRODUCTION

The development of a tractable computer based model to describe the entire blasting process from detonation to "rock on the ground" has been an endeavour that has only become partly possible in recent times due to the accessibility of affordable computing resources. To the authors' knowledge there does not exist a single code that can dynamically model the full three dimensional aspects of even a single hole blast by using a realistic mechanistic approach and there are very few codes that have been developed to tackle this complicated problem even in two dimensions.

A complete mechanistic simulation of a particular process would include simulations of all the known underlying physical events that contribute to that process. Naturally, the type of approach adopted must match the desired outcome; for example, there is little point in considering crystal lattice vibrations and phonons in order to model the ground vibration due to a large blast.

A truly mechanistic blasting model must include simulations of a series of macroscopic processes which work in concert. The overall model must include a description of detonation or some other high pressure–producing process that causes high amplitude stress waves to emanate from the blasthole wall into the rock fabric. There must be a model to track the stress/strain behaviour of the rock including the effects of insitu

stresses and a model that considers crack initiation, propagation and interaction with other cracks and existing geological features as well as a model that allows for cracking processes due to the high pressure gases produced in the blasthole. There should also be a kinetic model for the bulk motion of the fragmented rocks allowing for contact and settling. These processes should evolve dynamically and for multi–hole blasts the code should be capable of considering the effects of timing and blast geometry.

1.1 The importance of stress and gas

Until the early 1980's there were two schools of thought regarding the theories of rock fragmentation by blasting (Winzer & Ritter 1985). Differences centred on the relative importance of either the stress field or the high pressure gases to cracking and fragmentation. Although it was actually realised much earlier (Kutter & Fairhurst 1971; Fourney et al. 1983) that each process plays an important rôle the issue was not clearly resolved experimentally.

Reduced scale blasting tests carried out by Winzer and Ritter (1985) and Gur et al. (1984) identified the fact that the initial cracks at a free surface were caused by the tensile stresses resulting from the reflection of the p–wave as it travels back to the blasthole and were not due to cracks radiating from the blasthole as originally thought. Persson (1990) however, claims that these reflected tensile waves carry insufficient energy to cause damage, they simply promote the growth of the dominant radial cracks already emanating from around the blasthole.

Winzer and Ritter (1985) also claimed that the bulk of the fragmentation was due to longer term stress wave effects rather than the blasthole gases [which had vented]. Persson (1990) asserts that the blasthole gases act on the wedges of rock formed by the radial cracks between the blasthole and the free face resulting in bending stresses in the wedges that lead to fragmentation initiated mainly at existing joints, cracks and flaws.

Haghighi et al. (1988) reported results conducted by Britton in the early eighties that described measurements of gas pressures and relationships between the gas pressure and the work done on the rock mass for a range of decoupling ratios. It was found that the shock energy played an insignificant rôle especially for decoupled charges. Britton (1987) also showed the importance of dilatancy or volume strain induced by simple shear as an important bulking mechanism.

Haghighi et al. (1988) also developed a finite element model of a single blasthole bench geometry that incorporated several predefined radial vertical cracks which could be pressurised to different lengths. The results showed that beam bending and gas pressurisation were more important processes for rock breakage than shock or vibrational processes. Their results also showed that maximum displacements were achieved by allowing only the first two thirds of the cracks to be pressurised in the radial direction toward the free face.

In a series of computations and experiments involving steel lined and unlined blastholes in perspex McHugh (1983) clearly showed several important features: (a) short radial cracks due to the circumferential tensile stresses alone are generated close to the blasthole, (b) the primary effect of the explosive gases is to cause the crack length to increase by a factor of five to ten, (c) a secondary effect is a 50% increase in the number of small fractures, (d) in direct comparison with Winzer and Ritter's (1985) findings the gas acts dynamically and cooperatively with the stress field to produce cracks and, (e) the gases penetrate the cracks and reach the crack tip i.e. the gas pressurisation of the cracks causes the large crack extensions.

Similar comprehensive experiments were carried out by Brinkmann (1987, 1990) using unstemmed blastholes that were either unlined, or lined with aluminium or steel in a typical gold mine stoping environment. Brinkmann concluded that (a) initial fracturing close to the blasthole is caused by the shock loading and (b) fragmentation due to the action of the stress field alone is limited. With the lined blastholes the fragmentation was only 10% of that produced with the unlined blastholes. Similarly the heave velocities were reduced by a factor of between five and eight. These results clearly showed the dominance of the gas effects in producing both fragmentation and heave.

1.2 Blasting related codes and models

There have been several attempts to develop codes which consider some or all of the dominant blasting processes. It is not the intention of this paper to review the field, but some of the more relevant codes or models which deal with cracking, fragmentation and the effects of the gas in particular are discussed briefly. We are not concerned here with "engineering" codes like SABReX (Jorgenson & Chung 1987) and EXEN (Sarma 1994) since they are highly empirical and are neither dynamic nor truly mechanistic in the sense defined here.

FDM (Yang & Wang 1995) is a constitutive model that has been implemented in the finite element code PRONTO (Taylor & Flanagan 1987). This model has had an extended history and is based on the modified Sandia (Kipp & Grady 1979; Kuszmaul 1987) microcrack damage model. In FDM the damage is determined by the fractal dimensions for joints and cracks.

BFRACT (Simons et al. 1995) is a constitutive model that has been implemented in the finite element code DYNA2D (Hallquist 1982). It is a multiplane model capable of considering the heterogeneity of geologic materials and uses a fracture mechanics based stress intensity factor formulation with a viscous crack growth law in which the crack velocity is limited by a function of the shear velocity.

CAVS (Barbour et al. 1985) was developed as a 3D finite element constitutive model of tensile fracture initiation and growth. The porosity and permeability of the cracks were derived from the void strain due to crack opening and closing. Crack planes form normal to the principal tensile stress direction with the transient, laminar, viscous fluid flow being driven by crack segment pressure gradients providing a constant pressurisation over any given fracture segment.

Taylor and Preece (1989) have developed an elegant two dimensional distinct element code DMC in which the elements are disks derived from the cross–section of spheres, mainly for modelling the motion associated with blasting layered rock masses. The code does not model the fragmentation process at all but does employ a porosity based gas flow model (Preece 1993) and explosive blasthole loading based on non–ideal detonation.

PFC3D (Itasca 1995) is a three dimensional particle flow code in which an impermeable rock mass is modelled by aggregated sphere based distinct elements with soft contacts and compressive, shear and tensile contact "bond" breakage mechanisms. The dynamically coupled fluid "flow" is associated with penny shaped overlapping reservoirs produced at these "crack" locations interconnected via a pipe network of rectangular cross section.

A study of fracturing in microscopic random assemblies of dislocations has recently been carried out by Napier and Peirce (1995). They used the displacement discontinuity boundary element method which simulates distinct element processes for inertial systems that are strongly damped and that have small inelastic strains on individual discontinuities. The crack growth can extend from the nucleation sites according to specified growth rules.

Song and Kim (1995) have developed a Dynamic Lattice Network Model in which the rock mass is discretised into triangular elements which have particles situated at each vertex and massless springs connecting them. The mass of the particles is determined from the density and the heterogeneous spring constants are determined from the experimental range of the rock strength tests. A blasthole pressure profile is applied using a function that represents both the high strain component and the gas pressurisation phase. The creation and propagation of cracks is associated with the failure of the lattice springs.

Munjiza and co–workers (Munjiza 1992; Owen et al. 1992) developed a combined finite element – discrete element code that employs a simple fracture based softening Rankine plasticity model to generate cracks and a sophisticated contact detection scheme for handling the fragments. The interaction between the rock and detonation gas is handled using a dual discretisation technique.

2   MBM🜂: MECHANISTIC BLASTING MODEL

The challenge to develop a single computational model that is capable of dynamically modelling all the blasting processes mentioned in Section 1 in a realistic mechanistic fashion has only been achieved by one code to date, namely MBM🜂 (for Mechanistic Blasting Model). This code is an explicit finite element – discrete element code which has evolved from further work of Munjiza and co–workers (Munjiza 1992; Owen et al. 1992) and the present authors over the past two

years and is an extension of the ELFEN package (Rockfield Software) which among other things adds pre and post processing capabilities. The discrete element part of the code is currently restricted to two dimensional problems using plane strain or axisymmetric elements but the main thrust of current developments is towards the three dimensional version.

2.1   Finite elements

In the regular two dimensional finite element technique the problem domain $\mathcal{R}^2$ is mapped to the computational domain $\mathcal{C}^2$ as a series of polygonal elements e_i joined at nodes. Various types of elements have been used ranging from triangular to eight node quadrilaterals. We use triangular elements as the larger elements are computationally expensive and it is now generally accepted that with adequately fine meshing the triangular elements are more efficient and just as accurate. As an example, Figure 1 shows the radial displacement at the point (4.96, 5.04) in an axisymmetric half space measuring 10 m x 20 m. The mesh was triangular with 9664 elements and 4942 nodes and the purely elastic material with density 2500 kg/m^3 has a p–wave velocity of 2500 m/s and a Poisson's ratio of 0.25. The signal was produced by a travelling pressure loading of the form,

$$P = P_o t^n e^{-bt}$$

where $n = 6$, $b = 5000$ and P_o was 1 MPa applied on the line from (0.1,0.0) to (0.1,5.0) to replicate a velocity of detonation of 5500 m/s. The bottom and right boundaries were non–reflecting. The sampling interval was 50 μs and the output was not smoothed or altered. There is no noise in this signal and the filtering effects of the mesh are negligible; the signal was not altered by halving the number of elements.

Various boundary conditions or constraints can be applied to $\mathcal{C}^2$ to simulate $\mathcal{R}^2$; single nodes or a series of nodes constituting a line or surface can be constrained to move only in certain ways, for example, damping on certain exterior lines to prevent the reflection of waves back into the mesh (non–reflecting boundaries). Similarly, loads can be applied to $\mathcal{C}^2$, for example a pressure or face load might be applied to a series of nodes to simulate an impulse.

In essence then, an explicit finite element code solves for the nodal displacements over all e_i in $\mathcal{C}^2$ using spatial and temporal integration. The manner in which the nodes respond and move is determined by the material properties assigned to the related elements. For example if $\mathcal{R}^2$ is a purely elastic material then the stresses and strains generated throughout

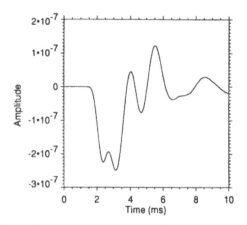

Fig. 1: The radial displacement at an interior point in an axisymmetric half space. Details are given in the text.

$\mathcal{C}^2$ due to the nodal displacements will be determined by the simple tensor based elasticity equations (Timoshenko & Goodier 1970).

2.2 Discrete elements

In this code a two dimensional discrete element d_j is a separate, distinct polygonal element made up of a collection of finite elements (at least one). Since the discrete elements are deformable a critical state of stress or strain can be reached where an element separates or fractures into two or more discrete elements or their boundaries can change if the fracture is only partial. This is different from the techniques used in the codes μDEC and $3DEC$ attributed to Cundall (Cundall 1980; Cundall & Hart 1988) which use two types of elements, one that is not discretised at all and is "simply deformable" and another type that is "fully deformable" and is discretised into finite difference zones. Neither of these element types can fracture.

Elaborate techniques have been devised to minimise the computational effort spent on contact detection i.e. establishing which elements, finite or discrete are in contact with the currently inspected element, and on the evaluation of contact forces (Owen et al. 1992). There is also an elaborate "book keeping" system associated with the memory management required for a system in which an unknown number of elements is created throughout the simulation.

2.3 Fracturing

We do not model the actual crack tip as is often done with codes that consider single crack propagation. It is common knowledge though that the speed of a moving crack is limited only by the rate at which stored elastic energy is delivered to the crack tip i.e. by the free fracture surface elastic wave velocity, the Rayleigh wave speed c_R. Experimental crack speeds are typically one third to one half c_R and recent theory (Ching 1994) suggests that steady state crack propagation above a critical speed of about $0.34c_R$ may be unstable.

Fracturing in the code is based on a fracture based softening Rankine plasticity model (Figure 2). In each element the fracture criterion is checked for at every integration point. Fracturing may occur if the tensile strength of the rock has been reached and if in the post peak process the stress has reduced to zero following the local softening gradient:

$$E_i^T = -\frac{\sigma_t^2}{2G_f}\sqrt{A_i} \qquad (2.1)$$

where σ_t is the tensile strength, G_f is the fracture energy release rate in tension and A_i is an area associated with the integration point of the element. Since the fracture rule involves a characteristic length for each element the cracking and fragment formation is independent of the mesh size.

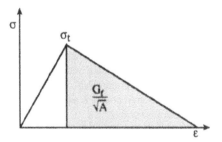

Fig. 2: Fracture based Rankine model

The accumulated effective plastic strain is monitored in the principal direction within each element. If the final stress state is non zero and on the softening branch a full breakage is deemed not to have occurred then the effective plastic strain is treated as a scalar variable for the next state of deformation.

At present we are unable to easily introduce cracks or joints into $\mathcal{C}^2$ to represent the existing structure. This is obviously a highly important feature which we are currently addressing. We are able to incorporate in situ gravitational stress using dynamic relaxation. Both the vertical and the horizontal components can be important especially at great depths since the stress gradient can be around 0.02 to 0.04 MPa per metre (Herget 1988).

3 MBM2D: GAS FLOW MODEL

The simultaneous solution of the rock fracturing (discrete element) equations and those governing the flow of detonation products in a highly fragmented rock is still a computationally intractable problem. For this reason we have chosen to model the gas flow independently of the rock dynamics and to couple the processes at every timestep. In this section we develop the model for the flow of the detonation products and describe the coupling with the fracturing code.

3.1 Detonation model

The detonation of the explosives in the blasthole is modeled by the non–ideal detonation code CPeX, (Kirby & Leiper 1985). This code models the reactive flow of the detonation products and the influence of the rock confinement and blasthole diameter. From CPeX we obtain the velocity of detonation (hence the duration of the detonation event) and the blasthole wall pressure profile. After detonation is complete it also gives us a polytropic adiabatic equation of state for the gas

$$p = c\rho^\gamma \tag{3.1}$$

where c and γ are constants and p and ρ are the gas pressure and density respectively.

3.2 1–D gas flow in a single crack

Many authors, including Nilson et al. (1995), Nilson (1988) and Paine and Please (1993, 1994) have studied gas fracturing processes in specialised geometries where the flow problem reduces to one dimension in a single crack. They then simultaneously solve the flow with a semi analytical solution for the rock response. The gas flow models are very similar and mainly differ in the empirical form adopted for the turbulent drag law and whether heat transfer effects are included or neglected. We have elected to use the form adopted by Paine and Please (1994) who have coupled the gas flow with a star crack subjected to an arbitrary distribution of pressure on the crack faces and a central blasthole pressure.

The gas flow model is developed from the fact that the Reynold's number is around 10^5 and the flow is turbulent. As is usually the case with turbulent flow its components are decomposed into mean and fluctuating parts and the Reynold's stress (arising from the interaction of fluctuating velocities) needs a drag law to mathematically close the equations. The flow equations are further simplified by assuming that the crack

height, h, is small so that lubrication theory applies and the Mach number, $\rho v^2 / p$, where v is the gas velocity, is small. Further, the wall heating is assumed to be small so that the gas expands adiabatically. This leads to the following mass and momentum conservation equations

$$\frac{\partial}{\partial t}(\rho h) + \frac{\partial}{\partial r}(\rho v h) = 0 \tag{3.2}$$

$$h\frac{\partial p}{\partial r} + \rho v|v|\psi = 0 \tag{3.3}$$

where t is time and r is distance along the crack. $\rho v|v|\psi$ is the drag law modelling the Reynold's stress and the empirical friction factor ψ takes the form

$$\psi = \frac{12\mu_t}{\rho|v|h}$$

for a constant turbulent eddy viscosity model. Here, μ_t is the coefficient of viscosity. We can solve equation (3.3) for v using (3.1) and substitute into (3.2) to give,

$$\frac{\partial}{\partial t}(\rho h) = \frac{1}{12\mu_t}\frac{\partial}{\partial r}\left(h^3\gamma c\rho^\gamma\frac{\partial \rho}{\partial r}\right) \tag{3.4}$$

a non–linear diffusion equation for the mass of gas per unit crack length, ρh. The equations for gas flow induced fracturing are completed in Paine and Please (1993) by relating h to p and hence ρ.

3.3 2–D gas flow in fractured rock

We now turn our attention to extending the 1–D gas flow model to 2–D and how we discretise and solve it. We form the 2–D analogue of (3.2) and (3.3) and substitute for ψ to give,

$$\frac{\partial}{\partial t}(\rho\alpha) + \frac{\partial}{\partial x}(\rho v_x \bar{h}_y) + \frac{\partial}{\partial y}(\rho v_y \bar{h}_x) = 0 \tag{3.5}$$

$$v_x = -\frac{\bar{h}_x^2}{12\mu_t}\frac{\partial p}{\partial x} \tag{3.6}$$

$$v_y = -\frac{\bar{h}_y^2}{12\mu_t}\frac{\partial p}{\partial y} \tag{3.7}$$

where x,y are rectangular space coordinates, α is the local void fraction, v_x, v_y are the x,y components of

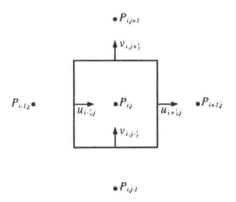

Fig. 3: Finite volume spatial discretisation

velocity and $\bar{h}_x, \bar{h}_y$ are the average crack heights per unit distance in the x and y directions respectively.

We now consider a uniform mesh of points with separations x, y in the x and y directions covering the whole of the blast geometry. We need to set up equations to advance density (and pressure via equation (3.1)) in time. We adopt the standard computational fluid dynamics technique of discretising the equations by finite volume on a staggered grid, i.e. we consider ρ and p to be evaluated at the mesh points and the velocities at the mid–points between them. We then integrate (3.5) about the control volume (see Figure 3) so that for the (i,j)th mesh point we obtain

$$\frac{\partial}{\partial t}(\rho_{i,j}\alpha_{i,j}) + \left[\rho v_x \bar{h}_y\right]_{x_i-\frac{x}{2},y_j}^{x_i+\frac{x}{2},y_j} + \left[\rho v_y \bar{h}_x\right]_{x_i,y_j-\frac{y}{2}}^{x_i,y_j+\frac{y}{2}} \quad (3.8)$$

The velocities are obtained by straightforward differencing of (3.6) and (3.7) to give

$$v_x(x_i + \frac{x}{2}, y_j) = -\frac{1}{12\mu_t}\bar{h}_y^2(x_i + \frac{x}{2}, y_j)\frac{p_{i+1,j} - p_{i,j}}{x}(3.9)$$

$$v_y(x_i, y_j + \frac{y}{2}) = -\frac{1}{12\mu_t}\bar{h}_x^2(x_i, y_j + \frac{y}{2})\frac{p_{i,j+1} - p_{i,j}}{x}(3.10)$$

For stability considerations $\rho(x_i + \frac{x}{2}, y_j)$ and $\rho(x_i, y_j + \frac{y}{2})$ take the upstream values rather than a local average.

The time integration of (3.8) is currently carried out by the forward Euler method.

3.4 Gas flow coupling with discrete elements

Let us consider how we couple the gas flow with the discrete element code in order to advance from time t_n to time t_{n+1}, as illustrated schematically in Figure 4. At time t_n we assume we know the gas pressures on

the flow mesh. We interpolate these pressures to derive loads for the exposed surfaces for input to the discrete element code. We then use the discrete element code to advance to t_{n+1} under this loading. These results are then processed to give us $\bar{h}_x(t_{n+1})$, $\bar{h}_y(t_{n+1})$ and $\alpha(t_{n+1})$ on the flow mesh. We re-evaluate the pressure at t_n due to the change in geometry. We finally use these values in the flow code to advance the gas and hence compute the pressure at t_{n+1}.

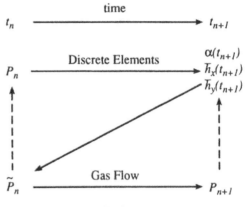

Fig. 4: Gas flow coupling

3.5 Heave model

Within the discrete element formulation the kinematic motion of the fragmented elements is handled naturally as is the stacking and settling of the fragments; these phases are not any different from the close contact phase of the calculations. The fragments will naturally fracture into polygons with a range of different shapes although small fragments will tend to be triangular. Due to their more realistic shapes there is no need to introduce artificial bulking or swell mechanisms (Preece 1990) to produce muckpiles with the correct packing densities.

With $MBM\hat{D}$ the stress field modelling, the fracturing, the gas flow and the bulk motions all occur essentially at the same time and throughout the entire calculation. Fracturing can also occur in the developing muckpile or through collisions provided the fracture criteria are met. With this code it is also possible to treat the discrete elements as rigid faceted elements (distinct elements); this feature was added as an option to facilitate more rapid calculation during the heave phase which is dominated by bulk motion rather than fracturing.

4 MBM*D* : EXAMPLES AND DISCUSSION

4.1 Early crack development

The early fracture development in a five hole bench blast is shown in Figure 5. This is a plan view of the bench with a burden of 2.0 m, spacing of 1.8 m, at 4.98 ms after firing the first hole. The blasthole diameter was 65 mm (nb. the holes are shown oversize for clarity). The delay timing is 4.5 ms. The rock has a p-wave velocity v_p of 4667 m/s, an s-wave velocity v_s of 2695 m/s, Young's modulus is 47 GPa, the tensile strength is 14.7 MPa. The bottom and side surfaces are non–reflecting (infinite) boundaries, the top surface is the crest of the bench. For this simulation we used a generic explosive which produced a peak pressure of 4.2 GPa with $\gamma = 2.87$.

   Cracks can be seen radiating from each of the blastholes with finer fragments appearing close to the holes. The simulation shows the initial stages of flexure at the crest line and clearly shows cracks that have originated at the crest coalescing with the cracks radiating outwards from the blastholes. On this figure we have superimposed four circles; the three smaller circles centred on the second blasthole in descending radii are approximate representations of the wavefronts for the p–wave, the s–wave and a hypothetical "crack front" respectively. The crack front velocity is assumed to be $0.34v_s$ (see Section 2.3). The largest circle is centred on the first blasthole and represents the hypothetical "crack front" produced from that blasthole. It can be clearly seen that the predicted crack lengths are of the correct order.

4.2 Effect of free faces on crack development

The fact that many cracks can and do originate from the free faces is due to the tensile stresses arising from the reflection of p–waves as they travel back to the blasthole. This mechanism is clearly shown in Figure 6. Contrary to the arguments of Persson (1990)

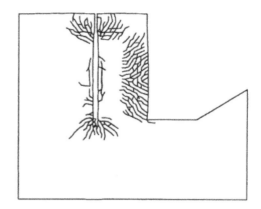

Fig. 6: Cracks developing from free surfaces

(also see Section 1.1) it would appear that there is sufficient energy in this process to cause damage. The simulation is an elevation view of a single hole bench blast with 6 m burden, 14.5 m blasthole and an 11.5 m column (2.5 m toe charge) of the generic explosive mentioned earlier fired instantaneously (i.e. infinite VoD). The blasthole diameter was 300 mm.

4.3 Effect of gas on crack development

To investigate the qualitative differences between the crack development due to stress effects alone and the effects of the gas penetration into the cracks we have modelled a single hole bench blast similar to that discussed above with and without stemming. In this case there was no toe charge resulting in an explosive column of 9 m. The results are shown in Figure 7 and Figure 8.

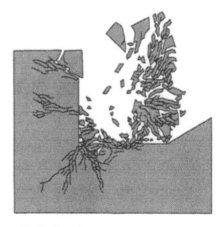

Fig. 7: Single hole bench blast – not stemmed

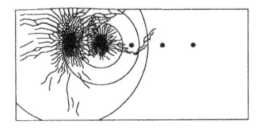

Fig. 5: Fracturing from two blastholes – plan view

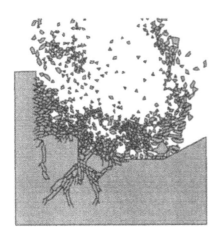

Fig. 8: Single hole bench blast – stemmed

It is immediately obvious that the fragmentation produced in the stemmed case (i.e. with gas) is much finer than the unstemmed case. In the unstemmed case large fragments are produced around the collar region and there is less overall throw. As mentioned there was no subdrill in either case and that has resulted in the uneven and rising floor. There is evidence of considerable damage behind the final wall and below the blastholes in both cases. (NB. the abrupt cutoff of the cracking region to the left and bottom of the problem is a restriction we imposed to prevent cracks running to the non reflecting boundaries). The equivalent results for a plan view of a bench blast with two blastholes are shown in Figures 9 and 10. In this simulation only the lefthand side blasthole has been detonated in both cases.

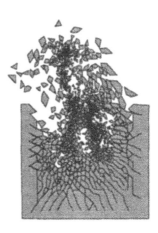

Fig. 10: Bench blast (plan) – stemmed

4.4 Computational details

A complex code like $MBM\oplus$ consumes considerable computing resources especially when attempting to simulate complete blasting events from initiation to rock on the ground.

As an example Figure 11 shows the results of the simulation discussed in Section 4.1 at 135 ms. The original mesh used 7337 elements, 3763 nodes. These numbers increase considerably throughout the calculation; in this particular case the number of elements almost doubled. In the original mesh the average element size was 0.4 m the smallest element being 0.02 m. Using a single processor Digital AlphaStation 250 (266 MHz (262 SPECfp92) with 128 MB RAM) running OpenVMS V6.2 this simulation ran at around 50 minutes cpu time per millisecond of problem time, the

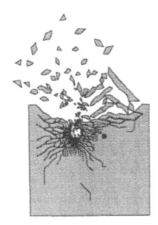

Fig. 9: Bench blast (plan) – not stemmed

Fig. 11: Bench blast at t=135 ms

whole simulation to this point taking nearly 5 days. As a consequence of these long runtimes we are investigating different specialised computing solutions for running the 3D version.

MBM2D will be used as an aid for understanding the complex blasting process. With the ability to model the dynamic and cooperative effects of the stress and gas and the ability to generate cracks and fragments we are in a position to analyse blasting dynamics in a more quantitative fashion than has hitherto been possible.

REFERENCES

Barbour T.G., K.K. Wahi & D.E. Maxwell 1985. Prediction of fragmentation using CAVS. *Fragmentation by Blasting*. W.L. Fourney, R.R. Boade & L.S. Costin (Editors). Society for Experimental Mechanics. 1st Edn. pp. 158–172.

Brinkmann J.R. 1987. Separating shock wave and gas expansion breakage mechanisms. *Proc. 2nd Int. Symp. on Fragmentation by Blasting*, Keystone, Colorado. pp. 6–15.

Brinkmann J.R. 1987. An experimental study of the effects of shock and gas penetration in blasting. *Proc. 3rd Int. Symp. on Fragmentation by Blasting*, Brisbane, Australia. pp. 55–66.

Ching E.S.C. 1994. Dynamic stresses at a moving crack tip in a model of fracture propagation. *Phys. Rev. E*. 49(4):3382–3388.

Cundall P.A. 1980. UDEC – a generalised distinct element program for modelling jointed rock. *Peter Cundall Assoc. Rep.* Rept. PCAR-1-80, European Research Office, US Army, Contract DAJA37-79-C-0548

Cundall P.A. 1988. Formulation of a three–dimensional distinct element model — Part I. A scheme to detect and represent contacts in a system composed of many polyhedral blocks. *Int. J. Rock Mech. Min. Sci.* 25:107–116

ELFEN/explicit 1995, *Rockfield Software*, Swansea, Wales,

Fourney W.L., D.B. Barker & D.C. Holloway 1983. *Proc. 1st Int. Symp. on Fragmentation by Blasting*, Luleå, Sweden. pp. 505–531.

Gur Y., Z. Jaeger & R. Englman 1984. Fragmentation of rock by geometrical simulation of crack motion – I *Engng. Fracture Mech.* 20(56):783–800.

Haghighi R., R.R. Britton & D. Skidmore 1988. Modelling gas pressure effects on explosive rock breakage. Int. J. Mining & Geological Engng. 6:73–79.

Hallquist J. 1982. User's manual for DYNA2D – an explicit two dimensional hydrodynamic finite element code with interactive rezoning and graphical display. *Univ. California, LLNL, Report* UCID-18756.

Herget G. 1988. *Stresses in rock*. Rotterdam: Balkema

Itasca Consulting Group Inc. 1995. PFC^{3D}: Particle flow code in 3 dimensions, Vols. I, II and III, Minneapolis, Minnesota: ICG.

Jorgenson G.K. & S.H. Chung 1987. Blast simulation – surface and underground with the SABREX model. *CIM Bulletin*, August, pp. 37–41.

Kipp M.E. & D.E. Grady 1979. Numerical studies of rock fragmentation. *Sandia Report* SAND79-1582.

Kirby I.J. & G.A. Leiper 1985. A small divergent detonation theory for intermolecular explosives. *8th Symposium (International) on Detonation*, Office of Naval Research, Report NSWC MP 86-194, P. 176.

Kuszmaul J.S. 1987. A new constitutive model for fragmentation of rock under dynamic loading. *Proc. 2nd Int. Symp. on Fragmentation by Blasting*, Keystone, Colorado. pp. 412–423.

Kutter H.K. & C. Fairhurst 1971. On the fracture process in blasting. *Int. J. Rock Mech. Min. Sci.* 8:181–202

McHugh S. 1983. Crack extension caused by internal gas pressure compared with extension caused by tensile stress. 1983. *Int. J. Fracture* 21:163–176.

Munjiza A. 1992. Discrete elements in transient dynamics of fractured media. *Ph.D. Thesis*. University of Wales, University College of Swansea, Wales, UK.

Napier J.A.L. & A.P. Peirce 1995. Simulation of extensive fracture formation and interaction in brittle materials. *Proc. 2nd Int. Conf. on the Mechanics of Jointed and Faulted Rock*, 10–14 April, Vienna, Austria, pp. 63–74.

Nilson R.H., W.J. Proffer & R.E. Duff 1985. Modelling of gas–driven fractures induced by propellant combustion within a borehole. *Int. J. Rock Mech. Min. Sci.* 22:3–19.

Nilson R.H. 1988. Similarity solutions for wedge–shaped hydraulic fractures driven into a permeable medium by a constant inlet pressure. *Int. J. Numer. Anal. Methods Geomech.* 12:477–495.

Nilson R.H. 1991. Dynamic modeling of explosively driven hydrofractures. *J. Geophys. Res.* 96, No. B11: 18.081–18.100.

Owen D.R.J., A. Munjiza & N. Bićanić 1992. A finite element – discrete element approach to the simulatiuon of rock blasting problems. *Proc. 11th Symp. on finite element methods in South Africa*, Centre for Research in Computational and Applied Mechanics, Cape Town, 15–17 Jan. pp. 39–58.

Paine A.S. & C.P. Please 1993. Asymptotic analysis of a star crack with a central hole. *Int. J. Engng. Sci.* 31:893–898.

Paine A.S. & C.P. Please 1994. A simple analytic model for the gas fracture process during rock blasting *Int. J. Rock Mech. Min. Sci.* 31:699–706.

Persson P-A. 1990. Fragmentation mechanics. *Proc. 3rd Int. Symp. on Fragmentation by Blasting*, Brisbane, Australia. pp. 101–107.

Preece D.S. & L.M. Taylor 1990. *Proc. 3rd Int. Symp. on Fragmentation by Blasting*, Brisbane, Australia. pp. 189–194.

Preece D.S. 1993. Momentum transfer from flowing explosive gases to spherical particles during computer simulation of blasting induced rock motion. *9th Ann. Symp. on Explosives & Blasting Res.* Society of Explosives Engineers, San Diego, California. pp. 251–260.

Sarma K.S. 1994. Models for assessing the blasting performance of explosives. *Ph.D. Thesis*. JKMRC, University of Queensland, QLD, Australia.

Simons J.W., D.R. Curran & T.H. Antoun 1995. A computer model for explosively induced rock fragmentation during mining operations. *68th Ann. Minnesota Section SME & 56th Univ. Minnesota Mining Symp.* Jan 24–26, Duluth, Minnesota. pp. 203–213.

Song J. & K. Kim 1995. Numerical simulation of the blasting induced disturbed rock zone using the dynamic lattice network model. *Proc. 2nd Int. Conf. on the Mechanics of Jointed and Faulted Rock*, 10–14 April, Vienna, Austria, pp. 755–761.

Taylor L.M. & D.P. Flanagan 1987. PRONTO2D – A two–dimensional transient solid dynamics program. *Sandia Report*. SAND86–0594.

Taylor L.M. & D.S. Preece 1989. DMC – a rigid body motion code for determining the interaction of multiple spherical particles. *Sandia Report*. SAND88-3482.

Timoshenko S.P. & J.N. Goodier 1970. *Theory of elasticity*. McGraw Hill. 3rd Edn

Winzer S.R. & A.P. Ritter 1985. Role of stress waves and discontinuities in rock fragmentation. *Fragmentation by Blasting*. W.L. Fourney, R.R. Boade & L.S. Costin (Editors).Society for Experimental Mechanics. 1st Edn. pp. 11–23.

Yang J. & S. Wang 1995. Study on fractal damage model of rock fragmentation. *2nd Int. Conf. on Engineering Blasting Technique*. Nov 7–10, Kunming, China. pp. 288–293.

Rock Fragmentation by Blasting, Mohanty (ed.) © 1996 Taylor & Francis. ISBN 90 5410 824 X

Effect of open joint on stress wave propagation

K.R.Y.Simha

Department of Mechanical Engineering, Indian Institute of Science, Bangalore, India

ABSTRACT : Stress wave characteristics are drastically altered by joints and other inhomogenities. This paper addresses the effect of an open joint on stress wave transmission. An elastodynamic analysis is developed to supplement and explain some recent observations by Fourney and Dick(1995) on open as well as filled joints. The analytical model developed here assuming spherical symmetry can be extended to filled joints between dissimilar media, but results are presented only for open joints separating identical materials. As a special case, stress wave transmission across a joint with no gap is also addressed.

1 INTRODUCTION

Generation and propagation of stress waves by explosions is a fundamental problem of geophysics. Close to the borehole, the magnitude and duration of a stress wave is contolled by the applied pressure. As the waves move away from the borehole, their characteristics are drastically altered by joints and other inhomogenities. This paper is concerned with the effect of open joints. The joint is modeled as a gap. The special case of a joint without any gap is also examined. This paper is motivated by the results presented by Fourney and Dick(1995), who examined the effect of a gap on stress wave transmission. In addition, they also examined the role of joints stuffed with some weak filler matereial. In the case of open joints, they showed that both magnitude and the pulse width of the transmitted stress wave are reduced depending on the extent of the gap. Transmission starts when the gap is closed. Beyond a certain critical gap there is no transmission. A remarkable feature about the results presented is that the wave terminates at the *same* instant for all gaps. This paper explains their results by examining the characteristics of the radial stress. Accordingly, wave transmission ends when the radial stress switches sign from compression to tension. For *filled* joints, the switching instant is likewise independent of the layer thickness.

2 ANALYSIS

Consider the detonation of a fully coupled spherical charge in a cavity of radius a. The elastic medium surrounding the cavity is characterised by c_1 and c_2 for P- and S-waves, respectively. The stress waves generated by the detonation encounter an open joint at a radius b. The open joint with a gap δ separates the spherical core from the surrounding material as shown in Figure 1.

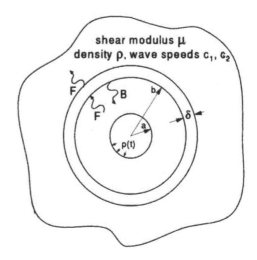

Figure 1: Geometry

It is shown in books on waves, Graff(1975) for instance, that the present problem can be analysed with the help of two functions F and B to account for diverging and converging waves, respectively. Diverging waves propagating radially outward generate converging waves upon reflecting at the open joint. The core displacement u_b at the joint situated at radius b is given by

$$u_b = -(F + B)/b^2 - (F' - B')/bc_1 \qquad (1)$$

where primes denote differentiation.

In a continuous unbounded medium, only diverging waves are present and B is absent. This is the case for the stress waves transmitted across the open gap. Wave transmission is possible only if u_b exceeds δ. Otherwise, the waves get trapped in the core itself leading to its ringing.

The forward wave function F is determined by solving the differential equation:

$$F'' + 4c_2^2 F'/ac_1 + 4c_2^2 F/a^2 = -pac_2^2/\mu \qquad (2)$$

where p is the transient pressure generated by the explosion and acting on the borehole wall; and, μ is the shear modulus.

Equation 2 is analogous to a forced spring-mass-damper system with a natural frequency ω_n equal to $2c_2/a$ and a damping ratio equal to c_2/c_1. The damping ratio is a function of Poisson's ratio(ν), and decreases steadily from $1/\sqrt{(2)}$ when $\nu = 0$ to zero when $\nu = 1/2$. At intermediate values of $\nu = 1/4$ and $1/3$, the damping ratios are $1/\sqrt{(3)}$ and $1/2$, respectively.

The radial stress σ_r at the joint is also given by equation 1 if a and $-p$ are replaced by b and σ_r, respectively. This analysis is valid when diverging P-waves generate the radial stress. At the free boundary adjacent to the gap, converging waves are also induced by reflection. In this case the wave function B must also be included to enforce $\sigma_r = 0$. This condition determines B as a function of F. For a plane boundary $B = -F$, but for a spherical boundary a small correction g solves the problem by taking $B = -F + g$, where g satisfies the equation

$$g'' + 4c_2^2 g'/bc_1 + 4c_2^2 g/b^2 = -8c_2^2 F'/bc_1 \qquad (3)$$

When b is large, $B = -F$ gives sufficient accuracy atleast as far as the early time motion of the joint

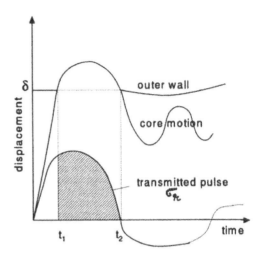

Figure 2: Transmission mechanics

free boundary is concerned. Referring to equation 1, this approximation gives $u_b = -2F'/bc_1$. It is also interesting to point out here that the core boundary at the joint moves out roughly *twice* as much as the displacement in a continuous medium without the joint when b is large. This feature is borne out by the results presented in the next section.

Returning to the mechanics of stress wave transmission across an open joint, the core boundary impacts against the outer medium after moving out by an amount δ at time t_1 reckoned after the wave first reaches the core boundary ie., after a lapse of time $(b - a)/c_1$. This impact impulsively generates a stress σ_r equal to the corresponding value in a joint-free *continuous* medium at the *same* instant in time. Following this initial rise, the stress and displacement histories follow the same variation as in a continuous medium until the radial stress switches sign from compression to tension at time t_2. At this instant, the contact is broken and the inner core reverts back to free ringing. The outer medium also returs to its initial configuration by executing damped oscillations at a much smaller frequency equal to $2c_2/b$. Due to the slower rebound of the outer medium, the likelyhood for secondary impacts is small when b is large. Even if they do occur, such secondary impacts are unlikely to be as energetic as the first one. These features of stress wave transmission across an open joint are schematically explained in Figure 2. The shaded zone represents the transmitted pulse across the open joint.

The duration of the transmitted pulse $t_2 - t_1$ is entirely governed by t_1 which in turn depends on the gap δ. The time t_2 is *fixed* for a given loading function and geometry. The same holds for the transmitted stress magnitude. An interesting special case of this problem results when $\delta = 0$. In this case the entire compressive front of σ_r is let through the joint and the tensile tail is filtered out. A similar situation prevails for fluid-filled joints as fluids cannot transmit tension. However, for a filled joint stuffed with some cohesive material, tensile tails can pass through as noted by Fourney and Dick(1995). This discussion helps understand the complex role played by joints to interpret field observations.

3 RESULTS

Farfield stress and displacement under explosive loading depend on the pressure magnitude and the elastic properties of the medium. All the results presented in this paper correspond to a medium with $c_1 = 2c_2$. Further, the borehole radius and P-wave speed c_1 equal unity. For these values, the spring-mass-damper analogue gives a natural frequency of unity and a damping ratio of $1/2$. This damping is typical of geological media with P-wave speed *twice* that of S-wave. The natural frequency of unity implies a stress wave period of 2π units. This sets up the time scale for interpreting the results. Stress magnitudes have to be scaled suitably to compare results of experiments in the field. Stress pulse transmission across an open joint depends on the joint gap δ, which is set equal to 0.02 units in all further discussion. The joint is assumed to be situated at a radius of 10 times the borehole radius.

3.1 STEP PULSE LOADING

Results are first presented for a step pulse $p = p_0$ of adequate duration. The pulse should last long enough to let the compressive part to go through but not long enough for the waves reflected from the core boundary to return after another reflection at the borehole wall after traveling a distance $2(b - a)$. Figure 3 compares joint wall motion u_b with that occuring in a joint-free continuous medium. It can be seen that the joint roughly doubles the displacements. Figure 4 comares the

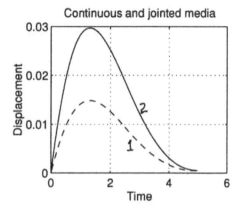

Figure 3: Displacement profiles

1 – continuous medium
2 – open joint

exact and approximate estimates of joint wall motion. It is seen that both estimates agree well initially when the core wall expands inside the gap. The agreement gets even better as the joint radius is increased. Once the joint gap is closed, the displacement corresponding to a continuous medium shown in Figure 3 takes over until the contact is broken by tension.

Stress pulse transmission across the open joint depends on δ. For $\delta = 0.02$, Figure 5 shows the transmitted pulse starting at a time $t_1 = 0.5$, and terminating at $t_2 = 1.3$. The terminating time is fixed for this cases while the starting time depends on δ. When the gap is reduced to zero, transmission starts at $t_1 = 0$. For δ greater than 0.03, there is no transmitted stress into the outer medium.

3.2 EXPONENTIAL PULSE: $p = p_0 t e^{-t/t_r}$

Explosively generated pressure is generally specified by a rise time, t_r and a corresponding peak magnitude p_m. The decay in pressure from this peak value generally follows an exponential variation depending on the properties of the surrounding medium as also the type of explosive deployed. Generally, harder rocks generate high p_m with low t_r while the vice-versa is true for softer formations. These pressure transients can be conveniently modeled using a function of the form $p_0 e^{-t/t_r}$. Such profiles have been examined by the author to formulate tailored pulse loading (TPL)

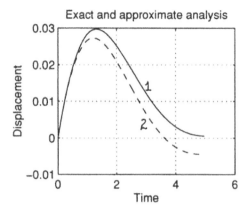

Figure 4: Displacement analyses

1 – exact
2 – approximate

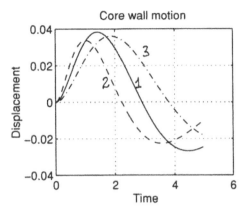

Figure 6: Core wall motion
1 – rise time $= 1$
2 – rise time $= 1/2$
3 – rise time $= 2$

concepts(Simha,1993). The peak value(p_m) for this profile is $p_0 t_r/e$. In the discussion to follow, the peak value is held fixed for different values of t_r.

Rise time plays a key role stress pulse transmission across an open joint as shown in Figures 6-11. In these results, joint wall motion is calculated using the approximate estimate for displacement u_b. Figures 6 and 7 depict the effect of rise time on displacements at the joint. Maximum response is observed when the rise time matches with the natural frequency ie., $t_r = \omega_n = 1$. As in the case of step loading, the core wall movement is approximately twice as much as in a joint-free continuous medium. Figure 6 sets the upper limit on $\mathcal{S}$ for

stress transmission across an open joint. Thus, $\delta = 0.035$ transmits no pulse into the outer medium when the rise time is 0.5, but not when the rise time is 1 or 2.

Predicting stress transmission across a joint under exponential pulse loading can follow the results in Figure 8. Unlike the case of a step pulse, the terminating time t_2 depends on the rise time t_r. Depending on the time taken by the core wall to close the gap, the transmission time extends to $t_2 \approx 1$ for the faster loading with $t_r = 1/2$ to as much as $t_2 \approx 2.6$ for the slower one. However, the maximum magnitude of stress transmission occurs under resonant loading when $t_r = \omega_n = 1$. Further predictions on stress transmission across an open

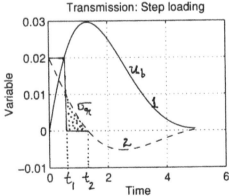

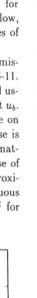

Figure 5: Stress transmission

1 – core wall motion
2 – radial stress

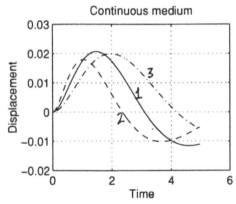

Figure 7: Displacement in continuous medium
1 – rise time $= 1$
2 – rise time $= 1/2$
3 – rise time $= 2$

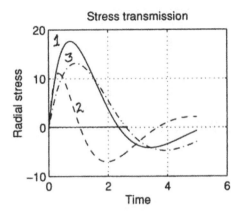

Figure 8: Stress transmission

1 – rise time = 1
2 – rise time = 1/2
3 – rise time = 2

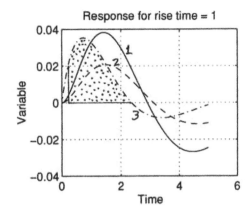

Figure 9: Response for rise time = 1

1 – core wall motion
2 – continuous medium
3 – radial stress
(dotted zone represents transmitted stress pulse)

joint with δ equal to 0.02 is delineated in Figures 9-11.

Figure 9 corresponds to resonant loading with rise time equal to unity. For a joint gap of $\delta = 0.02$, contact occurs at $t_1 \approx 0.6$ and the contact is broken at $t_2 \approx 2.4$. The transmitted stress pulse reaches its peak value immediately after the impact around $t \approx 0.7$. Figure 10 presents results for $t_r = 1/2$. Although the contact is made earlier at $t_1 \approx 0.5$, the transmitted stress magnitude is much more attenuated in both magnitude and duration when compared to the previous case. Finally, Figure 11 refers to the case of $t_r = 2$. The results are somewhat similar to Figure 9, with minor changes in magnitude and duration of the transmitted stress pulse. For a closed joint with $\delta = 0$, Figure 8 shows that the loading with $t_r = 1/2$ gives a transmitted pulse of duration 1 unit compared to a duration of 2.4 for $t_r = 1$.

4 CLOSURE

Joints play a key role in controlling stress wave transmission. The associated mechanics ma inly involves the joint geometry, and whether it is closed or open. Open joints with no filling material transmit only compression as do fluid-filled joints. However, joints stuffed with cohesive materials can transmit a part of the tensile tail of a stress wave

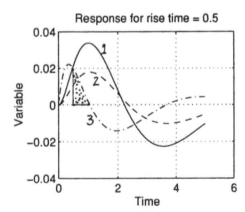

Figure 10: Response for rise time = 1/2

1 – core wall motion
2 – continuous medium
3 – radial stress
(dotted zone represents transmitted stress pulse)

incident on the joint. Formulating this problem in spherical space affords many advantages albeit the analysis becomes tedious for a general type of loading. However, for joints remote from the source, curvature effects become less important. Applying these analytical results to interpret experiments requires data on materials, loading and

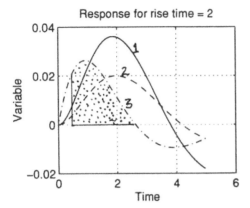

Figure 11: Response for rise time = 2

1 – core wall motion
2 – continuous medium
3 – radial stress
(dotted zone represents transmitted stress pulse)

geometric details of a specific field situation. The analysis can be extended to filled joints separating dissimilar materials. Stress wave attenuation in terms of magnitude and pulse width are often vaguely attributed to material factors rather than open joints. Strategically located joints have the potential to alleviate ground vibration and other undesirable effects of underground operations.

REFERENCES

Fourney, W. L. and R. D. Dick 1995. The utilization of explosive loading as anon-destructive evaluation tool in geological materials, *Int. J. Solids Structures*, v.32, no. 17/18, pp. 2511-22.

Graff, K. 1975. *Wave motion in elastic solids.* Clarendon Press, Oxford, pp. 298-304.

Simha. K. R. Y. 1993. Stress wave patterns in tailore pulse loading. *Proc. IV Int Symp on Rock Fragmentation by Blasting–FRAGBLAST-4*, Vienna, Austria, ed. H. P. Rossmanith, pp 79-85.

Rock Fragmentation by Blasting, Mohanty (ed.) © 1996 Taylor & Francis. ISBN 90 5410 824 X

Dynamic response by signal integration

W. L. Fourney & R. D. Dick
Mechanical Engineering Department, University of Maryland, College Park, Md., USA

T. A. Weaver
Group EES-3, Los Alamos National Laboratory, N. Mex., USA

ABSTRACT: An alternate way of examining the response of geologic materials to explosive loading is presented. The method provides a fast practical way of estimating the predominate frequency of response of any given material to an explosive source and means to determine how that frequency decreases with distance. Also provided is the ability to determine the particular type of rock in which the explosive was detonated and how much faster the velocity signal decays with distance than does the displacement. Results of small scale tests conducted at the University of Maryland along with results from model tests conducted at Stanford Research Institute and from nuclear testing at the Nevada Test Site are used to present the method.

INTRODUCTION:

The ability to measure close-in ground motions caused by the detonation of an explosive source have improved greatly over the past 10 years. Scientists from both Sandia National Labs and Los Alamos National Labs have made the measurement of acceleration and stress in the near vicinity of a nuclear detonation almost a routine activity. The results obtained from such measurements at this time provide good information about the response of the rock to the explosive source. At the same time the use of regional seismic instruments has been demonstrated to be successful in distinguishing large events (nuclear and chemical sources) from earthquakes. The science of blasting has developed greatly over the past 20 years and much more information is gleaned from the close examination of ground motion measurements than was previously obtained with more attention being paid to the frequency content of such measurements. It was with the intention of learning more about the response of geologic materials to explosive loading from close-in ground motion measurements that the signal integration technique was explored.

BACKGROUND:

We were interested in looking at what happened to

a ground motion pulse (in the time domain) as it propagates away from the explosive source. We took a very simple-minded approach to the problem and made a basic assumption that whatever changes the transmitted pulses might undergo that they should occur in a continuous manner. Figure 1 presents a schematic sketch of our thinking. We imagine in the figure that we are measuring particle

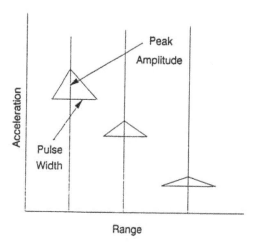

Figure 1. Changes in triangular shaped acceleration time pulse as distance from the source increases.

acceleration at various ranges from an explosive source. If, as shown, the pulse were triangular in form, as it propagates away from the source we would imagine that the amplitude would decrease and that the pulse width might increase. Our basic assumption would indicate that if no drastic changes in the geologic material occur (that is, no open joints or layers of intrusive material) that the form of the pulse would still be triangular. If the pulse were some other form, that characteristic shape would not change except for decreases in amplitude and perhaps an increase in pulse width. Any changes to the shape would be second order such as slight changes in elastic precursors, etc. and the pulse should be recognized as being the same shape with sight modifications.

In our testing we find it more convenient to measure particle velocity rather than acceleration. Figure 2 shows a typical velocity measurement obtained at a range of 25.4 mm from the a g PETN charge in Hydrocal (a fast setting gypsum cement). In looking for ways to determine if the shape of the propagating pulse were unchanged, we felt that the ratio of peak velocity to peak displacement might be a good indicator - where displacement is the area under the velocity time curve. Our results from Hydrocal tests showed that the graph of particle velocity versus particle displacement gave a linear relationship in log-log space. Figure 3 shows data obtained from model testing conducted in Hydrocal. The graph depicts results from three separate tests. The quantity plotted along the vertical axis is the

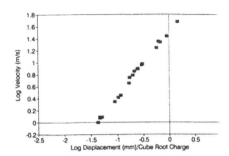

Figure 3. Peak particle velocity as a function of scaled displacement from model testing in Hydrocal.

maximum value of radial particle velocity (away from the explosive source), i.e. the peak of the first velocity pulse created by the explosive (approximately 22 m/s from Figure 2). The quantity plotted along the horizonal axis is the area under the first (positive) velocity pulse from the time at which the particle begins to move away from the explosive source up to the time that the velocity returns to zero (The area from 9 microseconds 67 microseconds in Figure 2). Electromagnetic velocity gages [1] were used to obtain the velocity versus time data for the explosively loaded material.

The observation of this relationship from our tests looked encouraging and gave an incentive to examine other experimental data. Figure 4 presents radial particle velocity versus radial particle displacement (again in log-log space) for three other materials - ash fall tuff, granite, and limestone. The results for granite and limestone are from testing conducted in models by investigators at Stanford

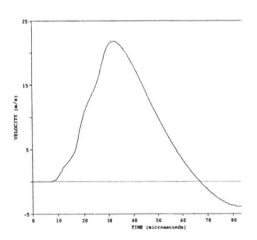

Figure 2. Velocity time measurement from Hydrocal model test.

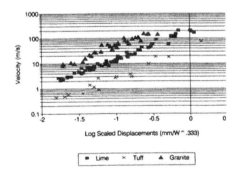

Figure 4. Peak particle velocity as a function of scaled displacement as measured in testing of granite, limestone, and tuff.

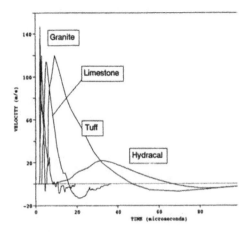

Figure 5. Radial particle velocity as a function of time as measured in granite, limestone, tuff, and Hydrocal.

Research Institute [2,3]. A few of the points shown for tuff are also from SRI model testing, but most are from nuclear tests conducted at the Nevada Test Site. All of the results from SRI were obtained from tests of models approximately the same size as the ones used in the Hydrocal tests conducted by the authors. The results from the nuclear testing involved charges in the several kiloton range. Figure 5 shows samples of velocity-time traces used to generate Figures 3 and 4. In Figure 5, the particle velocity for granite was measured 10 mm from a 3/8 g charge of PETN as was the limestone signal. For the tuff the measurement was made 19 mm from the 3/8 g explosive source. The Hydrocal result was measured 25 mm from a 1 g charge of PETN.

Notice that the displacements plotted in both Figures 3 and 4 are scaled values. This is due to the tremendous difference in size of the tests represented. The scaling used (charge size to the one third power) appears to work quite well and of utmost interest, the data separates quite nicely according to material types. We have indicated this separation in an earlier paper [4]. There is scatter in the data shown in Figures 3 and 4 which can be partially attributed to variations in the materials studied. For example, some of the granite was dry and some was saturated. In the case of both the dry and the saturated, some was tested in a frozen state and some at room temperature. The same is true for the limestone. Likewise some of the scatter in the tuff could be due to variations in either saturation or air filled voids of the material in place. The results

for the Hydrocal shown in Figure 3 are all for the same condition - dry and at room temperature.

We assumed that the data presented in Figures 3 and 4 was well represented by straight lines in log-log space and found the best fit by the least squares technique. We assumed a functional relationship between particle velocity and particle displacement of the form:

$$V = b \, (SD)^a \qquad (1)$$

where V is particle velocity, a and b are constants, and SD is scaled displacement. Table I gives the results from fitting the data in log-log space with straight lines. As can be seen from the Table the power "a" ranges from 1.01 for the saturated granite at room temperature to 1.17 for the tuff. The last row in the table gives R^2 values for the fits which are quite high in most cases. The constant "b" distinguishes the response of the four materials one from another and ranges from 1.52 for the Hydrocal to 3.06 for the dry granite tested at room temperature. This value is related to the slope of the velocity versus displacement curve in normal space. If the power "a" were unity (as with saturated granite at room temperature), "b" would be the slope of the straight line. When "a" is not equal to 1 the velocity versus displacement curve is nonlinear and the slope of the curve changes according to the formula:

$$\text{Slope} = a \, b \, (SD)^{(a-1)} \qquad (2)$$

TABLE I - Signal Integration Parameters

Material	a	b	R^2
Granite			
Frozen	1.07	2.922	0.929
Sat.(RT)	1.01	2.877	0.995
Dry (RT)	1.14	3.055	0.997
Limestone			
Dry (Frz.)	1.12	2.278	0.994
Tuff			
Insitu	1.17	1.644	0.959

We were curious to learn more about this manner of discriminating the response of various materials to explosive loading.

FREQUENCY CONTENT:

We were interested in looking at the response curves (Figure 5) in terms of frequency. We therefore conducted FFT analyses on the velocity time traces for the four different curves shown in Figure 5. We used the FFT option within MatLab for our analysis. The results the four rock types are shown in Figure 6. The amplitudes shown are not scaled with respect to one another and the results exhibit quite broad peaks that are not well defined with regard to the frequency at which the peak amplitude occurs. In all cases the maximum amplitude occurs at the second frequency position plotted and due to a rather large spacing between frequency points in the FFT analyses a better resolution on the exact frequency at which the maximum response occurs is not possible. The spacing in the frequency domain is determined from the length of the input data in the time domain and the time interval between measurement points. For the granite data the spacing between frequency points was 79,400 Hz (it had the shortest time record). For limestone the spacing was less - 39,200 Hz between points. For the tuff and the Hydrocal the spacing for the data presented was 13,300 Hz and 9,980 Hz respectively. We did not have access to the digital output from the tests conducted at SRI and obtained the information for the FFT analysis by digitizing hard copies of the first complete pulse of published results. Since we wanted the comparison among the various materials to be as unbiased as possible we only used information from the first pulse for the Hydrocal as well, even though

we had information over a much longer time frame. All that can be determined for certain from the FFT analysis performed is that the peak amplitude in each case occurs between 0 and twice the minimum spacing in the frequency spectrum.

It is clear, however, from looking at the results in the figure that the peak amplitude is at a higher frequency for the granite than for the limestone and that the limestone response frequency is higher than for either the tuff or the Hydrocal. The peak values for the tuff and Hydrocal appears to occur at about the same frequency. Our conclusion after performing the FFT analyses was that the differences in the response of different rock types can be detected from an FFT analysis but these results are not as well defined as we would hope. To obtain better results might require decreasing the interval between sampling times and collecting data over an extended period of time during the measurement of velocity.

VIBRATING BEAM RESULTS:

We wanted to investigate the significance of the velocity versus displacement results described above and felt that looking closely at a system that we understood better would be useful. Figure 7 presents a sketch of a clamped circular aluminum beam which has a concentrated mass at the end. We wanted to use this system to provide some insight into what we were seeing in the case of the response of rock to explosive loading. We placed a strain gage near the support on two such beams of different lengths (0.609m and 0.304m) and recorded the strain versus time signals while the beams were

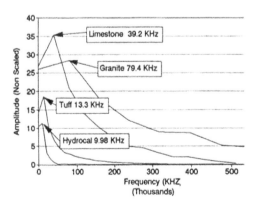

Figure 6. Results of Fast Fourier Analysis of velocity time signals from granite, limestone, tuff, and Hydrocal.

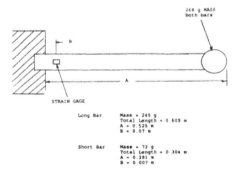

Figure 7. Experimental setup used in vibrating beam tests.

90

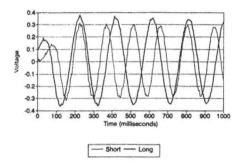

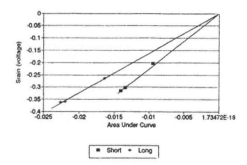

Figure 8. Strain versus time traces as measured in vibrating beam experiments.

Figure 10. Strain measured in two vibrating beams as a function of the area under the strain time curve.

vibrating freely after having the end mass displaced from equilibrium and released. In this case we are measuring the response of the beam based upon strain rather than velocity but for our purposes the results should be similar. Figure 8 shows the resulting wave forms obtained with the strain gages from the free vibration of the two beams. The longer beam has larger amplitude strains and is vibrating at a lower frequency, as expected. Figure 9 shows the results of an FFT analyses performed on the strain time signals shown in Figure 8. In this case the peaks defining the lowest natural frequencies are sharp and well defined. The longer bar has a natural frequency of about 5.2 Hz and the short bar's natural frequency was found to be 7.1 Hz.

We then selected at random a few of the strain pulses and integrated them to obtain the area under the selected pulses. We integrated a negative signal from the time when the strain was zero until the strain signal again returned to zero. This was done for 3 pulses from each of the two beam experiments. Figure 10 shows a plot of peak strain versus the area under the strain time curve (analogous to the

velocity versus displacement curve from the rock testing). Straight lines through the three points from each of the two tests extrapolate to the origin (as they should since at zero strain the area under the strain time curve will be zero). Negative numbers are shown in the figure since we only selected negative pulses for the integration process. The positive pulses would result in the extension of the two lines shown into the first quadrant.

A very simple minded way of determining the area under any curve would be to take the product of the height of the pulse, the width of the pulse, and a form factor. For example the triangular pulse in Figure 1 would have a form factor of 0.5. If the pulse were rectangular in shape the form factor would be 1.0. No matter what the shape of the pulse it is possible to find a number which when multiplied by the pulse width and the maximum amplitude will yield the correct area. If we assume that we will define the area in that fashion then the following relationship is valid.

$$A = a_p \, \lambda/2 \, f_f \qquad (3)$$

where A is the area under the strain time curve, a_p is the peak amplitude of the strain curve, $\lambda/2$ is half the pulse width of the strain time curve, and f_f is the form factor. For a pure sine wave the form factor is $2/\pi$ (0.637). The form factor is the same regardless of the amplitude or the pulse width and only depends upon pulse shape. For the strain time curves shown in Figure 8 the pulse shape is very close to being a sine wave. The slope of the straight lines shown in Figure 10 is the ratio of a_p/A (the amplitude divided by the area). Since the curves shown in Figure 10 are assumed to be straight lines, the ratio of the strain to the area under the strain time curve is constant and the ratio is given by the

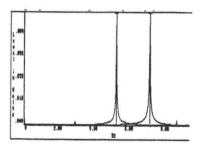

Figure 9. Results of FFT analysis of strain signals from long and short beams.

slope of the line. The slopes are therefore (from equation (3)); Slope = $1/(f_f \lambda/2)$. The pulse width is given by the reciprocal of the frequency of the wave form. Therefore the slopes are:

$$\text{Slope} = 2 * \text{frequency}/(f_f) \qquad (4)$$

The slope of the line in Figure 10 for the short beam is 22.14 per second. The slope for the line from the long beam was found to be 16/second. Using equation (4) and the natural frequencies determined from the FFT analyses the form factor for the long short beam is 0.641 and for the long beam 0.650. Both of these are acceptably close to the expected 0.637 value for a pure sine wave. It is likely that the difference between the values obtained and the expected value is due to the inability to read the slopes of the two lines in Figure 8 with greater accuracy.

We infer from this exercise that by plotting the velocity versus the displacement measured in ground motion studies we are providing an indication of the response of the medium to the explosive loading since the ratio of velocity to displacement is twice the predominate frequency divided by the form factor. The same would be true if we plotted the acceleration versus the velocity. Both give an indication of the predominate frequency of response of the rock to the loading imparted to the media.

RESULTS FROM EXPLOSIVE LOADING:

Looking at the results presented in Table I it is clear that the relationship between velocity and displacement is not linear in the case of explosive loading. The results are nearly linear with the largest power for the four materials shown being 1.17 for the tuff. The ratio of the velocity to the displacement still gives twice the frequency divided by the form factor but this ratio now changes with distance from the source. Very close to the charge (at large velocities) the slope is greater (the ratio is higher) than it is farther from the charge.

Finding a least squares linear fit to the data in normal space - even though there is a slight curvature to the relationship gives an estimate of the average dynamic response of the material over the range investigated. If this is done for the four materials under discussion the results are given in Table II.

TABLE II - Results of Least Squares Fit

For Hydrocal:
$V = -0.915 + 35.57 \delta$ with $R^2 = 0.996$

For Granite:
$V = 3.77 + 436.05 \delta$ with $R^2 = 0.910$

For Limestone:
$V = -0.938 + 192.32 \delta$ with $R^2 = 0.84$

For Tuff:
$V = -2.294 = 53.02 \delta$ with $R^2 = 0.96$

V is velocity and δ is displacement.

We were interested in finding how the form factor varies with range. Form factor is determined by dividing the peak displacement (the area under the velocity curve) by the velocity times the pulse width. For the four velocity pulses shown in Figure 4 the results are given in Table III which also presents some other properties of interest for the four materials.

TABLE III - Material Properties of Interest

Material	Form Factor	Modulus
Granite	0.285	76 GPa
Hydrocal	0.486	12 GPa
Tuff	0.435	12.5 GPa
Limestone	0.376	27.8 GPa

Form factor does vary with range. The greatest variation occurs at smaller ranges where the stress levels are quite high and the rock is being inelastically deformed. Figure 11 shows the form factor at various ranges as obtained from one of the SRI tests conducted with dry limestone. At a range of 10 mm the form factor is about 0.4. The form factor increases to a value of around 0.65 at a range of 40 mm and remains fairly constant at that value out to 80 mm. The form factor for granite varies between 0.2 and 0.32 over the range for which data is available. In other work we have found that the form factor obtained in testing in tuff at NTS was

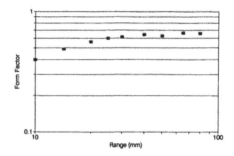

Figure 11. Form factor as a function of range from an SRI test with dry limestone.

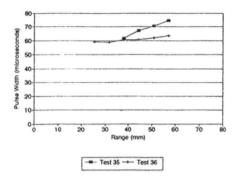

Figure 12a. Pulse width as a function of range from two tests in Hydrocal.

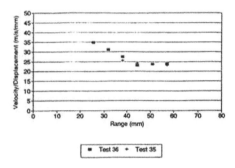

Figure 12b. Velocity displacement ratio as a function of range from two tests in Hydrocal.

between 0.47 and 0.50 over the range for which data was available. For Hydrocal the form factor varies from about 0.48 at 25.4 mm to 0.68 at 58 mm.

If the slopes of the velocity versus displacement curves in Table II and the form factors in Table III are used in Equation (4) the approximate natural frequencies for the response of the four materials are found as follows:

For Hydrocal ω = 8,644 Hz

For Granite ω = 62,137 Hz

For Limestone ω = 36,541 Hz

For Tuff ω = 11,537 Hz

These results are in agreement with results obtained from the FFT analysis and are felt to be more accurate than those results due to the problem with minimum frequency size discussed earlier.

Figure 12 shows results from two tests in Hydrocal. Figure 12a shows the pulse width as a function of range from Tests 35 and 36 which were intended to be identical in all aspects. The explosive performance, however, appeared to be different between the two tests. (The velocities recorded in Test 36 were significantly higher than in Test 35). Notice that the pulse widths measured in Test 35 are considerable larger than those measured in Test 36. Figure 12b shows the response frequency as determined by the signal integration technique for the two tests. Note that even though significant differences are seen in the wave characteristics between the two tests that the response frequencies at any given range agree.

The decrease in the velocity displacement ratio with range gives an indication of how the predominate response frequency changes with range - especially at ranges where the form factor is relatively constant. As the distance from the charge increases more of the high frequency content of the response is lost and the velocity displacement ratio decreases. Keep in mind that the form factor also varies with range as discussed earlier. The decrease in frequency with range overestimates the actual decrease at smaller ranges, since in that location the form factor itself is increasing.

There are two parameters obtained in the fitting of the data for velocity versus displacement, "a" and "b". The intercept value "a" has been shown to be very useful in distinguishing the response of one material from another when subjected to an explosive load and is tied to the predominate response frequency of the material. There appears to be a correlation of this number with modulus - as would be expected. Figure 13 shows a plot of this intercept value versus material modulus. The modulus values shown in the figure are a mixture of dynamic and static values. The materials

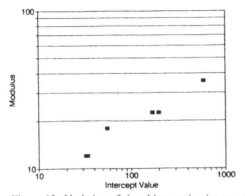

Figure 13. Variation of signal integration intercept value with elastic modulus.

represented by the data shown are granite, limestone, tuff, Hydrocal, and 2C4 (a rock matching grout).

The parameter "b" represents the exponent of the velocity-displacement curve. The range of this parameter for materials presented is from approximately 1 to 1.2. This parameter determines how the fall off in velocity magnitude compares to the fall off in displacement magnitude with range. For example, a value of b=1 would indicate that if the velocity falls off as range to the $1/R^2$ then the displacement also falls off at the same rate. A value greater than unity indicates that the velocity falls off at a greater rate than the displacement.

CONCLUSIONS:

It appears that a convenient method for determining the response of a material to an explosive loading and for discriminating among different materials is a plot of measured particle velocity versus particle displacement. In this presentation the particle displacement was obtained by integrating the velocity-time signal.

A vibrating beam was used to demonstrate that the ratio of the measured quantity to the integrated signal at any station is related to the predominate frequency of response of the media. In the case of the vibrating beam the ratio of strain to area under the strain curve was directly related to the natural frequency of the vibrating beam.

For rock the response to an explosive load is more complicated. From the data available it appears that the velocity/displacement ratio is consistent from test to test in the same media even though the pulse width of the velocity response is not.

The use of this ratio as a discriminate between material behavior appears to hold promise and deserves additional investigation. It could be especially helpful in organizing the response of various types of geologic materials to explosive loading.

ACKNOWLEDGMENT:

This work was funded by support from the Los Alamos National Laboratory under subcontract 9-XG3-6590H-1. The authors are grateful to the helpful discussions held with Steve Taylor and Fred App at Los Alamos National Laboratory.

REFERENCES:

Fourney, W.L. & R.D. Dick 1994. The Utilization of Explosive Loading as an NDE Tool in Geologic Materials. *International Journal of Solids and Structures.*

Miller, S.A. & A.L. Florence 1991. Laboratory Particle Velocity Experiments on Indiana Limestone and Sierra White Granite. PL-TR-91-2277, SRI Final Report to Phillips Laboratory, 56 pp.

Nagy, G. & A.L. Florence 1985. Particle Velocity and Stress Pulses Produced by Small Scale Explosions in Geologic Materials. In *Joint ASCE/ASME Mechanics Conference on Response of Geologic Materials to Blast Loading and Impact,* V 69, Ed. J.C. Cizec, American Society of Mechanical Engineers, pp. 105-118.

Young, C., B.C. Trent, N.C. Patti, & W.L. Fourney Aug. 1983. Electromagnetic Velocity Gauge Measurement of Rock Mass Motion in Explosive Fragmentation Tests. *1st International Symposium on Rock Fragmentation by Blasting,* Lulea, Sweden.

Rock Fragmentation by Blasting, Mohanty (ed.) © 1996 Taylor & Francis. ISBN 90 5410 824 X

A new constitutive model for rock fragmentation by blasting – Fractal damage model

Jun Yang
Department of Mechanics & Engineering, Beijing Institute of Technology, People's Republic of China

Shuren Wang
Beijing Graduate School, China University of Mining and Technology, People's Republic of China

ABSTRACT: Based on theoretical models and their developmental trends in rock blasting, a new constitutive model for rock fragmentation by blasting has been presented, by introducing fractal theory into the study on mechanism of rock blasting. In the new model, the natural damage is determined by the fractal dimension from joint, crack, fracture etc., which is basis of rock breakage   and the damage evolution is connected with energy disspation rate. The damage evolution process in rock by blasting is a fractal and the fractal dimension reflects the degree of rock damage and its   development. Thus, a simple equation system of the new damage model described with fractal and its evolution is constructed for rock fragmentation by blasting. The main point of view in the new model have been verified by the similarity material simulation, dynamic photo-elastic and practical tests.

The model is set into a two dimensional numerical calculating code with ALE metthod, which can be used to calculate in detail the rock damage evolution process by blasting. Based on the hypothesis of crater and the equation for predicting the size distribution, the crater profile (or broken range)and the fragmentation size distrbution can be predicted conveniently.

1 INTRODUCTION

The study on damage model of rock fragmentation by blasting is a hot topic for theoretical research of engineering blasting, which is important for realizing blasting optimizing design and predicting its effiect. The focus of the study is to describe constitutive functions of rock material under dynamic loading. Among all of the models including elastic model, fracture model, and so on, a damage model, describing the fracture that occurs as a result of the stress induced material deformation, has shown obvious advantage compared with other models above. Since 1985, some authors (Grady & Kipp , 1985; Taylor et al. , 1986; Kuszmaul, 1987) have established the constitutive model of irreversible evolution for a brittle rock. It is modified (Thorne, 1990) to extend it to large crack densities and allow finite element calculations that remain stable at late times. So the current damage model has developed a numerical model which can be put into program (PRONTO) to simulate rock fragment progress by dynamic load, and the experimental verification has been conducted.

However there are some shortages in the current damage model(Throne, 1990), it has not been used widely in common blast engineering. Thus, in this paper, by using the new achievements from fractal theory and its applications in rock damage mechanics area, the current model is improved and modified, and a new model, fractal damage model for rock fragmentation by blasting, is presented.

How to describe the damage in rock is a difficult problem in studying rock damage mechanics. It had been solved with crack activating hypothesis in the current model, but the original damage in rock material was not considered as a natural parameter. The fundamental assumption of the damage model is that rock is an isotropic material which is permeated by an array of randomly distributed and oriented micro cracks that extended under tensile loading. And the micro cracks density is not depended on the original damage in rock, but the loading, and some parameters caused by loading, such as volume tensile strains, the maximum strain rate and other material constants. In this way, it is ignore the important influence of joints, cracks, micro holes and other original flaws to blasting action. It is obviously disconformity to reality of rock blasting. As well known, rock, as material under geological stress, is unavoidable to keep a lot of macro and micro flaws named original damage. It is the original damage which grow and interacted with one another under dynamic loading that rock fragmentation satisfied certain distribution can be obtained in blasting. So the damage in rock before blasting cannot set at be "zero" and it must be a certain value, a parameter responded natural flaw in rock, as a base of evolution.

Besides this, the crack activating hypothesis neglects the fact that the original damage still influences the stress action, although the stress is compression. In fact, the original damage in rock determines the rock fragment size by blasting in large scale, which limited the largest size of the rock block after blasting (Zhao, 1985). Therefore, the action of original damage in rock must be fully considered in damage model. However, in the current damage model, the damage evolution is postulated as the function of bulk train, in which it is monotonous, and the fact is ignored that the more damage in some area, the more energy is needed for damage developing. In order to reflect the law of damage volution in rock fragmentation correctly, we must seek a new way of solving the problem.

2 A NEW DAMAGE MODEL FOR ROCK FRAGMENTATION BY BLASTING

Owing to the analysis above, a new model is presented, in which the original damage's fractal to be as main parameter and the energy dissipation rate during blasting to be regards the main basis of damage evolution. It is very complex of the distribution of the original fracture in rock and there is no final verdict about the definition and the survey of damage parameter D in damage mechanics. However, a new inspiration has been given from the outstanding application of fractal geometry in rock mechanics and geophysics. Why not try to throw away many factors during blasting physics process, and to consider only the crack fractal dimension and its evolution process to construct the damage model for rock fragmentation by blasting. This is undoubtedly a new daring attempt in blasting mechanics.

2.1 Fractal nature on rock damage

The damage evolution process of rock materials is a fractal, and has a good statistical self—similarity (Xie & Gao, 1991). The fractal dimension reflect the damage degree of rock material. According to fractal theory, the damage in rock can be regard as the reduction of the dimension of rock matter, and mechanical effect of the reduction is just the same as the change of effective module caused by damage.

Box dimension (box counting) is a fractal parameter which can reflect the fill degree of crack in rock(Xie,1990). It can be calculated with the function below:

$$D_f = dim_b F = \lim_{\delta \to 0} \frac{log N_\delta(F)}{-log(\delta)} \qquad (1)$$

here D_f is box dimension, $N_\delta(F)$ is least number of cubic box element covered the fractal F, and δ is the box size.

In calculating fractal dimension from crack image of rock material, counting the box number N_δ which hold crack with variable size scale δ, doing linear regression to the two term data in logarithmic coordinates, and the linear slope is the fractal dimension D_f. All the work above can be completed with micro—computer and a scanning image process unit.

From function (1), the crack number in unit volume with average size a is: $N = \beta a^{-D_f}$, introducing N into the crack density function (Thorne, 1990) $C_d = \beta N a^3$:

$$C_d = \beta a^{3-D_f} \qquad (2)$$

here a is the average size of ore crystal grain in rock material, and β is a proportionality constant.

According to the relationship of damage D and crack density C_d, have

$$D = \frac{16}{9} \frac{1-\nu_e^2}{1-2\nu_e} \beta a^{3-D_f} \qquad (3)$$

here, ν_e is effective Poisson modules, $\nu_e = \nu(1-16/9C_d)$, ν is the Poisson's ratio.

Thus the damage parameter can be express as a function of fractal dimension D_f.

2.2 The fractal evolution in rock fragmentation

According to the results studied (Xie, 1993; Yang & Wang, 1994), the damage evolution process in rock is always company with the crack fractal developing, and there is a linear relationship between the fractal dimension D_f and the damage energy dissipation rate:

$$D_f = D_o - kY \qquad (4)$$

where D_o is the fractal dimension of original crack, and k is material constant.

For the problem of isotropic thermo—elasticity damage mechanics, the damage energy dissipation rate is

$$Y = \rho \frac{\partial \varphi}{\partial D} \qquad (5)$$

where φ is free—energy potential and ρ is material density. Under common temperature, that is the linear elastic state

$$Y = -\frac{1}{2} \left[\lambda \, \varepsilon_{ii} \, \varepsilon_{jj} + 2\mu \, \varepsilon_{ij} \, \varepsilon_{ij} \right] \qquad (6)$$

where ε_{ij} is elastic strain tensor, λ and μ are lame' constants.

Introducing (6) to (5)

$$D_f = D_o + \frac{1}{2} k \left[\lambda \, \varepsilon_{ii} \, \varepsilon_{jj} + 2\mu \, \varepsilon_{ij} \, \varepsilon_{ij} \right] \qquad (7)$$

So it is obtained the fractal evolution equation described by elastic strain tensor.

2.3 The constitutive equations of the new damage model

The bulk response and deviator response of rock material are given below (Kuszmaul, 1987)

$$P = K_e \varepsilon \qquad (8)$$

$$S = 2\mu_e e \qquad (9)$$

where, P is bulk stress, ε is bulk strain, S is stress deviator tensor and e is strain deviator tensor. The effective bulk modules K_e and the effective shear modules μ_e are defining as follow:

$$K_e = (1 - D) K \qquad (10)$$

$$2\mu_e = \frac{3K_e (1 - 2\nu_e)}{1 + \nu_e} \qquad (11)$$

where K is bulk modules.

Introducing (10) (11) into (8) (9) and combining with (2) (3) (6), a simple equations system of the new damage model described with fractal and its evolution is constructed for rock fragmentation by blasting.

3 THE TEST VERIFICATION ON THE NEW MODEL

In order to prove the reasonableness of the new model described above, we have accomplished various kinds of tests such as simulated blasting (Yang & Wang, 1995).
The model material used in simulated test
is made of cement adding certain resin (5% or 10%). Because the resin is brittle and has a lower strength, a lot of flaws (damage) are produced throughout the material. The sample with $\Phi 300 \times 400 mm^3$ volume is poured in the material with various proportion of original flaws, so the results can reflect the influence of various original damage parameter on blast effect. The characteristic of the similarity material and the crater parameters in the tests are shown in Table 1.

The cumulative distribution curves for the tests in G-G-S regression equation of the pile are shown in figure 1. From the table and figures above it can be seen that the original damage degree has important influence over the blast effects. As shown in table 1, the crater volume relate to the fractal dimension of original damage.

Table 1. The characteristics of similarity material and the crater size in simulation tests.

Test №	Characteristics			Crater Size		
	Strength R(MPa)	Resin Proportion %	Fractal D_f	Radius (cm)	Depth (cm)	Volume (cm³)
T−1	19.62	0	1.3737	4.8	4.2	101.3
T−2	11.77	5	1.6519	5.6	4.5	147.8
T−3	10.79	10	1.7731	7.0	4.8	195.0

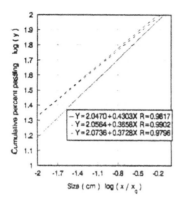

Fig. 1 Cumulative distribution curves
in G-G-S regression equation

The higher the crack density, the larger the diameter and the depth of crater blasted. The increase depth is greater than that of the diameter. Figure 1 shows the sieve size of the broken blocks in similar tests. It is obvious that initiation of new damage causes a decrease in block size. The G-G-S distribution regression results indicate that increase of the original crack in the material causes a decrease in the index n of G-G-S function.

For comparing the stress field in damaged material from that in undamaged material during blasting process, dynamic photo-elastic tests were carried out with polycarbonate containing various densities of man made micro cracks. The results (Yang, 1995) show that stress fringes grade in the sample contained original damage are higher than that in the sample not containing original damage under the same blast conditions.

4 THE MODEL IMPLEMENTATION AND VERIFICATION BY TEST IN SITU

A computation program based on the model described has been written and implemented in SHALE (Demuth et al. , 1985), which is a two dimensional finite difference code for simulating stress wave propagation and dynamic process in solid. It is implemented on the PC computer. The computer routine for FDM model is shown in figure 2.

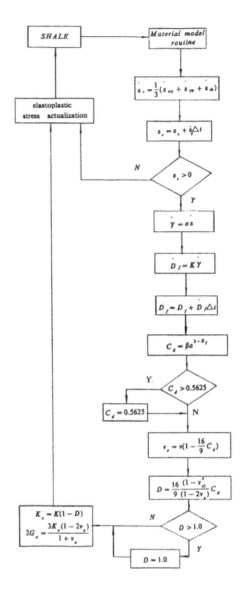

Fig. 2 Compute routine for FDM model

The program has been executed with the parameters in rock below: ρ=2.26 g/cm^3, c=4100 m/s, D_o=1.48, v=0.27, K-0.1298 MPa   The detonation of a vertical blast hole 1 m long and 42 mm in diameter with a column of 0.5 m of 2" rock AN-TNT explosive is analyzed. Figure 3 shows the damage parameter distribution 3 ms after the shot, when the damage phenomenon had already come to end. As shown in figure 3, a damaged area exits under an undamaged one by the surface. Severe damage appears in the close vicinity of the blast hole, which may be attributed to the radial cracks that typically grow around the blast hole after blast. In the new model, crack and damage grow under energy; this is why the damaged range is deeper than the results used in the current model.

Now let us attempt to define the crater range with the damage chart calcutated above and to predict the fragment distribution by blasting with the help of the fractal dimension within the crater. As research shown, blast crater contour in damage chart can be supposed as a quadritic curve (Yang, 1995), and there is a relationship between the G-G-S index n of fragment distribution by blasting and the fractal dimension D_f of crack in rock (Yang, Wang &. Wan, 1995):

$$n = 3 - D_f \tag{12}$$

So, we can predict the fragment distrbution with G-G-S function below:

$$y = (\frac{x}{x_0})^{3-D_f} \tag{13}$$

here x_0 is the size of the maxmium block.

Blasting tests in situ have been completed with blast hole 1 m long, 42 mm in diameter and with 0.6 kg of 2" rock AN-TNT explosive. The parameters of the rock and the explosive used in test are shown in table 2.

Figure 4 shows the comparative chart for crater contours of the polynomial regression result of measure date in test and the predicting one of numerical calculation. The predicting one is not beyond the distributed range of measure data, so it is practicable to predict the broken range (the crater)from damage calculating.

Figure 5 shows the comparative chart for the G-G-S regression curves of fragment distribution of sieving test in situ and predicting one based on calculating aveage D_f. This verify that the predicting result of blast effect with FDM model is able to meet the needs of practical production.

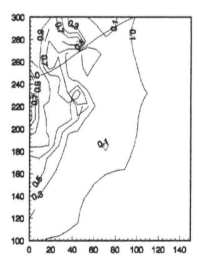

Fig. 3 Damage parameters of rock material. Distances in cm

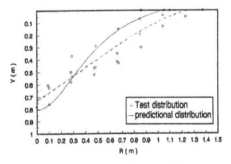

Fig. 4 Comparative chart for the crater contours of predicting result and test result

5. CONCLUSION

With applications of the fractal theory to describe the original damage in rock and to calculate the damage evolution during blasting process, FDM

Table 2. The parameters of rock material and explosive used in test in situ.

Rock					Explosive	
Density $\rho(g/cm^3)$	Wave Velocity $c(m/s)$	Yang'Moudule $E(MPa)$	Poisson ratio ν	Fractal D_f	Density $\rho_e(g/cm^3)$	velocity $D_e(m/s)$
2. 600	3600	34. 19	0. 28	1. 6490	1. 080	3600

model for rock fragmentation has emphasized the influence of natural crack in rock to blasting action, and improved the mechanics hypothesis in current model. The main point of view in the new model have been verified by the similarity material simulating, dynamic photo—elastic and practical tests, and the numerical calculating has successfully implemented in SHALE code with microcomputer. It is verified by tests in situ that the predictions about blasting effect of using the calculating results tally with the practice. However there are rooms to improve the new model.

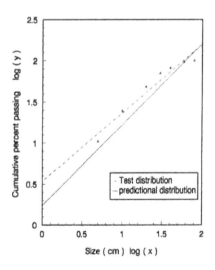

Fig. 5 Comparative chart for predicting distribution and test distribution of fragmentation

REFERENCES

Demuth, R. B. 1985. SHALE: a computer program for solid dynamics. LA — 10236 Los Alamos, New Mexico.

Grady, D. E. & Kipp, M, E. 1985. Mechanisms of dynamic fragmentation: fractors governing fragment size. SAND—84—2304C.

Kuszmaul, J. S. 1987. A new constitutive model for fragmentation of rock under dynamic loading. *2nd ISREFB*: 412—423. Coloradu U. S. A.

Tailor, L. M. Chen, E. P. &Kuszmaul, J. S. 1986. Microcrack—induced damage accumulation in brittle rock under dynamic loading. *Computer Method Appli. Mech. Eng.* 55: 301 —320.

Thorne, B. J. 1990. A damage model for fragmentation and comparision of calculations with blasting experiments in Granite. SAND —90—1389.

Xie, H. 1990. *Damage mechanics of rock & cement.* CUM Press. Xuzhou China.

Xie, H. & Gao, F. 1991. Fractal characteristic of damage evolution in rock material. *Rock Mech. & Eng.* 10: 74—82.

Xie, H. 1993. Fractal nature on damage evolution of rock materials. *Proc. ISACMRM*: 435 —441. Xian China.

Yang, J. & Wang, S. 1994. Study and analysis of the crack development of Hopkision bar under blasting loading. *Proc. FNCRD4*: 53 — 59.

Yang, J. & Wang, S. 1995. Study on the damage model of rock fragmentation by blasting. Proc. 2nd ICMJFR: 973—977. Rotterdan: Balkema.

Yang, J. 1995. Study on the fractal damage model of rock fragmentation by blasting. (Ph. D thesis) Beijing China.

Yang, J. Wang, S. & Wan, Y. 1995. The application of fractal geometry in predicting rock fragment distrbution by blasting. J. of Northeastern University 16(S): 107—111.

Zhao, b. 1985. Prediction of the distribution of rock fragmentation by blasting (Master thesis). Shengyang China.

Blast induced fractures and rock damage

Rock Fragmentation by Blasting, Mohanty (ed.) © 1996 Taylor & Francis. ISBN 90 5410 824 X

Borehole pressure measurements behind blast limits as an aid to determining the extent of rock damage

G. F. Brent & G. E. Smith
ICI Explosives Technical Centre, Kurri Kurri, N.S.W., Australia

ABSTRACT: Demands on blast design for the minimisation of damage are becoming more severe. In order to achieve control, the assessment of damage is vital. This paper discusses a technique which involves monitoring the pressure in sealed boreholes behind blast limits. The experimental technique is presented along with results from two sets of experiments. Underpressures ranging from -39 to -78kPa at one burden, 0 to -36kPa at two burdens and 0 to -20kPa at three burdens were recorded behind a series of bench blasts. After comparison with borehole video records and correlation with the dimensions of newly-formed cracks, the underpressures were attributed to the new volume created by the cracks. No evidence of penetration by high pressure gases was found, leading to the conclusion that the cracks were stress induced. Conversely, for the situation of fully confined crater blasts in a different rock type, peak overpressures in the range 5 to 280kPa were recorded at distances between 62 and 8 blasthole diameters. In this situation, gases were able to penetrate to the monitoring locations. Several conclusions are drawn from the work, most notably that the technique is a valuable aid to determining both the mechanisms and extents of blast-induced damage from different blast designs or in different rock types.

1 INTRODUCTION

The control of the extent of damage from blasting has assumed increasing importance as mining techniques have evolved. The minimisation of damage to the unblasted rock is of vital importance in both underground and surface mining where this rock forms part of the working environment or mine structure (Chitombo and Scott, 1990). In open cut mining, current trends are towards deeper pits and consequently higher walls which will require ever greater structural competence.

In order to control and reduce damage, assessment of the extent of damage is vital. Several techniques have been used for damage assessment behind blast limits. These include vibration measurements (which are generally related to various damage criteria), extensometers for gross movement, cross-hole seismic logging for determining crack density and, more recently, the direct viewing of cracks by borehole video cameras. This paper describes a system of recording the pressure-time history in monitoring holes behind blasts. Results from case

studies, including the use of this system in conjunction with a borehole video camera, are presented.

Originally, this type of system was employed for detecting high pressure explosive gases penetrating into the remaining rock. A technique for detecting the presence of high pressure explosive gases behind blast limits was first reported by Williamson and Armstrong (1986). They described a system of monitoring boreholes located at various positions behind a blast. Each hole was plugged at the collar and piezoelectric pressure transducers were fitted to tubes which sampled the air inside the holes. They adopted the convention of drilling the monitoring holes to the same depth and at the same inclination as the blastholes. Their results indicated that high pressure gases were able to penetrate back to the monitoring holes.

Later workers adopted this technique with varying results. In general, it has been employed in conjunction with some of the damage measurement techniques mentioned earlier. Bulow and Chapman

(1994) report its use together with vibration measurement, the use of extensometers and cross-hole seismics. Although they only used two monitoring holes behind blast limits, they report a decrease in the magnitude of pressures measured with distance behind the blasts. They detected only overpressures, (ie positive relative to the atmosphere), and concluded that these were evidence of gases penetrating along joints. Sarma (1994) also reports the use of this technique, but with highly variable results. Both Sarma (1994) and Williamson and Armstrong (1986) detected underpressures (ie negative relative to the atmosphere). They postulated that these were a result of the forward displacement of the burden.

Other recent work has shown more consistency in the results. Lejuge et al (1994) found that the initial signal constituted an underpressure which was sometimes followed by overpressures. Using the technique in conjunction with extensometers, they concluded that the underpressures were a result of ground dilation. They also concluded that the overpressures occurred closer to the blast limits while further back there was still evidence of underpressures due to ground dilation.

The most recently reported work is that of Ouchterlony (1995). In a series of tests involving 13 monitoring holes he reports the occurrence of only underpressures in 10 of these. Some overpressures occurred in monitoring holes close to blastholes, particularly where the blast was fully confined as in a presplit situation. After comparison to extensometer data he concluded, in agreement with Lejuge et al, that the underpressures were a result of ground dilation.

2 MEASUREMENT TECHNIQUE

The technique adopted in this work was a variation on those of earlier workers. Diaphragm pressure transducers were used in two configurations. The first comprised pressure transducers which referenced the atmosphere. These were fitted to tubes sampling the air inside the monitoring holes which had been sealed at the collar. The second comprised absolute pressure transducers which referenced an internal sealed vacuum chamber, and these were suspended directly into the sealed monitoring holes. It was confirmed in one of the bench blasts described in the following section that the pressure records

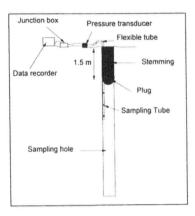

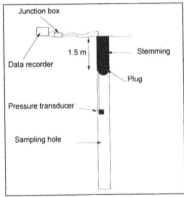

Figure 1. Two configurations for pressure monitoring holes.

from both configurations in the same monitoring hole were virtually identical.
The configurations are shown in Fig. 1. Generally, the top 1.5 m of the holes was sealed off , restricting the pressure chamber to the lower section of the hole.

Prior calibration of the pressure transducers was done against a conventional pressure gauge which was certified as accurate to within 2%. The calibration factors were set at 200kPa and an accurate response to within 5% was confirmed over the range 0-200kPa. This was considered adequate for the purposes of this work. Dynamic response was also tested by rapid depressurisation from 200kPa to 0kPa through a ball valve.

In this work the monitoring holes were drilled on staggered patterns behind the blast limits. This was done in order to avoid modification of the stress fields and/or gas flows by holes in front of each other. Preferential crack formation between holes would also be minimised. A further benefit of the staggered pattern was geometric as it slightly increased the distances between holes and allowed monitoring at different locations along the back of the blast. The convention of drilling the monitoring holes to the same depth and inclination as the production holes was also adopted.

3 SERIES OF BENCH BLASTS

The first series of experiments was undertaken at an open cut mine in the Hunter Valley, NSW, Australia. The rock was described as massive sandstone interburden. The mine required knowledge of damage mechanisms and the distances to which appreciable damage occurred behind production blasts. This was necessary in order to avoid damage as mining progressed toward an inground slurry wall. Borehole pressure monitoring using the technique

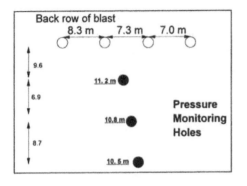

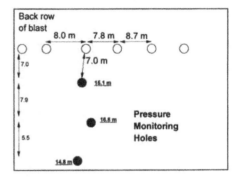

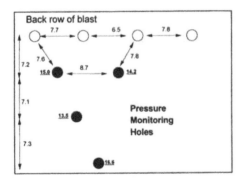

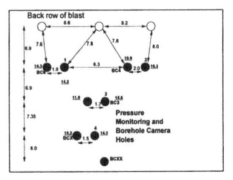

Figure 2. Plan views of monitoring locations for the bench blasts (blasts 1 - 4 clockwise from top left, hole depths underlined).

Table 1. Blast design and monitoring hole details for the series of bench blasts

Blast No	Hole Diameter (mm)	Burden x Spacing (m x m)	Powder Factor (kg/m3)	Explosive Type	Monitoring Hole Diameter (mm)	Monitoring Hole Distances behind Blast Limits (m)
1	200/270	7 X 8	0.45	ANFO(top)/HANFO(toe)	200	9.6, 16.5, 25.2
2	200/270	7 X 8	0.43	ANFO(top)/HANFO(toe)	270	7.0, 14.9, 20.4
3	200/270	7 X 8	0.44	ANFO(top)/HANFO(toe)	200	6.9, 13.8, 21.1
4	200/270	7 X 8	0.31	ANFO(top)/HANFO(toe)	200	7.2, 14.3, 21.6

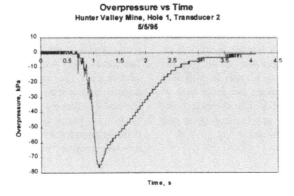

Figure 3. Typical pressure-time trace from a monitoring hole behind the bench blasts.

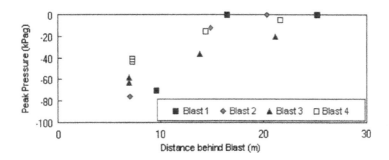

Figure 4. Summary of peak pressures measured from the series of bench blasts.

described earlier was employed behind four blasts, all of which were located within the same interburden regime. Details of the blasts and monitoring hole locations are shown in Table 1, while Fig. 2 schematically shows the plan views of the monitoring locations. The monitoring holes were all dry, vertical and drilled to approximately the same depth as the production holes. Generally, they were located at nominal distances of one, two and three blast burdens behind the limits. These distances correspond approximately to 35, 70 and 105 hole diameters. At the third blast, a proprietary borehole video camera was used to view borehole walls before and after the blast. The video camera holes were located closely adjacent to the pressure monitoring holes at between 1.5m and 2.0m away. These holes are labelled with the prefix BC in Fig. 2.

Underpressures were consistently recorded, with no evidence of high pressure gases penetrating to the monitoring hole locations. An example trace

showing the typical form can be seen in Fig. 3, while Fig. 4 summarises the data for all four blasts, showing the relationship between distance behind the blast and magnitudes of the measured peak underpressures.

Analysis of the borehole video records from the third blast revealed that there were no obvious visible structures or discontinuities in the borehole walls before the blast. After the blast, many large open cracks could be seen. These cracks were mostly horizontally orientated, however there were also several large vertical cracks and an example of these can be seen in Fig. 5 taken from the report by Raax Australia (1995). (The left image is the view across the circumference of the borehole wall with depth shown on the vertical axis, while the right image is a partial three dimensional representation of this view.) Figure 6 shows a map of the clearly visible cracks obtained through analysis of the video record. Maximum widths of the cracks (mm) are indicated.

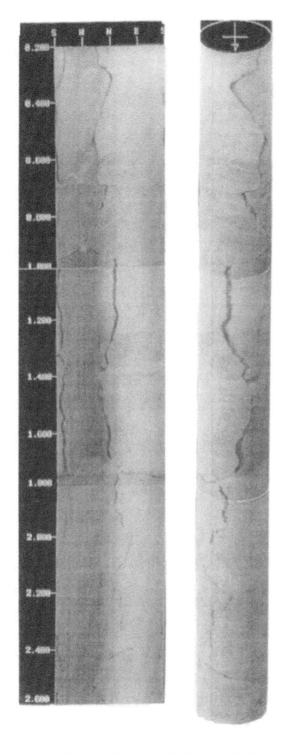

Figure 5. An example of cracking viewed by the borehole video camera.

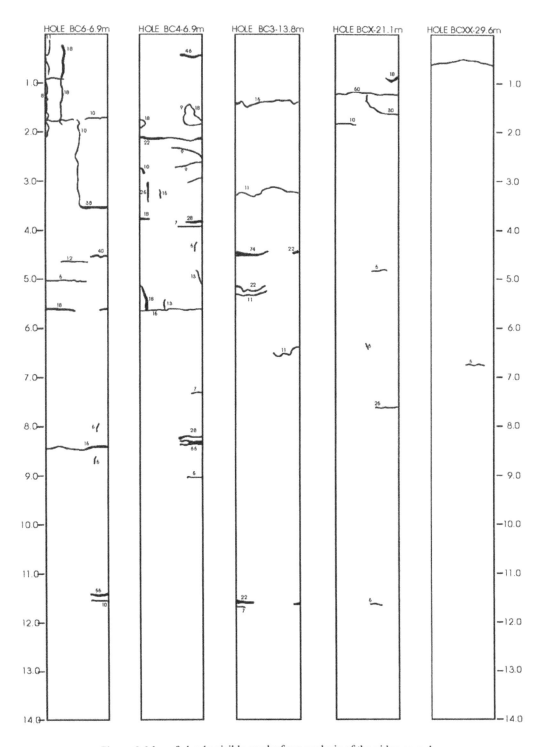

Figure 6. Map of clearly visible cracks from analysis of the video record.

108

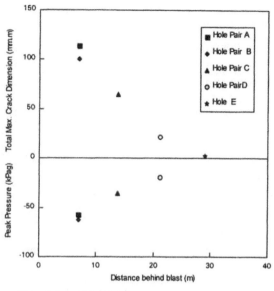

Only cracks below 1.5 m included.
Crack data for hole A only to 9.5m depth.

Figure 7. Plot of peak pressures and crack dimensions versus distance.

The vertical axis shows depth in metres. It can be seen that cracking is more pronounced nearer to the surface and that the extent of cracking decreases with distance behind the blast.

The maximum widths and the lengths of the individual cracks were measured from the video record. These dimensions were multiplied together and summed for each hole to obtain a maximum crack area dimension. These were plotted together with the peak pressures recorded from the corresponding adjacent monitoring hole as shown in Fig. 7. Note that only cracks below 1.5m were included, as the holes were sealed at this level. Borehole video data for one hole (BC4) was lost below 9.5m.

In order to test whether it was reasonable to conclude that the opening of new cracks was responsible for the underpressures, a rapid adiabatic expansion process for the air in the boreholes was assumed. Invoking the expansion law for an adiabatic process, the following relationship is derived:

$$P = P_{atm} \left(V_{hole} / \left(V_{hole} + V_{cracks} \right) \right)^{\gamma}$$

where :

P = measured peak pressure (atm)
P_{atm} = atmospheric pressure (1 atm)
V_{hole} = monitoring hole effective volume (m^3)
γ = adiabatic expansion co-efficient = 1.4 for air at ambient conditions
V_{cracks} = maximum volume of the cracks (m^3), which may be assumed to be a function of the crack area as follows:
V_{cracks} = $k (AREA_{cracks})$ where k represents some characteristic crack length and $AREA_{cracks}$ is the maximum crack area dimension (m^2).

The plot in Fig. 8 shows the fit of this relationship with the measured data and an imposed boundary condition of 1 atmosphere and zero crack area. Although there is limited data, the correlation co-efficient of 0.98 shows that the data is not inconsistent with the assumed expansion process. Note that the characteristic length from this relationship is of the order of metres, which is a reasonable result given the dimensions of the

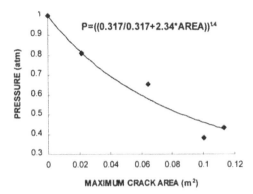

$$P = ((0.317/0.317 + 2.34 \cdot AREA))^{1.4}$$

Figure 8. Plot of pressure versus crack dimensions for the third bench blast.

monitoring station layouts. The characteristic crack length may be an indication of the effective length of the cracks which contribute to volume expansion, ie before communicating with, (and hence venting into), other cracks, boreholes or the surface.

It was thus concluded that the measured underpressures are caused by the volume increase of the air chamber due to the opening of new cracks. The very rapid opening of the cracks is reflected in the steep initial drop in pressure which is characteristically observed (Fig. 3), while the slower recovery of pressure over a typical timescale of 2-3 seconds may be attributed to air flow into the hole/crack volume from the surface or new free face. This conclusion is in general agreement with Lejuge et al and Ouchterlony, who ascribed the underpressures to ground dilation. Accepting this conclusion, the technique becomes an invaluable tool for damage assessment behind blast limits.

4 CONFINED CRATER BLASTS

The second set of experiments was done at an experimental site in the Hunter Valley. The rock at

this site comprised layers of mudstone, shale and sandstone. Previous fully confined experimental blasts had been done at this site and on several occasions explosive fumes had been observed to travel tens of metres from blastholes before emerging from the surface or the exposed highwall.

Two fully confined single hole crater blasts were monitored. The charges were just above the critical cratering depth as some cratering and surface expression was evident after the blasts. The blastholes were 129mm in diameter and each contained 80kg of ANFO. The monitoring holes were also 129mm in diameter and were located at distances corresponding approximately to 8, 16, 31 and 62 charge diameters. All holes were vertical, dry and approximately 10m deep. Figure 9 shows plan views of the monitoring hole locations.

The resultant traces from the first blast are shown in Fig. 10. The transducer in the first monitoring hole of the second blast was destroyed, however the peak pressures recorded in the other three monitoring holes were surprisingly similar to those of the first blast. It is clear that gases penetrated to all the monitoring holes in these cases. An initial underpressure is observable in some of the traces. This is again postulated to be due to new cracks opening in the borehole walls prior to the ingress of gases.

The conclusion drawn in this case was that the fully confined nature of the blast, together with the permeability of the rock mass, resulted in gas penetration to all the monitoring holes. While initial cracking may be due to stress phenomena, the gases could extend the cracking and dilate existing joints. It is observed that the extent of gas penetration as reflected by the measured peak pressures decays exponentially with distance as shown in Fig. 11, which is a plot of the peak pressures obtained from both crater blasts.

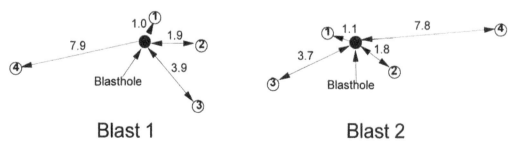

Figure 9. Monitoring hole locations for the fully confined crater blasts (distances shown in metres).

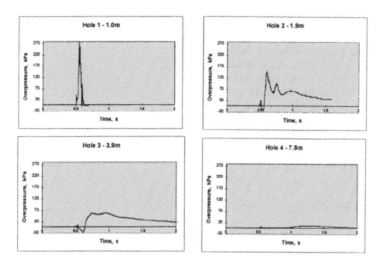

Figure 10. Traces obtained from the confined crater blast 1.

5 DISCUSSION AND CONCLUSIONS

There are several factors which may affect the pressure records obtained from this technique.
Communication between cracks which vent to the atmosphere or into nearby monitoring holes is one factor. Careful location and sufficient sealing height of the monitoring holes should minimise these effects.

It should be noted that this technique only samples events at the monitoring hole locations and will not detect gas flows which do not intersect these holes. However, given the surprising consistency in results at different blast locations, it is still reasonable to draw conclusions about the general phenomena in the region behind blast limits.

Another important point is that the invasive nature of the sampling holes modifies the rock structure, stress distributions and hence ultimate crack patterns (Hagan, 1995). This, however, is a drawback of all such sampling techniques (eg cross-hole seismics, borehole camera, etc). Nevertheless, it is still possible to draw valuable conclusions about the mechanisms and maximum extent of cracking behind blasts.

Despite these drawbacks, the technique has many important potential benefits:

1. A major use of the technique is to discriminate between mechanisms causing damage in different blast designs or rock types. In the series of bench blasts in the massive rock, damage appeared to be stress induced with no evidence of high pressure explosive gases reaching the monitoring locations. In the confined crater blasts in the layered rock, high pressure explosive gases were able to reach the monitoring locations and would be expected to cause dilation of existing cracks or joints.

2. Another important use is to determine the extent of damage behind various blast designs in order to optimise damage control blasting. Plots such as Fig. 4 and Fig. 11 may be used to reflect the extent of damage as a function of distance from the blast. A program of work examining various limits blast designs is currently in progress. Ouchterlony (1995) presents an example where this technique may be applied to determine whether presplitting reduces damage.

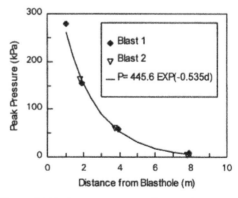

Figure 11. Peak pressure versus distance for both confined crater blasts.

3. The arrival times of the first cracks may also be estimated from the traces. Although a lag between crack opening and pressure reduction due to air outflow is to be expected, if the magnitude of this lag is reasonably constant from hole to hole, then the characteristic velocities of cracks may be found from monitoring holes at various distances. Furthermore, the arrival times of individual underpressure peaks within the pressure-time trace may be related back to blast hole or row firing times to determine the sources of damage.

6 ACKNOWLEDGEMENTS

The authors extend their appreciation to the management and personnel of Coal and Allied's Hunter Valley No. 1 Mine who provided excellent co-operation during the series of bench blasts for this work. In particular, Darren Stacey is thanked.

REFERENCES

Bulow, B.M. & J. Chapman 1994. Limit Blast Optimisation at Argyle Diamond Mine. *Proc. Open Pit Blasting Workshop '94:* 104-109. Perth: Curtin University.

Chitombo, G. & A. Scott 1990. An Approach to the Evaluation and Control of Blast Induced Damage. *Proc. 3rd Int. Symposium on Rock Fragmentation by Blasting:* 239-244. Brisbane: AusIMM.

Hagan, T.N. 1995. Personal Communication.

LeJuge, G.E., E. Lubber, D.A. Sandy & C.K. McKenzie 1994. Blast Damage Mechanisms in Open Cut Mining. *Proc.Open Pit Blasting Workshop '94:* 96 - 103. Perth: Curtin University.

Ouchterlony, F. 1995. Review of Rock Blasting and Explosives Engineering Research at SveBeFo. *Proc. Explo '95:* 133 - 146. Brisbane: AusIMM.

Raax Australia 1995. Blast Hole Borehole Image Processing System (BIPS) Study (Piercefield, Hunter Valley Mine), proprietary report prepared for Gutteridge Haskins & Davey Pty Ltd. Australia.

Sarma, K.S. 1994. Models for Assessing the Blasting Performance of Explosives. *PhD Thesis.* Brisbane: University of Queensland.

Williamson, S.R. & M.E. Armstrong 1986. The Measurement of Explosive Product Gas Penetration. *Proc. Large Open Pit Mining Conference:* 147 - 151. Melbourne:AusIMM.

Rock Fragmentation by Blasting, Mohanty (ed.)© 1996 Taylor & Francis. ISBN 90 5410 824 X

The relationship between strain energy, rock damage, fragmentation, and throw in rock blasting

Per-Anders Persson
RCEM/EMRTC, New Mexico Institute of Mining and Technology, Socorro, N.Mex., USA

ABSTRACT: Design of cast blasting rounds and optimization of the blasting process has taken on a new and much greater importance with the emergence of overburden casting operations involving thousands of tons of explosives in each blast. This paper addresses the fundamentals of blast design and optimization for such explosive casting. It shows that one key variable — the strain energy set up in the rock mass by the blast vibration — can be used to predict the extent of damage done to the remaining rock mass, the degree of fragmentation of the rock broken loose, and the velocity of the broken rock. It also shows how the dynamic strength of the rock mass, the fracture energy of its joints, and the geometry of the blast influence the fragment size distribution and the velocity of the broken rock.

The dynamic tensile and shear strength of the rock mass and its joints and weak planes may be measured in a new type of impact experiments. The paper gives examples of such strength measurements in one type of sandstone material. The strain energy distribution within the entire rock mass, including the rock being broken loose, can be predicted on the basis of a few accelerometer measurements, which provide calibration of the complicated interaction between the specific rock material and explosive used at a specific location.

A statistical approach is proposed for determining the local fragment size distributions and fragment throw trajectories from the strain energy field. The techniques outlined in this paper could form a framework for a statistical calculation — based on measured data — and an optimization of the entire rock blasting process with respect to rock damage, fragmentation and throw velocity.

1 INTRODUCTION

In modern cast blasting operations, such as those near Gillette, Wyoming and Grants, New Mexico in the United States and in the Hunter Valley, New South Wales in Australia, very large quantities of overburden material are set in motion in each blast. The volumes of overburden are approaching 25 million metric tons in a single blast using nearly five thousand metric tons of explosives. Explosive casting of the overburden reduces overall cost and increases productivity since it reduces the cost and time for overburden removal by the more costly and slow dragline. With blasting on this very large scale, even a small percentage increase in the distance of throw or improvement in fragmentation by explosive can be translated into large economic benefits.

The detonation of an explosive charge in a drillhole in rock exerts an initially very high pressure on the drillhole walls which sets the rock material in radial motion outwards. As the reaction product gases expand, the pressure on the hole walls decreases. The momentum being imparted to the rock by the gas pressure is transferred in the form of a radial wave motion to a larger and larger mass of rock material at lower and lower velocity. In the immediate vicinity of the drillhole, and especially in soft rock materials, large scale shear deformation under mainly compressive stresses occurs. In brittle rock, this leads to the formation of some fine crushed material, often called "fines" in mining. Further out from the hole, the peak stresses in the wave motion are compressive, and since most rock materials are strong in compression, no damage occurs until the wave motion begins to interact with nearby free surfaces, at which stage tensile stresses are created. These tensile stresses cause fragmentation, mainly by tensile fracturing of pre-existing planes of weakness, cracks, or joints filled with a weaker material. Commonly, three free surfaces contribute to generating tensile stress, namely the free surface from which the hole is drilled, the free surface parallel with the drillhole which determines the burden, and the new free surface created by the separation of the burden from the remaining rock. Even for a single hole, the wave motion of the rock mass and the resulting tensile fracturing is an extremely complex process, which can only be modeled numerically under very simplifying assumptions regarding the material's wave propagation and strength.

However, the wave motion of the rock mass can quite readily be measured, even within that part of the rock mass that will be removed and fragmented by the blast. Starting from such measurements, Holmberg and Persson (1978) derived a simple approximate method for calculating the peak particle velocity in the wave motion at any point around an extended charge. The calculated peak particle velocities within the rock left standing after the blast were found to be in good agreement with actual peak particle velocities derived from accelerometer measurements. Critical values of the peak particle velocity for incipient rock joint damage were determined from damage measurements. These measurements were made in hard Scandinavian bedrock. Subsequently, the Holmberg-Persson technique has been used extensively to predict and control damage to the remaining rock in other rock materials. The parameters in the Holmberg-Persson equations are then calibrated against actual measured peak particle velocities for new combinations of rock mass and explosive.

The present work extends the applicability of the Holmberg-Persson predictive technique to include predictions of fragmentation and cast velocity. It is shown how measured or calculated peak particle vibration velocities can be translated into strain energy values. (Strain energy, the product of strain and stress, is a more natural criterion for rock damage and fragmentation than peak particle vibration velocity, though for simple vibrations, one can be derived from the other). It is shown that not only the damage to the remaining rock, but also the degree of fragmentation and the throw velocity of the fragmented rock can be related to

the strain energy. Techniques for measuring the critical strain energy and fracture energy required for different levels of rock damage and fragmentation are proposed and tested in preliminary experiments on sandstone.

2 THEORY

The theoretical basis for the proposed techniques is outlined below. First, the fundamental properties of stress waves in rock are summarized. Second, the effects of the stress waves (in terms of rock damage, fragmentation, and throw) are related to measurable amplitude values of the wave motion.

2.1 Stress waves in rock

Consider the static extension of a test cylinder (drill core) of rock under an axial tensile stress σ. In tension, the cylinder will become longer as a result of the axial strain ϵ. (The test cylinder will also become thinner by an amount determined by the Poisson's ratio ν). The axial stress is related to the axial strain through Hooke's law

$$\epsilon = \frac{\sigma}{E} \qquad (1)$$

where E is the elastic modulus (Young's modulus) of the material. Equation 1, with σ replaced by $-p$, also expresses the compression of a test cylinder under a compressive stress p

$$\epsilon = -\frac{p}{E} \qquad (2)$$

In this case, the cylinder becomes shorter and slightly thicker.

If the compression or extension of the rod is by an elastic axial wave motion, the same equations (1) and (2) apply, relating the dynamic stress and strain to each other. (This is strictly true only if the wave length is sufficiently long for lateral expansion of the test sample to take place within the rise time of the stress). The lateral material motion to contract or expand the cylinder takes place by a reverberating shear wave which follows the front of the axial wave. The lateral vibratory motion

114

gradually abates and the stress system becomes essentially uniform at a point several cylinder diameters behind the axial wave front.

The peak axial strain ϵ in the stress wave in the rod is related to the peak particle velocity u_p and the longitudinal elastic wave velocity C through an equation similar to equation (1)

$$\epsilon = \frac{u_p}{C} \qquad (3)$$

C is a simple function of Young's modulus and the initial density ρ_o of the rock

$$C = \sqrt{\frac{E}{\rho_o}} \qquad (4)$$

Waves in other geometries than the straight rod propagate with velocities slightly different from those in the rod. A straight wave propagating in a thin plate, in a direction parallel to the plate's surfaces, for example, has the slightly higher velocity C_l, given by the equation

$$C_l = C \frac{\sqrt{1-2\nu}}{(1-\nu)} \qquad (5)$$

where ν is Poisson's ratio. Plane elastic waves in an infinite medium propagate with the slightly higher velocity C_p (see Persson, Holmberg, and Lee, Ref. (1994)),

$$C_p = C \sqrt{\frac{(1-\nu)}{(1-2\nu)(1+\nu)}} . \qquad (6)$$

A shear wave propagates with a considerably lower velocity C_S, given by the expression

$$C_s = \sqrt{\frac{G}{\rho_0}} = \sqrt{\frac{E}{2\rho_0(1+\nu)}} \qquad (7)$$

where G is the shear modulus. Finally, a Rayleigh or surface wave propagates with the velocity C_R, given by the approximate expression

$$C_R = C_s \frac{0.86 + 1.14\nu}{1+\nu} . \qquad (8)$$

For granite with Poisson's ratio $\nu = 0.21$, the ratios are $C_l/C = 0.964$, $C_p/C = 1.04$, $C_s/C = 0.796$, and $C_R/C = 0.723$.

Equation (3), derived for the peak strain associated with the longitudinal elastic compression wave with particle velocity u_p in a long rod, is approximately valid for the straight elastic wave in a plate and and is exact for a plane elastic wave in an infinite medium. If the axial wave (sound) velocity C in Equation 3 is replaced by the longitudinal wave velocity in the plate C_l, the plane compressive wave velocity in the infinite medium C_p, the shear wave velocity C_S, or the Rayleigh wave velocity C_R, we obtain values for the strain with varying degrees of approximation. In the general case of a three-dimensional elastic wave with a curved front, Equations (1) through (3) are only valid as approximations of the real, much more complicated relationships; however, for the purposes of the present work, the error in the predicted strain introduced by using C from Equation (4) is likely to be less than 10% of the predicted values.

The strain energy associated with a stress wave is the internal energy contained in the material in the form of elastic deformation energy. It is the product of stress and strain. It is also equal to the kinetic energy of the wave, and therefore can be written as

$$e_s = \frac{1}{2} \frac{\epsilon\sigma}{\rho_o} = \frac{1}{2} u_p^{\,2} \qquad (9)$$

where e_s is the strain energy per unit mass of rock. Combining Equation (9) with Equation (1), we arrive at the relationship

$$e_s = \frac{1}{2} \frac{\sigma^2}{\rho_0 E} \qquad (10)$$

which gives e_s per unit mass, for example in J/kg, the unit frequently used in rock crushing fragmentation studies. Equation (10) gives the stress σ as a function of initial density, Young's

Table 1. Damage and fragmentation effects in hard Scandinavian bedrock density $\rho_o = 2600$ kg/m^3, sound velocity $C_l = 4900$ m/s, Young's modulus E = 60,000 MPa.

Peak particle velocity (m/s)	Tensile stress (MPa)	Strain energy (J/kg)	Typical effect in hard Scandinavian bedrock
0.7	14	0.65	Incipient swelling
1	20	1.33	Incipient damage
2.5	50	8.3	Fragmentation
5	100	33	Good fragmentation
15	300	300	Crushing

modulus, and strain energy, and can be written in the form

$$\sigma = \sqrt{2\,\rho_0\,E\,e_s} \; . \tag{11}$$

The stress can also be written as

$$\sigma = \rho_o\,u_p\,C \tag{12}$$

in terms of the (peak) particle velocity u_p and the longitudinal wave velocity C.

2.2 Rock damage and fragmentation

Table 1 shows values of peak particle velocity, stress, and strain energy in blasting, using the above approximate relationships, together with a best estimate of the corresponding levels of rock damage and fragmentation from experience with blasting Scandinavian bedrock.

The values in Table 1 are only intended to serve as indications of the rough order of magnitude of particle velocity, tensile stress, and strain energy representative for the corresponding effects, and should not be used for any other purpose. However, the lowest levels of particle velocity, 0.7-1 m/s, are well established as indicating the beginnings of damage to the remaining rock after blasting in hard rock. Also, the highest value of strain energy, 300 J/kg, is in the range of strain energies known to be required for crushing of hard rock in a primary crusher.

2.3 A Statistical approach to fragmentation

Since most of the rock fragments in a blast are formed by tensile fracturing of existing weak planes or joints, the fragment size distribution obtained for a given applied strain energy depends on the statistical distributions of joint strength and joint spacing. Both of these can conveniently be represented by Weibull distributions. In most rock masses, there are a few, say three main sets of joints, with different mean directions. The joints of each set have a distribution of directions about its mean. There are thus 9 statistical distributions that determine the fragmentation. Each of these can be determined by two parameters in the Weibull distribution. The complete solution thus contains 18 variables. In a general way, however, we may assume that the joint spacing distribution with the smallest mean value will determine the 50% point on the 0 fragment size distribution, while that with the lowest mean strength will determine the strain energy level of that distribution.

Joint sets with very large spacings and/or high strengths will have less influence on the fragmentation. We can talk about a local fragment size distribution at a point a given distance away from the blasthole; the final fragment size distribution will then be the integrated mean of these local fragment size distributions. If we can solve the statistical problem, we will have a great instrument for predicting and controlling fragmentation in real rock. It is obvious, however, that some experimental information on the dynamic strength of joints and results from careful joint mapping will be needed before we have the solution in hand.

2.4 Rock throw velocity

In order to understand the mechanism of acceleration of the rock being broken loose by the blast, let us consider a stack of ice-hockey pucks made of very soft rubber. If the stack is placed vertically on a rigid floor and compressed, each puck will acquire the same strain energy. When the stack is suddenly released, a release wave will travel down the stack with sound velocity C, accelerating the material upwards. As the release wave reaches the floor, the entire stack is at zero stress and moving upwards, without separating, with a uniform velocity u_p. The strain energy e_s stored in the compressed rubber will have been transformed into kinetic energy, so that

$$u_p = \sqrt{2 e_s} . \qquad (13)$$

We can apply Equation 10 to find the throw velocity of a volume of jointed rock set in motion by release of a uniform compressive stress. The uniform throw velocity of the jointed rock corresponding to a strain energy of 33 J/kg using Equation 13 would be 14 m/s. Table 1 indicates that a strain energy of 33 J/kg would give good fragmentation in Scandinavian bedrock. As in the case of the stack of rubber disks, there would be no tendency for the rock mass to separate at the joints if the rock stress were uniform.

In reality, as shown by calculations using the Holmberg-Persson technique, the stress and the strain energy induced in a rock mass by the detonation of a charge in a drillhole is not uniform but decreases with increasing distance from the drill hole. The velocity corresponding to the strain energy of the rock at the free surface of the bench then becomes the initial, lower limit value for the throw velocity. The throw velocity may increase by 20-30 percent by transfer of momentum from the rock closer to the drillhole, and also somewhat due to the continued effect of gas pressure in the drillhole acting on crack surfaces.

This case is not amenable to a simple analytical solution; however, a simple numerical integration would give the entire throw velocity distribution within the burden.

When there is a gradient of velocity or strain energy, of course, the material will separate wherever there is a weak plane, or wherever the strain energy released exceeds the fracture energy.

The energy consumed in creating a single fracture might be expressed in units of energy per unit new surface area created by the fracture. For practical purposes, however, it is more convenient to express the overall fracture energy e_f in units of energy per unit mass of broken rock, to make it comparable to the strain energy. The fracture energy then becomes a function of the joint spacing, as detailed in the experimental section below. The throw velocity then would be reduced according to the equation

$$u_{throw} = \sqrt{2(e_s - e_f)} . \qquad (14)$$

3 EXPERIMENTS

3.1 Rock damage from blasting

Holmberg and Persson's measurements (1979) in hard Scandinavian bedrock showed that the damage in the rock left standing behind a bench blast is mainly in the form of swelling, i.e. a slight opening of existing cracks and joints. Only a limited number of new cracks are created, except close to the drillhole. The total swelling can amount to a few percent of the vertical expansion, resulting in a total lifting of the bench top surface of perhaps 0.15 m for a 15 m high bench. Considering that there may be as much as one hundred existing cracks and joints over that bench height (corresponding to an average joint spacing of the order of 0.15 m), the increase in average crack opening is limited to perhaps 1,5 millimeters. To understand how the fracturing is related to the strain energy, we need experimental information about the stress, strain, and strain energy needed to create one fracture, and the energy consumed by this fracturing process. Whether a given strain energy will result in fragmentation of a given rock mass or not is determined by the rock material's dynamic strength and fracture properties. The emphasis is on the word dynamic, since static tensile tests tell us little about the strength and energy absorbtion of the joints under the short duration stresses of blasting. The static compressive strength can be as little as one third of the dynamic compressive strength, and very little is known about the dynamic tensile strength of any rock materials, let alone joints.

3.2 Measurement of dynamic rock strength

A simple dynamic tensile fracture experiment was designed to measure the dynamic tensile strength of an individual joint and its dynamic fracture energy. For a homogeneous rock material, the experiment can also measure the dynamic tensile strength and the dynamic fracture energy absorbtion of the rock material itself. In this experiment, a homogenous tensile stress was generated in a straight cylinder of sandstone by the impact of a short striker cylinder of the same material. Figure 1 shows a schematic picture of the overall experimental setup, Figure 2 shows the idealized stresses in the two cylinders at different times after impact.

The striker and test cylinder both were cut out of the same 100 mm diameter drillcore of sandstone from the Alamo Indian Reservation near Magdalena, NM. A plastic tube with 100 mm outer diameter was glued on to one end of the striker, which was dropped through a vertical 105 mm inner diameter PVC tube, aligned so that

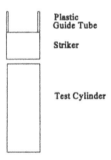

Figure 1. Schematic figure of the dropping striker impact arrangement.

the striker impacted the test cylinder accurately end-on-end.

In Figure 2, compressive stresses are shown above the zero line, tensile stresses below. At a sufficient impact velocity, a clean fracture occurs in the test cylinder at right angles to its axis (as if it had been cut by a knife) at a position which is one striker length from the end of the test cylinder (indicated by the dotted line at F). In reality, the stress wave fronts are not as sharp as shown in figure 2, partly due to

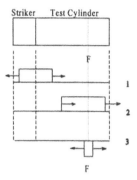

Figure 2. Idealized compressive and tensile stress waves in striker and test cylinder at three points in time (1, 2, and 3) following impact. Arrows indicate direction of wave front motion.

Table 2. Cylinder impact test results.

Drop height (m)	Impact velocity (m/s)	Result
5	9.8	Fracture
5	9.5	Partial fracture
4	8.8	No fracture

unavoidable minor misalignment between the two cylinders at impact, partly due to dispersion of shock stresses at the wave fronts. As shown in Table 2, the critical impact velocity of fracture is about 9.5 m/s. This corresponds to a stress wave particle velocity of $u_p = 4.75$ m/s. The corresponding strain energy (Equation 9) is 11.3 J/kg.

3.3 Fracture energy vs fracture toughness

The fracture energy per unit fracture surface created can be determined from these experiments in the following way:
If no energy were absorbed in creating the fracture, then the separated piece of the test cylinder, with a mass identical to that of the striker, would come off with a velocity equal to the impact velocity $u_0 = 9.5$ m/s. The kinetic energy of the separated piece would then have been

$$e_k = l \frac{\pi d^2}{4} \frac{\rho_0 u_0^2}{2} \qquad (15)$$

$$e_f = \frac{6 a w_f}{a^3 \rho_0} = \frac{6 w_f}{a^2 \rho_0} \qquad (18)$$

here l is the length of the striker, d its diameter, and ρ_0 the density of the material. In the experiment with impact velocity $u_0 = 9.5$ m/s only a partial fracture occurred. All of the kinetic energy in that experiment was apparently consumed within the sample. However, in the process of fracturing, only half of the impactor's kinetic energy can be used for deformation work at the fracture surfaces. The other half is transmitted back into the two separating parts in the form of elastic tensile waves. If we define the fracture energy per unit of new surface area as w_f J/m², we find

$$2 w_f \frac{\pi d^2}{4} = \frac{l}{2} \frac{\pi d^2}{4} \frac{\rho_0 u_0^2}{2} \qquad (16)$$

i.e.

$$w_f = \frac{l \rho_0 u_p^2}{8} = \frac{1}{2} l \rho_0 u_p^2 \quad . \qquad (17)$$

With the values $l = 0.075$ m, $\rho_0 = 2600$ kg/m³, and $u_p = 4.75$ m/s, the fracture energy becomes $w_f = 1974$ J/m². This value can be compared to literature values of the fracture toughness G_{IC}, which is the energy needed for slowly extending a crack through a material. Values of G_{IC} range from 1.2 J/m² for silicon glass to 10,000 J/m² for steel. For rock materials, values of G_{IC} range from 95 to 150 J/m² for different types of granite to 300 J/m² for sandstone (Reference 3). It is not surprising that the dynamic fracture energy for the sandstone we tested, 1974 J/m², is greater than the fracture toughness of sandstone, 300 J/m². It is, however, surprising that our value for the dynamic fracture energy is as much as 6.5 times greater than the fracture toughness.

For blast fragmentation, it is more convenient to know the fracture energy per unit rock mass. Assuming fragments of cubic shape with edge length a, we find

The fracture energy e_f is thus inversely proportional to a^2. Translated into rock blasting producing, say, uniform cubic fragments with an edge length of $a = 1$ m, the fracture energy per unit fragment mass is $e_f = 6*1947/2600 = 4.56$ J/kg. For a fragment size of 0.5 m edge length, the surface area per unit mass is doubled, and for a fragment size of 0.25 m it is quadrupled, i.e. the fracture energies are 9.11 and 18.22 J/kg, respectively. These values for sandstone (probably larger than those for granite), are of the same order of magnitude as the strain energies for fragmentation of Scandinavian bedrock according to Table 1, $e_s = 8.3$ J/kg for fragmentation, and $e_s = 33$ J/kg for good fragmentation.

3.4 Proposed field experiments

In the above sample calculation, we have compared experimental strain energies for fracture and fracture energies for homogeneous Alamo sandstone with predicted strain energy values for Scandinavian bedrock. To provide a better comparison, and to allow realistic predictions of fragmentation and throw velocities, we need to have the strain energy prediction and the rock strength measurement made with the same material. Preferably, to make use of the results where they can have maximum economic impact, these measurements should be made in the overburden material of an existing large cast blasting operation.

To provide a maximum of useful data, accelerometers should be placed in separate drillholes, with a diameter of, say 75-100 mm (3-4 in), situated between 3 and 15 m from the nearest drillhole, both in the burden forward of the hole and in the rock intended to be left standing behind the blast. Some measurements should be provided for a single blasthole of 250-375 mm (10-15 in) diameter, some for a hole near the back of the bench in a large multi-hole blast. The latter experiment would also provide information on the vibrations and rock motion associated with the explosions in the rest of the large blast. To recover the valuable accelerometers placed in the rock which will be fragmented and thrown, these accelerometers can be fastened securely inside schedule 40 steel tubes, several meters long, with markings on the outside to indicate where the accelerometers are. Because of their length, the tubes will be found in the rubble, though bent, and can be cut to allow the accelerometers to be recovered for re-use.

4 CONCLUSIONS, PRACTICAL SIGNIFICANCE OF THE RESULTS

The techniques outlined in this paper will allow prediction of three major effects of the blast, namely the damage to the remaining rock, the fragmentation of the rock removed, and the velocity of the cast, as functions of a single variable, the strain energy of the blast vibration. The strain energy and its distribution with distance from the blasthole can be predicted with good accuracy, using simple calibration experiments to take into account the properties of the specific rock material being blasted and the explosive used. The techniques therefore will provide a consistent framework against which to evaluate and optimize changes to the blast pattern and loading of explosive in cast blasting.

5 ACKNOWLEDGMENTS

The author is grateful to Professor Michael Hood, Director of the University of Queensland Cooperative Research Centre for Mining Techniques and Equipment (CMTE), Pinjarra Hills, Queensland, Australia and to New Mexico Institute of Mining and Technology, Socorro, NM, USA. CMTE and New Mexico Tech provided economic support during three months of sabbatical leave which the author spent at the University of Queensland's Julius Kruttschnitt Minerals Research Centre (JKMRC), Indooroopilly, Queensland. I am also grateful for technical discussions with Dr. Nenad Djordjevich, Dr. Andrew Scott, Dr. Gideon Chitombo, and Mr. David La Rosa. Dr. Don Mckee, Director of the JKMRC provided a scientifically stimulating climate and laboratory support for this research and took a personal active interest in the work, for which I am very grateful. Graduate student Petr Willem performed the impact experiments and prepared the drill core samples from the Alamo sandstone.

REFERENCES

Holmberg, R., and Persson, P.-A., "The Swedish Approach to Contour Blasting", Proceedings, 4th Conf. On Explosives and Blasting Technique, Soc. of Expl. Engineers, New Orleans, LA., February 1-3, 1978.

Holmberg, R., and Persson, P.-A., "Design of Tunnel Perimeter Blasthole Patterns to Prevent Rock Damage", Proceedings, Tunneling '79, Ed., Jones, M.J., Institution of Mining and Metallurgy, London, UK, March 12-16, 1979.

Persson, P.-A., Holmberg, R., and Lee, J., *Rock Blasting and Explosives Engineering*, CRC Press, Inc., Boca Raton, FL, 1994, p. 342.

Rock Fragmentation by Blasting, Mohanty (ed.)© 1996 Taylor & Francis. ISBN 90 5410 824 X

On the damage zone surrounding a single blasthole

Dane Blair & Alan Minchinton
ICI Explosives, Kurri Kurri, N.S.W., Australia

ABSTRACT: A critical analysis is given of vibration damage models based upon simple charge weight scaling laws that have been modified in an attempt to account for the near-field influence of charge length. It is shown that such simple models are significantly flawed since they do not correctly account for the charge length or realistically account for the known blast wave radiation pattern. For example, these models fail to predict a dominant vibration component which is a vertically polarised shear (SV) wave. An analytical vibration model is developed to account for the charge length and radiation pattern as well as a finite velocity of detonation. The results of the analytical model compare favourably with a Dynamic Finite Element Model (DFEM). Both models clearly show that it is invalid to assume any simple relationship between vibration and strain for waves radiating from a blasthole, even in the very far field. It is also shown that traditional Seed Waveform modelling employing vibration waveforms measured from short (elemental) charges cannot yield a realistic vibration output for the entire blasthole; an alternative method is suggested.

1 INTRODUCTION

A simple model for the assessment of damage surrounding a single blasthole has been recently reported by McKenzie et al. (1995). Although this model does not consider the influences of local geology or blasthole gases, there are at least five other assumptions implied by the model that warrant further investigation,

i) the assumption that the radiating blast wave obeys simple charge weight scaling laws,

ii) the assumption that the peak particle velocity due to each element of charge within the blasthole is numerically additive,

iii) the assumption of an infinite velocity of detonation (VOD) for the explosive column,

iv) the assumption that the complicated influence of the free surface(s) can be neglected,

v) the assumption that the particle velocity is proportional to dynamic strain.

In this study the term "vibration" is used to specifically describe the particle velocity. The particle displacement and acceleration are also included under the general definition, but are not considered here. Charge weight scaling laws are traditionally used to estimate surface vibration levels due to production blasts. However, it should be noted that such laws are not mechanistic insofar as they do not provide an insight into any blast

mechanisms (such as the separate attributes of p- and s-waves), and they do not predict vibration variations with time. These scaling laws only provide a simple engineering solution to the problem of estimating peak surface vibration levels (rather than underground damage levels) measured as a function of charge weight and distance, alone. In this regard, Blair (1990) highlights several problems associated with these laws which indicate that it might be ambitious to expect that they could also provide the basis of any blast damage model.

The model described by McKenzie et al. (1995) is founded on the well-known model of Holmberg and Persson (1979) who altered the standard charge weight scaling law to include the influence of the length of charge within the blasthole. It should be appreciated that this alteration is somewhat akin to adding a mechanistic component to a non-mechanistic model.

Figure 1 illustrates the relevant geometry; two blasthole elements are shown at A and B as well as a general element of length *dh*. According to charge weight scaling, the peak vibration, *v*, measured at a point P in the rock mass due to the element of charge *dh* is given by:

$$v = K w_e{}^\alpha R^{-\beta} \qquad (1.1)$$

where w_e is the charge weight of the element *dh*;

K, α and β are constants and R is the distance between P and the element dh. It must be appreciated that equation (1.1) implies a uniform vibration, decaying as $R^{-\beta}$, in all directions at any distance, R, from a fixed element of charge within the blasthole. Thus under the assumption of charge weight scaling, each element radiates a spherical vibration wave into the surrounding medium. According to this model, the total vibration from the blasthole, of length L, is given by a series of such spherically-radiating elements stacked along this length. However, the work of Baird et al. (1992) clearly showed that it is invalid to use spherically-radiating elements to simulate a long cylindrical column of explosive. This point is discussed later in some detail.

The Holmberg-Persson model further assumes that for any blasthole of length L the vibration peaks (such as v_1 and v_2) due to all elements (such as A and B) may be *numerically* added at the point P to yield the total peak vibration, v_T. Under this assumption they derive:

$$v_T = K\left(m\int_0^L \frac{dh}{\left(R_0^2 + (R_0\tan\phi_2 - h)^2\right)^{\beta/2\alpha}} \right)^{\alpha} \quad (1.2)$$

where m is the charge mass per unit length of the explosive column; other relevant variables are illustrated in Figure 1. However, in the near-field of the charge length the vibration peaks (such as v_1 and v_2) due to the elements cannot be numerically added. In this case the complete vibration time history for each element must firstly be resolved into horizontal and vertical components and the separate components added for all elements to yield total components. The vector peak, v_T, is then obtained using the total vertical and horizontal components. Clearly, since simple charge weight scaling laws do not incorporate any time dependence, they are not capable of

providing the correct near-field analysis.

In fact a direct consequence of the assumption that the peaks can be numerically added is that the point P in Figure 1 must lie in the far-field of the entire charge length for which ϕ_1 is approximately equal to ϕ_2. Under this condition, the variation of h with ϕ is negligible, so that $\left(R_0^2 + (R_0\tan\phi_2 - h)^2\right)^{\beta/2\alpha}$ simply reduces to $R^{\beta/\alpha}$ which is constant for any point P, irrespective of L. The integral of equation (1.2) then collapses to the scaling law of equation (1.1) with w_e replaced by the total charge weight W of the entire column. Thus nothing has been achieved. This unfortunate state of affairs is a direct consequence of attempting to use a simple, time-independent and non-mechanistic model to describe near-field vibration.

Blair and Jiang (1995) used a Dynamic Finite Element Model (DFEM) of a single blasthole to study the surface vibrations produced as a function of the blasthole explosive VOD. Although their model was restricted to a blasthole of 5 m length, it showed that assuming an infinite VOD and neglecting the free surface (as also implied by equation 1.2) can overestimate the true vibration by a factor of approximately five for distances greater than 2 m, and can underestimate the true vibration level by a factor of approximately two for distances less than 1.5 m. This finding, alone, suggests that any model which neglects the influences of VOD and the free surface may produce dubious estimates of blast damage.

McKenzie et al. (1995) also fully appreciated that assuming an infinite VOD could possibly produce an overestimation of the true vibration, and thus utilised the so-called Seed Waveform Model to account for a finite VOD. The Seed Waveform approach, which has enjoyed popularity for some time, firstly involves the selection of a specific Seed Waveform to represent a small element of charge within a blasthole. In this regard the technique is more powerful than charge weight scaling since it does, at least, introduce time dependent seed waveforms which are often obtained experimentally by monitoring the vibrations due to a small length of charge. The VOD influence is then modelled by adding an appropriate phase delay to a succession of such waveforms in order to simulate the total vibration due a series of stacked elements of total length, L. However, almost all Seed Waveform Models, by implication, incorrectly assume a *spherical* radiation pattern for all wave types emanating from each element, since the element vibration seed at any point is simply scaled by $R^{-\beta}$. It thus appears obvious that unless the radiation pattern of each element matches the known radiation pattern of a short *cylindrical* charge (Heelan, 1953, Abo-

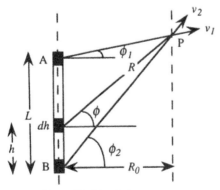

Fig. 1: The blasthole geometry.

Zena, 1977) then any Seed Waveform Model must be viewed with caution. The cylindrical charge pattern is also dependent upon the p- and s-wave velocities, (hence Poisson's ratio, ν) and the material density, ρ.

The generalised Hooke's Law in elementary elasticity theory shows that it is quite misleading to make the blanket statement that stress is proportional to strain for an elastic material. There are many components of stress and strain, so care must be exercised in specifying relevant components. This generalised law is given by equation (1.3), where σ_{ij} are the stress components and ε_{ij} are the strain components; λ and μ are the Lamé elastic constants and the volumetric strain $\Delta = \varepsilon_{xx} + \varepsilon_{yy} + \varepsilon_{zz}$, with reference to the x,y,z directions, and δ_{ij} is the Kronecker delta defined by $\delta_{ij} = 1$ if $i=j$, and $\delta_{ij} = 0$ otherwise (see, for example, Timoshenko and Goodier, 1971).

$$\sigma_{ij} = \lambda \Delta \delta_{ij} + 2\mu \varepsilon_{ij} \qquad (1.3)$$

For general wave propagation, the normal stress with respect to the x-direction is given by equation (1.3) as $\sigma_{xx} = \lambda \Delta + 2\mu \varepsilon_{xx}$. Thus the normal stress, σ_{xx}, is not simply proportional to the strain ε_{xx} since it also depends upon the strain components ε_{yy} and ε_{zz}.

However, if we consider planar p-wave propagation along the x-axis in an unbounded medium, then all the strain components vanish except for ε_{xx}; then equation (1.3) yields:

$$\sigma_{xx} = (\lambda + 2\mu)\,\varepsilon_{xx} = \rho v_p^2\,\varepsilon_{xx} \qquad (1.4)$$

where $v_p\;(= [(\lambda+2\mu)/\rho]^{1/2})$ is the p-wave velocity. Since λ and μ are both constants, then it is correct, in this instance, to state that stress is proportional to strain with respect to the wave propagation direction.

Let the planar p-wave displacement, $u_x(t)$, as a function of time, t, in the x-direction be given by:

$$u_x(t) = f(t - x/v_p) \qquad (1.5)$$

where f is a general function. If $v_x(t)$ represents the particle velocity, then:

$$v_x(t) = du_x(t)/dt = f'(t - x/v_p) \qquad (1.6)$$

also:

$$\varepsilon_{xx} = du_x(t)/dx = (-1/v_p)f'(t - x/v_p) \qquad (1.7)$$

Equations (1.4), (1.6) and (1.7) yield:

$$\sigma_{xx} = \rho v_p^2\,\varepsilon_{xx} = -\rho v_p v_x(t) \qquad (1.8)$$

Thus the particle velocity, $v_x(t)$, is proportional to dynamic strain, ε_{xx}, for planar p-wave propagation in an unbounded medium simply because the condition of equation (1.4) holds. Although it has become popular to assume otherwise, this proportionality is the exception, rather than the rule. For example, this proportionality does not hold for the propagation of the non-planar p-and s-waves radiating from a cylindrical blasthole, and it does not hold in the vicinity of any free faces. Unfortunately, since blast damage is basically concerned with radiation from cylindrical blastholes and free faces, the assumption of any simple proportionality between particle velocity and strain is dubious.

The vibrations induced at the base of a tall, thin structure, such as a smelter stack, provide a simple example of a situation when dynamic strain is clearly not proportional to vibration (i.e. particle velocity). If this structure is excited in its fundamental vibration mode, then the peak vibration is always at the top of the structure, yet in this region all components of dynamic strain are very small. In fact the peak dynamic strain (and possibly the region of potential damage onset) is close to the base of the structure. In this case, geometry (ie. the free sides of the structure) is one obvious reason that vibration is not proportional to strain. In a like manner, the geometry (free surface) of a half-space also destroys any proportionality between vibration and strain. For example, in the half-space problem, the surface vertical (z) vibration, $v_z(t)$, due to a normally incident plane wave is exactly twice the vertical vibration at depth below the surface. Yet in this case the surface normal strain, $\varepsilon_{zz}(t)$ (as opposed to the shear strains ε_{xz} or ε_{yz}) in the vertical direction is zero. Thus, clearly, the vibration $v_z(t)$ is not proportional to the strain $\varepsilon_{zz}(t)$.

The main aim of the present work is to use analytical solutions and DFEM for a single, axisymmetric blasthole in an infinite medium as well as DFEM in a medium with a nearby free surface to predict vibration and strain contours surrounding the blasthole. DFEM solves the elasticity problem in a rigorous manner, automatically handling all free surfaces and making no simplifying assumption between strain and vibration. These results will then be compared with the predictions based upon the model of Holmberg and Persson (1979) in order to assess the magnitude of any discrepancies. The starting point of this investigation is a comparison of the analytical and DFEM solutions for an element of charge in a blasthole. The solutions are then expanded to account for a finite VOD of the entire explosive column; finally, the DFEM is used to investigate the influence of a nearby free surface.

123

2 THE VIBRATION PRODUCED BY A SINGLE ELEMENT OF CHARGE

Heelan (1953) gave a theory for the radiation from a single element within a long, otherwise empty, blasthole for distances sufficiently removed from the source. Although both Jordan (1962) and Abo-Zena (1977) have criticised the mathematical detail of Heelan's solution, White (1983) noted that when Abo-Zena re-worked the mathematics then, in the far-field of the source, he produced a solution identical to that of Heelan. Furthermore White (1983) was able to reproduce Heelan's far-field solution by using a completely different approach based upon seismic reciprocity.

One of us (DB) has recently conducted a detailed comparison of the Heelan and Abo-Zena analytical solutions and finds that if the applied borehole pressure function is of the form $P = P_0 t^n e^{-bt}$ (P_0, n, b constants), then both far-field solutions agree if the length of the elemental charge approaches zero. In all the following models b was set at 10,000 with the time waveform having a sample interval of 25 μs.

Figure 2 (next page) shows the Abo-Zena predictions for the contours of vector peak particle velocity (vppv) surrounding a cylindrical element of charge of diameter 0.2 m and length 0.2 m. In the present study only pressure functions with n=6 are considered for both theories. All contour levels in this study are given for the range 0.0 to 2.0 in intervals of 0.2, and normalised to unity at the point 5 m horizontally out from the centre of the charge length. The material is assumed to have a p-wave velocity, v_p, of 5500 m/s, a Poisson's ratio, v, of 0.25 and a density ρ, of 2500 kg/m³. Unless stated otherwise, all boundaries extend to infinity; i.e. there are no reflections.

In this case of a 0.2 m charge length, there was insignificant difference between the predictions of the Heelan and Abo-Zena models. Thus only the Abo-Zena analytical results are shown in this study. The dominant vibration lobes at an angle of 45 degrees to the blasthole are due to the radiation of the vertically polarised shear (SV) wave, and the minor lobe along the horizontal axis is due to the p-wave. It is pertinent to note that neither the Holmberg-Persson nor traditional Seed Waveform Models predict this dominant SV component. Spherical wave radiation is implied by these simple models, which means that the contour plot of peak vibration would simply be represented as concentric circles centred on the charge element.

The dynamic strains were then calculated from the Abo-Zena displacements by evaluating the appropriate displacement derivatives in the cylindrical coordinate system (r, θ, z); in the present case of axisymmetry ε_{rr}, ε_{zz}, ε_{rz} and $\varepsilon_{\theta\theta}$ are the only non-zero components of strain.

In the present work, we have chosen to use two particular strain parameters which do not depend upon the rotation angle at any particular point in the rock mass. These chosen parameters are the absolute value of volumetric strain, Δ, as previously defined, and the maximum shear strain, τ_m, defined as

$$\tau_m = \{ \varepsilon_{rz}^2 + (\varepsilon_{rr} - \varepsilon_{zz})^2 \}^{1/2} \qquad (2.1)$$

The maximum shear strain is chosen because it is often associated with the strength of materials (Timoshenko and Goodier, 1971). It should be appreciated that one of the main aims of this work is to highlight the relationship between strain and vibration in an elastic material; it is beyond the present scope to investigate parameters (such as the tensile failure stress) that might be more relevant to rock damage.

Figure 3 shows contours of the peak volumetric strain, and Figure 4 shows contours of the peak value of the maximum shear strain.

A comparison of Figure 2 with Figures 3 and 4 clearly shows that the peak vibration is not proportional to either the peak volumetric strain or the peak maximum shear strain; this finding is not unexpected since the charge element radiates non-planar waves into the surrounding medium.

Figure 5 shows a DFEM solution for the vppv surrounding the element of charge within the long, otherwise empty, blasthole; Figure 6 shows the DFEM solution for the peak volumetric strain and Figure 7 shows the DFEM solution for the peak maximum shear strain; all four boundaries are non-reflecting.

There is a significant discrepancy between DFEM and the analytical models for all regions along the length of the blasthole lying within approximately ten times the blasthole radius. This is as expected since the analytical models are only far-field solutions, valid for distances given by $R_0/a > 10$ and $R/dh > 10$; where R_0 is the horizontal distance from the axis of symmetry, a is the charge radius, and dh the element length (Figure 1). However, it must be emphasised that this condition is not too restrictive for a single element of charge having $dh = 0.2$ m, since in the present case the analytical models only fail within 2 m or so from the charge column. The large lobe in the DFEM solution directly above the blasthole is due to the wave travelling along the borehole. However, due to the far-field approximations implied by the analytical models, this wave is not correctly modelled in either the Heelan or Abo-Zena solutions.

However, it is quite obvious that all three

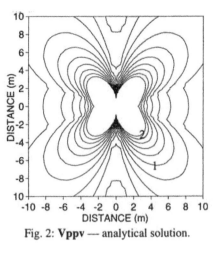

Fig. 2: **Vppv** — analytical solution.

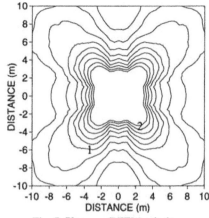

Fig. 5: **Vppv** — DFEM solution.

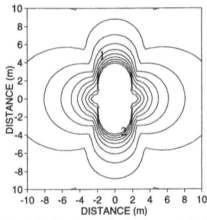

Fig. 3: **Peak volumetric strain**— analytical.

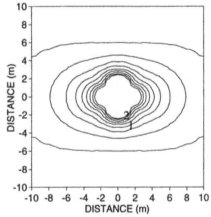

Fig. 6: **Peak volumetric strain** — DFEM.

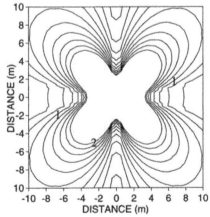

Fig. 4: **Peak maximum shear strain**— analytical.

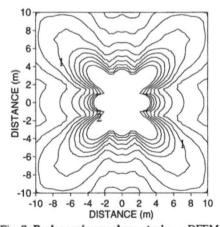

Fig. 7: **Peak maximum shear strain** — DFEM.

These figures show the specified contours surrounding a squat cylindrical element of charge

125

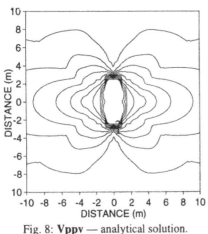

Fig. 8: **Vppv** — analytical solution.

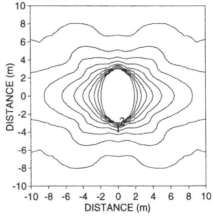

Fig. 11: **Vppv** — DFEM solution.

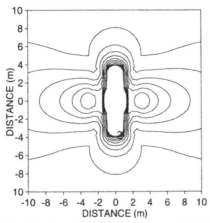

Fig. 9: **Peak volumetric strain** — analytical.

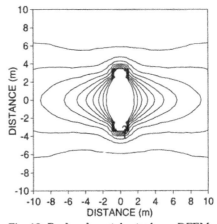

Fig. 12: **Peak volumetric strain** — DFEM.

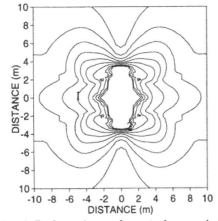

Fig. 10: **Peak maximum shear strain** — analytical.

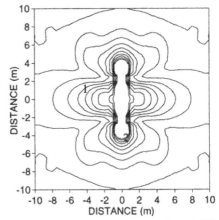

Fig. 13: **Peak maximum shear strain** — DFEM.

These figures show the specified contours surrounding a 5 m charge column having infinite VOD.

126

models (Heelan, Abo-Zena and DFEM) clearly show that the radiation pattern maintains its non-spherical characteristic for all distances into the far-field. Hence all these models clearly predict that waves radiating from a cylindrical blasthole will not become planar, even in the very-far field; they will always maintain their non-planar form. This characteristic of cylindrical charges was pointed out by Jordan (1960); it is also interesting to note that Nicholls et al. (1971) arrived at the same conclusion from an analysis of experimental data.

3 THE VIBRATION PRODUCED BY A COLUMN OF CHARGE

3.1 *The case for infinite VOD*

In order to model an extended charge a series of Abo-Zena elements were stacked vertically with the angles such as ϕ_1 and ϕ_2 (Figure 1) calculated for each element. It was found that the solution for a 5 m charge length gave acceptable convergence for as little as 11 elements; however 41 elements were used in the present study. The total horizontal and vertical waveforms components of velocity and appropriate strain were then summed for all elements, and the vppv, volumetric and maximum shear strains calculated for the total waveforms. This technique produces a solution that is only far-field in R_0/a, and not R/L, where L is the total length of the charge. This situation is different to that of the Holmberg-Persson model, which is far-field in R/L rather than R_0/a

Figure 8 shows the Abo-Zena predictions of the vppv surrounding a 5 m charge column, Figure 9 shows the peak volumetric strain and Figure 10 shows the peak maximum shear strain. As before,

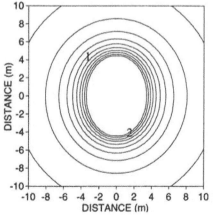

Fig. 14: Contours of **vppv** surrounding a 5 m charge column — Holmberg-Persson solution.

all boundaries are non-reflecting.

Figures 11 to 13 show the equivalent DFEM results for the 5 m charge column, and Figure 14 shows the vppv predictions for this charge length based upon the Holmberg-Persson model of equation (2.1). A direct comparison between Figures 8, 11 and 14 clearly shows that the Holmberg-Persson model does not predict the realistic distribution of vibration due to a column of charge, and the resulting patterns are too simplistic in both the near-field and the far-field. The Holmberg-Persson contours become circular in the far-field of the column because in this regime they are given by the simple charge weight scaling law. Unfortunately, in the near-field, the contours have no physical significance since they are based upon a flawed model that is clearly non-physical in this regime. Thus the Holmberg-Persson model does not even supply a first order approximation to either the near-field or far-field radiation from a blasthole

3.2 *The case for finite VOD*

Abo-Zena (1977) gave a treatment accounting for a finite VOD across the element; however his treatment did not account for variation in the angles such as ϕ_1 and ϕ_2 (Figure 1) and thus the stacked elements are still required. In this case each element is time delayed by dh/v_0, where v_0 is the VOD. However, despite the discrete summing of 41 elements, the VOD is essentially continuous throughout the entire column due to the VOD within each element which is internal to the Abo-Zena solution. Figure 15 shows the analytical vppv results for a base-initiated 5 m column having a VOD of 5500 m/s; in this case the base was located at (0,0).

Figure 16 shows the results for the peak volumetric strain and Figure 17 shows the results for the peak maximum shear strain.

Figure 18 shows the DFEM vppv results for the 5 m charge column, Figure 19 shows the DFEM solution for the peak volumetric strain and Figure 20 shows the results for the peak maximum shear strain. Again, in all cases the boundaries are non-reflecting.

3.3 *The influence of a free surface*

It is a formidable problem (perhaps impossible) to incorporate the influence of a free surface within the analytical model; thus only DFEM solutions are given. Figure 21 shows the vppv due to a 5 m column of explosive whose top is 5 m below a horizontal surface. Figure 22 shows the results for peak volumetric strain and Figure 23 shows the results for the peak maximum shear strain.

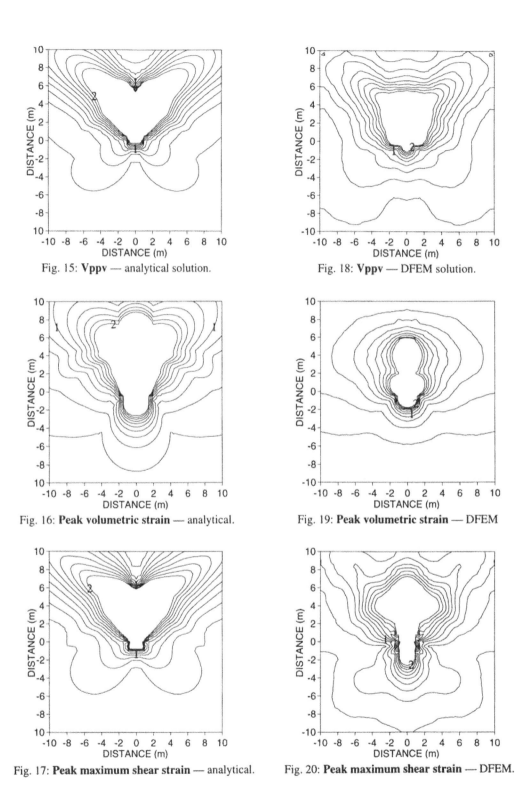

Fig. 15: **Vppv** — analytical solution.

Fig. 18: **Vppv** — DFEM solution.

Fig. 16: **Peak volumetric strain** — analytical.

Fig. 19: **Peak volumetric strain** — DFEM

Fig. 17: **Peak maximum shear strain** — analytical.

Fig. 20: **Peak maximum shear strain** — DFEM.

These figures show the specified contours for a 5 m charge column having a VOD of 5500 m/s.

It is clear from Figures 21 to 23 that the free surface has a complex and significant effect on the vppv and strain especially in regions close to the surface. From an analytical viewpoint, the influence of a horizontal free surface is to double all vertical velocities in a vertically-incident planar p-wave. Thus it is not surprising that the non-planar waves also have an increased vppv near the free surface. However, these figures also show that it is incorrect to account for the free surface by simply doubling all values of vppv on the surface. Unfortunately, this invalid technique has found its way into the blasting literature (see, for example, Persson, 1995). It is also clear from these figures that there is no simple relationship between strain and velocity for the realistic situation of a blasthole close to a free surface.

4 DISCUSSION AND CONCLUSIONS

There is good agreement between the analytical and DFEM far-field contours for the vppv in a single element (Figures 2, 5), the vppv in an extended column having infinite VOD (Figures 8, 11) and the vppv in a column with a finite VOD (Figures 15, 18). However, the agreement between the analytical and DFEM far-field contours for volumetric or maximum shear strain is not as good. The particle velocity is simply the time derivative of the appropriate displacement, whereas the strains are dependent upon spatial derivatives of the displacements. Thus any small differences between analytical and DFEM displacements throughout the contour grid may result in significant differences in respective strains.

The present work clearly shows that the Holmberg-Persson scaling model does not correctly describe the blast vibration due to a column of explosive. The major weaknesses of the scaling model are that it does not correctly account for the charge length or the radiation pattern in either the near-field or the far-field of the charge. Although the Abo-Zena model gives a good description of far-field vppv, and a reasonable description of far-field strain, it is not as realistic as the DFEM which correctly describes both the near- and far-fields of the charge assuming elastic theory. In this regard, all the DFEM contours show that there is no simple relationship between strain and vibration for an element of charge or an extended column of charge with or without a free surface. In fact there is no similarity in either the contour shapes or the contour gradients.

The Seed Waveform model attempts to account for the near-field influence of charge length by taking the elemental seed waveform measured from a short length of charge and summing all

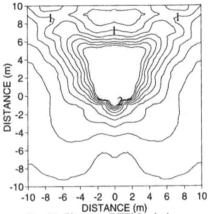

Fig. 21: **Vppv** — DFEM solution.

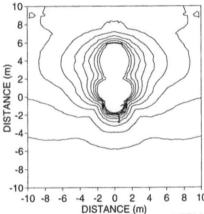

Fig. 22: **Peak volumetric strain** — DFEM.

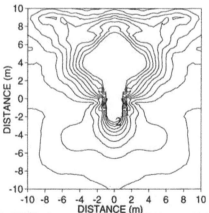

Fig. 23: **Peak maximum shear strain** — DFEM.

These figures are for a 5 m charge column whose top is 5 m below the surface, and whose VOD is 5500 m/s.

129

such elemental waveforms over the length of the blasthole. However, the present study on the radiation from an elemental charge clearly shows that the shape and magnitude of this measured seed will depend significantly upon the angles ϕ_1 and ϕ_2 (Figure 1) for which the measurement was conducted. Both these angles change markedly during any summation along the blasthole, and so a seed taken at any one location within the surrounding material cannot be used to represent each element.

A suggested alternative is to use a modified Seed Waveform approach in which only the radial component of particle velocity is measured in the far-field (greater than ten blasthole radii) along the line R_0 for an element of charge located at B (i.e. $\phi_2=0$, Figure 1). Under this condition it can be shown that the second derivative of the acting pressure function, P, is directly proportional to the measured radial velocity. The parameters n and b of the present source function $P=P_0 t^n e^{-bt}$ could be adjusted so that its second derivative is a least-squares fit to the onset shape of the measured elemental vibration waveform. Matching the measured peak of the radial velocity with the peak of the predicted radial velocity (using the determined values of n and b in either the analytical or DFEM solutions for a single element) would provide the required scaling factor. This scaling factor could then be input to the model of a full length hole. In this manner, it would be possible to predict meaningful contours of vppv and strain surrounding any blasthole.

Despite the present work it is by no means clear which parameter is the best indicator of damage. What is certain, however, is that typical iso-velocity and iso-fragment diagrams which appear in the literature (i.e. Holmberg and Persson, 1979; Persson, 1990; Persson, 1995) are seriously in error. In the present work, the maximum shear strain was one of several parameters employed because such a strain is often associated with the strength of materials (Timoshenko and Goodier, 1971). However, it is also possible to argue that in some regions of the rock mass the damage is induced by tensile failure (i.e. reflection spalling at a free surface), in other regions the damage is induced by compressive failure (i.e. close to the blasthole) and in yet other regions the damage might be induced by excessive vibration. Here the previous analogy of the smelter might be useful. If any damage occurred at the top of a resonating smelter then it would be vibration-induced, yet any damage that occurred at the base would be strain-induced. If $D(R)$ is the damage at a distance R from a blasthole, then a simple equation of the form

$$D(R) = F_1(R)*strain + F_2(R)*vppv \qquad (3.1)$$

might well be a more meaningful measure of damage than measures of either vppv or strain alone. In this equation $F_1(R)$ and $F_2(R)$ are general functions of distance but also chosen such that $D(R)$ is a dimensionless quantity. Furthermore, $F_1(R)$ would need to be short range in R (i.e. dominate near the blasthole), and $F_2(R)$ would need to be sensitive to the location of free surfaces, and thus be a dominant term in the medium-to-far field of the blasthole.

REFERENCES

Abo-Zena, A.M. (1977). Radiation from a finite cylindrical explosive source. *Geophysics*, 42, 1384-1393.

Baird, G.R., Blair, D.P. and Jiang, J.J. (1992). Particle motions on the surface of an elastic half-space due to a vertical column of explosives. *West Aust. Conf. Mining Geomech.*, 367-374, Kalgoorlie, Australia.

Blair, D.P. (1990). Some problems associated with standard charge weight vibration scaling laws. *Third Int. Symp. on Rock Fragmentation by Blasting*, 149-158, Brisbane, Australia.

Blair, D.P. and Jiang, J.J. (1995). Surface vibrations due to a vertical column of explosive. *Int. J. Rock Mech. Min. Sci & Geomech. Abstr.*, 32, 149-154.

Heelan, P.A. (1953). Radiation from a cylindrical source of finite length. *Geophysics*, 18, 685-696. Holmberg, R. and Persson, P.A. (1979). Design of tunnel perimeter blasthole patterns to prevent rock damage. *Proc. Tunnelling'79*, 2870-283 (IMM, London).

Jordan, D.W. (1962). The stress wave from a finite, cylindrical explosive source. *J. Math. Mech.*, 11, 503-551.

McKenzie, C.K., Scherpenisse, C.R., Arriagada, J. and Jones, J.P. (1995). Application of computer assisted modelling to final wall blast design. *EXPLO'95*, 285-292, Brisbane, Australia.

Nicholls, H.R., Johnson, C.F. and Duvall, W.I. (1971). Blasting vibrations and their effects on structures. *USBM Bull.* 656, 105 pp.

Persson, P.A. (1990). Fragmentation mechanics, *Third Int. Symp. on Rock Fragmentation by Blasting*, 149-158, Brisbane, Australia.

Persson, P.A. (1995). A new technique for predicting rock fragmentation in blasting. *EXPLO'95*, 421-425, Brisbane, Australia.

Timoshenko, S.P. and Goodier, J.N. (1971).*Theory of elasticity*. McGraw-Hill, Third Edition.

White, J.E. (1983). Underground Sound. *Application of seismic waves*. Elsevier.

Rock Fragmentation by Blasting, Mohanty (ed.) © 1996 Taylor & Francis. ISBN 90 5410 824 X

Measuring rock mass damage in drifting

M. Paventi – *INCO Limited, Manitoba Division, Thompson, Man., Canada*

Y. Lizotte – *Natural Resources Canada, CANMET Experimental Mine, Val d'Or, Que., Canada*

M. Scoble – *McGill University, Montreal, Que., Canada*

B. Mohanty – *ICI Explosives Canada, McMasterville, Que., Canada*

ABSTRACT: This paper is based upon field studies of damage associated with closely controlled drifting through contrasting rock masses at the Birchtree Mine. It reviews the development of a field procedure for damage monitoring through an empirical blast-induced damage index. The intensity and mechanisms of blast damage were found to relate to the inherent characteristics of each rock mass.

1. INTRODUCTION

A damage audit, monitoring the integrity of a rock mass, requires appropriate procedures and tools. Monitoring the response of the rock mass to the mining process and the damage inflicted should be part of the mine production control rationale. The extent of mining-induced damage reflects the nature and quality of the mining process. Controlling damage should assist in minimizing reconditioning, reducing costs and improving safety.

The *inherent damage* of a rock mass is its reduction in integrity as a result of natural processes. Damage arising from blasting, machine-mining or ground stress redistribution is termed here *mining-induced damage*.

This paper will focus on blast-induced damage resulting from drifting for underground mine development. It reviews studies aimed to identify and explain the differences in rock mass damage attributed to blasting, observed in the 83 orebody at INCO Limited's Birchtree Mine in Northern Manitoba. The damage associated with 5 x 4m drifting through three distinct rock masses was monitored. The blast design, explosives and accessories, and drilling and blasting personnel were maintained essentially constant during the studies.

An empirical mining-induced damage index is presented, which was developed through evaluating field parameters and their observed relationship to damage. The research also developed a classification of the mechanisms by which blast damage was observed to be propagated in each rock mass.

2. DAMAGE MEASUREMENT

Indices are commonly used in mining to empirically assess rock mass quality. These tend to consider only inherent discontinuities and ignore any mining-induced damage. Measurement is based on the quality of the in situ rock mass, a value that denotes the *inherent damage* produced as a natural result of geological evolution. Indices such as Deere's RQD (1964), Bieniawski's RMR (1976), Barton *et al.'s* Q (1974), and Laubscher's MRMR (1977) were not specifically intended to measure blast-induced damage. Cummings *et al.* (1982) and Kendorski *et al.* (1983) considered blast damage in a modified RMR classification, the Modified Basic RMR (MBR).

Geophysical tools (including seismic tomography, borehole acoustic, electromagnetic, radio imaging) and borehole instrumentation (including dilatometer, impression packer and others) may have future application to determine meso-features at the rock mass scale. Although many types of geophysical techniques exist to monitor structural defects or voids in rock, they are generally far too crude to detect small scale features including localized fracturing, meso-structures, and rock fabric. Other issues may relate to cost, site preparation, manpower, time and space requirements. In this study an approach was sought to quantify blast damage by observation with simple tools.

Inherent damage is the datum which is characteristic of the tectonic and regional history which has been imprinted onto a rock mass, Paventi (1995). *Mining-induced* damage is subsequently imposed onto the inherent rock mass damage. Mining-induced damage may take many forms: increased fracture frequency, degradation in discontinuity surfaces, change in the aperture of the discontinuities and sloughing around the excavation, as well as damage due to stress redistribution about the excavation.

This paper stems from work aimed to measure inherent and blast-induced damage in open stoping drawpoints. These were excavated through three distinct and contrasting rock masses at the Birchtree Mine in the Thompson Nickel Belt, Peredery (1982). These are metasedimentary (MD), massive sulphide (MSD) and serpentinized ultramafic (SUD) rock masses. The MD contains strong, foliated and abrasive rock, principally as gneisses, schists, iron formation and pegmatite rock units. Typical MD geomechanical properties are: unconfined compressive strength (UCS) 75-130MPa, static Young's modulus (E) 32-38GPa, volumetric joint count (Jv) 7 joints per m^3, Paventi *et al.* (1996). The MSD comprises sulphides with diverse rock inclusions, typically as follows: UCS 60-120 MPa, E 35-45GPa, Jv 11 jt/m^3. The SUD Rock units are more variable. They are generally foliated and low to moderate strength, typically as follows:UCS 30-100 MPa, E 20-35 GPa, Jv 10-18 jt/m^3.

Blast damage can be controlled by understanding the rock mass and how the blast pattern, number of drill holes, delays, spacing and burden determine the nature of the excavation with a given explosive. Ideally a perimeter controlled drift round will exhibit the following characteristics: minimal scaling of loose, ground sounds solid when struck with a scaling bar, a high percentage of half casts, minimal ground support, and minimal cycle time between mucking, supporting and drilling the next round.

Forsyth and Moss (1990) devised a method of quantifying blast-induced damage. Their Drift Condition Rating (DCR) system comprised two components: firstly, the drift back condition (related to the rock mass integrity and the percent of half casts visible); and secondly, the amount of overbreak. This empirical rating varies from 0 to 9 (minimal damage).

Singh (1992) investigated blast-induced damage mechanisms at the macroscopic scale by assessing the scaling time, overbreak, length of visible half casts and mucking time. His assessment of damage was based on blast vibration, percent of half casts and overbreak, seismic reflection, core logging and borehole camera mapping.

In the current study a device was built to obtain the horizontal and vertical orientation of drift face blastholes to determine the extent of deviation and its effect on the way the round broke. It was found that: deviation either created a larger damaged zone at the toe or caused the neighbouring holes to misfire; and holes with a look-out in excess of 7° tended to misfire. Fifty four drift rounds on the 1900 and 2000 levels in the 83 orebody were monitored over three months in 1993. This period was spent in the company of the crews and was devoted to observing the daily routine of development excavation. Monitoring commenced as soon as the jumbo operator had finished drilling the prescribed drill pattern. Working from the muckpile after the blast, a loading stick was used to help survey the relevant features evident in the drift. In all cases the geology was mapped and any structural features recorded. The muckpiles for most drift rounds and other pertinent drift features were photographed for image analysis. Since no guidelines existed regarding data collection for such monitoring, a standardized data collection procedure was adopted, see Table 1.

Table 1. Data Collection for Damage Assessment.

---

1) Pre-blast data:
- Hole: diameter, location, length, orientation
- Drill pattern, perimeter holes and sequence
- Quantity & type of explosive and delays
- Hole loading
- Face geology and structure
- Face geomechanical data
- Face geometry

2) Blast data:
- Blast vibration & firing sequence

3) Post-blast data:
- Length and condition of half barrels
- Crushed zone & bootlegs
- Misfires, unbroken portions of blastholes
- Round profile, over- and underbreak
- Sloughing geometry
- Discontinuities controlling slough cavities
- Scaling: location and time
- Sounding with scaling bar
- Exposed round geology & structure
- Round geomechanical data
- Paint and map structural features along round
- Fracture mapping, geophysics or photography
- Face survey position
- Powder & energy factors

4) Fragmentation:
- Size distribution
- Rock type & structure displayed in fragments

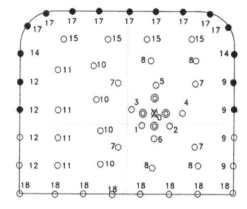

O Pneumatically loaded with AMEX II

● PRIMAFLEX pneumatically anchored with AMEX II

◎ Hole reamed to 7.6 cm

⊠ SUPERFRAC 4000 the remainder pneumatically loaded
with AMEX II

(HCF was only calculated for those holes with PRIMAFLEX)

Figure 1. 5x4m Drift Blast Design.

Post-blast data, related to geology/structure, geomechanical, drift morphology and fragmentation features, was collected after the ground was scaled, bolted and screened. The standard drift blasting design used throughout this study is illustrated in Figure 1.

3. MINING-INDUCED DAMAGE INDEX, D_M

This index aims to characterize the extent of damage arising solely from the mining process. Five parameters are summed: the reduction in intact rock strength; post-scaling half cast factor; drift condition (sounding the back and walls with a scaling bar); normalized scaling time; and the orientation of dominant discontinuities. These parameters were considered to account for all aspects of damage, as well as being accessible with simple tools.

Rock strength reduction is brought about by stresses related to blasting, mechanical excavation and ground stress redistribution. A good correlation was achieved in the field between Schmidt Hammer Rebound values and UCS, and was used to monitor strength reduction.

Half casts define the datum outline of the excavation created. If most are visible after the ground has been supported, then this indicates that little, if any, scaling

was necessary and also that fracturation beyond the perimeter holes in the round was limited. A low half cast factor (HCF) value implies that damage was sufficient to allow sloughing or scaling of loose.

Sounding of the back and walls with a scaling bar can be subjective. The results may differ depending on the skill and experience of the individual, Forsyth and Moss (1990). Scaling time was determined by adding the time taken to inspect, sound and scale the round. It was normalized to reflect the size of the excavation, assuming that the scaling time was linearly dependent on the surface area represented by the round. The scaling time component of D_M, like the rock sounding component, can be influenced by the experience of the miner and his knowledge of the behaviour of the ground. During the time frame of this study, the contract crews were solely responsible for developing and rehabilitating the 83 Complex over a period of five years.

The relationship between the HCF, drift condition rating and normalized scaling time was statistically evaluated. The highest correlation obtained was only 0.41. This increased significantly, however, when the data was segregated according to rock mass. In many cases the removal of the data within the transition zones between the rock masses improved the correlation further. Statistical analysis confirmed the relationship that exists between these parameters, and justified their evaluation according to the rock mass.

The final D_M component was the orientation of the drift axis with respect to the direction of the prevailing meso- and macro-structure. This component accounts for the most persistent and abundant meso-structures encountered in the drifts. It is calculated by addition of the cosines of the angle between the drift direction and the structure involved. Damage was observed to increase as structures approached the drift direction. Studies by Worsey et al. (1981) showed that a relationship exists between discontinuity orientation and fracture propagation. This was found to apply with the meso- and macro-structures at Birchtree. There are four types of meso-structure to consider: foliation, cleavage and jointing for metamorphic, lineation and jointing for igneous rocks and bedding, cleavage and possibly jointing (if < 2 m spacing) for sedimentary rocks.

The following is the sequence by which the D_M index component ratings are computed:

(I) Consider the reduction in intact rock strength due to micro-fracturing.

Strength Reduction (% UCS)	D_M Rating
No Reduction	1.00
< 5%	1.25
5 to 10%	1.50
10 to 15%	1.75
>15%	2.00

(II) Evaluate the extent of the exposed excavation surface area remaining in place, using the post-scaling half cast factor, HCF, expressed as % of total area.

HCF (%)	D_M	HCF (%)	D_M	HCF (%)	D_M
>80	1.0	50-60	2.5	20-30	4.0
70-80	1.5	40-50	3.0	10-20	4.5
60-70	2.0	30-40	3.5	<10	5.0

(III) Determine the drift condition component of D_M by assessing the drumminess of the back with a scaling bar, as follows:

Description	D_M Rating
Sounds solid before scaling	1.0
Sounds solid after scaling	1.5
Sounds drummy in places	2.0
Sounds drummy	2.5

IV) Account for the amount of scaling arising from damage using the normalized scaling time (minutes).

(V) Finally, consider the direction of structure with respect to the drift direction, to account for the anisotropy potentially caused by structural features at the meso- and macro-scale. The most dominant and persistent structures (including foliation) are considered by summing the cosines of the angle between the drift axis and each structure direction.

(V_a) Meso-structure: including foliation, joints, and shears.

(V_b) Macro-structure: mine-wide, including faults and shears.

D_M is then calculated according to:

$$D_M = I * II * III * IV * (V_a + V_b)$$

4. DAMAGE ANALYSIS

Fifty four rounds were monitored in the 83 orebody on the 1900 and 2000 levels, with the procedure reviewed earlier. Mining-induced damage takes full advantage of the inherent weaknesses, therefore, the more inherently weak the rock mass the easier it will be to damage. The least inherently damaged of the three rock masses was the metasedimentary domain (MD), followed by the massive sulphide domain (MSD). It is logical that more energy is required to damage a stronger rock mass than a weaker one, all other factors being constant. The same can be said for both the transition zones; for example, T2 between the weaker rock masses exhibited a higher D_M than T1 between the inherently stronger rock masses.

The SUD mass is inherently a heterogeneous and highly damaged rock mass, requiring little effort to reduce its integrity. The study shows that, whether sheared or unsheared, D_M is in excess of 2.5 times its value in the other rock masses.

Table 2 shows a statistical summary of all fifty four drift rounds monitored in the three rock masses. The MD and SUM masses were categorized into sheared and unsheared rock masses. There is a clear distinction in the D_M value for these categories.

5. DAMAGE MECHANISMS

Six types of blast damage mechanism were observed at the Birchtree Mine. These can be grouped into three damage categories: dilated fabric, dilated structural, and dilated lithological, as shown in Figure 5. Two types of dilated fabric were observed: firstly, fracturing along pre-existing planes of weakness created by the alignment of weak minerals, e.g. foliation in a biotite schist, referred to as Type I, see Plate 1; and secondly, Type II damage, involving the creation of new micro-fractures at some angle to the widespread fabric, primarily caused by blasting, see Plate 2.

Dilated structural damage refers to breaking or extending pre-existing discontinuities present in the rock mass. The distinction between Type III and Type IV is the inherent condition of the discontinuity. A tight or annealed discontinuity is associated with a Type III dilated structural damage mechanism, Plate 3. Open joints, or those with weak infill, perhaps

Table 2. Variation of D_M between rock masses.

Rock Mass		n	mean D_M	δ	cv
MD	All Data	19	22.4	8.6	38.2
	Sheared	10	23.5	7.5	32.1
	Unsheared	9	21.2	9.9	46.7
MSD	All Data	19	54.6	27.2	49.8
	Sheared	7	76.0	20.4	26.9
	Unsheared	12	42.1	22.7	54.1
SUD	All Data	10	112.8	26.3	23.3
	Sheared	9	127.8	12.0	9.4
	Unsheared	1	77.8	-	-
	T1	3	18.8	15.9	84.5
	T2	3	84.4	48.6	57.6

n = number of rounds

δ = standard deviation

cv = coefficient of variation (%)

transition zone: MD to MSD

transition zone: MSD to SUD

slickensided (or shear-like at the meso-scale) are typical of Type IV, Plate 4.

Dilated lithological damage refers to the fracturing at the interface of two or more lithological units. The condition at the contact can be weak resulting from the presence of one or a combination of platy minerals including biotite, chlorite, and graphite. Sheared and slickensided contacts (macro-scale) were not uncommon in the case of MD-MSD or MSD-SUD interfaces. These surface conditions are typical of the Type V dilated lithological contact shown in Plate 5. Type VI, unlike Type V, is a stronger contact that requires more energy to break, particularly if the boundary does not contain mica. This damage mechanism is generally restricted to lithological units that possess similar physical-mechanical properties, e.g. biotite gneiss and biotite schist in the MD.

The mechanisms observed for the three rock masses are a combination of the various types. Type II is part of the dilated fabric damage originated at micro-scale (inter-, intragranular cracks or along grain boundaries) where a number of these micro-fractures can coalesce into larger fractures that may ultimately promote instability. This mechanism type is common to all rock units and rock masses.

Some types of damage mechanism were observed to be particular to certain rock masses:

- the MD generally exhibits the Types I and VI mechanism.
- the MSD was characterized by Type V. Type IV, if present, would be observed in large (>2 m wide) SUD inclusions within the SMD.
- the SUD mechanisms were dominated by Type IV, with subordinate Type II only.

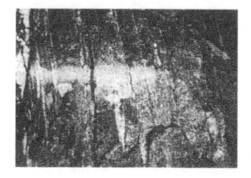

Plate 1. Type I - dilated fabric damage: breaking along foliation in biotite schist, drawpoint 816, 1900 level.

Plate 2. Type II - dilated fabric damage: creation or extending fractures in peridotite, 1500 level.

Plate 5 Type V - fracture along weak boundary, massive sulphide and peridotite inclusion, 1500 level.

Plate 3.Type III - breaking-extending discontinuities in biotite schist, drawpoint 830, 2000 level.

Plate 6 Type VI - fracture in well banded gneiss and massive white pegmatite, Thompson Open Pit.

6. DAMAGE CONTROL

In drifting, the energy of the explosive, blast pattern, delays, number of blast holes, their burden and spacing and other blasting related factors will control the amount of damage inflicted on the remaining rock mass and the consequent ground support requirements. Traditionally, these blast design factors have been considered globally, using such parameters as powder factor, only from the viewpoint of their influence over fragmentation, rather than rock mass damage.

This study has demonstrated that when the above blasting factors are kept constant within three distinctively different rock masses, then the consequent differences in rock mass damage are evident in the following data: scaling time, drift profile, overbreak, half cast factor HCF, drift

Plate 4. Type IV - shear discontinuities in peridotite, slickensided and infilled with talc, 1500 level.

morphology and fracture density. Clearly the two most fundamental differences in the three rock masses in this study at the Birchtree Mine have been the bulk rock composition and the structure.

Mining-induced damage, particularly blast damage was seen to increase in the weaker rock masses. The SUD mass, at the Birchtree Mine, is seen to be inherently weak and this was more susceptible to damage than the other two domains. Moreover, SUM masses may frequently form stope hanging wall exposures in the 83 orebody. SUD inclusions are also a common occurrence in the massive sulphide domain; their size, number and area can be observed to control the stability of the stope excavations. Hence, it is crucial to understand the rock mass properties and its susceptibility to damage before formulating a blast design to minimize damage and ore dilution. Blast design should aim to minimize damage by considering geology rather than pursue the traditional trial-and-error approach in controlled blast design.

In controlled blasting the energy of the explosive is harnessed in such a way as to fragment and displace the rock and control overbreak. The ability to control overbreak was demonstrated to vary significantly within the three rock mass domains studied at Birchtree. Overbreak at the stope scale was further exacerbated by the presence of macro-structure, such as shears and faults. These were observed to traverse the different rock masses and influence damage by kinematic control and modification of ground stress distribution. Their relative orientation to the excavation was important, together with the direction of the drift in governing overbreak, see Plate 7.

Plate 7. Overbreak due to intersection with a sub-parallel shear.

7. CONCLUSION

This work attempted to bring an understanding of the rock mass to the study and control of damage resulting from blasting. It was considered important to understand the types of damage observed and then to link the observed damage mechanisms to the mining process and rock mass characteristics.

The methodology established for recognizing and monitoring mining-induced damage met the characterization needs for monitoring development drifting. A system for the quantification of mining-induced damage with simple tools was developed based on field observation and testing. Damage was clearly reflected in the D_M index, derived from the reduction in intact rock strength, post-scaling HCF, drift condition rating, normalized scaling time, and direction of meso- and macro- structure with respect to the orientation of the drift.

Mining-induced damage intensity could be termed low (D_M less than 25), moderate (D_M between 25 and 50) and high (D_M greater than 50). The type of support observed for a given D_M varied depending on the overbreak, historical behaviour of the rock mass upon exposure, and the proximity to macro-structures.

The damage types observed at the Birchtree Mine were grouped into three categories: (1) dilated fabric damage, which comprises foliation (Type I) and fracturing (Type III); (2) dilated structural damage, exploiting discontinuities (Type II) and shears (Type IV), (3) dilated lithological damage, consisting of parting along weak, well defined boundaries between two different rock units (Type V) and breaking or parting along lithological boundaries between similar rock types (Type VI).

D_M indicates that mining-induced damage is influenced not only by the mining method, but also by the inherent nature of the rock mass. The presence of macro-structures, namely shears and their condition, also plays a significant role in the generation of both inherent- and mining-induced damage. D_M should permit future extrapolation to other geological sites, mining methods or excavation processes. The challenge remains, however, to integrate this work into ground control practice by identifying and accounting for damage also generated by stress redistribution after excavation. One example is the recent evolution in studies to characterize damage mechanisms related to rockbursting, Jesenak et al. (1993). At this time it is not possible to claim that any of the damage considered in this work was not related to this cause. The data, however, was captured immediately after each blast and so limited time was available for such damage to develop. A further challenge is to expand this foundation work to stoping and dilution control. This will require the integration of stope morphology, blasthole drilling precision and petrological-structural survey data.

ACKNOWLEDGMENT

The authors wish to thank INCO Limited, Manitoba Division, for allowing the publication of this paper and assistance with this work. The views expressed are entirely those of the authors.

REFERENCES

Barton, N., Lien, R. and Lunde, J. 1974. Engineering classification of rock masses for the design of tunnel support. Rock Mechanics, Vol.6, pp. 189-236.

Bieniawski, Z.T., 1976. Rock Mass Classification in Engineering. Proc. Symp. Exploration for Rock, Johannesburg. pp. 97-106.

Cummings, R.A., Kendorski, F.S. and Bieniawski, Z.T. 1982. Caving mine rock mass classification and support estimation, U.S.B.M. Contract J010010, Eng. Int., pp. 195.

Deere, D.U. 1964. Technical Description of Rock Cores For Engineering Purposes. Rock Mech. Engng. Geol., Vol 1, no. 1. pp. 17-22.

Forsyth, W.W and Moss, A.E., 1990. Observations on Blasting and Damage Around Development Openings, 92nd Can. Inst. Min. Metall. Annual General Meeting, Ottawa.

Jesenak, P., Kaiser, P.K. and R.K. Brummer 1993. Rockburst damage potential assessment - an update. Proc. 3rd. Int. Symp. On Rockbursts and Seismicity in Mines, Balkema, Rotterdam, pp. 81-85.

Kendorski,F.S., Cummings, R.A., Bieniawski, Z.T. and Skinner, E.H. 1983. Rock mass classification for block caving mine drift support. Proc., 15th Int. Congr. Rock Mech., ISRM, Melbourne, pp. B101-113.

Laubscher, D.H. 1977. Geomechanics classification of jointed rock masses - Mining application. Trans. Instn. Min. Metal. (sect. A) vol. 86, pp. A1-A8.

Paventi, M., 1995. Rock Mass Characteristics and Damage at the Birchtree Mine. Unpubl. PhD Thesis, McGill University, Montreal, Canada.

Paventi, M., Scoble, M. and D. Stead 1996. Characteristics of a serpentinized ultramafic rock mass at the Birchtree Mine, Manitoba. Proc. 2nd North American Rock Mechanics Symposium, Montreal, in press.

Peredery, W.V., 1982. Geology and Nickel Sulphide Deposits of the Thompson Belt, Manitoba. In Precambrian Sulphide Deposits, H.S. Robinson Memorial Volume, R.W. Hutchinson, ed. GAC Special Paper 25, pp. 165 - 209.

Singh, S.P., 1992. Investigation of Blast Damage Mechanisms in Underground Mines. Report to the Mining Research Directorate. Laurentian University.

Worsey, P.N. et al. 1981. The Mechanics of Pre-splitting in Discontinuous Rocks. 22nd U.S. Symp. on Rock Mech., M.I.T., pp. 205-210.

Rock Fragmentation by Blasting, Mohanty (ed.) © 1996 Taylor & Francis. ISBN 90 5410 824 X

Evaluation of rock mass damage using acoustic emission technique in the laboratory

M. Seto
Research Development Corporation of Japan (JRDC), Tokyo, Japan (Presently: University of New South Wales, Sydney, N.S.W., Australia)

D. K. Nag
Monash University, Churchill, Australia

V. S. Vutukuri
The University of New South Wales, Sydney, N.S.W., Australia

ABSTRACT:
The damage of the rock is associated with the growth of microcracks and brittle fracturing, and the analysis of acoustic emission (AE) signals is well suited to study this phenomenon. In this paper, potential use of AE and the Kaiser effect in detecting and assessing the amount of rock damages has been investigated in the laboratory study using Gosford sandstone under uniaxial cyclic and triaxial loading conditions. The influence of time, stress state on the Kaiser effect has also been investigated. The experimental results show that Kaiser effect occurred reliably, and that rock damage behaviour can be assessed with a new suggested parameter described here in. Damage evaluation under triaxially reloaded rock has also been assessed and reported.

1. INTRODUCTION

Acoustic emission (AE) is a burst of high frequency elastic wave emitted by a local failure such as microcracking or pore collapsing in rock. It is observed in rocks that AE is detected during the cyclic loading in which the maximum stress of each cycle is higher compared to the previous cycle. If the previously applied stress state is exceeded, the AE events are again observed. This is so-called Kaiser effect. It is formally described as the absence of detectable AE events until the load imposed on the specimen exceeds the previously applied level.

The Kaiser effect was first observed in metals by Kaiser (1950). The investigations of the Kaiser effect in rocks started to be reported by Kanagawa et al. (1979) and Kurita and Fujii (1979). The motivation for the study of the Kaiser effect in rocks at the beginning was to estimate geostresses. It seems, however, that the theoretical and technical problems of estimating geostresses by means of the Kaiser effect have not been successfully solved so far.

The Kaiser effect is a actually a measure of damage which has been developed in a material subjected to a load. In describing brittle rocks where deformation and failure are due to the growth and coalescence of microcracks, the concept of damage is useful. It is known that AE is associated with the development of damage in rocks (microcracking and pore collapsing). The Kaiser effect may be used to detect and assess the amount of damage that has been developed in rocks (Holocomb and Costin, 1986; Holocomb et al., 1990; Li and Nordlund, 1993).

The objective of this study is to conduct a laboratory investigation of the acoustic emission in rocks and explore the possibility of its application to evaluate the damage in rock mass. The emphasis of this paper will be on the verification of the Kaiser effect in different stress levels and the influences of stress state on the Kaiser effect. And, we suggest a new parameter, which was produced by AE measurements of differently damaged rocks, in order to evaluate the rock damage.

2. EXPERIMENT FACILITIES

The AE instrument, MISTRAS-2001 system, was employed in the tests. The MISTRAS-2001 systems was a computerised AE system that performed AE signal measurements and stored, displayed and analysed the resulting data. AE signals were amplified by a pre-amplifier (Gain: 40 dB, frequency filter: 50-1200 kHz) and a post amplifier inside the system (Gain: 20 dB). The threshold could be set in the screen set-up menu of the test running code. Each AE signal was described in terms of its counts, energy, amplitude, rise time and duration in the record.

In the screen set-up menu of the MISTRAS-2001, there were three other parameters, besides gain and threshold, which had to be set before testing. They were PDT (Peak Defining Time), HDT (Hit Definition Time) and HLT (Hit Lockout Time). A proper setting of the PDT ensured correct identification of the signal peak for rise time measurements. A proper setting of the HDT ensured that each AE signal was reported as one and only one hit. With the proper setting of HLT, spurious measurements during the signal decay were avoided. In our tests, the values for the corresponding set-up parameters in the MISTRAS-2001 were chosen as follows:

Threshold : 40-50 dB,
PDT   : 50 μsec,
HDT   : 200 μsec
HLT   :1000 μsec

We employed two different types of piezoelectric transducer (PAC nano-30 (sensor 1) and NF AE-901S (sensor 2)). The resonant frequencies of sensor 1 and sensor 2 are 500 kHz and 120 kHz, respectively. In general, piezoelectric transducers cannot have a flat frequency response over a wide band of frequencies: they have multiple sensitivity peaks. Some of these peaks were accentuated through the use of active band filters to create narrow bandwidth windows through which acoustic emissions were monitored. We conduct a statistical comparison of the emission event rate monitored through a lower frequency window to that simultaneously detected through a higher frequency window. Lower frequency emissions were monitored by a sensor 2 through a lower frequency band from 50 to 200 kHz, and higher frequency emissions were monitored by sensor 1 through a higher frequency window from 200 to 1200 kHz. Two of these transducers were attached to a specimen and cemented with electron wax. In the present experiment the two transducers were placed adjacent to each other to minimise the difference in propagation path from emission sources between the two transducers.

The specimen tested, Gosford sandstone produced in Australia, were cored from a cubic rock block (30×30×30 cm) and prepared to 45 mm in diameter and 100 mm in length. Rock specimens were uniaxially or triaxially loaded under compression by means of a servo-controlled hydraulic testing machine, Schenck TREBEL. The rock specimens were loaded and unloaded repeatedly. Displacement control was employed. The loading rate under displacement control were

50 and 100 μm/min. Average uniaxial compressive strength of the Gosford sandstone was 40 MPa.

3. ACOUSTIC EMISSION BEHAVIOURS IN CYCLIC LOADINGS

A specimen was loaded and unloaded eight times until it failed. Figs.1 to 3 show the cyclic loading pattern and AE event rate signatures in the cyclic loadings. The arrows in the figure mark the maximum stresses of the corresponding previous loading cycles. AE continuously occurred during the entire process in the first loading up to 5.52 MPa. In the second loading up to 10.56 MPa, however, AE was absent until the load reached the maximum stress level of the first cycle. At that point, the continuous AE started again. This is the Kaiser effect. It was noticed that, at a relatively higher stress level, onset of the continuous AE did not occur exactly at the maximum stress level of the previous cycle, rather a bit before it. This is the so-called felicity effect. The breakdown of the Kaiser effect can be represented quantitatively by the

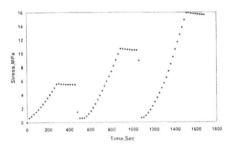

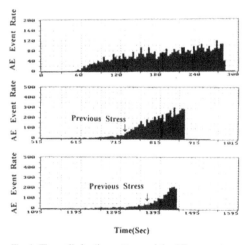

Fig. 1 The cyclic loading pattern and the AE event rate signatures from the 1st to 3rd loading.

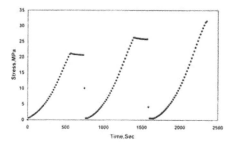

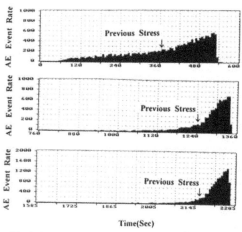

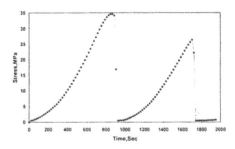

Fig. 2 The cyclic loading pattern and the AE event rate signatures from the 4th to 6th loading.

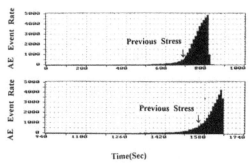

Time(Sec)

Fig. 3 The cyclic loading pattern and the AE event rate signatures from the 7th to 8th loading.

felicity ratio that is defined as the ratio of the AE-onset stress to the maximum stress of the previous cycle. A high felicity ratio means that the rock is of good quality (Li and Nordlund; 1993). The curve of the felicity ratio for a Gosford sandstone specimen is illustrated in Fig.4. The felicity ratio remains at the level of near 0.9 from the beginning of loading to the stress level of about 80 %, but it decreases rapidly to around 0.4 when the stress approaches to the level of the strength.

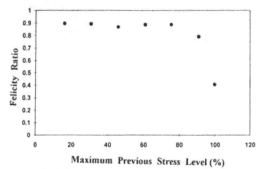

Fig. 4 The relation between felicity ratio and the maximum previous stress.

We also investigated the variation of m-value and lower frequency contents of the emissions in the individual loading. A m-value can be derived from the analysis of cumulative AE's amplitude distribution. The amplitude distributions were analysed using the power law

$$n(a)=(a/a^*)^{-m} \qquad (1)$$

where $n(a)$ is the fraction of the emission population whose peak exceeds an amplitude a, a^* is the lowest detectable amplitude and the exponent m is a constant which characterise the amplitude distribution.

The variation of lower frequency contents of AE was analysed by using a parameter $AE(L)/(AE(L)+AE(H)$ which is indicative of the relative change in the frequency content of emission waves. It is true that the count level itself depends upon the sensitivity and the threshold level set for each channel. But these sensitivity and threshold level were fixed throughout the experiment. Accordingly, the ratio $AE(L)/(AE(L)+AE(H)$ is a useful parameter for checking frequency dependence of the emission rate.

As a typical example, Fig.5 shows the variations of m-value and the frequency ratio $AE(L)/(AEL)+AE(H))$ in the 6th loading shown in Fig.2, up to the stress level of 91 % of the strength. A m-value remains constant at the level of about 1.3

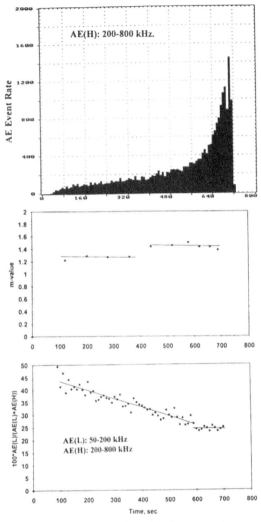

Fig.5 The variations of m-value and the frequency ratio AE(L)/(AE(L)+AE(H)) in the 6th loading up to 31.3 MPa with the AE(H) signature.

from the beginning until a certain stress level. Beyond this stress level nearly corresponding to the maximum previous stress, it increases to the level of about 1.5 with applied stress. The frequency ratio AE(L)/(AE(L)+AE(H)) decreases with stress from the initial stress level to the stress at which AE starts to increase, then remains constant at the lower value. It means that higher frequency contents of the emission waves increases with stress. From the results of m-value and the ratio AE(L)/(AE(L)+AE(H)), it can be concluded that smaller and higher frequency emissions are predominant beyond the maximum previous stress in the subsequent loading.

4. ROCK DAMAGE PARAMETER (χ_{RDP})

As shown above, the felicity ratio is likely to be one of the effective parameters to evaluate the damage of rock, if it is possible to know the maximum stress previously applied to the rock sample. In practical situations, however, the information which we can obtain from the rock sample, even if the AE measurements can be done, are only "rock strength", "applied stress in the subsequent loading(testing)" and "the onset of AE increase" during the loading. We first investigated, therefore, the onset at which AE starts to increase, and calculated the ratio of the AE-onset-stress(σ_{AE}) to the strength(Sc) (AE-onset ratio). Fig. 6 shows the relation between AE-onset ratio and the maximum previous stress level. As shown in Fig. 6, AE-onset ratio increases linearly with the maximum previous stress level from the beginning to near the level of 80 %, then decreases rapidly when the previous stress level approaches the level of the strength. The problem with AE-onset ratio is that non-damaged rock has a similar value to the one of highly damaged one, as indicated in the figure. It is stressed here that the recognition of the AE-onset is dependent on the applied load in the subsequent loading. As a typical example, Fig. 7 is given to indicate the evidence by using the AE signatures of the rock that was previously loaded up to 29.5 MPa (79 % level of the strength). In three AE signatures shown in Fig.7, the top one is in the loading from 0 to 1.4 MPa, the middle one from 0 to 14.9 MPa, and the bottom one from 0 to 25.1 MPa, respectively. We can recognise the different AE-take-off point in every three signatures, which is individually marked by arrow. Thus, we have to consider the effect of applied stress in the subsequent loading in order to evaluate the rock mass damage using a certain indices.

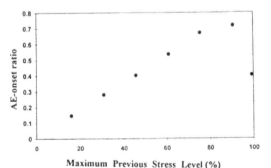

Fig. 6 The relation between AE-onset ratio and the maximum previous stress.

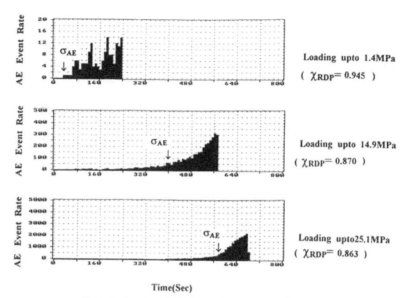

Loading upto 1.4MPa

(χ_{RDP}= 0.945)

Loading upto 14.9MPa

(χ_{RDP}= 0.870)

Loading upto25.1MPa

(χ_{RDP}= 0.863)

Time(Sec)

Fig. 7 Comparison of AE signatures up to three different
stress levels. A different AE-onset can be recognised
from each AE signature despite loading to the same
damaged specimen.

Here, we suggest the rock damage parameter (χ_{RDP}) to evaluate the rock damage The rock damage parameter is expressed by the following equation:

$$\chi_{RDP} = (\sigma_{AE} \times S_C)/\sigma_{Applied}^2 \quad (2)$$

where,

σ_{AE} : stress of AE onset,

S_C : uniaxial compressive strength of the specimen,

$\sigma_{Applied}$: applied stress in the testing.

Calculating the rock damage parameter of the rock shown in Fig.7, the parameters in three AE signatures are 0.945, 0.876, and 0.863, respectively. From this result, the rock damage parameter prove to be independent of the applied stress in the subsequent loading. In exception, if the applied stress is extremely low (1.4 MPa shown in Fig.7), the rock damage parameter is likely to be overestimated. If the rock fails at the lower stress than expected, the fracture stress should be used as Sc in Eq. (2) to calculate the rock damage parameter.

Fig.8 indicates the variation of χ_{RDP} with the maximum previous stress level. The rock damage parameter decreases dramatically with the maximum stress level. When the maximum previous stress is less than 50 % of the strength, which means it is within elastic range, χ_{RDP} is higher than 1.0. A high

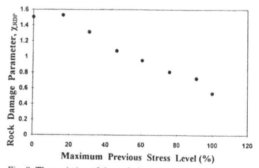

Fig. 8 The variation of the rock damage parameter. χ_{RDP}
with the maximum previous stress.

χ_{RDP} , higher than 1.0, means that the rock is of good quality. Beyond the stress level of 60 %, χ_{RDP} decreases rapidly when the maximum stress level approaches the level of the strength. If we define a magnitude of the rock damage parameter (χ_{RDP}), such as χ_{RDP} =1.0 which is chosen arbitrarily here for Gosford sandstone, as a criterion, we can obtain the information from the rock damage parameter (χ_{RDP}) to judge whether rock mass is being or was subjected to high stress concentration due to inadequate mining, structural instability, and geological irregularities etc., or blasting shock wave. In practical situation, the rock damage parameter (χ_{RDP}) , therefore, should be a better indices to evaluate the rock mass damage than felicity ratio which has been reported by Li and Nordlund (1993).

5. DAMAGE EVALUATION OF THE ROCK TRIAXILLY PRELOADED

In order to evaluate the rock mass damage at delay time, retention aspects of rock stress level, and the effect of confining pressure in evaluating the rock damage, a test program was designed. In this investigation, four sandstone core samples (uniaxial compressive strength, Sc=50 MPa) of 45 mm dia × 90 mm length were selected. All samples were previously stressed under triaxial stress condition,

confining pressure 10 MPa, and axial stress 20 MPa. The strength of the tested sandstone under confining pressure of 20 MPa was 149 MPa. The individual core specimen was loaded in a combination of a Hoek triaxial cell and servo-controlled testing machine, at the loading rate of 100 μm/min. At the 3 days' delay after the triaxial preloading, which is shown in Fig. 9, all samples were loaded again up to 30 MPa under uniaxial compression condition at the same loading rate of 100 μm/min. A typical examples of the resulting AE data are shown in Fig.10 which gives detail of AE event rate vs time of higher frequency AE rate, AE(H), and lower frequency AE rate, AE(L),

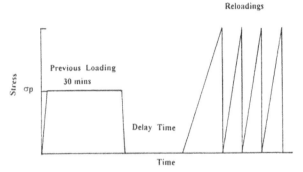

Fig.9 The patterns of previous loading and subsequent loadings under triaxial condition. Delay time varied from 10 min. to 3days.

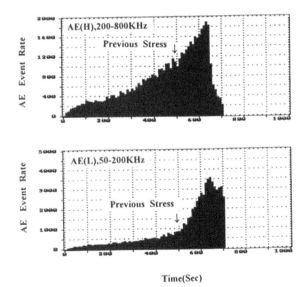

Fig. 10 A typical example of AE event rate (AE(L) and AE(H)) versus time, 3 days after the triaxial previous loading.

during the whole loading.

Close observations of the recorded AE data indicated that the previous axial stress level of 20 MPa was easily detectable in both frequency ranges. There was no adverse effect of delay time within 3 days, and the previous axial stress level could be recognised quite well. In case of lower frequency level, ie. lower frequency AE event rate, AE(L), was more comparable to higher frequency rate, AE(H), and highly sensitive to the previous axial stress level. From this experiment, it can be concluded that short term time delay does not have significant effect in evaluating the rock damage, even when the rock was subjected to triaxial stress condition. We applied the rock damage parameter(χ_{RDP}) to the rock triaxially preloaded. Using equation (2) described above, rock damage parameter(χ_{RDP}) is 1.11, which means that rock is of good quality and in the elastic stage. The evaluated result is consistent with the previous loading condition, axial stress 20 MPa and confining pressure 10 MPa, which is within elastic region.

6. CONCLUSIONS

This study gives better understanding of the behaviour of rock under different stress conditions, eventually this can be used to evaluate the rock mass damage. The following conclusions can be drawn:

(1) AE study provides important information to assess the rock mass damage.

(2) The new suggested rock damage parameter, χ_{RDP}, can be successfully used to evaluate the quality of rock.

(3) There is strong correlation of AE with the applied stress and a well defined Kaiser effect.

(4) Short term time delay does not deter in evaluating the rock damage, even when the rock is subjected to triaxial stress condition.

(5) In the present study, we only examined sandstone samples, and further investigations will be required for other types of rock.

REFERENCES

Holocomb,D.J., Costin, L.S. (1986): Detection of damage surfaces in brittle materials using acoustic emission. Trans. ASME 53, 536-544.

Holocomb, D.J., Stone, C.M., Costin, L.S. (1990): Combining acoustic emission locations and a microcrack damage model to study development of damage in brittle materials. In: Hustrulid,W.A., Johnson, G.A. (eds) Proc. 31st U.S. Symposium, Rock Mechanics Contributions and Challenges. Balkema, Rotterdam, 645-651.

Kaiser, E.J. (1950): A study of acoustic phenomena in tensile test. Doctoral Thesis, Technische Hochshule Munchen.

Kanagawa, T., M. Hayashi, Nasaka, H. (1976): Estimation of spatial geostress in rock samples using the Kaiser effect, Rep. No.375017, Central Res. Inst. of Electrical Power Industry, Abiko, Japan.

Kurita, K. Fujii, N. (1979): Stress memory of crystalline rocks in acoustic emission. Geophys. Res. Lett. 6 (1), 9-12.

Li, C., Nordlund, E. (1993): Experimental verification of the Kaiser effect in rocks. Rock Mech. Rock Engng. 26 (4), 333-351.

Rock Fragmentation by Blasting, Mohanty (ed.) © 1996 Taylor & Francis. ISBN 90 5410 824 X

A study of detonation timing and fragmentation using 3-D finite element techniques and a damage constitutive model*

Dale S. Preece & Billy Joe Thorne
Sandia National Laboratories, Albuquerque, N. Mex., USA

ABSTRACT: The transient dynamics finite element computer program, PRONTO-3D, has been used in conjunction with a damage constitutive model to study the influence of detonation timing on rock fragmentation during blasting. The primary motivation of this study is to investigate the effectiveness of precise detonators in improving fragmentation. PRONTO-3D simulations show that a delay time of 0.0 sec between adjacent blastholes results in significantly more fragmentation that a 0.5 ms delay.

1 INTRODUCTION

One of the advantages of precise detonators (microsecond accuracy) appears to be enhanced fragmentation. The reasons for this are not clearly understood but it has been surmised that complimentary wave interaction in the region between two detonating blastholes results in better fragmentation. To address this question a study has been undertaken using the 3-D transient dynamic computer code PRONTO in conjunction with a damage material constitutive model.

PRONTO-3D (Taylor and Flanagan, 1989) has been continuously evolving at Sandia National Laboratories for many years. As an explicit transient dynamics finite element code, PRONTO is capable of addressing a large variety of dynamic simulations including: impact, metal forming and explosive/structure interaction. A wide range of material responses and large deformations and strains can be treated during any simulation. The damage constitutive model used in this study was specifically developed for predicting blast induced fragmentation in rock and has been exercised on several closely controlled crater field experiments with reasonable results (Thorne, 1990a&b, 1991).

The configuration examined in this study consists of two blastholes (crater style) located relatively close to on another. The geometry is similar to a single blasthole crater experiment that was used to qualify the damage constitutive model but with two blastholes instead of one. Two different delay times, 0.0 s and 0.5 ms, between the blastholes are simulated and the rock is assumed to be granite.

2 DAMAGE CONSTITUTIVE MODEL

The damage model is intended to simulate the dynamic fracture of brittle rock. It is based on work started by Kipp and Grady, 1980 and continued by Taylor, Chen and Kuszmaul, 1986 and Kuszmaul 1987a. It was modified (Thorne, 1991) to extend it to large crack densities as suggested by Englman and Jaeger, 1987. Its essential feature is the treatment of the dynamic fracture process as a continuous accrual of damage in tension due to microcracking in the rock. The fundamental assumption of the model is that the rock is isotropic and permeated by an array of randomly distributed and oriented microcracks. These microcracks grow and interact with one another under tensile loading. A complete derivation of the damage model is given in Preece et al, 1994. A brief summary is presented in this paper as an aid to understanding the computational results.

Englman and Jaeger, 1987, introduce a regularized damage parameter, F, which is related to Budiansky and O'Connell's, 1976, crack density, C_d, but takes into account the overlap between the damage vol-

* This work performed at Sandia National Laboratories supported by the U.S. Department of Energy under contract no. DE-AC04-94AL85000 and also supported by ICI Explosives USA.

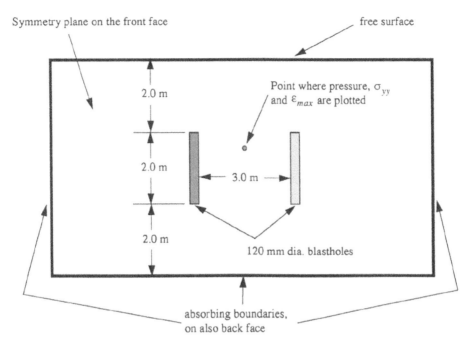

Figure 1: Two Blasthole Crater Geometry

Labels in figure:
Symmetry plane on the front face

free surface

2.0 m

Point where pressure, σ_{yy} and ε_{max} are plotted

2.0 m

3.0 m

2.0 m

120 mm dia. blastholes

absorbing boundaries, on also back face

umes of different cracks. To this end they define F by

$$F = 1 - exp(-\alpha C_d) \qquad 2.1$$

where $\alpha = 16/9$.

In order to relate stress to strain we will generate a system of equations which can be solved for the effective elastic moduli of the cracked medium. It is convenient to introduce a damage parameter, D, defined by

$$D = f(v_e)F \qquad 2.2$$

where v_e is the effective Poisson's ratio. K_e is the effective bulk modulus of a cracked medium and is given in terms of the undamaged bulk modulus K by

$$K_e = (1-D)K \qquad 2.3$$

The crack density, C_d, can be related to an average flaw size, a, by

$$C_d = \Upsilon N a^3 \qquad 2.4$$

where N is the number of active cracks and Υ is a proportionality ratio.

At this point, it should be noted that there is variety of assumptions which can be made as to the form of N, and there is almost no agreement as to the proper form for a. Kipp and Grady, 1980, and Kuszmaul, 1987a, assume that the number of cracks activated at a volumetric strain ε is described by a Weibull distribution of the form.

$$N = k\varepsilon^m \qquad 2.5$$

where k and m are material dependent constants and the volumetric strain, ε, is one third of the time integral of the trace of the deformation tensor, d with ε being positive in tension. Equations 2.5 and 2.4 imply

$$C_d = k\varepsilon^m a^3 \qquad 2.6$$

Based on energy considerations at high strain rates, Grady, 1983 derives the following expression for the nominal fragment radius, r, for dynamic fragmentation of a brittle material

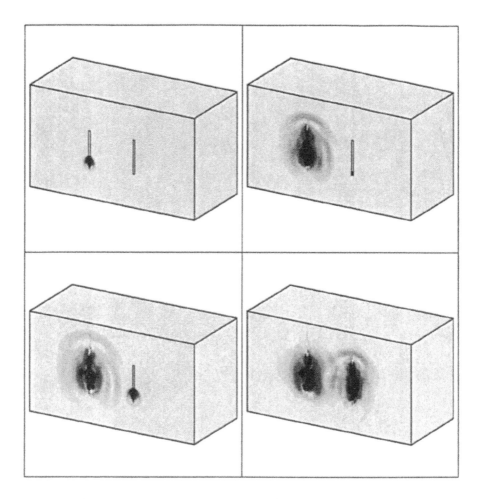

Figure 2: Pressure distribution at times 80 μs, 0.5 ms, 0.6 ms and 1.0 ms for a delay
time of 0.5 ms. The gray-scale (light-to-dark) pressure range is from 0.0 to
-400.0 Mpa. Pressures above and below the range are the same shade as the
upper or lower limits. Tension is positive.

$$r = \frac{1}{2}\left[\frac{\sqrt{20}K_{IC}}{\rho cR}\right]^{2/3} \qquad 2.7$$

Here K_{IC}, ρ and c are the fracture toughness, density and sound speed of the undamaged material and R is the strain rate, which is assumed in the derivation to be both constant and large. This is an average fragment radius for the global response of a uniformly expanding sphere. We will assume that the local average flaw size, a, is proportional to the value of r appropriate to the local strain rate.

In order to apply equation 2.7 to the case where the strain rate is not constant, Taylor, Chen and Kuszmaul, 1986, replace the constant strain rate, R, in equation 2.7 with the maximum strain rate, R_{max}, which the material has experienced. Making some assumptions about the maximum strain rate and combining equations 2.6 and 2.7 yields an expression for C_d based on measurable material parameters.

$$C_d = \frac{5k\varepsilon^m}{2}\left[\frac{K_{IC}}{\rho cR_{max}}\right]^2 \qquad 2.8$$

149

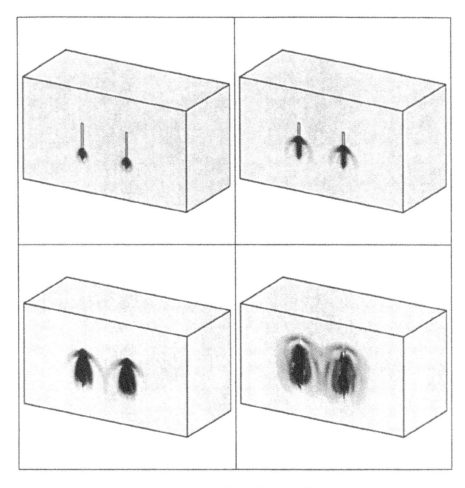

Figure 3: Pressure distribution at times 80 μs, 0.2 ms, 0.32 ms and 0.48 ms for a delay time of 0.0 ms. The gray-scale (light-to-dark) pressure range is from 0.0 to -400.0 Mpa. Pressures above and below the range are the same shade as the upper or lower limits. Tension is positive.

In the constitutive model implementation, equations 2.1 through 2.8 form the basis for derivation of a coupled system of ordinary differential equations which can be integrated to define the response of the damaged material.

The material parameters for granite were measured by Olsson, 1989, and Chong et al, 1988, are listed in Table 2.1.

Table 2.1: Granite Material Properties

Density	$\rho = 2680\ kg/m^3$
Youngs' Modulus	$E = 62.8$ GPa
Poissons' Ratio	$\nu = 0.29$
k	$k = 5.3 \times 10^{26}/m^3$
m	6.0
Fracture Toughness	$K_{IC} = 1.68$ MPa$\sqrt{m}$

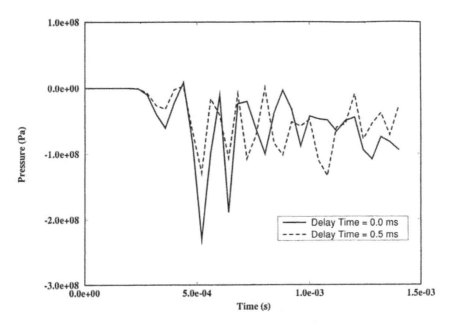

Figure 4: Pressure versus time at the point indicated in Figure 1 for
delay times of 0.0 *ms* and 0.5 *ms*. Tension is positive.

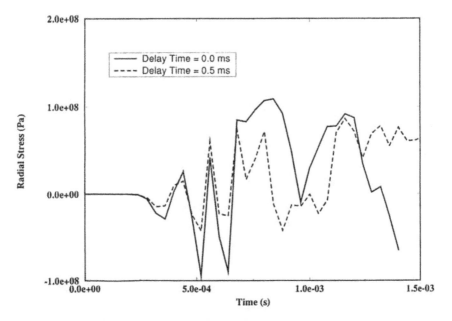

Figure 5: σ_{yy} versus time at the point indicated in Figure 1 for
delay times of 0.0 *ms* and 0.5 *ms*. σ_{yy} at the point indicated is also the circumferential stress. Tension is positive.

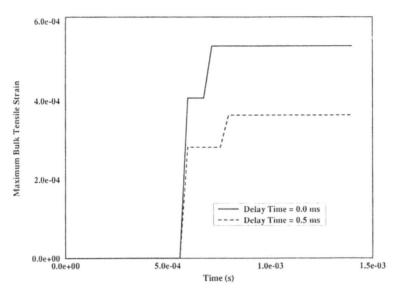

Figure 6: ε_{max} versus time at the point indicated in Figure 1 for delay times of 0.0 *ms* and 0.5 *ms*. Tension is positive

3 FINITE ELEMENT MODEL

The cross-sectional geometry of the 3-D finite element model is shown in Figure 1. This simulation models two explosive columns that have a diameter of 120 mm, a height of 2.0 m, a separation of 3.0 m, buried 2.0 m deep and filled with 28.27 *kg* of emulsion explosive. Full coupling is assumed between the explosive and the surrounding granite. Detonation begins at the bottom of each explosive column and is modeled with a controlled burn based on a specified detonation velocity.

The finite element model employed here has 41760 3-D hexahedral elements and 45933 nodes. The spatial resolution of this model is too fine to be drawn in this paper. Detonation delay times of 0.0 s and 0.5 ms were treated in two separate calculations. Each calculation required approximately 4 days of cpu time on a SUN SPARCstation 10-41 workstation. The databases produced by these calculations contained time steps saved every 40 μs. The zero-delay-time database contained 36 time steps and had a size of 144 megabytes while the 0.5 ms delay-time-database contained 48 time steps and occupied 191 megabytes.

4 COMPUTATIONAL RESULTS

Figures 2 and 3 show the pressure distribution as a function of time for the two different delay times in this study. A time history of the pressure halfway between the two blastholes and 1/4 of the blasthole height from the top (see Figure 1) is given in Figure 4. This graph indicates a significant increase in the bulk pressure $(\sigma_{xx} + \sigma_{yy} + \sigma_{zz})/3$ in the case of the simultaneous detonation that is less obvious from Figures 2 and 3. Although the bulk pressure is generally compressive, the circumferential (hoop) component (in cylindrical coordinates) will be tensile at least part of the time during the passage of the wave. This is illustrated in Figure 5 which shows σ_{yy} at the point of interest on the front symmetry plane. At this location σ_{yy} is equal to the cicumferential stress and is tensile part of the time. It is this tensile stress that results in tensile volumetric strain and consequently damage in the material. In studying Figures 4 and 5 one should keep in mind that they represents the pressure and stress in the model at specific output times. As discussed in section 3, the model is so large that the number of output time steps had to be limited.

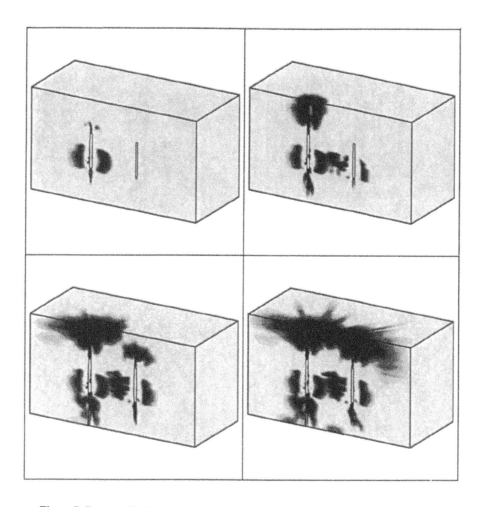

Figure 7: Damage distribution at times 0.5 *ms*, 0.8 *ms*, 1.2 *ms* and 1.88 *ms* for a delay time of 0.5 *ms*. The gray-scale (light-to-dark) damage range is from 0.0 to 1.0.

Figure 6 shows the maximum bulk tensile strain, ε_{max}, as a function of time and at the location indicated in Figure 1. ε_{max} is the maximum tensile volumetric strain at that point up to the time of interest and can only increase with time. Significantly higher values of ε_{max} are indicated in the simultaneous detonation case. The crack density, C_d, in equation 2.6 is an exponential function of the volumetric strain ε where the exponent, m, in this case (and for most rocks) is 6. Exponentiation to a power of 6 results in a large difference in the crack density and the damage produced by the two different delay times. These results are typical of the pressures and volu-

metric strains along a line of symmetry between the two blastholes and also for some distance on either side of that line.

The calculated spatial distribution of the damage at four different times is shown in Figure 7 for a 0.5 ms delay between detonations. A damage or crack density threshold value for fragmentation has been analyzed by Kuszmaul, 1987b and Thorne, 1990a&b. For the purposes of this study fragmentation is assumed to occur in the dark regions where the damage is close to one. The same plot is given for the simultaneous detonation in Figure 8. Comparison of

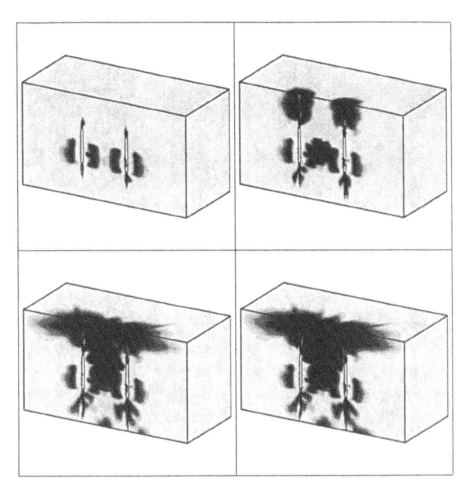

Figure 8: Damage distribution at times 0.5 *ms*, 0.8 *ms*, 1.2 *ms* and 1.4 *ms* for a delay time of 0.0 *ms*. The gray-scale (light-to-dark) damage range is from 0.0 to 1.0.

these two Figures indicates a significant improvement in the damage and fragmentation with the simultaneous detonation case being superior. A weakness of this damage constitutive model is its inability to predict damage in compression. Thus the undamaged regions adjacent to the blastholes would, in reality, be damaged and fragmented.

5 CONCLUSIONS

The 3-D transient dynamics finite element computer program PRONTO has been employed along with a damage constitutive model to study the influence of delay timing on the fragmentation of rock during blasting. This constitutive model accumulates damage based on episodes of tensile volumetric strain. These simulations show that simultaneous detonation significantly improves the fragmentation between blastholes. The reason for this improvement is the positive reinforcement of waves from the two blastholes arriving at the same location at the same time. This positive reinforcement manifests itself in significant increases in pressure and volumetric strain which results in improvements in damage and fragmentation when the blastholes are detonated simultaneously.

This study has implications for the value of precision detonators which can deliver μs accuracy in the delay time between blastholes. This study indicates that improvements to fragmentation due to precision timing may well be worth the additional cost of the detonator.

This paper has demonstrated a numerical capability that can be used to study the influence of precision timing on fragmentation. The effects on timing and fragmentation of many other parameters such as blasthole spacing and depth as well as rock and explosive types can also be studied using the techniques presented in this paper.

ACKNOWLEDGMENTS

The authors wish to acknowledge the contributions of personnel from ICI Explosives USA who have supported this work both financially and through field experimentation.

REFERENCES

Budiansky, B. and O'Connell, R. J., 1976, "Elastic Moduli of a Cracked Solid," Computer Methods in Applied Mechanics and Engineering, vol. 12, pp. 81-97.

Chong, K. P., Basham, K. D., Wang, D. Q. and Estes, R. J., 1988, "Fracture Toughness Characterization of Eastern Basalt and Gneiss," KPC & Associates report to Sandia National Laboratories on contract No. 55-5698, Laramie WY.

Englman, R. and Jaeger, Z., 1987, "Theoretical Aids for Improvement of Blasting Efficiencies in Oil Shale and Rocks," AP-TR-12/87, Soreq Nuclear Research Center, Yavne, Israel.

Grady, D., 1983, "The Mechanics of Fracture Under High-Rate Stress Loading," in William Prager Symposium on Mechanics of Geomaterials: Rocks, Concretes and Soils, (Bazant, Z. P., ed).

Kipp, M. E. and Grady, D. E., 1980, "Numerical Studies of Rock Fragmentation," SAND79-1582, Sandia National Laboratories, Albuquerque, NM.

Kuszmaul, J. S., 1987a,"A New Constitutive Model for Fragmentation of Rock Under Dynamic Loading," Proceedings of the Second International Symposium on Fragmentation by Blasting, Keystone CO, pp 412-423.

Kuszmaul, J. S., 1987b,"A Technique for Predicting Fragmentation and Fragment Sizes Resulting From Rock Blasting," Proceeding of the 28th U. S. Symposium on Rock Mechanics, Tucson, Arizona.

Olsson, W. A., 1989, "Quasi-Static and Dynamic Mechanical Properties of a Granite and a Sandstone," SAND89-1197, Sandia National Laboratories, Albuquerque, NM.

Preece, D. S., Thorne, B. J., Baer, M. R. and Swegle, J. W., 1994, "Computer Simulation of Rock Blasting: A Summary of Work From 1987 Through 1993," SAND92-1027, Sandia National Laboratories, Albuquerque, NM.

Taylor, L. M., Chen, E. P. and Kuszmaul, J. S., 1986, "Microcrack-Induced Damage Accumulation in Brittle Rock Under Dynamic Loading," Computer Methods in Applied Mechanics and Engineering, vol. 55, no.3, pp. 301-320.

Taylor, L. M. and Flanagan, D. P., 1987, "PRONTO 3D A Three-Dimensional Transient Solid Dynamics Program," SAND87-1912, Sandia National Laboratories, Albuquerque, NM.

Thorne, B. J., Hommert, P. J. and Brown, B., 1990a, "Experimental and Computational Investigation of the Fundamental Mechanisms of Cratering," Proceedings of the Third International Symposium on Fragmentation by Blasting, Brisbane, Queensland, Australia.

Thorne, B. J., 1990b, "A Damage Model for Rock Fragmentation and Comparison of Calculations With Blasting Experiments in Granite," SAND 90-1389, Sandia National Laboratories, Albuquerque, NM.

Thorne, B. J., 1991, "Application of a Damage Model for Rock Fragmentation to the Straight Creek Mine Blast Experiments," SAND 91-0867, Sandia National Laboratories, Albuquerque, NM.

Rock Fragmentation by Blasting, Mohanty (ed.)© 1996 Taylor & Francis. ISBN 90 5410 824 X

Micro-sequential contour blasting – Theoretical and empirical approaches

P.A. Rustan
Division of Mining Engineering, Luleå University of Technology, Sweden

ABSTRACT: The optimal delay time between the contour holes in rock blasting has been studied by theoretical and empirical research in Sweden, regarding ground vibrations, increase in crack frequency, radial crack length and finally overbreak (half cast factor). The model test presented in this paper concerns controlled contour blasting in tunnelling and the full-scale blasts concern tunnelling, road cutting, and dimensional stone quarrying. The results indicate that the micro-sequential contour blasting technique (contour holes fired in sequence and with a delay in the order of 1-2 ms) is superior to simultaneous initiation both regarding blast-induced ground vibrations and crack frequency increase in the rock mass. Both these evaluation methods reflects the conditions deeper in the remaining rock mass. Simultaneous initiation, however, is superior to micro-sequential contour blasting both regarding the half cast factor* and the length of radial cracks emanating from the blastholes. These two parameters are more related to the surface conditions after blasting. The industrial applications of this new knowledge are the use of micro-sequential contour blasting when ground vibrations are of greater concern than the contour, e.g. in trench blasting or quarrying in urban areas, and the use of simultaneous initiation when an even rock surface is of high priority.

1 INTRODUCTION

An important aim in all rock excavations is to reach stable contours. The most common technique to blast contours today is to use the smooth blasting technique, in which the contour holes are charged with a low linear charge concentration and are initiated on the last delay and simultaneously. Usually HS-detonators (half-second detonators) are used for the contour holes and they normally have a scatter in nominal timing of about ± 200 ms, according to Langefors (1967). Often the contour holes in the walls are initiated on a separate delay and those in the roof on the next delay.

The smooth blasting technology was developed during the 1960's. However, due to the large scatter in timing for HS-detonators commonly used for the initiation of the contour holes, it is not possible to reach a simultaneous initiation with pyrotechnical detonators. To make a really smooth blast it is

therefore necessary to use precise delaying technique, e.g. electronic detonators, or to use a certain length of shock tube or detonating cord as the time delay between the contour holes.

When simultaneous initiation is used, the blast waves from neighbouring holes will meet half way between the contour holes and superimpose, which means that the amplitude of the blast waves will double, and this is expected to increase the blast-induced damage due to ground vibrations.

During the 1980's researchers at Luleå University of Technology developed the new idea of investigating whether there could be an optimal time delay in contour blasting which would reduce the following four parameters: blast-induced ground vibrations, increase in crack frequency, radial crack length and overbreak of rock behind the theoretical contour (half cast factor). The research methods used were laboratory tests followed by theoretical work, and continued by full-scale tests undertaken both underground and later on by other research teams on the surface, trying to verify the hypothesis that there exists an optimal delay time between the contour holes.

---

*) Half cast factor is defined as the percentage of the visible length of identifiable blasthole "barrels" or "half casts" over the total number of perimeter holes drilled.

Theoretical work at Luleå University of Technology during the 1980's, see Rustan et al (1985), indicated that micro-sequential contour blasting using delay times of about 1,5 ms between the contour holes during drifting should be optimal at a burden of 0,5 m and a 1,0 m spacing of the contour holes. A full-scale test during drifting in magnetite ore at LKAB in Malmberget indicated a reduction of the blast-induced vibration to 1/6 in the near field (6-7 m) compared with HS-initiation of the contour holes. No reduction in overbreak could, however, be found using the micro-sequential blasting technique, because the half cast factor was almost the same, 85,7% for micro-sequential blasting and 86,0% for HS-initiation.

Since then, more tests have been undertaken in Sweden to show the effect of using micro-sequential delays between contour holes. From our field tests and also from other tests in Sweden, it is now verified that the micro-sequential blasting technique reduces the ground vibrations considerably, ~1/6 to 1/4 has been measured. It also reduces the blast-induced crack frequency in the remaining rock mass. However, the half cast factor was found to be equal or larger, 0% to 17%, with simultaneous initiation, and the radial crack length was less than with simultaneous initiation.

The industrial application of this new knowledge is therefore the use of micro-sequential initiation of contour holes, for example in trench blasting in populated areas or in the vicinity of stone quarries where there is a need for reduced ground vibrations.

2 LITERATURE REVIEW

2.1 Introduction to the development of the micro-sequential blasting formula

The initial smooth blasting theory was developed at Nitroglycerin AB's Detonic Research Laboratory in Stockholm by Lundborg and reported by Langefors (1953) and (1959). The technique was based on laboratory tests using plexiglass and full-scale blasts. The influence of the delay time between the contour holes on the radial crack length and fragmentation is shown for model blast tests using plexiglass in Fig. 1.

Measuring the longest radial crack from each blasthole and calculating the mean of these values at each delay time gives the following result. Short delay blasting produced a 33% longer mean maximum radial crack length from each blasthole compared with simultaneous initiation and shot-by-shot initiation a 13% longer mean maximum radial crack length from each blasthole compared with simultaneous initiation, see Fig. 2.

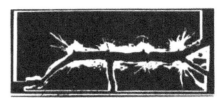

a) Simultaneous initiation (0 delay time)

b) Short-delay blasting

c) Shot by shot (∞ delay time)

Fig. 1. Blasting a row of four holes (experiment using plexiglass) varying the delay time between holes; a) simultaneous initiation (0 delay time), b) short delay blasting c) shot-by-shot blasting (∞ delay time). Langefors (1953).

Relative mean maximum radial crack length for each blasthole

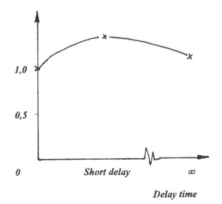

Fig. 2 Influence of delay time between contour holes on mean maximum radial crack length from each blasthole. Langefors (1953).

158

This result is consistent with the Vånga blast tests showed futher on, see Fig 10.

Langefors makes the statement that "The delay time between the blastholes should be as small as possible without getting the blast-induced damage caused by simultaneous initiation. In practice this means short delay blasting with delays of one or a few milliseconds at the most". Langefors therefore suggests a delay time close to that delay time which in this paper is called micro-sequential initiation (1-2 ms). Langefors was expecting the smallest blast-induced damage at that delay time, but as we will learn further on from the full-scale blasting tests carried out so far, an increase in the half cast factor or reduction in radial crack length cannot be obtained with micro-sequential blasting. Our results only show that ground vibrations and the increase in crack frequency could be reduced.

Another well-known effect, which can be seen in Fig. 1, is that simultaneous initiation produces a more course fragmentation.

Brown (1968) also showed by model blast tests using plexiglass, that simultaneous initiation of the contour holes is very important for the development of the radial cracks in the direction of the contour of the blast. In the same paper it was also shown by high speed photography of birefringement material, that the initial cracks between the contour holes occur very early, already 5 μs after initiation, and that the cracks are elongated by the sustained (quasi-static) pressure in the blastholes. Still at 130 μs after initiation, the radial cracks have not joined between the blastholes.

Experiments by Dally et al (1978) using transparent polyester, known commercially as Homalite 100, showed a development of cracks between the contour holes at ≈ 5-40 μs after initiation. (Homalite 100 is a polymeric material and it is extremely brittle as evidenced by its extremely low critical fracture toughness $K_{Ic} = 0,440$ MN/m$^{3/2}$.

No cracks were, however, started in the middle between the two instantaneously initiated blastholes, where the strain waves superimpose, as pointed out by some authors. The radial crack extension between the blastholes was found to be due to the quasi-static pressure in the two blastholes.

Rustan et al (1985) reported results from a model test on the scale of 1:36, simulating a contour tunnel blast in a hardened mixture of quartz sand and epoxy. The test was performed at Luleå University of Technology. A low strength detonating cord (1 g/m)

was used as the column charge simulating Gurit pipe charges 17 mm in diameter and in blastholes 1,5 mm in diameter. The burden was 27 mm and spacing 16 mm. Three different time delays were used and numbered 1-3, with a 27 ms delay time between the delays, see Fig. 3.

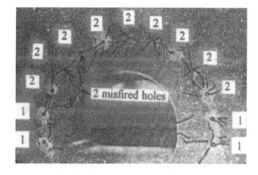

Fig. 3. Radial crack system around contour holes when the burden is too large and no breakage occurs (presplitting case). Two holes out of thirteen misfired. Naarttijärvi et al (1980).

Because of too large a burden, a presplit was formed between all contour holes, and a long radial crack developed next to the two holes which did not fire on the left side of the drift, see Fig. 3.

It is believed that the long time delay between the neighbouring holes was the cause of the long crack in the left wall. The time delay can be regarded as infinite when a neighbouring hole does not initiate as planned.

The primary conclusion, despite very little experimental data, was that a long delay time between the contour holes can create a larger blast-induced damage to the rock mass. This finding is similar to that reported by Langefors (1953) and (1959). Simultaneous initiation of the blastholes will therefore promote cracking between the holes and hinder the cracking into the remaining rock mass. The cracks are generally propagating in the direction of the largest stress.

The new question was therefore raised as to whether there exists an optimal delay time between the contour holes, as pointed out by Langefors, where overbreak and radial crack length could be minimized. The derivation of this optimal delay time will be shown in the next section.

---

*) In the paper by Dally et al (1978) an incorrect transformation from $K_{Ic} = 400$ psi(inch)$^{1/2}$ to $K_{Ic} = 360$ MN/m$^{3/2}$ has been done.

2.2 Development of the microsequential blasting formula

The mathematical formula was based on the hypothesis that the time delay between the contour holes should be as small as possible according to Langefors (1953) and (1959) and Brown (1968), to improve the splitting effect between the contour holes and at the same time it should be long enough to avoid the large vibrations caused by simultaneous initiation.

The goal was therefore to find a delay time where the stress waves (the longitudinal "P-wave" and transverse "S-wave") created upon detonation in one blasthole would not cooperate with the stress waves coming from a neighbouring hole. The hypothesis was therefore that, if both the P- and S-waves were to pass the neighbouring hole before that hole is initiated, even after reflection at the free surface created by the holes next to the contour holes, this would reduce the ground vibrations and blast-induced damage to the remaining rock mass, see Fig. 4

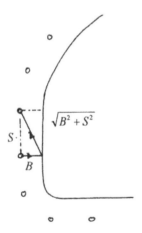

Fig.4 Schematic view of the basic set up for the micro-sequential initiation theory

The longest propagation distance for the wave will be the sum of the burden distance (B) plus the square root of the burden squared plus the spacing squared ($\sqrt{B^2 + S^2}$). All other wave traces within the area defined by the burden times the spacing will be shorter. By delaying the neighbouring holes the energy superposition of the blast waves between the blastholes will be avoided and the vibration level should theoretically be reduced to half of that created in simultaneous initiation. Because of the longer travel time for the reflected S-wave (the propagation

velocity is roughly half of the P-wave velocity), this time will be the most critical. The mathematical equation is based on geometry and wave velocities, and the optimal micro-sequential blasting delay time will therefore read as follows:

$$t_o = k \left(\frac{B + \sqrt{B^2 + S^2}}{c_s} \right) 10^3 \qquad (1)$$

where t_O is the optimal delay time between two adjacent contour holes in (ms), B is the burden of the contour holes in (m), S is the spacing between the contour blastholes in (m), c_s is the S- wave velocity in (m/s) and k is a safety factor which may be used when the S-wave velocity is not very well known. The selection of k should be larger than 1. This formula was first published by Rustan et al (1979) in Swedish and later on by Rustan et al (1985) in English.

The hypothesis was verified by full-scale tests in drifting at LKAB in Malmberget. Originally the method was given the name "Cutblasting" to distinguish it from smooth blasting and presplitting. The name is a direct translation of the Swedish term "Skärsprängningsteknik". This rock blasting technique could be most simply explained by comparing its effect to cutting a bread with a knife, see Fig. 5.

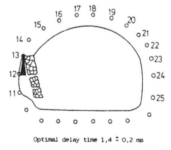

Optimal delay time 1,4 ± 0,2 ms

Fig. 5 The micro-sequential blasting principle. Rustan et al (1985).

The name cutblasting may be mixed up with the term cutblasting used in tunnelling, meaning the blasting of the first holes in a tunnel round, and therefore the name was later on changed to micro-sequential contour blasting technique.

3 VERIFICATIONS OF THE MICRO-SEQUENTIAL BLASTING HYPOTHESIS

The micro-sequential blasting theory has now been tested in three different working environments:

tunnelling in magnetite ore at LKAB in Malmberget and Kiruna underground mines, tunnelling in waste rock at LKAB in Kiruna underground mine, in a road cut at Svalbo, and finally in a dimensional stone quarry at Vånga. In three of the places, Malmberget, Svalbo and Kiruna, the half cast factor was determined, in two of the places, Malmberget and Svalbo, the ground vibrations and increase in crack frequency behind the blastholes were measured, and in one test at Vånga, the radial crack length was determined.

3.1 Field tests at LKAB in Malmberget underground mine

In the field tests performed at LKAB in Malmberget, the micro-sequential contour blasting technique was compared with conventional initiation with HS-detonators when drifting in magnetite ore having a uniaxial compressive strength varying from 20-111 MPa, and a Young's modulus of 27-108 GPa. The HS-detonators had a scatter in timing of about ± 200 ms. Five roof holes with a diameter of 45 mm and length of 3,5 m were blasted in a separate blast after all other holes had been blasted. The bottom charge had a length of 0,5 m, consisted of Dynamex B (DxB), and had a weight of 0,57 kg, while the column charge consisted of detonating cord with a linear charge concentration equivalent to 140 grams DxB /m. Detonating cord (1 g/m) was used as a delayer for the micro-sequential contour blasting initiation, and the optimal delay time was calculated to be 1,3 ms at 0,5 m burden and 1,0 m spacing for the contour holes, see Fig. 6.

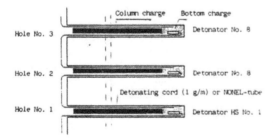

Fig. 6 Initiation sequence set up to test the micro-sequential blasting principle. Detonating cord with a linear charge concentration of 1 g/m was used to create the delay time between the contour holes. Rustan et al (1985).

The blast-induced damage was measured by borehole viewing, 0-1 m above the roof, accelerometers in the near field, 6-7 m from the roof holes, and geophones

at a larger distance of ~20 m from the roof holes. The acceleration sensors were mounted in boreholes above the roof and the geophones at the wall of a drift situated above the roof. The increase in crack frequency was studied in boreholes drilled in angled holes above the roof. A borehole viewer was used to observe the cracking. If the accelerometers are mounted in a proper way in the boreholes, the measurements of vibration are usually regarded as more reliable than borehole viewing. If the rock has many open airfield cracks, however, the results from vibration measurements can be misleading.

One test round was fired for each technique, round No. 10 (HS-initiation) and round 12 (micro-sequential initiation).

The results indicated that when the micro-sequential blasting technique was used instead of HS-initiation, the ground vibrations were reduced to 1/6. The blast-induced damage zone (defined by more than 3 new cracks/m created at the blast) was reduced to half for the micro-sequential blasting technique (0,4 m compared with 0,75 m for HS-initiation) according to the borehole viewing. The half cast factor was almost the same, 85,7% for micro-sequential contour blasting and 86,0% for HS-initiation of the contour holes.

From literature studies and the field tests at LKAB in Malmberget it was concluded that the half cast factor is not a good indicator of blast-induced damage deeper into the rock mass. The engineers in the LKAB Kiruna mine had the experience that a high half cast factor could suddenly result in rock fall, especially in poor rock conditions, and the half cast factor was therefore not regarded as a good indicator of a good controlled contour blast in poor rock.

Because only one test round was fired for each technique, the conclusion was that the micro-sequential contour blasting technique had to be tested more. However, the micro-sequential initiation was superior to HS-initiation both regarding ground vibrations and crack increase in the remaining rock, although there was no reduction in half cast factor as expected.

3.2 Field tests at Svalbo road cut

Within the "Swedish Mining 2000 Project", controlled contour blasting tests were performed during the blasting of a road cut at Svalbo along Highway No. 60, 15 km south of Lindesberg in the middle Sweden. The micro-sequential contour blasting technique was tested here by the Swedish Detonic Research Foundation in co-operation with

Nitro Nobel, Boliden Mindeco, LKAB, Vielle Montagne, the Swedish Road Administration and Luleå University of Technology, see Niklasson and Karlsson (1991).

The rock consisted of an unweathered leptite with fine and medium grains. The bench height was 4-5 m. Three rounds were used to test the micro-sequential blasting technique, two of the rounds with 21 holes in one row in each round and one round with 10 holes in a row. Simultaneous initiation was tested in two rounds with 21 holes in each row and one round with 10 holes in one row. HS-initiation was tested in one round with 21 holes in a row and one round with 11 holes in a row. Electronic detonators from Nitro Nobel with very high precision in timing (scatter 0,5 - 1,0‰ from the nominal detonation time) were used for simultaneous and micro-sequential initiation and Nitro Nobel GT/T No. 55 detonators for HS-initiation.

The optimal delay time calculated by the micro-sequential blasting formula (1) was 1,6 ms. To be sure that the scatter in timing of the electronic detonators (1,5 - 2,0 ms) would not cause any overlapping, a delay time of 2 ms was selected as the micro-sequential contour blasting delay time. The first delay in the contour was initiated at 3000 ms, which would be the nominal delay time in a tunnel or drift round. High speed filming of the blasts showed that there was no overlapping between the blastholes (which means that no later delay number was initiated before an earlier delay number).

The results from the vibration measurement on the surface 10 m behind the blast front are shown in Fig. 7.

Acceleration (g) and peak particle velocity (mm/s)

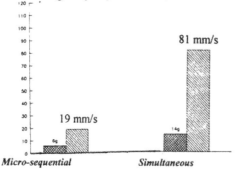

Micro-sequential   Simultaneous

▨ *Acceleration*   ▧ *Peak particle velocity*

Fig. 7 Acceleration measured on the surface with accelerometers 10 m behind the blastholes in the Svalbo road cut tests and calculated peak particle velocities. Niklasson and Karlsson (1991).

The micro-sequential contour blasting technique, if compared with simultaneous initiation induced about half or 57 % lower acceleration values and about 1/4 or 76% lower peak particle velocity values. This verifies the micro-sequential blasting hypothesis. Unfortunately the vibration measurements of the HS-initiation failed.

The increase in the half cast factor using the micro-sequential blasting technique could, however, not be verified by the Svalbo tests, see Fig. 8.

Half cast factor (%)

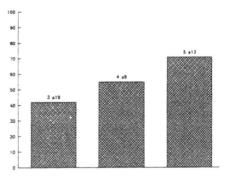

HS-initiation   Micro-sequential   Simultaneous

HS-initiation   = *half-second initiation by Nitro Nobel GT/T detonators*
Micro-sequen.   = *micro-sequential initiation with electronic detonators*
Simultaneous   = *simultaneous initiation with electronic detonators*

Fig. 8 Modified half cast factor for contour holes (visible half cast related to charged borehole length) in a road cut at Svalbo. Niklasson and Karlsson (1991).

The modified half cast factor (visible half cast related to the charged borehole length) was much higher with simultaneous initiation, 71%, compared with 55% using micro-sequential contour blasting initiation.

One a priori explanation of the fact that the micro-sequential blasting hypothesis cannot be verified regarding half casts may be that simultaneous initiation gives a more equal load on the remaining rock mass compared with the micro-sequential contour blasting technique. An even load means a higher lateral confinement at the instant when the blast waves propagate through the rock, and it is a well-known fact that the strength of material increases with an increase in confinement.

The reason why the micro-sequential contour blasting technique produced less half casts compared with simultaneous initiation may also be the fact that the real delay time used was larger than the optimal. Therefore more shear forces than necessary were induced into the rock mass. This effect may have been reduced, if the safety factor k in formula (1) had been set to 1, thereby reducing the optimal delay time from 1,6 to 1,1 ms or ~ 1 ms. Compared with the 2 ms delay time used, this means therefore a reduction of 100%.

The increase in crack frequency was measured by borehole viewing in six holes directed slightly upwards and drilled into the bench face through the whole test area before the test started. The raw data in the report has been used by the author of this paper to draw Fig. 9.

Number of new cracks per metre (No./m)

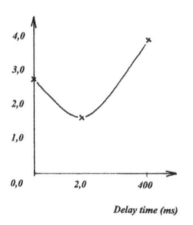

Delay time (ms)

Fig. 9 Influence of delay time on new cracks observed by borehole viewing in the first metre of rock behind the blast front. Svalbo test in leptite. Raw data from Niklasson and Karlsson (1991).

The micro-sequential initiation hypothesis was therefore verified by the Svalbo test both regarding the ground vibration and the crack frequency measurement. A lower half cast factor could, however, not be verified. More tests should be done using a shorter delay time, 1 ms, for simulation of micro-sequential contour blasting. The accuracy of the electronic detonators has to be increased further, if they are going to be used for micro-sequential contour blasting initiation with 1 ms delays.

3.3 Field tests at LKAB in Kiruna underground mine

Micro-sequential contour blasting and simultaneous

initiation were also compared in drifting at LKAB in Kiruna during 1990-1991 in the Sofia project, see Niklasson and Keisu (1991). Parallel hole tunnel rounds with 64 mm diameter holes and a depth of 4 or 7 m were drilled in waste rock or magnetite ore. The half cast factor for twelve rounds with simultaneous initiation was 38% compared with nine micro-sequential contour blasting rounds where the half cast factor was 33%. The micro-sequential contour blasting hypothesis was therefore not valid using the half cast factor as a blast damage criteria.

3.4 Field tests at Vånga dimensional stone quarry

In the Swedish granite dimensional stone quarry, Vånga, belonging to Nilsson & Söner AB in Västervik and situated about 19 km northeast of Kristianstad in south Sweden, the influence of delay time on radial cracking behind blastholes was studied by Olsson (1994). The granite is medium to coarse grained and has a uniaxial compressive strength of 200 MPa and a tensile strength of 12 MPa. The blasting was carried out using the Nitro Nobel Gurit pipe charges (17 mm in diameter) in boreholes 51 mm in diameter. The bench was vertical and the height 5 m, the burden 0,5 m and spacing 0,5 m. Electronic detonators (with a scatter in timing of ± 0,1 ms) from Nitro Nobel AB/Rinobel were used in the tests. After the blasting, huge rock blocks were taken out from the rock mass behind the blastholes, and the blocks were diamond sawed at three different horizontal levels above the bottom of the holes.

For five holes in a row and simultaneous initiation the maximum radial crack length was 6 cm compared with 55 cm for micro-sequential blasting (1 ms delay time between neighbouring holes). In another test with 10 holes in a row the mean value in all cuts showed a mean maximum crack length of 16 cm for micro-sequential blasting (1 ms delay time between neighbouring holes) and 12 cm for simultaneous initiation.

Fig. 10 shows how the radial crack length depends on the delay time.
The scatter in radial crack length is so large that in a specific situation with little data it will be difficult to determine which method is the best. A tendency could however be seen of an increase in radial crack length at increased delay time from ~ 10 cm in simultaneous initiation to 25 cm at 3 ms delay time.

From studies of fragmentation in model and half- and full-scale tests it is a well-known fact that the maximum size of fragments is achieved (few and short radial cracks in the removed rock) when the blastholes in a row are initiated simultaneously, see Langefors (1953) and (1965), Rustan (1970),

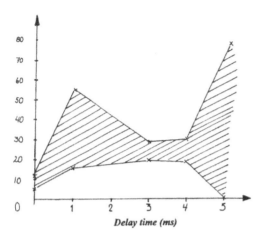

Radial crack lengths (Mean or maximum) (cm)

Fig. 10 The influence of delay time between contour holes on the mean or maximum radial crack length into the remaining rock mass. Raw data taken from Olsson, (1994).

Bergman and et al (1974), Fourney and Barker (1979), and Norell (1985). It may therefore be logical to assume that the radial crack length in the remaining rock also should be low in simultaneous. This has now been verified by the tests at Vånga.

The conclusion from the Vånga field test is therefore that simultaneous initiation produces shorter radial crack lengths.

4 CONCLUSIONS

According to the vibration measurements and the measurement of increase in crack frequency behind the theoretical contour of the blast performed at both LKAB in Malmberget and Svalbo, the micro-sequential blasting hypothesis is confirmed. These two evaluation criteria are the best, according to the authors opinion, regarding the conditions deeper in the remaining rock mass.

The hypothesis could, however, not be verified regarding the half cast factor studied in the LKAB tests in Malmberget and Kiruna and the Svalbo tests. Here simultaneous initiation produces a higher half case factor. The hypothesis could also not be verified regarding radial crack length in the Vånga tests which showed that simultaneous initiation produces shorter radial crack length's compared with micro-sequential contour blasting.

The total result is summarized in Table 1.

The following conceptual model of controlled contour blasting has been established on the experiences gained by this paper. The radial cracks are important for the overbreak since, if a radial crack crosses a weakness plane in rock, there is a high probability of rock outfall. This will affect the overbreak and also the half cast factor, see Fig 11. Deeper in the rock mass the damage is governed mainly by opening of weakness planes already existing in the rock mass before blasting. These cracks are therefore parallel to the dominating fracture systems in the rock mass. The amplitude of blast vibrations determines the extent of this damage. The micro-sequential blasting technique results in a lower level of ground vibration and less increase in crack frequency, see Fig. 11.

ACKNOWLEDGEMENT

Associate Professor Kou Shao Quan is hereby acknowledged for his kindness in checking the manuscript before submission.

Table 1 Confirmation of the micro-sequential contour blasting hypothesis is marked by yes. The two new criteria, not originally included in the hypothesis, do not however verify the hypothesis and are therefore marked with no.

Damage criterion	LKAB/Malmberget tunnel blast (1985)	Svalbo road cut (1990)	LKAB/Kiruna, tunnel blast (1991)	Vånga dimensional stone quarry (1994)
Ground vibrations	Yes	Yes	-	-
Increase in crack frequency	Yes	Yes	-	-
Half cast factor	-	No	No	-
Mean maximum radial crack length	-	-	-	No

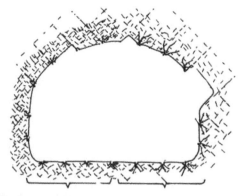

Simultaneous initiation

Advantage

-More half casts
-Shorter radial cracks

Disadvantages

-Higher ground vibrations
-More induced cracks
 parallel to the weakness
 planes

Micro-sequential initiation

Advantage

-Lower ground vibrations
-Less induced cracks
 parallel to the weakness
 planes

Disadvantages

-Less half casts
-Longer radial cracks

Fig. 11 A conceptual comparison of simultaneous and micro-sequential initiation. The purpose of the blast will determine what method should be selected.

REFERENCES

Bergman, O.R., Wu, F.C. and Edl, J. W. (1974). Model rock blasting measures effect of delays and hole on rock fragmentation. *Engineering and Mining Journal*, June.

Brown, A.N. (1968). Notes on an investigation into the basic fracture mechanisms encountered in controlled blasting. *Journal of the South African Institute of Mining and Metallurgy*, Oct. 1968.

Dally, J.W., Fourney, W.L., and Ladegaard Pedersen, A. (1978). A dynamic photo-elastic evaluation of some current practices in smooth wall blasting. *Society of Mining Engineers, Mining Engineering*, Feb.

Fourney, W.L. and Barker, D.B. (1979). Effect of time delay on fragmentation in a jointed model. *Mechanical Engineering Department, University of Maryland*, College Park Campus, Aug.

Langefors, U. (1953). Slätsprängning. (Smooth blasting), pp. 2-7. *Jernkontorets Annaler* Vol. 137, 1953 pp. 436-441. (In Swedish).

Langefors, U. (1959). Smooth blasting, pp. 1-7. *Water Power*, May.

Langefors, U. (1965). Fragmentation in rock blasting. *VI Symp. on Rock Mechanics*. The Pennsylvania State University, University Park, Pennsylvania, June 14-16.

Langefors, U. (1967). *The modern technique of rock blasting*. Almqvist & Wiksell/Gebergs förlag AB, Stockholm.

Naarttijärvi, T., Rustan, A., Öqvist, J. and Ludvig, B. (1980). Laboratorieförsök i försiktig sprängning (Laboratory tests in controlled contour blasting). Swedish Work Environment Fund Part report No. 7, Project No. 76/218:2. *Forskningsrapport TULEA 1980:22*, Luleå University of Technology. (In Swedish).

Niklasson, B. and Karlsson, L. (1991). Intervalltidens inverkan på slätsprängningsresultatet. (The influence of delay time on the result of smooth blasting). *Swedish Detonic Research Foundation Report No. DS 1991:6G*. (In Swedish).

Niklasson, B. and Keisú, M. (1991). Ny teknik för ort- och tunneldrivning- Sofiaprojektet. (New technique for drifting and tunneling-the Sofia Project). *Swedish Detonic Research report No. DS 1991:10*. (In Swedish).

Norell, B. (1985). Intervalltidens inverkan på fragmenteringen. *Swedish Detonic Foundation report DS 1985:1*. (In Swedish).

Olsson, M. (1993). Sprickutbredning vid skonsam sprängning. *Swedish Rock Engineering Research Report No. 3*. (In Swedish).

Olsson, M. (1994). Sprickutbredning vid flerhålssprängning. (Crack propagation in multiple hole blasting). *Swedish Rock Engineering Research Report No. 18*, pp. 1-38. (In Swedish).

Rustan, A. (1970). Mätmetod för bestämning av malmhalt hos en blandning av malm och gråberg - kinematik, svällning, och styckefall i försättningen vid sprängningn mot berg i pall- och skivrasmodeller. *Teknisk Licentiatavhandling, Royal Inst. of Technology*, Stockholm. (In Swedish).

Rustan, A., Naarttijärvi, T. and Ludvig, B. (1979). Försiktig sprängning i dåligt berg - modellstudier. (Controlled contour blasting in poor rock - model studies). *Bergsprängningskommitténs diskussionsmöte i Stockholm*, Feb. 8th, pp. 113-150. (In Swedish).

Rustan, A, Naarttijärvi, T. and Ludvig, B. (1985). Controlled blasting in hard intense jointed rock in tunnels. CIM Bulletin, Dec., pp. 63-68.

Rock Fragmentation by Blasting, Mohanty (ed.) © 1996 Taylor & Francis. ISBN 90 5410 824 X

Monitoring of large open cut rounds by VOD, PPV and gas pressure measurements

F. Ouchterlony, S. Nie, U. Nyberg & J. Deng
Swedish Rock Engineering Research, SveBeFo, Stockholm, Sweden

ABSTRACT: A project with the goal to minimize the blast damage to the remaining pit walls has been carried out at the Aitik open pit mine in North Sweden. Factors like confinement during the blast, blast direction and size of blast holes in the contour were systematically changed. During the blasts VOD in the blast holes plus PPV and gas pressure in gauge holes behind the contour were monitored.

The VOD measurements were used to check the explosive performance, to obtain the real initiation sequence and to identify the sources of the PPV pulses in the composite acceleration records. The PPV levels were used to establish a scaling law for the test area which later was used for blast damage assessments.

The gas pressure measurements show that the ordinary rounds, in which the blast holes are unstemmed, don't force any significant amounts of high pressure blast fumes into the walls. This is probably explained by a shock wave initiated dynamic swelling movement that opens up fracture planes. Those that are connected with the gauge hole increase its volume and reduce its pressure. The measured under-pressures correlate relatively well with the measured residual swell values. Further a shot presplit line didn't transmit direct shock waves or blast fumes into the walls. The presplit blast holes themselves though, despite being unstemmed, did however force pressurized blast fumes into the rock. This may be a source of blast damage.

1 INTRODUCTION

The Boliden Mineral AB, Aitik open pit copper mine in North Sweden produces about 15 Mton of 0,4% grade ore annually. The slope conditions are of vital importance to the mine economy, hence the strong requirements to reach the planned inter-ramp and bench face angles. Even if geology determines the stability to some extent, the blast damage may be quite important for the individual bench crests and have large economic consequences.

The mine uses double 15 m benches with a catchment berm every other level. The berm has to be at least 11 m wide over 90% of its length for it to function properly. This was not always so and projects were initiated to deal with it, the first in 1991 (Sognfors 1994). Despite a steepening of the bench angles, the blast damage to the walls still seemed to be large. Thus a more pronounced cautious blasting of the final slope seemed motivated.

Part 2 started in 1993 with Boliden as coordinator and with SveBeFo and Nitro Nobel as partners. Field tests were made during late fall and early winter of 1993.

Two production rounds were divided into smaller rounds while varying the following factors;
- the degree of confinement of the round, i.e. the number of rows fired,
- the direction of mass movement through variations in the drilling and ignition plan, and
- the use of either the larger production holes ($12\frac{1}{4}$" = 311 mm) or the smaller contour holes ($5\frac{1}{2}$" = 140 mm) in the last row of the round.

Presplitting of the contour was also tried as a complement to the ordinary smooth blasting.

A monitoring program was set up with the purpose of showing how back-break and other forms of blast damage are influenced both by the blasting itself and by the geology. It had three parts;
- careful documentation of drilling, charging and the laying out of the firing lines,
- measurement of VOD, PPV and air pressure in the slope behind the contour holes and
- measurement of the back break and crest profiles in the contour plus the vertical swelling of the bench and the round.

An evaluation of the tests later led to recommendations to the mine how to conduct its contour blasting.

Project "Cautious Blasting of Final Slope"

Instrumentation profiles:

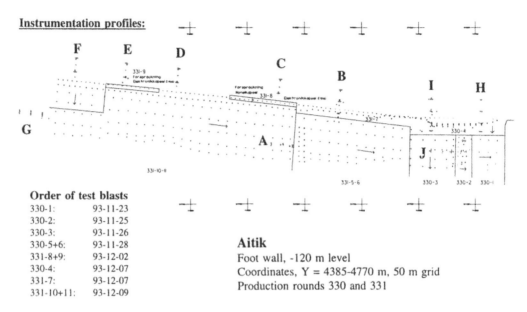

Order of test blasts

330-1:	93-11-23
330-2:	93-11-25
330-3:	93-11-26
330-5+6:	93-11-28
331-8+9:	93-12-02
330-4:	93-12-07
331-7:	93-12-07
331-10+11:	93-12-09

Aitik

Foot wall, -120 m level
Coordinates, Y = 4385-4770 m, 50 m grid
Production rounds 330 and 331

Figure 1: Lay out of test blasts and instrumentation profiles in Aitik mine.

This paper focuses on the blast monitoring. Its purpose was to check that the explosive in the blast holes detonates with the correct VOD, in the right sequence and with the proper delay. Furthermore

1. The VOD-records were used to identify which blasthole was the source of what PPV-pulse.
2. The PPV-records were used to construct the site scaling laws needed for damage zone evaluations.
3. The measurement of gas/air pressure in empty bore holes was used both to determine the contribution of the blast fumes to the damage in the remaining bench face and to study the effect of

the presplit crack in protecting the face from PPV pulses and gas penetration. The work by LeJuge et al. (1994) was a very helpful starting point.

Of these the pressure measurements have been presented to an international audience (Ouchterlony, 1995), the VOD and PPV ones only nationally.

2 MEASUREMENTS

2.1 Lay-out of test blasts

The layout of the field tests is shown in Figure 1 and Table 1. Eleven different blasts were fired, using 279 tons of Emulan 7500, a gassed heavy ANFO type emulsion, to break 316200 m^3 of rock.

A typical drilling and charging plan is shown in Figure 2. Note that none of the holes are stemmed. The presplits (331-8 and 331-9) were made with about 90 kg of 10 m long decoupled Emulan charges in Ø 100 mm plastic pipes, initiated partly in a 1 ms sequence by EDD detonators and partly "simultaneously" by Nonel Unidet U 500 detonators.

The initiation system used in the other rounds was Nonel Unidet with U 500 detonators in 0,5 kg bottom primers together with UB 176 surface delay elements between rows and UB 17 elements in the rows. See Figure 3.

Table 1: Lay out of test blasts.

Round no	No of holes	Wall-back Ø mm	Confinement/ direction of throw
330-1	11	311-311	free / along wall
330-2	18	311-140	free / along wall
330-3	22	311-311	free / along wall
331-5+6	66	311-140	conf. / along wall
331-8+9	27	140-...	conf. / across wall
330-4 331-7	100 both	140-...	free / across wall
331-10 331-11	199 both	140-311 140-311	conf. / along wall conf. / across wall

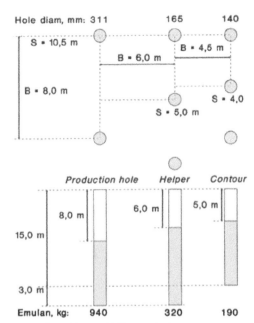

Hole diam, mm: 311   165   140

S = 10,5 m

B = 6,0 m

B = 4,5 m

B = 8,0 m

S = 4,0

S = 5,0 m

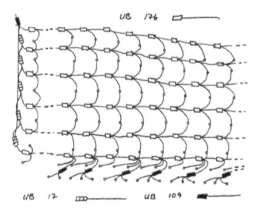

Production hole   Helper   Contour

8,0 m   6,0 m   5,0 m

15,0 m

3,0 m

Emulan, kg: 940   320   190

Figure 2: Ordinary drilling and charging pattern.

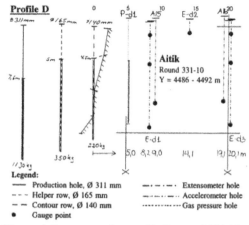

Profile D

Aitik
Round 331-10
Y = 4486 - 4492 m

Legend:
—— Production hole, Ø 311 mm   —·—·— Extensometer hole
– – – Helper row, Ø 165 mm   —··—··— Accelerometer hole
– – Contour row, Ø 140 mm   ·········· Gas pressure hole
● Gauge point

Figure 4: Instrumentation profile D, round 331-10.

UB 176

UB 17   UB 109

Figure 3: Nonel Unidet initiation on foot wall.

The test site lay in the north and middle parts of the foot wall between the 120-135 m levels. The rock was mainly a biotite gneiss with zones of feldspar and epidote parallel to the foot wall. These zones coincide with the structurally dominating feature, a steeply dipping, about 70°, foliation system which strikes about 10°E from the wall. This anisotropy was considered in the test layout.

The test layout may be described as a number of contour sections, final or temporary, where different contour blasting techniques were tried. In 13 out of 16 test sections an instrumentation profile had been prepared, see Figure 4, where extensometer, PPV and pressure measurements were made.

2.2 VOD measurements

The VOD measurements were made with the VODR-1 instrument from EG&G (Chiappetta 1993). It uses a coaxial cable which runs through the charge. It records the movement of a short circuit caused by the detonation front by sending down a pulse roughly every 10 µs and measuring the travel time for the returning echo. If the initiation sequence is known, the cable can be threaded through several holes and the actual initiation delays measured. There are 2 channels which can record up to 10 holes each.

Figure 5 shows a recording of the position of the detonation front with time. The pigtail part AB is instantaneously shorted by the primer, the slope of part BC gives the VOD in the charge and CD shows the movement of a pressure wave or the blast fumes through the empty unstemmed part of the bore hole.

Figure 6 shows how the instrument was used to measure the relative initiation times of holes in 3 trunk lines and the contour row.

2.3 PPV measurements

The relevant part of the ground vibration event is the peak particle velocity at the front of the shock wave which emanates from a detonating charge. This value, the PPV, is often used as an engineering measure of the loading to which the material is subjected and consequently as a means to evaluate the blast damage (Holmberg and Persson 1979).

Close to a blast hole, it is our experience that the frequencies are well above 1 kHz so we prefer to use accelerometers. At Aitik the ground vibrations were measured at the bottom of 12 m deep Ø 165 mm

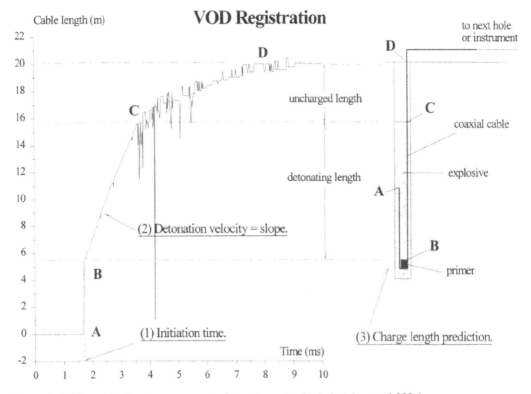

Figure 5: VOD registration, i.e. movement of shorting point for hole 1 in round 330-1.

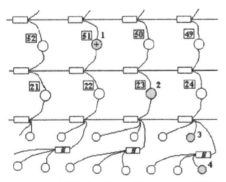

Figure 6: VOD holes in profile D, round 331-10.

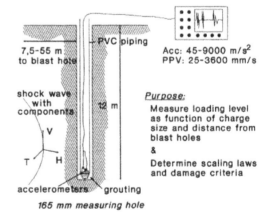

Figure 7: Gauge hole for PPV measurements.

holes, approximately at the same level as the center of gravity of the production charges, see Figure 4.

The gauges were mounted on a base plate with a threaded stud, see Figure 7, which was attached to the end of a string of Ø 63 mm PVC pipes. Thus we could insert the gauges into the hole before measurements and remove them afterwards while protecting them from moisture. The base plate stud fits into a conical foot which had been securely grouted in place with expanding, quick curing and freeze

resistant cement. The mount in turn was protected by a PVC pipe sleeve which made it possible to screw the base plate to the mount from the surface.

The measuring system consisted of accelerometers, charge amplifiers and recording units for a maximum of 16 channels. The accelerometers were of type Brüel&Kjaer 2635 or 2626 with a 12,5 kHz bandwidth. The recording units were one 14 channel FM tape recorder TEAC XR-510 with 20 kHz bandwidth and one 6 channel digital recording unit with a 6 kHz bandwidth and a 10 s storage window. Whenever possible channels were doubled and in each instrumentation profile there were at least two sets of 2 or 3D accelerometers, see Figure 4.

A trace is given in Figure 8. It shows that it may be quite difficult to match each pulse in the round with the correct source. We used both ionization probes and the VOD records to do the matching.

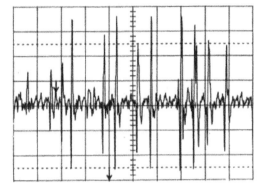

Figure 8: Vertical acceleration, hole A-20 in profile G, round 331-10. 100 ms/square.

2.4 Air pressure in bore holes measurements

All air pressure measurement holes except one were vertical, 15 m deep and had Ø 165 mm. The damaged upper part was sealed off by a Ø 110 mm, 6 m long capped plastic pipe, Figure 9.

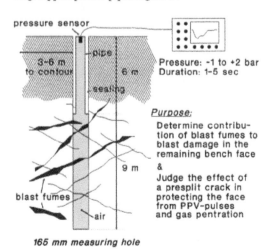

Figure 9: Sealed holes for gas/air pressure meas.

The gauge was mounted in the cap under extra protection. In the gauge output, a DC level represented atmospheric pressure, so there was no mistaking a nonresponding live gauge for a dead one.

The pressure gauges were of foil strain gauge type, Haenni model ED 517/314.211/075. They had a range of 0-10 bars because we expected mainly overpressures from the pressurized blast fumes that would penetrate the cracks in the slope. The holes were positioned 3-25 m behind the contour. Pressure measurements were made in 9 of the instrumentation profiles, using 1-3 holes. Signals were obtained from 17 out of 18 holes and from all profiles.

3 RESULTS AND INTERPRETATIONS

3.1 VOD results

The VODR-1 worked well even in the wet subfreezing conditions. Data are given in Table 2. From 46 instrumented holes, 44 initiation times were obtained and from these 35 good VOD versus time histories were obtained. Four of the missing ones came to the pre-split rounds 331-8+9 where the delay was 1 ms.

The VOD values were as expected, being highest near the primer and decaying towards the top of the holes, as the density of the explosive decreased. The VOD values of the production holes were about 10%

Table 2: Summary of VOD-values.

Round no	Ø mm + no holes	VOD bottom m/s	VOD top m/s
330-1	311-3	5750-6180	4650-5380
330-2	311-1	6550	5880
	165-2	5530-5800	4250-5090
	140-2	5440-5600	4210-4570
330-3	311-4	6060-6110	4700-5590
331-5+6	311-6	5940-6740	4160-5650
	165-1	5620	3950
	140-1	6320	3580
330-4+ 331-7	165-1 140-2	5810 5870-5890	5170 4930-4980
331-8+9	140-2	4930-5130	4200-5160
331-10+11	311-7	5680-6710	4770-5420
	165-3	5420-5690	3600-5560
	140-1	5690	5030
Total	311 140-165	6200±300 5600±300	5200±400 4600±600

higher than for the smaller holes. The differences between the VOD values of the Ø 140 and Ø 165 mm holes were however too small to be significant compared to the scatter, which was about 5-10%. This shows that the explosive held a consistent quality during our tests. The VOD recordings could also be used to judge the charge length with acceptable accuracy, but not the uncharged length.

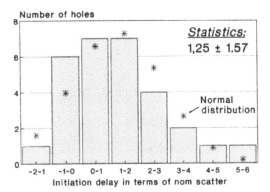

Figure 10: Relative delay in initiation of VOD measurement holes in rounds 330 and 331.

Figure 10 shows the relative delay of the initiation time of the holes along a VOD cable with respect to the previous hole, divided by the nominal scatter of the relevant delay elements and caps. The values should simplistically be normally distributed with a standard deviation of 1 around the average 0.

The actual average 1,25 can be explained by the additional 12,5 ms delay of the initiation impulse as it propagates down about 25 m of shock tubing at 2000 m/s. The actual standard deviation 1,57 is probably an RMS build up of individual scatters.

Figure 11 shows the VOD holes outside instru-

mentation profile D. The underlined numbers <u>177</u>, <u>392</u> and <u>520</u> show how late after hole no 51 the real initiation occurred. The number 1(∓10) for hole no 22 is a back calculation from hole no 23, considering the initiation time scatter of delay element and cap.

According to Figure 6, hole no 22 should have been initiated 17 ms after no 51 but in reality they were initiated nearly simultaneously. The same then was true for hole pairs 52-21 etc. This shows the difficulty to identify the true sources of the shock wave pulses in a recording like Figure 8, using only the initiation pattern. Despite the example above, there were no specific complaints on the initiation sequences of the test rounds by production people.

3.2 PPV results

PPV was measured in 11 of the 16 instrumentation profiles using 2 gage holes each. Three dimensional gauges were used in all rounds except the last one where the vertical and horizontal components along the profile were measured. Signals were obtained from 17 of the 22 gauge holes and from all profiles. The scatter in the amplitudes and the absence of a dominating component made vector evaluation unnecessary so all results refer to velocity components.

These components were plotted in synthetic PPV profiles like the one for instrumentation profile D in Figure 12. Here all identifiable Ø 140 mm sources have been moved to the contour position, all identifiable Ø 160 mm sources to the helper row position 4,5 m out and all Ø 311 mm sources to the first production row another 6 m out.

The quality of the results was sufficient to determine site scaling laws. The swelling in the bench could however not be obtained from the vertical

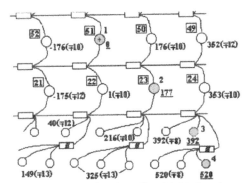

Figure 11: Measured and predicted initiation times in round 331-10, relative to hole no 51.

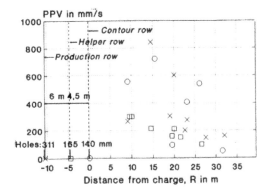

Figure 12: Synthetic PPV profile for instrumentation profile D, nominal burdens 4,5 and 6 m.

acceleration through a double integration with sufficient accuracy.

3.3 *Air pressure results*

The air pressure measurements are summarized in Table 3. A clean pressure signal from an ordinary round is shown in Figure 13. Here the pressure usually starts to drop before the closest contour charges detonate. We had expected overpressures to occur but in 10 out of 13 holes there are only underpressures like in the figure. In one case a cable was cut and from the two farthest holes, 22 and 25 m we measured insignificant pressure variations.

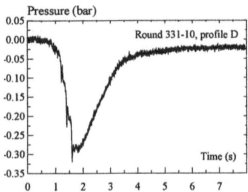

Figure 13: Air pressure in gauge hole P-d1, 5 m behind contour in round 331-10.

This underpressure is probably caused by the vertical movement in the rock which the shock fronts from the charges initiate and which is amplified by the reflections from the top surface of the bench. This swelling opens up fractures connected to the measuring hole, increasing its effective volume. The pressure drops when the air in the hole is sucked into the fractures. The hole and the fractures behave roughly like an "accordion" together.

The pressure equilibrium is restored as air and maybe blasting fumes flow into the hole. We can not exclude the possibility of the blast fumes from entering the hole, but if so, their pressure effects are negligible. The duration of these pulses are 1-5 s and the maximum underpressure measured was about 0.6 bars 3-4 m behind a charge. This would require a doubling of the original hole volume.

During the presplits two gauge holes at roughly 3 and 6 m behind the charge line were used in each of the profiles C and E. Three of them returned measurable levels, the remaining one a negligible level. All

Table 3: Air/gas pressure measurement data; distances, pressures, durations and arrival times.

Round/ profile no	Dist. to hole m	Press. atm	Durat. s	Arriv. ms
330-3/I	11.7	-0.28	3.7	-12
330-3/I	21.8	±0	-	-
330-3/I	24.6	±0	-	-
330-4/I	2.9	-0.62	1.3	-
330-4/I	5.7	-0.39	2.6	-
330-7/B	4.6	-0.62	1.1	147
330-7/B	2.6	-0.21	>0.5	87
330-7/B	6.3	-0.22	5.3	70
331-8/C	3.0	+1.77	1.3	-3
331-8/C	5.6	±0	-	-
331-9/E	2.8	+1.60	4.5	-7
331-9/E	6.4	+0.19	12.2	-9
331-10/D	5.0	-0.31	2.5	2420
331-10/E	5.1	±0.08	-	-
331-10/E	7.7	-	-	-
331-10/G	6.1	-0.27	1.3	288
331-10/G	12.3	-0.37	1.2	17
331-11/F	3.5	-0.47	1.8	344

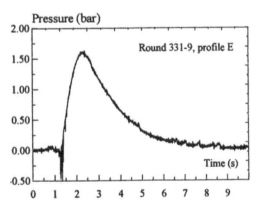

Figure 14: Pressure in gauge hole P-e2, 2,8 m behind presplit in presplit round 331-9.

three show the same behavior, see Figure 14. After a short negative pulse (< 0.2 bar except for noise spikes) the pressure rises fast to a substantial overpressure, between 1.6-1.8 bar 3 m behind the split.

What sets the presplit apart from ordinary rounds is the confinement of the surrounding rock and that the 1 ms delay in charge initiation allows the charges in adjacent holes to cooperate. This appears to be enough to force blast fumes into the remaining rock.

The effect of the presplit was to be judged by 2 holes in profile E but the only signal came from 1 m behind the uncharged presplit or about 5 m behind the charged Ø 165 mm holes.

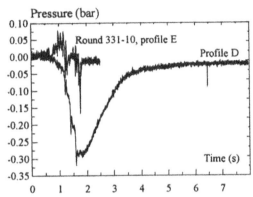

Figure 15: Pressure in gauge hole with (E1) or without a presplit crack covering the charges.

Figure 15 shows that the presplit effectively cuts off the pressure, compared with hole D1 which is situated just as far behind a charged contour row.

The PPV measurements indicated the same effect on the incident shock wave. However, the damage may already be there since pressurized blast fumes have penetrated the rock mass during the presplit.

The vertical swelling, h, in the bench after blasting can go about 20-25 m in from the blast holes. If we assume that it is due mainly to an irreversible dilation of the fractures during the round we can relate it to the measured underpressures. Since the upper 6 m of the pressure holes were sealed off, it follows that the extensometer readings from the 6 m level are a measure of the fracture widths accumulated below it. The active hole length $H \approx 9\text{-}10$ m.

For simplicity assume that these fractures can be estimated by a number of circular fractures of radius R and that the air in the hole follows a polytropic gas law. Then it follows that

$$p_{min} \approx 1/[1+4(R/\emptyset)^2 \cdot (h/H)]^{1.4}-1. \qquad (1)$$

Like the PPV-data, the swelling data scatters much but the envelopes for the two main hole sizes, Ø 140 and 311 mm, give upper limits. The right magnitude for 2R would be somewhat larger than the average distance between fracture planes, 0.9-1.7 m. With 2R = 1.8 m we get the results in Table 4 and Figure 16.

Table 4: Estimated underpressures in gauge holes.

Swelling mm	Pressure atm	Distance to Ø 140 m	Distance to Ø 311 m
70	-0.57	3.0	4.5
50	-0.48	5.0	7.1
30	-0.35	8.2	11.5
20	-0.26	11.0	15.1
15	-0.21	13.0	17.2
10	-0.15	15.2	20.1
5	-0.08	18.2	25.0

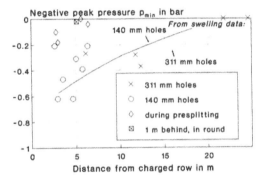

Figure 16: Measured underpressures and estimates based on equation 1 and data in Table 4.

Since swelling envelopes were used, the curves should envelope the measured values. This they do reasonably well so the agreement is acceptable and this supports the hypothesis that the measured underpressures in the holes are due mainly to the opening of fracture planes in the rock mass.

Our observations are largely in agreement with those of LeJuge et al. (1994) from the Rössing mine. They attribute the underpressure to the same mechanism as we and they didn't detect any gas flow across the presplit crack either.

Behind a stemmed presplit blast they measured a signal much like Figure 13 at 2 m, but with a shorter duration. At 4 m the overpressure has decayed substantially and at 6 m the initial negative pulse dominates completely. The difference might be caused by their more lightly charged presplit, about 18 kg per hole vs our 90 kg, and a more fractured rock which both exaggerates the fracture dilation effect and shortens the pressure restoration time.

The pressure record they show from a Ø 381 mm, 15 m deep stemmed production hole is different from Figure 13 though. At 5 m they recorded a substantial overpressure, something which we never found behind our unstemmed Aitik holes.

We can probably attribute this to the pressure retaining effect of the stemming at Rössing. It then follows that an unstemmed blast hole is important in venting the blast fumes and avoiding gas penetration into the rock mass. For the Aitik mine at least we were able to conclude that the blast fumes from an ordinary unstemmed blast in all probability don't give a significant contribution to the blast damage.

However, it doesn't necessarily follow that;
• the blast damage would be measurably smaller if there were no stemming, or that
• the stemming effect which contains the blast fumes would significantly add to the damage.

Firstly, the swelling goes considerably deeper than the gas penetration and all fracture dilation should be considered as damage once it has reached a certain level, no matter what caused it. Secondly, we have no comparative experiments where the stemming is the only test variable.

3.4 The site scaling laws

The synthetic PPV profiles, like in Figure 12, give a good idea of how the measured PPV-values depend on distance R, charge size Q and charge length H. They show how far into the wall waves of a certain amplitude can penetrate. The present Aitik data could be divided into two groups, mainly the foot wall profiles and those parallel to the wall.

A site scaling law was fitted to the complete set of foot wall data,

$$PPV = 160 \cdot (\sqrt{Q}/R)^{1.42} \text{ in mm/s.} \qquad (2)$$

The prefactor for our profiles varied between 130 and 220. The uncertainty in A could also be expressed through the standard deviation factor 2,03. The equation and some of the data are shown in Figure 17 together with an upper limit line. Previously measured PPV data follow this trend well too.

The different data for instrumentation profiles oriented along and across the foot wall made us estimate the wave propagation velocity parallel and perpendicular to the foot wall. We combined the VOD initiation time and the PPV arrival time data with the distance between source and receiving gauge to obtain a P-wave speed of 4850 m/s independent of direction.

Scaling laws of the form (2) predict the average PPV value as a function of charge size and distance. In estimating the depth of a blast damage zone we

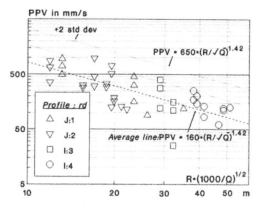

Figure 17: PPV data from profiles I and J.

are perhaps more interested in obtaining engineering estimates of their maximum penetration. Ouchterlony et al. (1993) handled this by using a double deviation factor for A. If the errors are random and normally distributed, 98% of the measured values should fall below this limit.

Close in, only part of the charge is expected to contribute to the PPV value (Holmberg and Persson 1979). Adding this charge length correction to equation 2 we obtained the following limiting curve for the PPV values in the foot wall

$$PPV = 650 \cdot [\sqrt{(fQ)}/R]^{1.42} \text{ in mm/s with} \qquad (3a)$$

$$f = [atan(H/2R)/(H/2R)]. \qquad (3b)$$

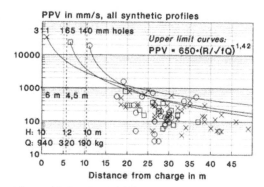

Figure 18: Synthetic PPV profile for foot wall.

In Figure 18 all the foot wall data are shown together with the limiting curves for the different holes. Close to the final wall, the contour holes dominate the loading but already 8-9 m in the production holes start to dominate. The Ø 165 mm holes seem to be relatively harmless though.

CONCLUSIONS

Based on these monitoring results the following conclusions could be drawn:

- VOD is an excellent means of checking the performance of explosives and initiation system in a blast. At Aitik both of these worked well.
- At Aitik the VOD records were instrumental in pin pointing 10-20 individual PPV pulse sources in each round.
- The charges in ordinary rounds, in which the blast holes are unstemmed, didn't force any significant amounts of high pressure blast fumes into the walls since only subatmospheric pressures were measured.
- A shock wave initiated dynamic swelling movement opens up fracture planes. Those fractures that are connected with the gauge hole increase its volume and reduce its pressure. The measured under-pressures correlate relatively well with the measured residual swell values.
- The shot presplit line didn't transmit direct shock waves or blast fumes into the slope. The presplit holes themselves did however, despite being unstemmed, force pressurized blast fumes into the rock.

Despite their relatively confined conditions, the blast fumes from normal blast holes could thus be eliminated as a direct source of blast damage. This doesn't rule them out as a considerable source of damage if the holes are stemmed so that the blast fumes are prevented from venting axially.

The blast damage assessment could thus focus on the PPV levels. Measurement of back break and swelling made it possible to quantify the blast damage and, through the scaling law, to relate it to the blasting patterns. This has been the basis for suggested designs of cautious blasting that the mine has been trying out. The results of this part of the project will be reported later.

ACKNOWLEDGEMENTS

The authors are grateful to Boliden Mineral AB for the permission to publish this work. The staff of the Aitik mine and the project group under the leadership of Per-Olof Sognfors is also gratefully acknowledged.

REFERENCES

Chiappetta, R F 1993. Continuous velocity of detonation measurements in full scale blast environments. In P Weber (ed), *Rock blasting, an international summer seminar*, p 27-74, Alès: Ecole des Mines.

Holmberg, R & P-A Persson 1979. Design of tunnel perimeter blasthole patterns to prevent rock damage. In *Proc Tunnelling '79*, London: Inst Mining and Metallurgy.

LeJuge, G E, L Jubber, D A Sandy & C K McKenzie 1994. Blast damage mechanisms in open cut mining. In T N Little (ed), *Proc Open pit blasting workshop 94*, p 96-103. Perth WA: Curtin Univ.

Ouchterlony, F, C Sjöberg & B A Jonsson 1993. Blast damage predictions from vibration measurements at the SKB underground laboratories at Äspö in Sweden. In *Proc 9th annual symposium on explosives and blasting research*, p 189-197, Cleveland OH: ISEE.

Ouchterlony, F 1995. Review of rock blasting and explosives engineering research at SveBeFo. In *Proc Explo 95 conference*, p 133-146, Carlton VIC: AusIMM.

Sognfors, P-O 1993. Cautious blasting of final slopes at the Aitik mine. In *Proc Blasting conference*, paper no 13, Gyttorp: Nitro Nobel. In Swedish.

Vibration measurements in the damage zone in tunnel blasting

Ingvar Bogdanoff
Geotechnical Peak Technology AB and Chalmers University of Technology, Göteborg, Sweden

ABSTRACT: During blasting of an access tunnel in a rock cavern in Stockholm city, vibration measurements were done at distances between 0,25 and 1m outside the tunnel perimeter holes. The perimeter and the perimeter helpers were charged with conventional smooth -blasting explosives. Electronic detonators were used with 1 ms delay time between the perimeter holes. This initiation technique is known, as "shear blasting". The measurements gave the duration -times for the high level vibrations in the damage zone. The duration - time is closely related to the principal frequencies and must be evaluated when very short delay times are used. The peak particle vibration levels (PPV) were 2 - 2,5 times higher than the normally expected value, at the distances for normally expected rock damages. The problems of resonance between the accelerometer probe and the resonance in the rock were solved by small accelerometer probes and digital filtering technique.

1 INTRODUCTION

The damage zone outside the contour has been a subject of several research projects. Among direct methods are investigations with core drilling and bore hole telescopes been used. An indirect method was given by Holmberg and Persson (1979). They proposed that the damaged zone is directly proportional to the peak particle velocity (PPV). They made measurements in relatively large holes 64 - 250 mm, at distances down to two meters from the ANFO charges. The damage zone for contour blasting explosives, with low charge concentrations, was extrapolated. The damage is a result of induced strain, ε, which for an elastic medium is given for a sine wave approximation as:

$$\varepsilon = PPV / c_p \qquad (1)$$

c_p = the compressional wave velocity

The PPV range for damages was found to be 0,7 - 1,0 m/s.

Theoretical calculation can give a wide range for calculated PPV in the damage range. Forsyth (1993) calculated that the tensile strength for low quality rock should be at 2400 mm/s and for High quality rock at 1275 mm/s.

Nyberg and Ouctherlony (1991) measured vibrations down to 3,8 m from the contour. The expected damage zone here was 0,2 - 0,5 m from the contour, a normally expected damage range for cautious tunnel blasting with smooth blasting explosives.

Yang *et al.* (1993) reported PPVs from 2 - 4 m from heavy charges in 100 mm bore holes. Damaged was found at 2 m from the constricted hole. The PPV was measured to 6 m/s. At 4 m the PPV was 0,9 m/s.

Rustan *et al.* (1985) measured vibrations from contour blasts with low VOD pipe charge (0,18 kg/ m), detonating cord (0,14 kg/m) and ANFO with plastic beads (0,26 kg/m relative dynamite). The PPV at the lowest measured range, 2m from the charge, was 0,3 - 0,9 m/s for explosives commonly used for smooth blasting. The lowest PPV was recorded from detonating cord. An extrapolation for the 0,5 m range gives PPVs around 1 - 3 m/s. This is considerably higher than the often referred range for damages, 0,7 - 1,0 m/s. The calculations with 0,7 m/s gave damage 0,1 m range for detonating cord and 0,35 m 17 mm Gurit . The observed damage ranges by direct methods are 0,4 m and 0,5 m respectively. This work suggests that the PPV for damages can be higher than 0,7 - 1,0 m/s.

The calculating process for the damage zone is questionable because the lack of measurements in the damage zone. Classifying the level for the damage is also difficult. In most cases, the criteria of damages are visible new cracks and widened old cracks.

1.1 The smooth blasting concept

The concept for smooth tunnel blasting is well known and is used in construction tunnelling and in high rock cuts. The pipe-charge is 0,10 - 0,2 kg/m . Common explosives are detonating cord and low VOD pipe charges. The detonating cord is used because of its stabile detonation and simple charging work. Both types of explosives are however commonly used in Sweden for similar rock conditions. The initiation of the perimeter is usually done with as many holes as possible within the same delay. Groups of 5 - 10 holes are sometimes initiated simultaneously with detonating cord.

1.2 Shear blasting

It is well known to Rock Blasters that simultaneous initiation of a row with blasting holes gives the best breakage and throw for lowest consumption of explosives. The drawback is that vibration level increases. Two adjacent holes can also interact. The wave fronts from the holes can meet halfway between the holes and superimpose on each other. The increased stress may then cause a longer damage zone. If the holes in the contour are given a small time delay, the superimposing effect can be avoided. This technique (shear blasting) has been investigated by Rustan et al.(1985). They proposed that the optimal time delay t_0 between two adjacent perimeter holes can be calculated as.

$$t_0 = k \left[\ B + (B^2 + \ S^2)^{0,5} \right] / c_s \ 1000 \qquad (2)$$

B = burden (m); S = hole spacing (m); c_s = the shear wave velocity (m/s); k = safety factor to give time for the wave to pass the area between the holes The safety factor was proposed to 1,5.

If this formula is applied to the smooth blasting concept for hard rock conditions with burden = 0,7 m, hole spacing = 0,55 m and shear wave velocity = 2300 m/s, the t_0 will be 1,0 ms.

The superimposing effect is however also dependent on the vibration frequency. The risk for superimposing increases with low frequency. This effect is not taken in account in formula 2.

It should also be noted that the compressional

waves are mostly used in formula 1 meanwhile the calculation in formula 2 uses shear wave velocity.

Nicklasson and Karlsson (1991) compared damage from open cut blasts. The rounds were ignited simultaneously, with electronic delays with 2 ms delay and with conventional pyrotechnical detonators. The rock damage from shear blasting and simultaneous initiation was the same, meanwhile increased damages was observed from the pyrotechnical detonators, with random initiation in the same delay.

2 FIELD TESTS

2.1 Instrumentation

The intentions of these tests was to measure the vibration level in the range where visible cracks still can arise. This is expected for smooth blasting explosives with low charge concentrations of 0,15 - 0,2 kg/m to occur up to 0,2 - 0,5 m from the perimeter holes, for gneiss and granites.

At each round, 3 - 4 instrumentation holes were drilled in the roof outside the tunnel perimeter. The depth of the holes was 1,4 m. See Fig1. The blast holes were drilled 3,6 m.

Accelerometers were used as sensors and positioned inside an aluminum probe. The probe consisted of two parts where one part was bonded to the bore hole with expanding cement. See Fig 2. The other part could be retrieved The two parts had a slight conical shape and were pressed together. After the blasting, the part with the accelerometer was retrieved with a bar that could be fastened to the probe`s screw. The cables and the accelerometers in the measuring hole were protected against water with a thin plastic pipe. Outside the holes, the cables were protected with heavy polythene pipes. To get similar acoustic impedance between the probe and the rock, aluminum was used in the probes.

An important factor in the analysis process of vibrations is knowledge of the measuring systems resonance frequencies. The resonance frequency of the probe with an installed accelerometer was determined by a light blow against the probe while the signal from the accelerometer was recorded. Fig 3 shows a recording from a test blow. The resonance frequency is 8,5 kHz for this probe. It was assumed that the resonance frequency could increase when the probe was cast in a bore hole with expanding cement. However, this could be not tested.

In many measuring situations 3 perpendicular accelerometers are used. This requires a bigger and a

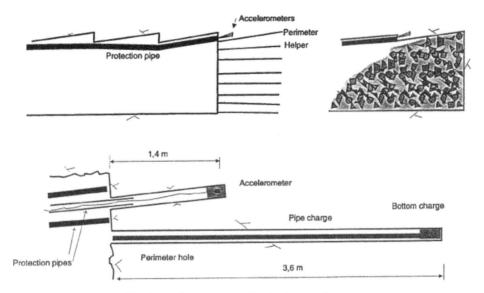

Fig 1. Accelerometer position in the tunnel

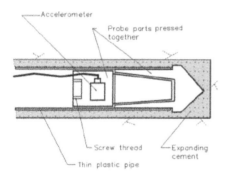

Fig 2. The accelerometer probe in the 55 mm bore hole

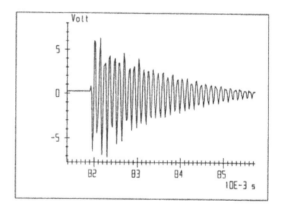

more complex probe. To make the probe as small as possible, only one accelerometer was used in each probe. It was also assumed that the dominating vibrations should be perpendicular to the perimeter holes.

Two types of accelerometers have been used, 17 g Endevco and 2,5 g Bryel & Kjaer. The resonance frequency for the Endevco accelerometers is 32 kHz and 52 kHz for the smaller Bryel & Kjaer sensors. Before installing, a calibration test was done with a handheld field calibrator. This test was also done after each test. None of the retrieved accelerometers was damaged during the test.

The data acquisition was performed with a PC. The sampling rate / channel was 25 kHz in introductory tests and later, it was increased to 40 and 50 kHz.

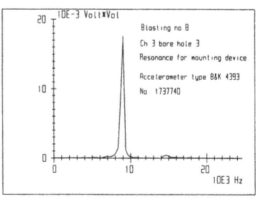

Fig 3. Resonance frequency of an accelerometer probe before positioning in the hole

179

2.2 Explosives

The pipe charge explosives were:

Perimeter	Helpers	Stoping
Detocord 80	22 mm Gurit	ANFO
(0,1 kg/m)	(0,3 kg/m)	(1,45 kg/m)
VOD = 6000 m/s	VOD = 2200 m/s	Hole dia. = 45 mm
17 mm Gurit		VOD = 2500 m/s
(0, 18 kg/m)		

The charge concentrations are given relative to dynamite with VOD = 5000 m/s. This dynamite is in charge calculations in Sweden set to 110 % of ANFO.

In the tests, electronic detonators from Nitro Nobel were used. In these detonators the pyrotechnical elements have been replaced with an electronic timer. The delay time can be set in the range 1- 6250 ms. A common delay time in the tests was 1 ms between the perimeter holes and 2 ms between the perimeter helpers. The longer delay for the helpers was motivated by the larger burden and hole spacing.

2.3 Signal Analysis

The accelerations and PPV were analysed from the time - history events. Frequency analyses were performed for all acceleration and PPV events and were mainly used as a tool, for estimating the signal quality.

A typical analysis process was:
- Frequency analysis of the accelerations for comparing signals from different accelerometers
- Low -Pass (LP) filtering to get rid of resonance's in the probes, usually in the 8 - 12,5 kHz range
- High - Pass (HP) filtering. Very low frequencies and offsets were filtered out
- Integration of the acceleration to PPV
- Frequency analysis

3 OBSERVATIONS FROM THE TESTS

3.1 Observations from blast five

Eight rounds were prepared for measuring. Five of them could be analysed.

Blast five was ignited with conventional, pyrotechnical detonators. Fig 4 shows the hole pattern with end position of the measuring holes. The distance to position 1 was 0,25 m from the closest perimeter hole and 0,8 m from the nearest helper.

The perimeter holes were charged with Detocord 80 and the perimeter helpers with 22 mm Gurit.

Only the holes with the highest vibration levels were analysed from the perimeter and the helper holes. It was assumed that these signals were produced from the holes nearest to the measuring holes. The positions of the other perimeter holes in the time-history events cannot be determined since they are ignited randomly within the same delay.

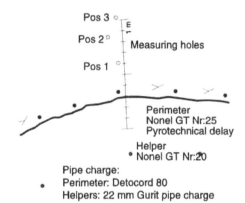

Fig 4. The hole pattern at round five. All holes were ignited with pyrotechnical delays. The measuring hole positions were determined at the hole bottom 1,4 m behind the tunnel face

The raw signals from 3 measuring holes are shown in Fig 5. The signal content is dominated by the 6 kHz component. After 0,4 ms after the first arrival in Pos 1 and 0,5 ms in Pos 3 a new wave arrives. It is assumed that this is a reflecting shear wave from the rock face.

Fig 6 shows the PPV signals. The signals are LP - filtered to 7 kHz. This eliminates the high frequency components from resonance's in the probes. HP-filtering was then performed to extinguish the offsets caused most likely from the high forces acting upon the probe and the accelerometer. The slight bend before the signals in position 1 and 3 depends on the HP - filters. The ringing in position 2 and 3 is 1,5 kHz and is assumed to be the resonance frequency of the rock. This frequency may fluctuate since the constriction of the rock fluctuates, due to existing cracks in the rock and the actual perimeter shape. Important for the shear blasting concept is the duration time for the signal. Obviously, the high level duration for a perimeter hole is about 1 ms at this range. It means that shorter delays between perimeter holes will interfere with each other and randomly magnify or extinguish the signal . The duration time can consist of the time for the direct P - and S - wave and their reflecting waves.

The signals in both the time and frequency domain are similar. The filtered PPV values are considered true since the signals are well within the frequency ranges for the measuring system. From Fig 6 the PPV is obviously still 1,5 - 1,6 m/s at 0,65 m from the perimeter. The damage range for this blasting set up is often given as 0,2 -0,3 m. The PPV at this distance is 2,5 m/s.

The PPV from the helper was 1,6 m/s at position 3. It is the same level as for the perimeter hole. This verifies the good choice of the drill pattern and charges between the perimeter and helpers.

3.2 Observations from blast eight

Fig 7 shows the holes in a round where all holes were charged with 17 mm Gurit. The perimeter and helper holes were blasted after the main round. The ignition was done with electronic detonators with 1 ms delay.

Fig 8 shows the acceleration from the helpers. Position 1 is dominated by a 15 kHz component and 11 kHz at position 3. The resonance's are probably caused in the probes. These high frequencies are superimposed upon the low frequency wave.

Fig 9 shows the filtered acceleration and PPV signals. The integration of the signal from acceleration can be compared with LP - filtering. High frequencies are filtered out. In this case the PPVs were the same as without LP filtering. This means that if the resonance frequencies are high enough they will not affect the PPV values.

Fig 10 shows the PPVs from the perimeter. The PPVs are 1,1-1,2 m/s for both the perimeter and the helpers in position 1. It can be due to the different measuring direction of the dominating perimeter and helper holes. However, it is well known that the vibrations are dependent on the constriction of the holes. More constricted holes produce higher vibrations. The holes in this blasting must be regarded to have relatively low degree of constriction since it was blasted as stoping blasting, two rounds after the main round.

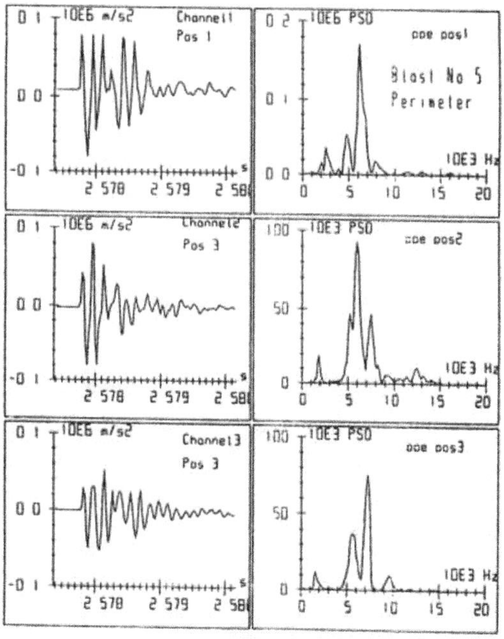

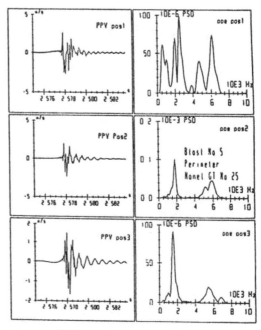

Fig 5. Raw signals from the nearest perimeter hole in blast five. The distance between the charge and the measuring spot is 0,25 m. The frequency diagrams are given as Power Spectrum Density (PSD)

Fig 6. The PPV signals from the perimeter after filtering. The PPV from detonating cord is 2,5 m/s at distances of 0,25 m and 1,6 m/s at 0,65 m

181

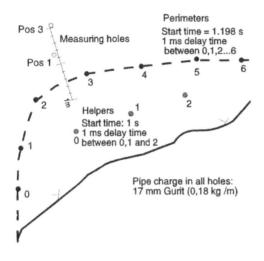

Pos 3

Measuring holes

Pos 1

Perimeters
Start time = 1.198 s
1 ms delay time
between 0,1,2...6

3   4   5   6

2

3

Helpers   1

Start time: 1 s
1 ms delay time
between 0,1 and 2

1

2

0

Pipe charge in all holes:
17 mm Gurit (0,18 kg /m)

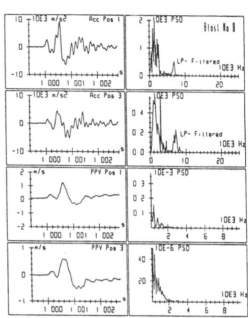

Fig 7. The hole pattern at blast no.8. It was performed
as a stoping blasting the tunnel roof. Electronic
delays with 1 ms delay time were used

Fig 9. LP- filtered signals from the helpers in Fig 7.
The PPV is 1,1 m/s at 1,0 m and 0,6 m/s at 1,4 m
from the charges. The PPV in pos 3 may be
suppressed somewhat due to interference

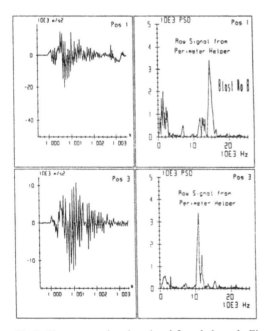

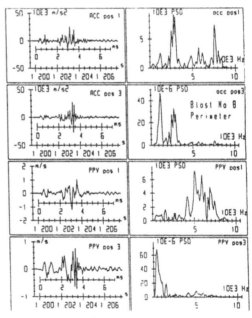

Fig 8. The raw acceleration signal from helpers in Fig
7 (blast no.8). The high frequencies originate from
the accelerometer probes. These high frequencies
(15; 11 kHz) have no influence on the PPVs since
they are filtered out in the integration process

Fig 10. The PPVs from perimeter holes in blast no.8;
Fig 7. The PPV in pos 1 is almost the same as for the
helpers. This may be due to low constriction or to
differing in the measuring direction between the
perimeter delay no 3 and the helper delay no. 0

3.3 Observations from blast three

In this round seven electronic 1 ms delays were used in the perimeter and eight 2 ms delays in the helpers. The rest of the round was ignited with pyrotechnical delays. Fig 11 shows the hole pattern. The shortest distance between the measuring hole and the perimeter hole was 0,35 m and 0,76 m between the helper and the measuring hole.

Fig 12 shows the PPV signal from position 1 - 3. A low frequency signal, 600 - 800 Hz, conforms to the fundamental signal. The high frequency signals are superimposed on this and are randomly magnifying or lowering the PPV signal. The fundamental signal is about 25% of the total PPV value. The signal in position 3 is dominated by the 2,5 kHz signal.

The delay number two gives a weak response in the time history. An explanation can be that the explosive detonates essentially in the air without breaking rock and creates therefore no vibrations. At some blasting rounds with electronic detonators it was noted in the routine sound pressure measurements, that higher sound pressure levels occurred than usual. The sharp bangs, which can commonly be heard from tunnel blasts can depend on explosives detonating almost in the air without, breaking rock.

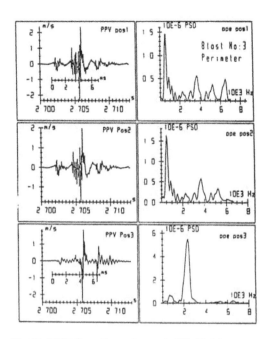

Fig 12. PPVs from the perimeter in Fig 11, blast no.3. The delay is 1 ms between the holes. There is unexpected gap at the position for delay no 2

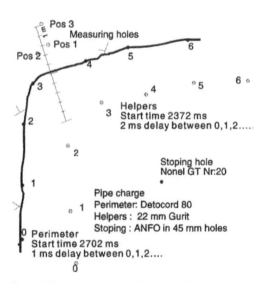

Fig 11. The hole pattern at blasting with a 1ms delay between the perimeter and 2 ms between the helper holes. The stoping was initiated with a pyrotechnical delay

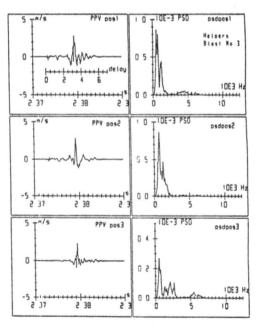

Fig 13. PPVs from the helpers in Fig 10. The signals are similar in all three measuring positions

The distance to the perimeter is 0,25 m and 0,4 m to the nearest perimeter hole. The PPV is 2,5 m/s at this distance. The signals from position 1 and 2 are very similar, which verifies the signal quality. The PPV from the nearest helper was 2,5 m/s, Fig 13. The distance was 0,8 m.

The stoping hole ignited with pyrotechnical delay was charged with ANFO. The PPV was 1,2 m/s at position 3 and 3,7 -3,9 m/s at position 1 and 2. The difference is too large according to the distances. A possible explanation can be interference between direct and reflecting waves. However the signals were very similar and well in the range for the measuring system.

4 CONCLUSIONS

At these short distances only a part of the charge gives a contribution to the PPV at the measuring point. Calculations of the efficient charge length is left out of this report. The usefulness of calculating the efficient charge is that different explosives can be compared at different distances against a PPV value. This investigation covers the PPVs in the damage range for common charge concentrations for smooth blasting and for ANFO charged in 52 mm holes.

The PPV as a function of the low VOD explosive 22 mm Gurit is plotted in Fig 14. This explosive is commonly used in the perimeter helpers 0,7 - 1,0 m from the perimeter. The PPV in this range is 2,5 - 3 m/s.

The PPVs from 17 mm Gurit and detonating cord Detocord 80 are plotted in Fig 15. These explosives have for a long time been used in the perimeter holes in construction tunnels and rock cuts. The PPV in the assumed damage range is 2 - 2,5 m/s, during hard rock conditions.

The LP and HP - filtering process filters out resonance and offset signals. The measured values are then within the frequency range for the measuring system. There is a risk for filtering out high incoming frequencies if they have the same frequencies as the probes. By comparing several signals, the high frequencies, can in most cases be identified. Their contribution to the PPV values after integration of the acceleration signal are however, damped and must be regarded in the aspect of the normal scattering of vibrations from blastings at close range.

The duration time of the high level vibration from a single hole is important if short delays 1 - 2 ms are used between blasting holes. In these tests the duration time which may increase the PPV more than normal scattering, was 1 ms for the perimeter,

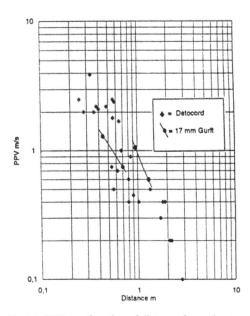

Fig 14. PPV as a function of distance for perimeter holes charged with detonating cord (0,1 kg/m) and 17 mm Gurit (0,18 kg/m relative dynamite)

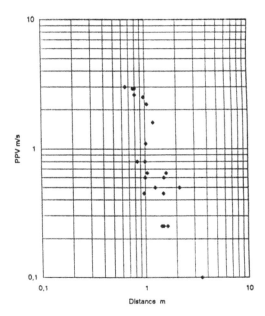

Fig 15. PPV as a function of distance for perimeter helpers charged with 22 Gurit (0,3 kg/m relative dynamite)

approximately 1,5 ms for the helpers and 2 ms for the stopings next to helpers. The duration times are only valid in the close range. Since the duration time is dependent of the distance between the charge and the measuring spot, much longer delays must be used to avoid high vibrations on constructions close the blasting site.

4.2 Smooth Blasting Results

The electronic detonators were arranged for shear blasting in the left tunnel wall and roof, meanwhile the rest of the tunnel was ignited with conventional pyrotechnical detonators with a random ignition of the detonators with the same delay. The contour of the tunnel became mostly very smooth and no visible difference between the two ignition systems could be observed.

The damage zones were not investigated by direct methods. Such an investigation can however, be done in the future since the measuring holes are marked in the tunnel and there is access to the tunnel.

5 REFERENCES

Forsyth W, 1993
  A discussion of blast-induced overbreak
  around underground excavations
  FRAGBLAST - 4, p. 161-166.
Holmberg R, Persson PA, 1979
  Design of tunnel perimeter blasthole patterns
  to prevent rock damage. Tunnelling "79.
Ouchterlony, Nyberg, Sjöberg, Johansson, 1991
  Äspö hard rock laboratory, Progress report 25-91-
  14, no:3. Damage zone assessment.
  by vibration measurements.
Rustan, Naarttijärvi, Ludwig 1985
  Controlled blasting in hard intense jointed
  rock in tunnels. CIM Bulletin, Dec1985, p63-68.
Karlsson L, Niklasson B, 1991
  Intervalltidens inverkan på sprängningsresultatet
vid slätsprängning. Swedish detonic research
  foundation report DS 1991: 6G.
Yang R.L, Rocque P, Katsabanis P & BawdenW.F
  Blast damage study by measurements of
  blast vibration in the area adjacent to blast hole.
  FRAGBLAST 4, p 137-144.

Rock Fragmentation by Blasting, Mohanty (ed.) © 1996 Taylor & Francis. ISBN 90 5410 824 X

Crack lengths from explosives in multiple hole blasting

Mats Olsson
SveBeFo, Stockholm, Sweden

Ingvar Bergqvist
Nitro Nobel AB / Dyno Explosives Group, Gyttorp, Sweden

ABSTRACT: Crack lengths from explosives in multiple hole blasting have been studied by cutting blocks and spraying them with penetrants. The research includes test blasting of totally some 150 holes. Different explosives, hole patterns, decoupling, initiation time and stemming were some of the examined parameters. It was found that decoupling and simultaneous initiation were the factors that reduced crack lengths most.

1 INTRODUCTION

At Fragblast 93 in Vienna we presented some results from a project looking into crack lengths in remaining rock from single hole blastings, (Olsson, Bergqvist 1993). Within the SveBeFo research program this project has now been extended to include also multiple hole blastings.

The object of the project is to study cracks in remaining rock from different combinations of explosives, hole diameters, patterns, timing, etc to get a better understanding of the mechanism behind this and consequently also to get better possibilities to minimise the cracks. In the end this means that we will get a better tool to design blastings of final contours, both above and under ground, for the purpose of achieving a higher quality of the remaining rock surfaces.

The project is still in progress, but some of the results obtained so far will be presented here.

2 FIELD TESTS

2.1 Test Site

The tests are carried out in a quarry for dimension stone in southern Sweden. The rock consists of fine grained granite with very few natural cracks. The granite is very competent with a compressive strength of about 200 MPa and a tensile strength of about 12 MPa. This means that the rock is of a very good quality, which in turn means that the disturbance from natural cracks is minimised and that the results are more comparative to each other.

2.2 Method

In a bench, with a height of about 5 metres, up to 20 vertical holes at a time are drilled mainly using hole diameters of 51 mm and/or 64 mm. Burdens as well as hole spacings are varied according to Table 3. The holes are charged with explosives according to section 2.3 below, together with a 50 g primer. They are charged to a height of about 4.5 metres, normally without any stemming material. In order to get as an exact detonation time as possible, electronic detonators from Nitro Nobel are used in most tests. The holes are divided into groups of minimum 3 holes, in all respects being equal to each other, see Figure 1. After blasting, another row is drilled about 2 metres behind the first row, using spacings of about 0.25 metres and blasted very carefully by 20 gram detonating cord, resulting in large blocks. When needed these blocks are then divided into smaller blocks by wedges.

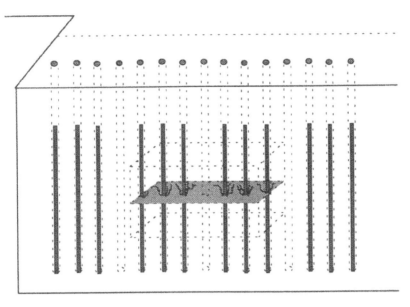

Figure 1 Test technique

The blocks are cut perpendicular to the remaining drill holes, by a rotary diamond saw. One of the surfaces created is then sprayed with a penetrant, which enhances the crack pattern from the holes, making it possible to study and measure the cracks.

2.3 Explosives

Most of the explosives commonly used in Sweden for smooth blasting have been tested. They are listed in Table 1 below.

Gurit: A nitroglycerine/nitroglycole sensitized explosive in plastic pipe cartridges

Kimulux 42: An emulsion type explosive in plastic pipe cartridges
Emulet 20: A low density ANFO type bulk explosive
Detonex 80: Detonating cord (80 g/m)

When using plastic pipe charges, Gurit and Kimulux, decoupling ratios according to Table 2 are achieved

Table 2. Decoupling ratios of Gurit and Kimulux 42.

Hole Diameter, mm	24	38	51	64
Charge Diameter, mm	22	17	17	22
Decoupling Ratio	0.92	0.45	0.33	0.34

Table 1. Specification of tested explosives and charge concentrations in different hole diameters.

Explosive Type	Explosive Diameter (mm)	Hole Diameter (mm)	Density (kg/litre)	VOD (m/sec)	Gas-volume (l/kg)	Energy (MJ/kg)	Charge concentration (kg/m)
Gurit	17	38, 51	1.00	2000	930	3.40	0.21
Gurit	22	24, 51, 64	1.00	2000	930	3.40	0.40
Emulet 20	Bulk	51	0.25	1800	1117	2.6	0.51
Kimulux 42	22	64	1.15	4800	903	3.20	0.37
Detonex 80	11	51	1.05	6500	780	5.95	0.08

2.4 Test Program

The following tests have been carried out so far (November 1995):

- Influence of type of Explosive
- Influence of Hole Pattern
- Influence of Decoupling
- Influence of Timing
- Influence of Stemming

The parameters which have been studied are found in Table 3, which also shows the number of holes tested, burdens (B), hole spacings (S) and time interval between the holes, (Δt). Δt = 0 means that electronic detonators of the same period number are used, resulting in scatter less than 0.1 ms. Δt ≠ 0 means that scatter is simulated by electronic detonators or that Nonel LP period number 50 (5000 ms) is used.

3 RESULTS

Generally all test blasts turned out very well with full breakage. The wall was very smooth with half casts from almost every hole. The exeption was blasting with fully coupled holes, where the half-casts were damaged. However, half-casts do not always indicate

that the remaining rock is undamaged, which will be shown clearly later in this paper.
The method of showing cracks with the help of penetrants in cut blocks works very well and catches preexisting as well as blast-created cracks in the rock.

3.1 Influence of Explosives (simultaneous initiation)

The crack analysis shows, as for the former single blasted holes, a significant difference due to the explosive. Every explosive seems to have it´s own signature. Some typical exampels are shown in Figure 2 and 3.

Figure 2 shows the crack pattern from 17 mm Gurit in a Ø 51 mm hole, S=B= 0.5 m. Here the cracks were very short, less than 0.05 m. An increase of the amount of explosive, like 22 mm Gurit in a Ø 64 mm hole (same decoupling), will increase the crack length.

The picture in Figure 3 shows a completely different crack pattern. The explosive here is an emulsion, 22 mm Kimulux in a Ø 64 mm hole, S=B= 0.5 m. There are many short cracks and also a pattern of longer radial cracks with the longest crack of about 0.25 m. We believe that the high frequency of short cracks is due to the high VOD of this explosive.

Table 3. Investigated parameters and number of holes.

Timing	Δt ≠ 0	Δt = 0	Δt = 0	Δt = 0	Δt = 0	Σ
B x S (m)	0.5 x 0.5	0.5 x 0.5	0.8 x 0.5	0.8 x 0.8	1.0 x 0.8	Holes
17 mm GURIT						
38 mm					3	3
51 mm	23	19	4	4	7	57
51 mm / stemming					3	3
22 mm GURIT						
24 mm					3	3
51 mm		1				1
64 mm		8	4	3	7	22
64 mm / stemming					3	3
22 mm KIMULUX						
64 mm		9			7	16
64 mm / stemming					3	3
EMULET 20						
51 mm		8			5	13
DETONEX 80						
51 mm		10			7	17
Σ Holes	23	55	8	7	48	141

Figure 2 Cracks from 17 mm Gurit in a Ø 51 mm hole

Figure 5 Arcs between holes charged with 22 mm Gurit

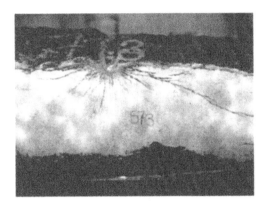

Figure 3 Cracks from 22 mm Kimulux in a Ø 64 mm hole

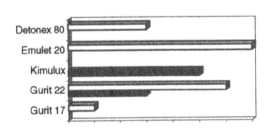

Maximum crack length (cm)

Figure 4 Maximum crack length from different explosives (black columns in Ø 64 mm holes the others in Ø 51 mm holes), S=B= 0.5 m

A bulk explosive which fills the entire hole (fully coupled), here Emulet 20, results in long cracks.

In Figure 4 the results from some of the tested explosives are presented.

3.2 Influence of Hole Pattern (simultaneous initiation)

Here we had expected that an increase of spacing and burden should give longer cracks. However, the result was not so unambiguous. For example, 17 mm Gurit in Ø 51 mm holes, did not show any change in crack length with increased B and S. For some of the other explosives there were an increase in crack lengths. One interesting effect, which we believe was due to increased spacing, was the creation of arcs or "bananas" between the holes. The effect seemed to be strongest for 22 mm Gurit, see Figure 5. On the surface the rock is virtually undamaged and the half cast ratio very high, but scaling of such a surface could bring down a substantial amount of rock. This shows that the half-cast ratio can be misleading as a measure of blast damage. However, more tests have to be done to state the influence of hole pattern.

3.3 Influence of Decoupling (simultaneous initiation)

Early test results with both single and multi-hole blasts have shown that the crack length increases with an increase of the decoupling ratio ($\emptyset_{charge}/\emptyset_{hole}$). To confirm this a test was done with 22 mm Gurit charged in Ø 24 mm holes with a decoupling ratio of 0.92, i.e almost fully coupled. Figure 6 shows the result in a very obvious way with a substantial number of long cracks. The longest crack is 0.9 m. Compare that to 22 mm Gurit charged in Ø 64 mm holes with the same pattern as above. Decoupling ratio is 0.34 and this results in considerably shorter cracks, see Figure 7. Here the longest crack is 0.3 m.

Figure 6 Crack pattern from fully coupled holes

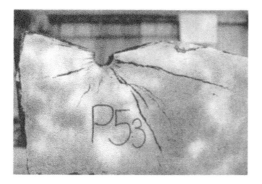

Figure 7 Crack pattern from decoupled holes

3.4 Influence of Timing

Influence of timing has been tested several times, using electronic detonators and 17 mm Gurit in Ø 51 mm holes.

We started the tests by a simulation of random initiation of holes with pyrotechnic deonators of the equal numbers. The spread in initiation time was 0-22 ms with the shortest interval between two adjacent holes of 3-5 ms. This resulted in cracks lengths of more than 0.2 m, which stated that the time difference was too long.

Tests were now made with shorter time difference. We tried both 1, 2, 3, 4 , 5 ms and 5, 8, 11, 14 ms but long cracks were still obtained, more than 0.2 m. Trying an alternating sequence was better. The time difference was 1 ms and 3 ms with sequences of 15, 16, 15 ,16 ms and 17, 20, 17, 20 ms. This resulted in crack lengths less than 0.05 m, consequently like instantaneous initiation.

The best way to obtain short cracks is to use

instantaneous initiation by electronic detonators. An other important thing with instantaneous initiation is the absence of undetonating cartridges. Ordinary contour blasting with LP-detonators often end up with a lot of undetonating cartridges in the muck pile. A test with 17 mm Gurit and 22 mm Gurit initiated with ordinary Nonel LP detonators no 50 (5 sek delay and a scatter of > 150 ms) resulted in lots of undetonated explosive cartridges (25 %) and substantially longer cracks than similar tests with instantaneous initiation.

4. CONCLUSIONS

The method to excavate blocks behind the drill holes and study the cracks, after cutting by a rotary diamond saw, has shown to be very successful. Compared to other methods it shows exactly what happens in the remaining rock behind the contour holes. In this case we have studied a rock of a very good quality with few natural cracks. The results may not be transfered in all details to other types of rock, but will still give a good picture of differences in crack lengths due to different conditions.

So far the conclusions of the test could be summarised as:
- The crack length is reduced when decoupling is increased
- Instantaneous firing reduces the crack length. Even small deviations (1 ms) in delay time between neighbour holes may ruin the result
- Fully coupled holes show a more complex pattern of cracks than decoupled holes
- A high VOD creates a high frequency of fine cracks in the vicinity of the drill hole
- The crack length is increased when the charge concentration is increased

Our intension is to further study how hole pattern and stemming influence the crack length, and also to study the effect on crack lengths of different explosive properties. Finally we also plan to look at other rock types.

REFERENCES

Olsson, M. and Bergqvist, I. Crack lengths from explosives in small diameter boreholes. 4 th Int Symp. on Rock Fragmentation by Blasting, Wienna 1993, p 193-196. Balkema, Rotterdam

Olsson, M. and Bergqvist, I. Sprickutbredning vid flerhålssprängning. Research report no 18. SveBeFo, Stockholm, Sweden. In Swedish.

Fracture control blasts

Rock Fragmentation by Blasting, Mohanty (ed.)© 1996 Taylor & Francis. ISBN 90 5410 824 X

Mechanism of smooth blasting and its modelling

Gregory Szuladzinski
Analytical Service Company, Sydney, N.S.W., Australia

Ali Saleh
Sydney University of Technology, N.S.W., Australia

ABSTRACT: The term 'smooth blasting' designates an operation intended to create a crack connecting a row of boreholes. The paper presents two quantitative approaches, which allow to determine the necessary explosive charges. The first is a quasistatic calculation related to the 'heave' action of the exploding charge. Static stress field around both boreholes (source and target) is analyzed with due attention to cracking involved. The dynamics approach uses an FEA method, in which the explosive is simulated by a set of concentric, preloaded springs. The dynamic effect is quantified in terms of the peak hoop stress around the target hole. One of inescapable conclusions of this paper is the importance of the tensile strength of rock material for a rational approach to the problem. Paradoxically, this property is often not even mentioned in investigation reports.

1 STATIC STRESS LEVELS IN THE VICINITY OF A BOREHOLE

A borehole of radius 'a' is subjected to pressure p_b is shown in Fig.1. A point at radius 'r' away from the center of hole has the following radial and hoop stress components:

$$S_r = -p_h \left(\frac{a}{r}\right)^2 \quad S_h = p_h \left(\frac{a}{r}\right)^2 \quad (1)$$

Next, consider a hole in a continuous medium subjected to stretching in one direction with stress of magnitude S_o, as shown in Fig.2. Any text dealing with stress concentrations will show that the peak stress in points A and B is $3S_o$, (tension), while at C and D there is a compressive stress with the magnitude of S_o. Returning now to Fig.1 we place a small hole in point G, while keeping r sufficiently large so as to be able to treat this new hole as only a

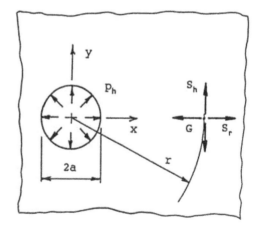

Fig.1 Stress field in a medium with pressurized hole

local disturbance to the over-all stress field. With this in mind we can designate by S_o the hoop stress at this radius and note that the radial component has the same magnitude, but opposite sign. The resultant stress field around the

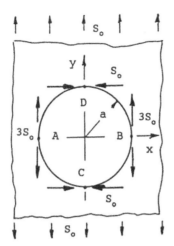

Fig.2 A hole in a uniaxial stress field

hole at G is a superposition of the effect of the hoop stress (shown in Fig.2) and that of radial stress. The total hoop stress at the edge of the hole is therefore given by

$$\sigma_A = \sigma_B = 4S_o \qquad \sigma_C = \sigma_D = -4S_o \quad (2)$$

(Note that σ stands for the local effect at the hole under consideration while S designates the continuous-field magnitude at the same radius from the source.) In a normal smooth blasting process the exploding hole pressure (and therefore the hoop tension) is much larger than the tensile strenth of rock material F_t. As a result a pattern of radial cracks appears around the hole called the "rose of cracks". Ideally, all cracks reach the same radius R. The region within the radius can be regarded as an equivalent hole, as the cracked material offers no hoop resistance to the pressure acting inside. At the radius R the hoop stress is equal to F_t. The applied radial stress has the same absolute value for the edge of hole in an infinite medium, therefore we conclude that the applied pressure inside the equivalent hole is also equal to F_t. As a result, we are able to use Eqs.1, originally derived for a hole in the intact material, also outside the rose of cracks:

$$S_r = -F_t\left(\frac{R}{r}\right)^2 \qquad S_h = F_t\left(\frac{R}{r}\right)^2 \qquad (3)$$

2   CONDITIONS FOR QUASISTATIC CRACKING

In Figure 3 there is a fragment of a long string of identical holes, at an even spacing 's', typical for smooth blasting. The center one of the three holes is the object of interest. At least three cases may be distinguished here: I) If hole 1 is detonated, hole 3 goes off later with such a delay that there is no superposition of stress fields. II) Holes 1 and 3 are detonated simultaneously, or nearly so. III) All three holes are charged and detonated at once or with a delay. In either case the objective is to create a continuous fracture which passes through the centers, as marked with a dashed line.

Out of these three Case I is the main object of interest, although II will be occasionally mentioned. Case III, being the 'easiest' will not be considered.

Case II is much more effective in creating a break line along the row of holes than Case I. Keeping in mind that spacing is an order of magnitude larger than the hole diameter, there is little difference between stress at points A and B of Fig.3 with only hole 1 fired. For this reason Case II produces (practically) twice the stress due to Case I, assuming of course that the explosive charges are identical.

If, in Case I, the hole pressure grows to F_t, the circumference of the hole is on the verge of cracking. A frequently used spacing s, equal to 12 hole diameters, results in the hoop stress in the adjacent hole of only 1/576 of the above value, per Eqs.1. Even if the relative spacing of holes is reduced, the next hole will not be significantly influenced. This brings us to a rather obvious conclusion that the hole, which was fired must begin cracking in order to have a chance to crack the next hole.

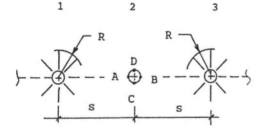

Fig.3 Repeating fragment of a long string of holes

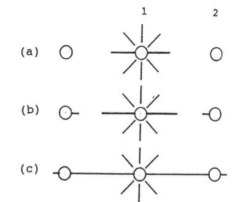

Fig.4 Sequence of events in creating a continuous crack

Let us now take the equivalent hole with radius R into account, this being hole 1 in Fig.3. In order to induce at point A of hole 2 the tensile stress of F_t, we need S_o of only a quarter of that, according to Eqs.2. If this is substituted, along with r=s in the second of Eqs.2, our cracking condition becomes:

$$R = \frac{s}{2} \qquad (4)$$

This means that if the radius R of the rose of cracks calculated from quasi-static gas-rock balance is large enough to fulfil Eq.4, cracking of hole 2 will be initiated. This, of course, is for Case I. If Case II is involved, R = 0.354s will suffice, according to Eq.3.

The following sequence of the actual event is proposed, and illustrated in Fig.4. At first the rapidity of the dynamic action makes the rose of cracks grow reasonably uniformly. After some time static action will let itself be known in that the direction of higher over-all stress will be associated with the preferred direction of cracking. In Fig.3 the tensile stress around hole 1 will be higher across the dashed line than across any other direction, because of the removed material from the boreholes. Once the radius of the rose is large enough to initiate cracking in A, the radial crack growing from hole 2 will soon join an already elongated crack emanating from hole 1.

It should be clear by now that the effectiveness of an explosive charge can be judged by the cracking radius R that would be induced in the infinite medium by said charge. The following must be mentioned to keep the calculation of R in a proper perspective: (a) There is an ample direct and indirect evidence that failure of rock can take place in either crushing mode (Mode I) or gas penetration mode (Mode II). The difference between the two is a zone of crushed material around the borehole acting like a semi-liquid during the explosion and preventing the explosive gas from penetrating radial cracks forming outside, (b) for smooth blasting Mode II is applicable, (c) the exploding gas may be treated as a nonlinearly-elastic body, pre-compressed to the initial pressure, then suddenly released. A detailed presentation of this approach is given by Szuladzinski (1993), (d) the mentioned paper outlines a procedure for determining R for Mode I, while for Mode II the process is simpler, (e) both modes have been quantified by this writer by manageable sets of equations, but the work has not yet been published.

Whatever the computational approach to rock-gas interaction resulting in a certain cracking radius R, the procedure begins with initial pressure, or the release pressure mentioned above, consistent with the internal energy of explosive. An example of the results obtained is quoted below.

Consider presplitting of granite with the following: E = 61600 MPa, ν = 0.22, F_t = 18 MPa. The explosive is ANFO of 12.7 mm radius with the total energy of 3.72 MJ/kg, VOD of 2000 m/s and density of 800 kg/m³. The boreholes are of 51 mm radius spaced at 1200 mm and stemmed. After the in-hole expansion and prior to cracking the initial pressure was found to be p_h = 21.17 MPa. (This was determined by a formula acknowledging the dynamic character of the process. Quasistatic expansion would have yielded artificially low pressure.) Since this pressure is only slightly larger than the tensile strength, the radius of cracking due to further expansion to the balance point was only R = 72.5 mm, much less than R = 600 mm required by Eq.4. The result is partially due to a low VOD value arising out such a small diameter charge.

The radius of the explosive in that example was the lower bound of what would be used in actual field conditions and the presplitting shot under these circumstances would be unsuccesful, not only in Case I, but also in Case II of firing arrangement. A more realistic charging of a hole would be to use a 19 mm radius of the explosive. All the remaining parameters quoted previously remain unchanged except for the VOD, which is now 3600 m/s. The release pressure for the cracking process is now 138.8 MPa, which results in cracking radius at balance of 2023 mm, in excess of the hole spacing and more than adequate under the conditions. One should note a strongly nonlinear relationship between radius of the explosive and the cracking radius.

3 DYNAMIC EVALUATION

After the explosive pressure is applied, a stress wave is emanated and, while travelling with the speed of sound, this wave passes the target hole. After some oscillations take place, the stress field stabilizes to what was previously described in the

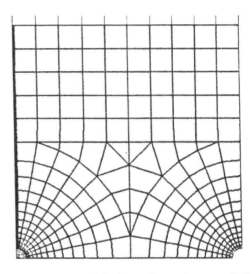

Fig.5   Finite-element   model simulating stress wave action

quasistatic analysis. During the preceding transient event the stress level at most points becomes, for a very short time, much higher than the quasistatic level. Our purpose is to examine to what extent can the stress wave crack the rock around the target hole.

The situation is much harder to grasp than before, because there is a spatial as well as temporal variation. For example, our equations (1), (2) and (3) are no longer valid and the magnitude of radial stress is not the same as of the hoop component. (In fact, the latter decays much more rapidly than the former, as the observation point moves away from the source hole.) The finite-element model, which had been constructed for this investigation, is a practical way to monitor stress levels at the locations of interest.

Suppose that one wants to model what is depicted in Fig.3, including a simultaneous firing of holes 1 and 3. The horizontal line through the holes is a line of symmetry. So is a line normal to it, through points D an C of hole 2. Thus, due to double symmetry, it is sufficient to model only one quarter of the medium. The outline of the actual model is in Fig.5, its horizontal base corresponding to 1-2. (These boundary conditions imply an infinite line with alternating source and target holes, but in practice the results are applicable to what is shown in Fig.3)

The upper boundary of the model simulates the remainder of the continuous medium, through the use of elastic and viscous elements.

The explosive in the source hole 1 (left side of the model) is simulated by a set of concentric springs, preloaded with radial forces, which are equivalent to the initial pressure described before. A sudden release of the preload is the onset of explosion. The same 19mm charge is used and the holes are separated by the same distance 's' as in the static investigation. On passage of the stress wave over the target hole the following peak stress levels were noted:

σ_D = -31.18 MPa; σ_A = 8.811 Mpa

Let us define the continuous-field stress for this hole in the same manner as done for statics before. This will be stressing due to the same explosion at the source hole arising at distance 's' in a continuous medium, i.e. at a good angular distance from the target hole. Those stress levels can be found from a very simple, axisymmetric model and in this case we have

S_D = -13.582 MPa; S_A = 2.228 MPa

(Note that the above is a doubled response of an axisymmetric model. This is to be consistent with the main model, which was set up for the simultaneous discharging of holes 1 and 3.)

The above calculation was done for the intact rock model, which we know to underestimate the response. It must be viewed as the first step only, done mainly for a reference, because in a more realistic approach a model must be made capable of cracking, at least around the source hole. This was implemented in the next model, which can crack within 1000mm radius. The following results were now obtained:

σ_D = -71.59 MPa; σ_A = 25.87 MPa

The continuous-field stress also shows an increase:

S_D = -27.12 MPa; S_A = 6.686 MPa

The allowed cracking is only about a half of 2023 mm radius calculated for this hole before. To extend the rose of cracks any further was undesirable because the target hole would be gradually placed between two radial cracks, which is not a setting planned in this investigation. Since increasing the cracked radius also increases the target response, our dynamic results are somewhat underestimated.

The most important stress component is the hoop tension, σ_A, because tension is responsible for the onset of cracking. Notice that this component increases very significantly when one changes from intact to a cracking medium, while the radial compression σ_D undergoes a smaller increase. The peak tensile stress of nearly 26 MPa at point A exceeded the tensile strength of 18 MPa, we can therefore say that cracking will take place at the target hole. At the same time we have to recognize that the tensile pulse is of a short duration and that we can expect only the initiation of cracking, rather than a continuous process. In this case the part of the pulse, which exceeded 18 MPa lasted for 0.193 ms, which translates to about 1000mm of the effective pulse length. The speed of cracking is usually estimated at one quarter of the p-wave speed, the latter being 5200 m/s. This gives the estimated crack length of 250mm, but this is an oversimplified estimate and most likely it is on the high side. Introduction of cracking capability in the model at the target hole would answer this question more decisively.

Another point to remember is that F_t is a static strength and should be viewed as a lower bound of the dynamic strength. Depending on material sensitivity, strain rate and duration of loading the dynamic strength may turn out to be several times larger. If our response figures are accurate, then only a minor variation of the tensile strength would prevent cracking from being initiated by the stress wave.

4 SUMMARY AND CONCLUSIONS

When a typical presplit blasting configuration is investigated, it is found that while the stress wave has a chance for initiating cracking near the target hole, it is not certain that this will indeed happen, due to dynamic stress not being large enough in magnitude and duration.

When a quasistatic approach is used, the criterion for creating the crack line was comfortably fulfilled.

Statics alone predicts the appearance of the desired cracking line with proper orientation near the target hole. The stress wave (shock) action, when of sufficient magnitude may "precondition" the rock material by precracking it.

The ratio σ/S of a peak stress at the edge of the target hole to the continuous-field stress is a stress concentration factor. For static conditions this factor is 4.0 for both compression (point D in Fig.3) as well as tension (point A). The dynamic simulation gave 2.30 in compression and 3.96 in tension for the intact rock case. The corresponding values were 2.64 and 3.87 for the cracking model. The reduction in stress concentration, especially for the compressive point D, is largely due to the disturbance and delay experienced by the radial impulse in travelling around the edge of hole.

The investigation was carried out assuming isotropic medium, beginning at some distance outside the hole. The influence of discontinuities, if strong enough, may invalidate the above conclusions. A joint oriented radially with respect to a source hole will not allow transmission of hoop stress in its vicinity. The effect of a circumferential joint may be to arrest both hoop and radial wave components. Strong enough discontinuities away from the intended crack line may make the line irregular.

REFERENCES

Szuladzinski, G. 1993. Response of rock medium to explosive borehole pressure. The fourth international symposium on rock fragmentation by blasting. Vienna, Austria, 1993. (FRAGBLAST-4)

Rock Fragmentation by Blasting, Mohanty (ed.)© 1996 Taylor & Francis. ISBN 90 5410 824 X

Study of pre-split blasting using fracture mechanics

J.J.Jiang

Department of Mining Engineering, WA School of Mines, Kalgoorlie, W.A., Australia

ABSTRACT: Pre-split technology has been widely used for highwall formation in surface mines and for extraction of dimensional stones in quarries. Although the mechanism of the initiation and formation of cracks along the pre-split line has been studied widely, the pre-split design is still carried out using "trial and error" or traditional methods. With the increasing importance of the pre-split blasting further research is required to improve the understanding of the mechanism and design.

This paper discusses the mechanism of the crack initiation around a blasthole and the critical pressure required to initiate the cracks using fracture mechanics. The formation mechanism of the cracks along the pre-split line is also studied. Although the rock fails under tension stress, the crack is formed gradually by crack extension and propagation. Stress concentration plays an important role in the control of the orientation of crack initiation and propagation. A formula to determine the blasthole spacing is derived using the concept of stress intensity factor and fracture toughness. There is good agreement between the spacing calculated by the formula developed in this paper and the spacing currently used in the pre-split blasting.

1 INTRODUCTION

Pre-split blasting technology has been applied to wall protection in surface mines and dimensional stone extraction for many years, and the mechanism of the pre-split formation has been studied widely by numerous researchers (Kutter and Fairhurst, 1971; Wei and Wang, 1987; Dunn and Cocker, 1995; Brent, 1995). However, there is no agreement concerning the "shock wave action" and "gas action" for the pre-split formation (Dunn et al, 1995). Furthermore, different criteria are used to determine the blasthole pressure and the blasthole spacing in a pre-split blast. As open pit depth increases, the importance of a high quality wall for slope stability also increases. Hence, there is a clear requirement of improving the pre-split blast quality.

A good pre-split design is one in which upon firing "half the blasthole" is still visible, there is no damage to the wall, and the blasthole spacing is maximised. Generally a pre-split design includes the blasthole diameter, blasthole spacing and blasthole pressure which should be determined according to the rock properties. In this regard, a good understanding of the crack initiation around a blasthole and the crack formation along the pre-split line are essential for a good pre-split blast design.

2 THE INITIATION OF A CRACK AROUND A BLAST HOLE AND ITS PRESSURE

It has been suggested (Jaeger and Cook, 1979) that in the rock mass there are micro cracks (flaws) whose length is in the magnitude order of the length of the maximum grain size. Since the cracks are randomly distributed in the rock mass it will be acceptable to assume that around a blasthole there are numerous edge radial cracks uniformly.

For a single blast hole under constant internal pressure standard analysis of the stresses around the blast hole yields:

$$\sigma_r = - \frac{R^2}{r^2} P , \tag{1}$$

$$\sigma_\theta = \frac{R^2}{r^2} P \tag{2}$$

where, σ_r and σ_θ are radial and tangential stresses respectively, R is the blasthole radius, r is the radial distance from the centre of the blasthole and P is the blasthole pressure.

If we assume that the radial cracks around a blasthole are uniformly distributed, then the stress intensity factor K_I of the blasthole under the internal pressure P is determined by (Whittaker, Singh and Sun, 1992):

$$K_I = FP\sqrt{\pi a} \qquad (3)$$

where a is the average length of the radial cracks and F is a constant dependent on the loading condition of the cracks. If the surfaces of the cracks are free of load, F is taken to be 1.12; but if the surfaces of the cracks are loaded at the same pressure as the blasthole pressure, then F is taken to be 2.24. Presumably, if the blasthole fails under "shock wave action", then one should take F to be 1.12, while if the blasthole fails under "gas action", then one should take F to be 2.24. The critical blasthole pressure required to make an existing "crack" propagate is given by

$$P_c = K_{IC}/F\sqrt{\pi a} \qquad (4)$$

where, P_c is the critical pressure and K_{IC} is the fracture toughness of the material.

For example, if the K_{IC} is 2 MPa√m̄, a is 1 mm and F is taken to be 2.24, then Pc will be 16 MPa. It should be appreciated that the stress intensity factor will vary with the crack propagation. In fact, Tweed and Rooke (1976) have shown that the stress intensity factor increases with crack length when the crack length is relatively short and then decreases with the crack length under constant pressure. In this regard the critical pressure P_c will not only initiate the crack but also be able to extend the crack to a certain length. Since this critical pressure is only dependent on the crack length and the fracture toughness of the material, there is lack of theoretical support for the claim that the blasthole pressure must exceed the tensile strength of the material (Dunn et al, 1995; Brent, 1995).

The critical pressure may be greater than or indeed less than the tensile strength. In many cases, however, the critical pressure is approximately the tensile strength of the material. In the static case Jiang (1984) measured the critical pressure to be 11 MPa for marble with K_{IC} equal to 0.94 MPa√m̄ and tensile stress 10 MPa. On other hand, Persson, Holmberg and Lee (1993) estimated the critical pressure to be in the order of 2 to 4 times higher than the uniaxial tensile strength.

3 A BLASTHOLE WITH A PAIR OF RADIAL EDGE CRACKS OF DIFFERENT LENGTHS

It should be appreciated that the analysis of section 2 only applies to the case where the blasthole is surrounded by numerous uniformly distributed micro edge cracks. The case in whick the blasthole has only a pair of radial edge cracks is, see Figure 1, quite different (Tweed et al, 1976).

In fact, Tweed et al (1976) have shown that:
1. The shortest crack has the largest stress intensity factor;

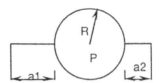

Figure 1. A blasthole with a pair of edge cracks.

2. If the length of one crack is fixed, then the stress intensity factor of the other crack decreases with length;
3. If the length of one crack is fixed, then the stress intensity factor of the crack of fixed length increases with the length of the other crack.

Hence, if a single crack, ie. $a_2 = 0$, is initiated from the blasthole and propagates to a certain length, the propagation will cease since the stress intensity factor decreases with crack length. Let us suppose a second micro crack now develops, ie. $a_2 \neq 0$. Since the stress intensity factor of a crack increases with increasing length of the other crack, the longer crack can induce crack propagation in the micro crack. With the propagation of the "micro" crack the stress intensity factor of the first crack will increase and this will further promote the first crack to propagate. Because the "micro" crack has larger stress intensity factor the "micro" crack will propagate faster than the first crack until finally both cracks will have a approximately the same length. This phenomenon has been observed by Jiang (1984) in the static case. If the internal pressure does not increase the crack propagation will cease since the stress intensity factor decreases with the crack length (Newman, 1971).

4 THE CRACKS FORMATION OF A ROW OF BLASTHOLES

The crack distribution around a blasthole is random. It follows that cracks can be initiated at any point of the blasthole wall under the blasthole pressure. A number of methods to control the orientation of the crack propagation are summarised by Persson et al (1993). The use of notches to promote fracture has been known for some time. It is possible to control the cracks by using a special tool to notch the blastholes along most of its length. The crack pattern can also be obtained by cutting two slits into a brass tube that is lowered into a blasthole and filled with explosive. The brass tubes protect the holes at all points except where fractures are required. However, it is difficult to implement notching techniques for a full-scale open pit pre-split blasting because of the difficulties of making the notches in a blasthole, particularly for deep blastholes. In fact, even in the quarries for dimensional stone extraction, the notching method is seldom used. Shaped charge (Persson et al, 1993) is another method to make notches and then cracks in required directions . The application of this method to full-scale blasting in surface mining will be dependent upon the efficient manufacture of the shaped charges and how to load the charge to blastholes without too much difficulty.

A more acceptable approach to control the crack pattern is to utilise the the stress concentration effect from adjacent blastholes. In fact it is the stress concentration effect that controls the crack propagation along the pre-split line in pre-split blasting. However, it seems to the author that the importance of the stress concentration is not fully appreciated. In this section the stress concentration effect will be discussed. The stress field around a blasthole due to the pressures from its adjacent two blastholes is diagrammed in the Figure 2.

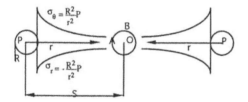

Figure 2. Static stress field around a blasthole due to pressures from its adjacent two blastholes.

The centre blasthole is in the stress field of the two pressurised adjacent blastholes. In the absence of the centre blasthole, the stress at the point "O", by the superposition principle, is given by

$$\sigma_r = -2\frac{R^2}{S^2}P \quad \text{and} \quad \sigma_\theta = 2\frac{R^2}{S^2}P \qquad (5)$$

In the presence of the centre blasthole, the tangential stress σ_θ at the point "A", is approximately given by

$$\sigma_\theta \approx 3\frac{2R^2}{(S)^2}P - \frac{-2R^2}{(S)^2}P = 8\frac{R^2}{S^2}P \qquad (6)$$

In deriving this approximation we have used Kirsch equations and assumed that the centre blasthole is under an uniform horizontal stress and vertical stress of

$$-2\frac{R^2}{S^2}P \text{ and } 2\frac{R^2}{S^2}P$$

respectively. Similarly the tangential stress σ_θ at the point "B", is approximately given by

$$\sigma_\theta \approx -8\frac{R^2}{S^2}P \qquad (7)$$

Of course the radial stress at boundary of the blasthole is zero.

The equations 6 and 7 show that at point A the blasthole surface is under approximately tensile stress 4 times higher than that if the hole is absent, while stress at point B is under compression stress 4 times higher than that if the hole is absent. The change from tensile at "A" to compressive at "B" gives a very high relative stress concentration effect around the blasthole. If the spacing is 6 times the blasthole diameter and the centre blasthole has the same pressure value P, then the tangential tensile stress at point A is P+0.055P while at point B is P-0.055P, this will give a 11% difference for the tensile stress from point B to A. For spacing equal to 10 times the blasthole diameter, the difference is 4%. However, a numerical modelling using finite element method for a row of blastholes showed about another 50% higher stress concentration. For spacing equal to 6 times the blasthole diameter, a 16% relative stress concentration was obtained and 6% for spacing equal to 10 times the blasthole diameter. The reason behind this discrepancy between the superposition method and the numerical modelling is, partly, due to the stresses from the next adjacent blastholes. If the rock properties and crack distribution are reasonably uniform, then we will expect the cracks to be initiated along the pre-split line, ie. the line connecting the centres of the blastholes.

5 THE DETERMINATION OF BLASTHOLE SPACING

The current method to determine the blasthole spacing is based on the "static tensile stress analysis" (Brent, 1995) as described as:

$$S = 2R (P + T)/T \qquad (8)$$

where S is the blasthole spacing and T is the uniaxial tensile strength of the material. The hypothesis behind this method is that the force required to fail the rock between two blastholes by tension is equal to the force from the blasthole pressure which is the product of the pressure P and the diameter (2R). Because the stress applied to the rock between the two blastholes is far from uniform, the reliability of the "static tensile stress analysis", ie. Equation 8, is quite questionable. The work of Kutter et al (1971) showed that the pre-split cracks are likely to be initiated at the surface of the blasthole. In fact, this phenomenon has been observed by Jiang (1984) in the static case.

If we assume the uniaxial tensile strength of the material is 1/10th the compressive strength and the blasthole pressure is of the magnitude of the compressive strength, the calculated blasthole spacing by Equation 8 is only 11 times the blasthole diameter. In fact, most of the pre-split blasting employed a much lower blasthole pressure but with a much larger spacing.

The blasthole spacing is the distance between the two blastholes which the cracks from both holes can break through. Since the blastholes break starting at surfaces of the blastholes and the rock fails by cracks propagation, it should be more sensible to apply the fracture mechanics principles to the determination of the blasthole spacing.

Regarding the mechanism of the cracks initiation around a blasthole and formation of the entire crack between two adjacent blastholes, many researchers have put forward their arguments about the "shock wave" mechanism and "gas" mechanism. A good summary of these arguments can be found from Dunn et al (1995). It has been understood that the gas effect increases with the decoupling extent. If a blasthole is fully coupled then the shock wave will be responsible for initiating the cracks around the blasthole. In this case it will be difficult to imagine that stress concentration can influence the direction of the crack initiation since a stress concentration field has not yet been built up. In other words the stress concentration will have little effect on the "shock wave action". In fact, numerous cracks will be initiated around the blasthole. However, if the blastholes are closely placed as some of the pre-split designs are, cracks will still be formed between the blastholes. This can be explained as although cracks are initiated in all directions, only the cracks along the pre-split line get extended and merged under the stress concentration effect built up later. In this case the cracks will remain to the half-blastholes (half barrels). Of course, if the blastholes are over charged the rock around the blastholes will be extensively crushed which should be avoided for a pre-split blasting.

In most cases pre-split blastholes are laterally decoupled. With increasing decoupling effect, the "gas action" will dominate the crack formation. If the "shock wave action" is not strong enough to initiate cracks around the blastholes, the "gas action" will be responsible for initiating cracks around blastholes. It should be appreciated that the duration of "gas action" is much longer than the "shock wave action". In this case it will be logical to say that only cracks along the pre-split lines will be initiated under the stress concentration effect.

To determine the blasthole spacing, we assume that the cracks could be initiated in any direction from a blasthole whether under "shock wave action" or "gas action" dependent upon the charge condition, but only the cracks along the pre-split line will propagate and join together. Since rock is not uniform and the perfect simultaneous initiation of the charges is not achieved in reality, the determination of the spacing will be complicated and presumably depend on the interpretation of the crack formation mechanism. Here the determination of the blasthole spacing will

be discussed under so called unfavourable and favourable conditions.

The unfavourable case is that the considered blasthole has two symmetrical edge cracks, but no cracks have been initiated in the adjacent holes as shown in Figure 3. Both cracks will propagate forward under the stress field generated by the pressures from those three blastholes. Since there are no cracks from both adjacent holes, the stress intensity factor of the crack(s) is relatively small and the cracks are relatively difficult to propagate. This can be seen later from the favourable case where each blasthole has a pair of edge radial cracks.

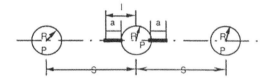

Figure 3. Unfavourable case.

The spacing is determined as each crack is able to propagate to the centre point of two adjacent blastholes; in other words, the spacing S is defined as twice the crack length. The process is to determine the crack length by comparing the stress intensity factor with fracture toughness under the stress condition. By the superposition principle of stress fields and superposition principle of stress intensity factors, the simplified stress intensity factor of the crack(s) can be derived as:

$$K_I = \frac{2PR}{\sqrt{\pi l}} + \frac{4\sqrt{\pi l}\,R^2P}{3\sqrt{3}\,l^2} \qquad (9)$$

When $K_I = K_{IC}$, the crack length, l, can be determined which is considered to be half of the blasthole spacing. Substituting K_{IC} for K_I and the blasthole spacing S can be derived from the equation as:

$$S = 2\left[\sqrt[3]{-\frac{q}{2}+\sqrt{\left(\frac{q}{2}\right)^2+\left(\frac{p}{3}\right)^3}} + \sqrt[3]{-\frac{q}{2}-\sqrt{\left(\frac{q}{2}\right)^2+\left(\frac{p}{3}\right)^3}} - t\right]^2 \qquad (10)$$

where,

$$q = \frac{27A^2C + 2B^3}{27A^3}, \quad p = \frac{-B^2}{3A^2}, \quad t = \frac{-B}{3A}, \text{ and}$$

$$A = 3\sqrt{3\pi}K_{IC}, \ B = -6\sqrt{3}PR, \ C = -4\pi PR^2.$$

The blasthole pressure P is generally estimated by calculating the CJ pressure and then the coupled (or decoupled) blasthole pressure. For a decoupled charge, as normally used in pre-split blasting, the blasthole pressure can be estimated by (Persson et al 1993):

$$P = \frac{\rho_0 D^2}{8}\left(\frac{d_0}{d}\right)^3 \qquad (11)$$

where ρ_0 is the density of the explosive, D is the explosive velocity of detonation, d is the blasthole diameter and d_0 is the explosive charge diameter.

A case study is a gold mine in Kalgoorlie with the following conditions:

Blasthole diameter = 89 mm (3 1/2"),
Charge diameter = 23 mm,
VOD = 4000 m/s,
Density = 1.2 g/cm³.

If we assume the fracture toughness is 2.5 MPa√m and apply the above parameters to equation 11 and then 10, a blasthole pressure 41 MPa is obtained and a 1.7 m blasthole spacing is defined. The actual used blasthole spacing is 1.4 m which seems to be the optimal in terms of the smooth face. However, it has been noted that a 2 metres stemming was used for the 10 metres long blastholes.

The previous study is the so called unfavourable case. The favourable case is defined as all the blastholes are initiated at same time and cracks are initiated and propagate simultaneously for all the blastholes as shown in Figure 4.

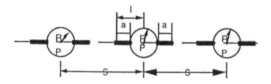

Figure 4. Favourable case.

The simplified stress intensity factor of the crack(s) is be given by

$$K_I = \frac{2}{\pi} P \sqrt{S \tan\frac{\pi(R+a)}{S} \, \sin^{-1}\frac{\sin\frac{\pi R}{S}}{\sin\frac{\pi(R+a)}{S}}} \qquad (12)$$

Numerical analysis of the Equation 12 reveals that, for a given blasthole radius R, the stress intensity factor K_I possesses a minimum when the crack length, a, varies from 0 to 0.25S. In other words, the crack length a can be expressed as a function of spacing when the K_I is minimum. Substituting K_{IC} for K_I minimum with crack length a as a function of spacing, the spacing S can be determined from Equation 12 for given blasthole pressure P. Since this spacing is derived from the favourable case, the determined spacing is much larger than the blasthole spacing used in the pre-split blasting currently. Hence the spacing determined from Equation 12 does

not have a practical meaning; but it does show the potential to use the blasthole pressure efficiently.

6 DISCUSSION AND CONCLUSIONS

It seems reasonable to determine the blasthole spacing using the proposed fracture mechanics model. Although there is no direct comparison between the fracture mechanics model and the "static tensile stress analysis" because of lack of the required rock properties, the examples from Brent (1995) seem to give much smaller blasthole spacings than that determined by fracture mechanics model for given blasthole pressures. It should be appreciated that by the fracture mechanics model the rock between two blastholes is assumed to fail progressively, but by the "static tensile stress analysis" the rock is assumed to fail simultaneously.

This research has shown that the control factor of the blasthole spacing is the requirement of smooth face rather than the blasthole pressure. In other words, the spacing is limited by the stress concentration effect which is attributed to the smooth face formation. In fact, a well below UCS (uniaxial compressive strength) blasthole pressure can give a much larger blasthole spacing than that determined in terms of the smooth face. Further research, such as the utilisation of satellite holes, should be carried out to optimise the use of both stress concentration effect and blasthole pressure. Further research is also needed to validate the blasthole spacing predicated by both models. Rock properties such as fracture toughness and uniaxial tensile strength are essential for the model validation.

The initiation and propagation of the cracks along the pre-split line is explained by the stress concentration effect. Although a simultaneous initiation of the blastholes is not required, a "pseudo-uniform stress field" from blastholes is assumed. This assumption implies a relatively long duration of gas pressure in each blasthole, and also a very small initiation delay between the blastholes. In fact, with a blasthole spacing of 15 times the blasthole diameter as in the case study, the stress concentration effect is relatively small. The stress concentration effect should be further examined closely.

Generally the fracture mechanics model predicates a critical pressure about the uniaxial tensile strength. It should be appreciated, in terms of the tensile failure criterion, the blasthole should not fail under the internal pressure of the magnitude of tensile strength because of the geometrical difference.

In order to improve the pre-split blasting results economically and technically, blasthole diameter, blasthole pressure and blasthole spacing should be carefully examined and then determined. The blasthole diameter affects the blasthole pressure, gas pressure duration and the stress concentration effect.

Adequate blasthole pressure enables the cracks to propagate to the required length. Blasthole spacing should be adequately determined according to the blasthole diameter, blasthole pressure and the rock properties. Since the drilling cost is relatively high, it is the blasthole spacing that dominates the cost of the pre-split blasting.

Fracture mechanics possesses the potential to be used to investigate the mechanism of the initiation and formation of the cracks along the pre-split line, and furthermore to guide the design of the pre-split blasting.

REFERENCES

Brent, G.F. 1995. The design of pre-split blasts. *Proc. EXPLO '95 Conference:* 299-305. Brisbane.

Dunn, P. & A. Cocker 1995. The design of pre-split blasts. *Proc. EXPLO '95 Conference:* 307-314. Brisbane.

Fourney, W.L. 1993. Mechanisms of rock fragmentation by blasting. *Comprehensive Rock Engineering, principles, practice & Projects* 4: 40-69. Ed. Hudson, J. A., Pergamon, Oxford.

Jaeger, J.C. & G.W. Cook 1979. *Fundamentals of rock mechanics.* Chapman and Hall, London.

Jiang, J.J. 1984. *The study of the static blasting agent and its application.* Master thesis. Wuhan University of Technology, Wuhan, China.

Jiang, J.J. 1985. The static blasting agent and its application to the marble exploitation. *Journal of Wuhan Institute of Building Materials* 2: 170-175.

Kutter, H.K. & C.F. Fairhurst 1971. On the fracture process in blasting. *Int. J. Rock Mech. Min. Sci.* 8: 181 - 202.

Newman, J.C. 1971. An improved method of collocation for the stress analysis of cracked plates with various shaped boundaries. *NASA,* TN D-6376.

Persson, P., R. Holmberg & J. Lee 1993. *Rock Blasting and Explosives Engineering.* CRC, Florida.

Tweed, J. & D.P. Rooke 1976. The elastic problem for an infinite solid containing a circular hole with a pair of radial edge cracks of different lengths. *Int. J. Engineering Science* 14: 925 - 931.

Wei, Y. & S. Wang 1987. A new method of fracture plane control and its application. *Proc. 2nd Int. Symp. on Rock Frag. by Blasting.* Keystone, Colorado.

Whittaker, B.N., R.N. Singh & G. Sun 1992. *Rock Fracture Mechanics, Principles, Design and Applications.* Elsevier, Netherlands.

Rock Fragmentation by Blasting, Mohanty (ed.) © 1996 Taylor & Francis. ISBN 90 5410 824 X

Role of discontinuity on stress field in wall-control blasting

S. H. Khoshrou
Mining and Metallurgical Engineering, McGill University, Montreal, Que., Canada

B. Mohanty
Mining and Metallurgical Engineering, McGill University, Montreal & ICI Explosives Canada, McMasterville, Que., Canada

ABSTRACT: The effect of discontinuities in the rock mass on wall-control blast results has been investigated in detail from a theoretical stand-point. The nature of the stress field around pressurized boreholes is evaluated by means of joint element modelling. The various conditions investigated in this study are the following: i) the effect of a free face, ii) narrow joint plane parallel and perpendicular to free face, and iii) joint plane of various widths. The regions of fracturation in the rock mass as well as the direction of preferential crack growth are predicted in this analysis, in term of the geometry and the orientation of the joint planes. The theoretical predictions are also compared with current engineering practice, and appropriate guidelines for wall-control blast design are developed.

1 INTRODUCTION

In blasting practice, many existing problems are extremely difficult or impossible to solve by analytical methods. One possibility is to simplify the problems to the point where analytical solutions can be used effectively. In some cases this procedure works, but in many cases it is not possible to obtain closed-form solutions; without relying on finite element methods.

A number of studies has been carried out to analyze the stress distribution around a hole loaded with explosives in the presence of a horizontal free face. Most of these studies discussed and modeled the in situ fragmentation for oil shale fragmentation (Young et al., 1985; McHugh et al., 1985; Shaffer et al., 1987). The stress distribution through the hole in a vertical section and the effect of bench height and burden in a three dimensional bench have been studied with a two dimensional finite element program by Ash (1973) and Smith (1976) respectively. The effect of borehole diameter on the distribution around a pressurized hole at constant burden was discussed by Ghosh (1990). Song and Kim (1995) have also attempted to model the smooth blasting process. These studies largely deal with the condition of stresses in normal production blasting, except for Song and Kim (1995). None of these studies however consider the effect of a free face or discontinuities on the stress field around pressurized

holes, especially on the centreline between the holes, which is a critical line for the wall-control blasting process (Khoshrou, 1996).

In the simple analytical approach, the stresses depend only upon the borehole pressure and the width of the material surrounding the borehole. It ignores strength properties of the latter, the influence of free face and any type of discontinuity. However, treatment of the latter is crucial to understand the mechanism of fracturation in rock under realistic conditions. The effects of these parameters on the stress distribution can be studied with a numerical approach and checked against observations. To illustrate this case, two-dimensional finite element modelling has been carried out to determine the effect of the presence of a free face, and a parallel and normal joint or weak plane to the free face on the stress field around a two pressurized boreholes.

2 NUMERICAL MODELLING

In this analysis, a finite element program (*I-DEAS*) has been used to analyze the stress distributions in the regions of interest. *I-DEAS* is an integrated software tool which has been developed by Structural Dynamic Research Corporation, SDRC (Lawry, 1991; I-DEAS, 1990). A complete finite element model can be built by *I-DEAS*, including physical and material properties, loads and boundary conditions. The finite element analysis consists of three steps: preprocessing, solution and postprocessing.

Preprocessing is graphically complete. It includes the process of developing the geometry of a model, creating of mesh, entering three physical and material properties, describing the boundary conditions and loads, and checking the model. The triangular and quadrilateral elements are available in the library of element types. These elements have two or three nodes along each edge and are known as linear and parabolic elements, respectively.

In the finite element procedure, the coordinate values of nodal points of each element are arranged into a matrix $\{D\}_n$, and the elastic properties of the material set up into the constitutive matrix [E]. The element stiffness matrix in a local coordinate system $[K_m]_n$ is obtained from

$$[K_m]_n = \int_{v^e} [B]^T [E] [B] \, dv^e \qquad (1)$$

in which [B] is the strain-displacement matrix, [E] is the constitutive matrix and v^e is the element volume.

The above integration is evaluated using one of the numerical integration schemes such as the Gauss-quadrature procedure. The transformation of the element stiffness matrix from the local to the global axes is performed by:

$$[K]_n = [T]^T [K_m]_n [T] \qquad (2)$$

in which $[K]_n$ is the element stiffness matrix in the global coordinate system and [T] is the transformation matrix. The global stiffness matrix of the whole structure [K] is obtained using the summation of the element stiffness matrices. This process can be represented symbolically by:

$$[K] = \sum_{n=1}^{N} [k]_n \qquad (3)$$

where N is number of elements in the domain.
For linear analysis, equilibrium equations can be expressed as:

$$\{F\} = [K] \{D\} \qquad (4)$$

where [K] is the stiffness matrix of the structure, {F} is the total force vector due to in situ stress, gravity and boundary pressure, and {D} is the nodal displacement vector, all in the global coordinate system. The displacement at any point of the element can be evaluated in terms of the displacements of the nodal points on the boundaries, or within the element as:

$$\{u\} = [N] \{U\}_n \qquad (5)$$

where {u} is the displacement at any point of the element, [N] is the matrix of shape function and $\{U\}_n$ is the element nodal displacement vector in the element local coordinate system.

The strain at any point of the element, $\{\varepsilon\}$, is related to the element nodal displacement, $\{U\}_n$, by the following equation:

$$\{\varepsilon\} = [B] \{U\}_n \qquad (6)$$

where [B] is the strain-displacement matrix and can be obtained as:

$$\{B\} = [L] [N] \qquad (7)$$

in which [L] is the differential operator matrix and is defined such that:

$$\{\varepsilon\} = [L] \{U\} \qquad (8)$$

Based on the stress-strain relations, stress at any point of the element can be found as:

$$\{\sigma\} = [D] (\{\varepsilon\} - \{\varepsilon^*\}) + \{\sigma^*\} \qquad (9)$$

where $\{\varepsilon^*\}$ is the vector of initial strain, [D] is the material property or material stiffness matrix and σ^* is the in situ stress.

For the present investigation, a horizontal section normal to the axis of the pressurized hole has been modeled using the above program. In all cases, a 100 mm diameter borehole is pressurized at 2000 MPa. This corresponds approximately to the explosion pressure generated by detonation of a fully-coupled ANFO (Ammonium Nitrate/Fuel Oil: 94/6) explosive charge. The concept of weak planes with varying compliances is used to simulate joints, faults and foliation planes in rock and their effect on the stress field. The material properties of the rock and weak plane are presented in Table 1. The properties of the two media represent an average rock and a weak cementing or gouge material for a weak plane. The results are normalized in terms of borehole pressure and radius to facilitate extrapolation of the results to other geometries and explosives.

Table 1. Material properties of the rock and the weak plane.

Material Properties	Unit	Rock	Weak Plane
Modulus of Elasticity	(GPa)	40	0.4
Poisson's Ratio	-	0.15	0.4
Density	(g/cm³)	2.5	2.1

2.1 Free Face

The spacing between two holes in the models is 10 and 15 times the hole diameter. These values are realistic for different types of wall control blasting methods used in mines, quarries and road cuts. The dimensions of burden are 2.5 to 20 times the borehole diameter, and that of spacing 15 and 10 times the borehole diameter respectively. The boundaries of the models are fixed in the x-direction on the right and the left sides and in the y-direction at the back of the holes. For various burdens, stresses are calculated at x and y directions as well as principal stresses in the solution regions. Figure 1(a) shows the simulated model for this analysis.

2.2 Weak Plane Parallel to the Free Face

The effect of weak planes, such as open clay-filled joint or thin clay-layers, parallel to the free face on the stress field is analyzed. It consists of placing a weak plane parallel to the free face but at varying distances from the boreholes at a constant burden and spacing. These weak planes are located in the front of two and three hole configurations (Fig. 1(b)).

2.3 Weak Plane Normal to Free Face

A weak plane is placed between the holes normal to the free face for two burden distances. The principal, compression and tensile stresses are calculated between the holes. The final results are compared with the same model but in the absence of any weak planes. Figure 1(c) shows the simulated model for this analysis.

2.4 Weak Planes of Varying Widths

The discontinuities in the rock may be filled or open. The width of discontinuities in a rock mass differs. In this section, the effect of weak planes of varying width will be discussed. For the purpose of modelling, the former are represented by weak planes of varying width and filled with gouge materials and named weak planes. These materials are much weaker than the rock. The width of the weak plane is varied from 1 to 20 mm, as shown in Fig. 1(d). The effect of similar weak planes normal to the free face (located at $2.5 \times D$ and $5 \times D$) on the stress distribution is investigated for constant burden and spacing ($10 \times D$), D being the borehole diameter.

3 RESULTS AND DISCUSSIONS

3.1 Effect of Free Face

The principal stresses as well as stresses in the co-linear and perpendicular directions are calculated at each node of the elements. These stresses are plotted along the direction normal to free face from the centre of the hole and the midpoint, as well as on the centreline, Fig. 2.

Figures 3 and 4 show the calculated maximum principal stress for different burdens at a constant spacing ($15 \times D$). The maximum principal stress at the midpoint for an infinite burden is 2 times greater than the stress at the same point for a burden of $5 \times D$. The tensile stresses and maximum principal stresses are identical along a direction normal to the free face and connecting the midpoints and centres of holes (N1 and N2). These stresses are reduced as the burden is decreased, and are about zero for a simulated model for a burden and spacing equal to $2.5 \times D$ and $15 \times D$, respectively.

On the other hand, the tensile stress increases dramatically in the burden region which is close to the face at the smallest distance, along a direction normal to the free face and connecting the pressurized boreholes. The maximum principal stress for a small burden ($2.5 \times D$) is 6 times greater than the infinite burden for a constant spacing ($15 \times D$), (Fig. 3(b)). The effect of a free face on the stress field for a burden greater than $15 \times D$ is negligible. A burden can be characterized as an infinite burden when the ratio of that to the spacing exceeds unit, and the critical size of the burden for pre-splitting is defined as a distance equal to spacing. In an isotropic and homogeneous medium, a spacing

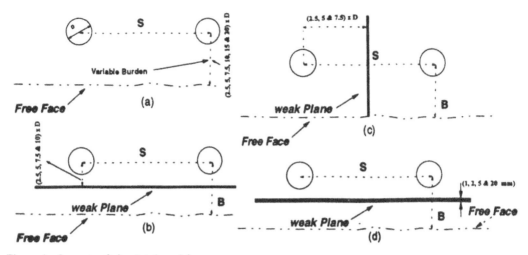

Figure 1 : Layouts of simulated models.

between 10 ×D and 15 × D should be acceptable, but the exact value would depend on the rock properties. As the burden decreases, the tangential stresses also decrease on the line normal to the free face from the midpoint (Fig. 3(a)). Conversely, the stresses on the line normal to the free face from the hole centre increase as the burden decreases. Consequently, the form of the fracture zone changes from elliptical to an approximately circular shape for each hole. It should be noted that the shape of this fracture zone is similar to an ellipse for an optimum spacing in an infinite burden.

For a ratio of burden to spacing of about 0.8, a fracture zone can be achieved between the holes, and to a lesser extent, in the burden regions. The effect of a free face at 0.8 times spacing on the stresses along the centreline and the line normal to it through the midpoint is almost negligible. In contrast, the stresses in front of each hole is increased (Fig. 3). This description is analogous to the concept of cushion blasting method. As stated before, in this technique the aim is to create a narrow fracture zone at the perimeter of the excavation, trimming or slashing the excess material from the final walls. The ratio obtained by numerical analysis (B/S=0.8) is also similar to the recommended ratio in common practice in cushion blasting.

The same analysis can be used to explain the recommended ratio of the burden to spacing ranges from 0.8 to 1.2 in buffer blasting. For a ratio smaller than 0.8, an irregular face (pillar left

between the holes) is predicted; for ratios greater than 1.2, large muck and cratering can be expected.

3.2 Effect of Weak plane Parallel to the Face

In this section, the effect of distance of a weak plane (with 20 mm width) from the borehole wall is analyzed for multiple pressurized holes. The weak plane is fixed at 2.5 × D, 7.5 × D, and 12.5 × D for a constant burden and spacing. As shown in Fig. 5(a), a weak plane with a distance greater than 7.5

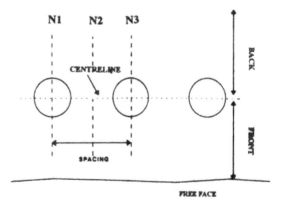

Figure 2 : Imaginary lines for plotting stresses around boreholes.

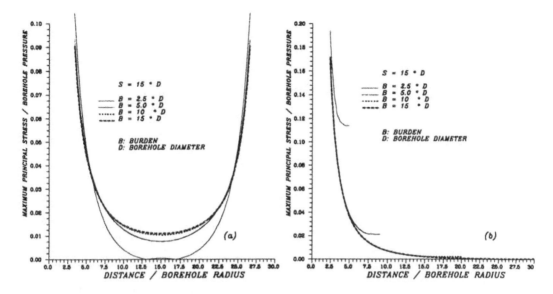

Figure 3 : Maximum principal stresses along the centreline and along line normal to face from hole centre for different burdens at a constant spacing.

× D times the borehole diameter does not affect the distribution of stresses on the centreline at a constant burden. This value has a direct relationship to the spacing, and decreases to 5 × D, where the borehole separation decreases to 10 × D with the same burden. Therefore, the influence of a fixed weak plane parallel to the face at 2.5 × D and 5 × D from the hole centre on the stress field should be the same if the spacing is decreased to 5 borehole diameters. As a result a fixed discontinuity parallel to the free face at half the spacing should not have any effect on the stresses along the centreline.

For a constant burden and spacing, the value obtained for maximum principal stress near the weak plane, on the lines normal to the face from the hole's centre, is 3 times greater than the stress at the same point for similar models without the weak plane (Fig. 5(b)).

The role of a filled discontinuity parallel to the centreline at a constant burden and spacing is approximately analogous to a free face which is located at the same distance from the hole wall. Overbreak around the holes and the pillar between holes are thus predicted by numerical analysis for a weak plane (20 mm wide) parallel to the free face with a distance smaller than half the spacing from boreholes wall. In all cases, the presence of a wide weak plane causes the tensile stress to drop

immediately to zero on the weak plane. The recovery distance is found to be inversely proportional to the proximity of the weak plane to

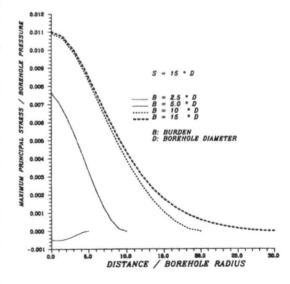

Figure 4 : Maximum Principal stresses for different burdens at a constant spacing, along line normal to free face from midpoint.

211

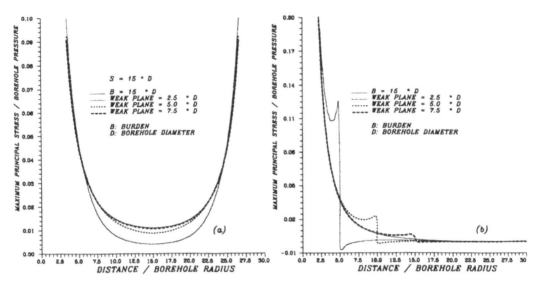

Figure 5 : Maximum principal stresses for different weak planes parallel to free the face along the centreline and along a direction normal to the face and connecting the hole for a constant spacing and burden.

the borehole. Therefore, fractures should not cross the wide weak plane (over 20 mm) or an open discontinuity. In all cases, fractures will start to develop from the borehole wall and extend to the parallel weak plane along the direction normal to the latter. The rock should remain intact between the holes, especially around the mid-region.

3.3 Effect of Perpendicular Weak plane to the Face

Several models have been simulated to examine the effects of weak planes normal to the free face on the stress distribution. The conditions of these models are exactly the same as those for two pressurized holes with a parallel weak plane. In this case, the parallel weak plane is rotated 90 degrees, and fixed at $2.5 \times D$, $5 \times D$ and $7.5 \times D$ at constant spacing equal to $15 \times D$ with two different burdens ($10 \times D$ and $15 \times D$).

Analysis of the stress distribution using the model with normal weak planes shows characteristics which are considered very important to the degree of success of wall control blasts. Normalized maximum principal stresses on the centreline are shown in Fig. 6(a). The magnitude of the maximum principal stresses increases as the distance of the weak plane to the borehole wall decreases. A circular tensile

zone is developed between the weak plane and the nearest borehole wall, when the distance of the weak plane from the hole centre is smaller than $5 \times D$. This zone is similar for all simulated models. The value of tensile stresses reaches its maximum near the boundary of the weak plane and presents the same condition for all models. As shown in Figs. 6(a), the calculated maximum principal stress for a model with a fixed plane at $2.5 \times D$ borehole diameter is about 2.5 times greater than the real value. The presence of a weak plane perpendicular to the centreline causes the stress to drop to zero immediately at the boundary of the plane. The stresses start to recover quickly after the weak plane, but the recovery rate depends on the distance of the weak plane from the borehole wall as well as the width of the discontinuity. It reaches its real value at half spacing and continues to the other holes normally.

As Fig. 6 shows, the effect of the weak plane normal to the free face from the middle of the centreline on the stress distribution is negligible. The results illustrate that the plots for two different cases, with weak planes at the middle of the spacing and without weak plane, correspond to each other. A comparison between the final results of finite element models and the experimental results (Belland, 1966; Worsey, 1984) shows that the

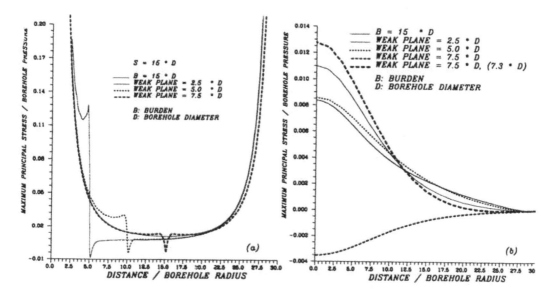

Figure 6 : Maximum principal stresses for weak plane normal to the face for a constant spacing and burden, along the centreline and along the line normal to the free face from midpoint.

numerical results are in good agreement with the field results. In a field study at the Carol Lake Mine, Belland observed that blasting across the major joint close to normal had caused a vertical face in the back of blastholes. Based on a laboratory investigation and field observation, Worsey concluded that the discontinuity normal to the final line of pre-split has little effect on the blast results.

The calculated stresses for a constant spacing and burden are shown in Fig. 6(b), along directions normal to the free face and connecting the midpoint on the centreline. The results illustrate that the tensile stress at midpoint is a function of weak plane distance from the borehole wall. As the distance is reduced, the stresses also decrease. Optimum results can be obtained when the weak plane is located at mid-point between two holes. Therefore, a weak plane with an intersection angle equal to 90 degrees to the line of pre-split at half spacing has minimal effect on the results of the blast.

3.4 Effect of the Width of a Weak Plane

The influence of the width of the discontinuity is analyzed for planes both parallel and perpendicular to the face. For each case, stresses in x and y directions and maximum principal stresses have been

calculated, and the final results plotted along the centreline and along directions normal to the free face and connecting the pressurized boreholes and the midpoint (N1 and N2). The width of the weak plane has been varied keeping the burden and spacing constant to discuss stress distribution around the pressurized holes for a weak plane with a fixed distance to the hole. The effect of the width of a weak plane parallel to the free face on the tensile stresses along the centreline is illustrated in Fig. 7(a). For a fixed plane close to the hole centre (2.5 × D), the maximum principal stresses are augmented as the width of the weak plane increases. The stresses reach the maximum value near the boundary of the weak plane and drop to zero on the discontinuity. This value approaches the real one (without a weak plane) for a narrow plane (1 mm). The amplitude of stress, immediately before the boundary, varies directly with the width of the weak plane. Also, a very narrow weak plane causes an immediate recovery from the stress. The recovery distance depends on the width of the weak plane for a fixed weak plane, and it is much greater for a larger width, Fig. 7.

At the midpoint, both tensile and compressive stresses are functions of the width of the weak plane. For an infinite burden, the magnitude of tensile stresses for a 1 mm wide weak plane is 1.25

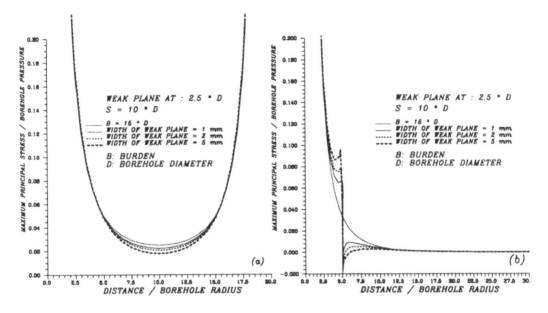

Figure 7 : Maximum principal stresses for weak planes of varying widths parallel to the free face for a constant burden and spacing, along the centreline and along the line normal to the free face from hole centre.

times greater than the stresses for the similar model with a 5 mm wide weak plane, Fig 7(b). The results predict that the difference between the magnitude of stresses for varying widths (1, 2 and 5 mm) at constant spacing and various burden will be almost equal.

Consequently, a very narrow discontinuity or an open discontinuity cemented with a strong material would have little effect on the final results of wall control blasting.

According to the results, the weak planes with a width smaller than 5 mm, and a distance greater than 5 borehole diameters do not have any effect on the stress distribution. For distances smaller than this, the rate of stress increase largely depends upon the width of the weak plane, and for a very narrow discontinuity (less than 1 mm) it approaches the value calculated from the same model without any weak plane.

The maximum principal stress between the holes and the boundary of a normal weak plane with various widths is equal, when the plane is located at middle of spacing between two holes. The magnitude of the stresses are variable close to the boundary of the weak plane, the narrowest weak plane, but the difference between them is not significant. Similarly, a parallel plane (1 mm wide) does not have any effect on the stress distribution,

and the influence of a weak plane width smaller than 5 mm is negligible.

As the distance of a discontinuity perpendicular to the borehole decreases, the magnitude of tensile stresses increases close to the boundary of the weak plane. For a weak plane 5 mm wide and located at $2.5 \times D$, this value is 2 times greater than that calculated for similar models without a weak plane. This reduces to 1.5 times when the width of the weak plane is decreased to 1 mm. However, a discontinuity with a distance smaller than $5 \times D$ always produces higher stresses in a zone between the borehole and the discontinuity. The magnitude of the stresses in this zone depends on the weak plane width and its distance from the borehole wall. It is expected that several tensile fractures would form in this zone. Consequently, backbreak should result in the area between the plane and the pressurized hole. The extent of overbreak in the field should be higher than those numerically calculated, due to the penetration of the explosion gases into the fractures and finally along the discontinuity.

4 CONCLUSIONS

The analysis shows that hole separation would range up to 15 borehole diameter for pre-split blast (infinite burden) in an isotropic and homogeneous material. For an optimum spacing, the shape of the fracture zone for each hole is approximately elliptical. The major axes of these ellipses coincide with the centreline between the holes. As the burden or spacing decreases, the fracture zones change from elliptical to circular or conical shape.

The study also shows that a ratio of burden to spacing up to 0.8 for cushion blasting and between 0.8 and 1.2 for buffer blasting would be applicable. With these ratios, a dominant fracture plane would be created between the holes, along with some fractures in the burden region. The analysis predicts "pillars" between the holes when the ratio is less than 0.8, and large blocks and extensive backbreak when the ratio is greater than 1.2. These ratios are similar to those in current engineering practice.

The presence of a relatively wide (parallel, normal and inclined) weak plane (of the order of 10 mm) causes the tensile stress to reach the maximum value at the boundary of the weak plane; the stress drops immediately to zero on the weak plane. It does recover, however, to its original value without the weak plane, but only at a considerable distance from the weak plane. The effect of a weak plane wider than 60 mm is virtually identical to that due to a free face.

Excessive overbreak in the direction of the centreline and pillars between the holes are also predicted by the model for a wide weak plane parallel to the free face with a distance smaller than half the spacing from the borehole wall. In the presence of a weak plane behind the boreholes, this weak plane will become the final wall provided the distance of the weak plane is less than half the spacing.

A weak plane located at the midpoint between the holes and lying normal to the centreline is shown to have only minimal effect on the results of the wall-control blast. It is also seen that when the distance of the perpendicular weak plane is changed by moving it closer to one of the holes, the tensile stress at this plane not only increases, as would be expected, but also considerable overbreak would be seen to ensue near the closer hole. The model shows that at 1/4 the spacing, the presence of this weak plane would cause the stress field between it and the borehole to assume a more circular shape. Therefore, more intense fracturing would take place around this hole, which would adversely affect the degree of fracture-plane control.

Although a high degree of correlation between these theoretical predictions and actual experimental results has been demonstrated (Khoshrou and Mohanty, 1996), the theoretical treatment does have some limitations. For example, isotropic and homogenous materials are selected as the media, and elastic solution is used to calculate the stresses in the regions of interest by a two-dimensional model. A concept of weak plane is used to present any type of discontinuities in the rock mass. Simultaneous detonation of the two holes is assumed, and the role of explosion gases inside the open cracks is ignored. The latter would significantly affect the final outcome, resulting in rather conservative estimates of burden and spacing.

ACKNOWLEDGMENTS

The authors are grateful to Profs. F.P. Hassani, H.S. Mitri for many useful discussions, and profs. L. Lessard and J.A. Nemes for assistance with the *I-DEAS* code.

REFERENCES

Ash, R.L., 1973, "The Influence of Geological Discontinuities on Rock Blasting," Ph.D. Thesis, Dept. Min. Eng., University of Minnesota.

Belland, J.M., 1965, "Structure as a Control in Rock Fragmentation," *The Canadian Mining and Metallurgical Bulletin*, pp. 323-328.

Ghosh, A., 1990, "Fractal and Numerical Models of Explosive Rock Fragmentation," Ph.D. Thesis, Dept. Min. Eng., University of Arizona.

I-DEAS, 1990, " Finite Element Modelling," Structural Dynamics Research Corporation, Milford, Ohio, U.S.A.

Khoshrou, S.H., 1996, "Theoretical and Experimental Investigation of Wall-Control Blasting Methods," Ph.D. Thesis, McGill University, Montreal, Canada.

Khoshrou, S.H. and Mohanty, B., 1996, "Some Critical Features of Wall-Control Blasts in the Presence of Discontinuities,"Int. J. Rock Mech. and Min. Sci. (in preparation).

Lawry, M.H., 1991, "I-DEAS™ Student Guide," Structural Dynamics Research Corporation, Milford, Ohio, U.S.A.

McHugh, S.L., Curran, D.R. and Seaman, L., 1985, "The NAG-FRAG Computational Fracture Model and Its Use for Simulating Fracture and Fragmentation," Society of Experimental Mechanics, Connecticut, pp. 173-185.

Shaffer, R.J., Heuze, F.E. and Nilson, R.H., 1987, "Finite Element Models of Hydrofracturing and

Gas Fracturing in Jointed Media," *Proceedings of the 28th U. S. Symposium on Rock Mechanics*, University of Arizona, pp. 797-408.

Smith, N.S., 1976, "Burden-Rock Stiffness and Its Effect on Fragmentation in Bench Blasting," Ph.D. Dissertation, University of Missouri-Rolla.

Song, J. and Kim, K. 1995, "Micro-Mechanical Modelling to Study Smooth Blasting," *SME Annual Meeting*, Denver, Colorado.

Worsey, P.N., 1984, "The Effect of Discontinuity Orientation on the Success of Pre-Split Blasting," *Proceedings of the Tenth Conference on Explosives and Blasting Technique*, Lake Buena Vista, Florida, pp. 197-217.

Young, C., Barbour, T.G. and Trent, B.C., 1985, "Geology Control of Oil Shale Fragmentation," Society of Experimental Mechanics, Massachusetts, pp. 93-99.

Rock Fragmentation by Blasting, Mohanty (ed.) © 1996 Taylor & Francis. ISBN 90 5410 824 X

Influence of discontinuities on presplitting effectiveness

Thomas Lewandowski, V. K. Luan Mai & Richard E. Danell

BHP Research, Newcastle Laboratories, Wallsend, N.S.W., Australia

ABSTRACT: The impact of joint frequency and relative joint orientation to the presplit have been examined in medium strong overburden rock from 6 pits of a large open cut coal mine in Central Queensland, Australia. The results of this study have shown for this rock type that there is a significant decrease in the presplit quality with increased joint frequencies and that presplit quality decreases with an increase of relative orientation from 5 to 30 degrees. These findings are consistent with theory presented in the literature.

1 INTRODUCTION

Open cut mining operations commonly use presplitting for wall control. Within BHP Australia Coal Pty Ltd (BHPAC), whose operations include seven open cut mines in the Bowen Basin of Central Queensland, Australia, presplitting is used mainly for short term highwall control, especially in throw blasting areas. In these operations, presplit is required to be of sufficient quality to provide three major benefits:

• To reduce the likelihood of hazards for mining underneath the face.

• To provide a face that leads to faster dragline recovery during off-line key digging or chop cutting.

• To enable a consistent front row burden in the next bench for engineering control in throw blasting.

In some cases a higher quality presplit is required at final pit walls in association with subsequent highwall mining of coal. Here the focus is solely on the reduction of hazards to people and machinery working at the base of the wall.

BHP Research and BHPAC have conducted a project to improve their understanding of the presplitting process, develop or modify engineering design tools and practices, so as to ensure that the business requirements listed above are achieved. In particular, the effect of geotechnical features such as joints was recognised as one that has a significant impact on presplit results, yet almost no current design procedures for BHPAC presplits explicitly consider jointing. This paper discusses a study to quantify the effects of jointing on presplit quality and to provide a basis for a practical method to include their effects into presplit design.

To put the effect of geotechnical factors in context, an idealised presplit design process is outlined in Figure 1. The presplit design process would contain the following elements:

1. Geological parameters: The strength of the rock, both in compressive and tensile fracture, is critical to the successful propagation of a presplit. Design processes have been designed around these factors and are expected to work well where massive rock is involved (Calder 1977, Chiappetta 1991).

2. Geotechnical factors: In addition to jointing, major structures such as faults require identification and specific attention. The effect of jointing is considered here as a modification to the behaviour of the presplit from that predicted from rock strength. Qualitative adjustments are made in some circumstances. A quality design requires a quantitative approach, based on the actual jointing characteristics of the site. This is the goal of the work presented in this paper. This work does *NOT* deal with the inherent stability of the face. Joints can lead to wedge and other failures in highwall faces and this should be assessed early in the mine design process.

3. Drilling parameters: As the rock is drilled, there is an opportunity to confirm the rock strength and to identify any local variations that can impact on presplit quality. For example, a soft band coinciding with an explosive deck can lead to adverse and costly results. The use of a drill monitoring system can identify the depth and relative strength of any anomalies (Danell 1989).

4. Equipment assessment: The available drilling equipment and its characteristics, such as minimum hole spacing, available drilling diameters and

accuracy play a significant role in the design process.

5. Economic assessment: Economic factors such as the purpose of the presplit and expected benefits need to be balanced against the costs, which largely revolve around drilling and explosives.

6. Site characteristics: The engineering history of blasts in the area and assessment of other effects such as the presence of water also have a role in the design process.

The first two geology/geotechnical assessments, shown in Figure 1, are used to produce an idealised design, which is compared and modified with the 'practical' design processes of the latter two assessment categories. Engineering judgement is used to resolve conflicts. Current BHPAC design practices for presplits rely more heavily on the latter three assessment categories. This is due to a lack of a design methodology to include jointing and insufficient databases on rock strength and jointing characteristics. While the approach based on historical blasting data has generally been effective, it has led periodically to unsatisfactory results and reductions in dragline and mining productivity. Elimination of these productivity losses through measurement and application of geological and geotechnical consideration is the main driver of this project.

2 THEORY

Presplit blasting is designed to generate a single crack along the row of blastholes which are initiated prior to the production blast. The aim is to break the rock in tension, while avoiding excessive compressive failure around the blastholes. Tensile failure is achieved as the compressive rock strength is several times greater than the tensile strength. When the blasthole wall pressure is adjusted to be approximately equal to the rock compressive strength the presplit hole spacing is maximised without causing damage to the rock near the explosive charge.

A theoretical examination of the production of tensile presplit cracks through interaction between rock strength and stress waves from explosives was conducted by Sanden (1974). This theory has been further studied by Calder (1977) and Chiappetta (1991). The presplit spacing based on the above consideration is given by:

$$S \leq 2rh \, (Pb+T)/T$$
where
S - spacing between the boreholes (inches)
rh - borehole radius (inches)
Pb - borehole pressure (psi)
T - tensile strength of rock (psi)

However, these methods deal with intact rock strength only, they do not include geotechnical aspects of the strata.

Literature discusses three predominant aspects of joints that affect presplit quality. These are the joint frequency, the joint angle in relation to the presplit line, and type of opening and infilling. Usage literature contains qualitative design comments with respect to geotechnical factors. These include:

1. Calder (1977) provided a general approach to the analysis of structures and their influence on the pit wall stability. The author distinguished between the impact of joints for almost parallel cases, at 45 degrees, and at 90 degrees to the face. It was concluded that the orientation of these structures in relation to pit wall direction have a major influence on the quality of the wall especially in terms of backbreak and face loss.

2. Chiappetta (1991) reported that frequency of joints in the range of 3-4 joint planes per spacing will produce adverse effect on presplit conditions.

3. Similar recommendations have been made by Workman and Calder (1991). They acknowledged the significance of joint spacing and their influence on presplit control. For example, hole spacing should not exceed 2-3 times the predominant joint separation.

4. Dunn and Cocker (1995) have noted that rules of thumb such as "the hole spacing should not exceed twice the joint spacing" provide some assistance to presplit design. As an alternative the authors produced a conceptual set of co-operation curves, which take into account structure orientation, frequency and strength. These curves are in line with the experimental results presented in subsequent sections of this paper.

Theoretically, where there is a joint the induced stress waves can be expected to be more attenuated. The attenuation will depend on frictional properties of the joint. Reinhart (1964) presents a mathematical examination of the interaction. He demonstrated that the attenuation of the wave increases with decreased joint frictional properties. He also demonstrated that the attenuation of the stress wave transmitted through the joint is a function of the angle of incidence of the stress wave onto the joint face. The attenuation is minimum when the angle of incidence is parallel or perpendicular to the face and increases to a maximum when the angle is between 15 and 45 degrees.

Following from this, to obtain optimum presplit wall quality the radial fractures should be in the plane of the presplit wall. This implies that the crack will proceed with minimum attenuation when the relative angle of jointing with respect to the presplit line is parallel, nearly parallel or normal. For oblique relative angles wave attenuation is

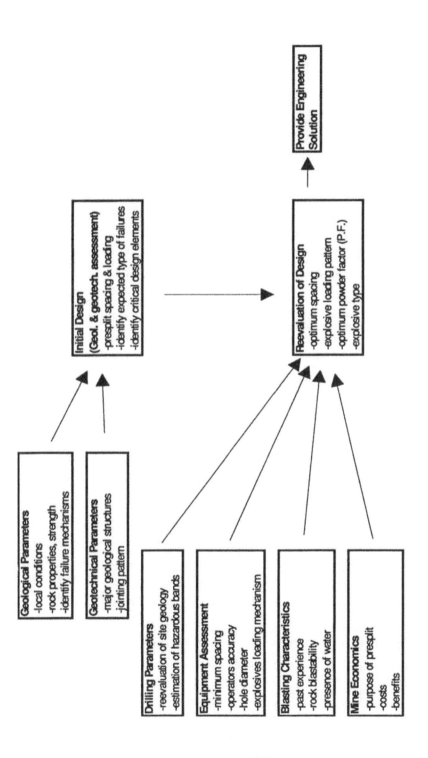

Figure 1. A methodology for the design of presplit blasts. The relative weighting of various elements depends on the quantity and quality of information from each element and their perceived impact. The purpose of the work desribed here is to provide a means for increasing the quality of information from geotechnical factors and to provide a quantitative method for assessing thier impact.

significantly increased and hence the presplit line formation is more difficult.

The effect of the wave attenuation is more pronounced in the presence of multiple joints. In this case stress wave attenuation will accumulate. In extreme cases the stress wave intensity will be reduced to level where the presplit crack is stopped. It can be inferred that with an increased level of joint frequency the probability of proper crack formation is decreased, resulting in inferior wall quality. Experimental testing with blasts in plexiglass confirmed incomplete presplitting in the presence of pre-existing joints (Kutter & Kulozik 1990).

These theoretical aspects can be summarised by the following points:

1. The orientation of the joints relative to the presplit wall will have a significant effect on presplit effectiveness. Presplits perpendicular to the jointing are little affected by jointing and some authors suggest that performance of presplits parallel to jointing may be enhanced. Presplits at other angles will have varying degrees of impaired effectiveness.

2. The frictional properties of the jointing will affect presplit effectiveness, with lower friction joints having increasing deleterious effects.

3. The frequency of joints will affect the number of joints each presplit must cross and the impairment of the stress wave is expected to be additive for each joint.

These points provide the basis for this study and experimental data is discussed with respect to these hypotheses in section 4.

A final point is the handling of multiple joint sets. The most common case is orthogonal joints. Each of the principal joint, J1, and its orthogonal counterpart joint, J2, will have the same impact on the presplit crack formation. This is due to the 90 degrees difference between the angle of incidence of J1 and J2. The attenuation effects are similar for each of these incidence angles. In the following sections the definition of the relative joint angle is the angle between principal joint direction J1 and pit direction (presplit line). Orthogonal joints are counted towards the total joint frequency. Any other joint sets are considered on a case by case basis.

3 SITE DESCRIPTION

Of the seven open pit operations of BHPAC one, Moura Mine, had assembled in the past five years an extensive database of jointing, wall quality and blasting information that was suitable for this study. As a result, Moura was chosen as the pilot study site.

In the areas of Moura considered here there is very little differentiation in overburden strata geology. About 80 -100% of the overburden height

in each case was classified as a medium strong to strong rock (50 - 150 MPa UCS). Also, only one to a maximum of three strata units are present in the overburden of each of the considered strips. The effect of bedding planes and their influence on the quality of presplits was not quantified in this study.

Even though there was limited geological variation some differences in presplit quality with rock strength have been observed. The most stable presplit walls are obtained for strata with 100 - 150 MPa (UCS). For 50 - 100 MPa (UCS) presplit walls are classified as stable but of lower quality than for the previous range. Strata with 0 - 50 MPa (UCS) produced deleterious effects on the wall quality. A variety of localised problems were observed. They ranged from slope sliding to acceleration of toppling failure originating from soft mudstone bands.

The focus of this study was to investigate the influence of joint properties on the presplit wall quality. An extensive geotechnical study was undertaken in 1993-95 by G. Leblang and Associates. These studies included detailed geotechnical mapping for all occurring structures (Godfrey and Thrift, 1995).

A typical jointing system at Moura mine consists of sub-vertical, orthogonal two joint system (J1, J2). The principal joint direction of J1 is N-S, while J2 is trending E-W. There is also some joint rotation from pit to pit. In addition a third joint set, J3, occurs in some areas and is mainly related to the predominant faulting system. For details of the joint occurrence for each particular pit refer to Table 1.

In all analysed cases only the systematic joints J1, J2 and sometimes J3 were analysed. The presence of J3 in the case of strip 18BL11S was omitted, as J3 is almost normal to the pit wall. Therefore, the expected level of attenuation is low and J3 can be neglected. For strip 13BL5N all 3 joints made oblique angles to the wall, and thus all of them were included for analysis.

Table 1 Summary of Joint Occurrences.

Pit Ref.	J1 dir.	J2 dir.	J3 dir.	Pit dir.
	(degs)	(degs)	(degs)	(degs)
13BL5N	336 (P)	66 (S)	305 (W)	345, 4, 356
16DU5N	330 (P)	69 (S)	-	337
16DU5S	330 (P)	70 (S)	-	335
17DU6N	337 (P)	59 (S)	-	345
17DU7S	333 (P)	64 (W)	-	346, 5
18BL11S	29 (W)	-	140 (W)	45

P - principal
S - strongly developed
W - weakly developed

Other structure data such as the length of joints, and the type and properties of infilling have been neglected in this analysis. Insufficient information was available to include these factors. These and other factors such as minor changes in the local geology, planning and execution of blast will provide significant amounts of variation within each data set.

The presplit blast design at Moura Mine is classified as large diameter, air-deck style presplit blasting. Typically 311mm diameter blastholes, with the exception of 270mm for strip 17DU6N, were used. The standoff distance between the presplit line and the production blast was 3.5 - 4m. In each case the production blast was a throw blast with sufficient inter-row delays to allow adequate relief, reducing the amount of backbreak from the production blast itself. Table 2 contains the presplit design parameters for each area considered in this analysis. Design variations in these areas are largely due to local observations from the previous strip, that is the Blasting Characteristics design step discussed in Section 1.

4 DATA ANALYSIS

The theory as presented in Section 2 indicated that there were at least 3 significant properties of joints that could affect presplit formation. These were the joint frequency, joint orientation relative to the presplit crack and the degree of frictional coupling across the joint. Data on the first two points were available from the database compiled at Moura and the analysis concentrated on these factors.

Jointing database was classified as to the frequency of joints and their orientation to the presplit line. The process first involved mapping each joint on a photograph of the highwall face. In this investigation only joints that occurred systematically were considered. This included joints J1, J2, and sometimes J3. On average, it was found that a set of systematic joints produced only from 50 to 70 percent of the total number of joints for each analysed strip. The relative orientation of each joint then followed from the set they were associated with and the difference in orientation of the pit wall and the joint direction.

A typical vertical face map of the 17DU6N pit is provided in Figure 2. In this case only J1 and J2 were plotted. Overall these joints produced 54 % and 18 % of the total number of joints. Note that each joint is identified with a point according to the relative height of the pit. These maps enable a high quality analysis of the joint frequency, as every exposed joint is mapped.

To obtain a measure of the presplit quality a second type of analysis was conducted, which involved the estimation of the half-cast factor. The half-cast factor is a ratio of the length of half holes of the presplit line that are visible in the final wall, divided by the total presplit blasthole length. The estimation of half-cast factor was based on high quality mosaic photos of each highwall face.

The third stage of the analysis considered the relationship between joint frequency and presplit quality, as measured by the half-cast factor. An example of the analysis method is provided in Figure 2, where presplit holes as identified in the blast plan are plotted onto the map of joint exposure in the presplit face. The face was subdivided into 20 - 50m long blocks. For each block the joint frequency as well as half-cast factor were measured. The joint frequency and half-cast factors for each block are then plotted on a scatter plot as shown in Figure 3.

The joint frequency analysis considered the effects for constant joint orientation. The analysis included data from 6 different pits, with a total of 9 case studies, when changes of pit direction were considered. The data obtained for each case produced a strong to reasonable correlation between joint frequency and the half-cast factor. These correlations confirmed the hypothesis that increasing

Table 2 Presplit Design Information

Strip Ref.	Spacing (m)	Powder Factor (kg/sq. m)	Depth (m)	Stemming (m)	Loading
13BL5N	5.5	0.76	60	5	2 decks, 125kg each
16DU5N	4	1	46 - 55	2	2 decks, 100kg each
16DU5S	4.5	1	39 - 42	2	2 decks, 130kg each
17DU6N	4 - 4.5	0.9	53	2	2 and 3 decks
17DU7S	6	0.6	45	4	1 deck, 160kg
18BL11S	4	1	30 - 35	4	2 decks, 75 and 50kg

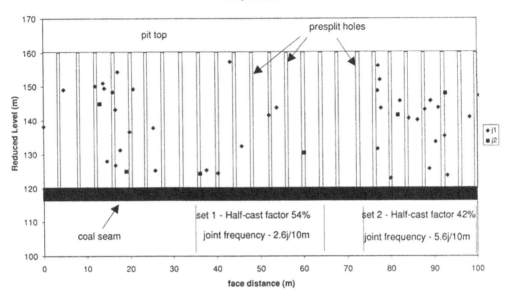

Figure 2. Face map of pit 17, D upper seam overburden, strip 6 north. The map indicates the position of presplit holes and the leading position of each exposed joint of sets J1 and J2. The joint frequency for each of two sections of the face are indicated, along with the half-cast factors (the length of the presplit holes that were visible after the blast).

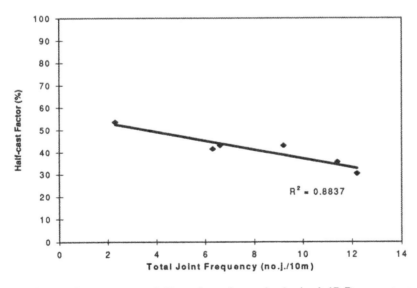

Figure 3. Joint frequency versus half-cast factor for overburden in pit 17, D upper seam, strip 6 north. The regression line and correlation coefficient are indicated.

joint frequency leads to a reduction in presplit quality, all other factors being equal. It was not expected that correlations will always be strong as some of these walls extend for a kilometre or more. Geological variations, blast loading techniques and slight changes in joint set orientation are some of the factors expected to contribute to additional variation in presplit effectiveness across a highwall.

Figure 4 contains a compilation of the data from a number of presplit walls in a similar geologic environment of the mine. This figure illustrates a decrease in quality of the presplit wall not only with joint frequency in each case, but also as the relative orientation of the jointing to the pit face changed from near parallel to approximately 30 degrees. This is consistent with the theoretical relation between relative joint orientation and presplit quality.

Figure 4 illustrates the requirement to consider both joint frequency and joint orientation in presplit design. The regression lines of joint frequency and half-cast factor, shown in Figure 4, were used to produce contours of joint frequency versus both relative orientation and half-cast factor. First, from each of the cases of relative joint angle the half-cast factor for 2 joints / 10 metres was estimated via interpolation or extrapolation using the regression line. This value was then used in a scatter plot of relative joint angle versus half-cast factor. This produced seven points of half-cast factor versus relative joint angle for a joint frequency of 2j/10m. A regression curve was fitted to these points. The exercise was repeated for 4, 6, 8 and 10 j/10m. This produced a family of curves as shown in Figure 5.

Figure 5 provides an overall picture of the inter-relationship between relative joint orientation, joint frequency and presplit quality for the range of rock types and blasting practices at Moura Mine. There are a number of similarities between Figure 5 and the more general conceptual 'co-operation curves' produced by Dunn and Cocker (1995). However, they vary in one particular respect. The conceptual curves indicate that joints nearly parallel to the wall actually enhance presplit formation. While the data in Figure 5 does not extend below relative angle of 5 degrees, there is little evidence of the significant improvement of half-cast factor that would be required to support the enhancement hypothesis. Examination of the wall condition indicate that the wall quality will continue to be limited for relative angle less than 5 degrees due to toppling failure along bedding planes.

Within the range of data considered, Figure 5 provides a basis for a technique to adjust the presplit design determined by rock strength (as discussed in Section 2) to accommodate the observed jointing. As an example, consider a case where the medium strong rock calls for a presplit hole spacing of 5 metres. This is typical of the rock strength and presplit designs of the areas used to produce Figure 5, so the Figure should be representative. The principal joint direction, J1, for the future presplit wall is 220 degrees. The two joint sets, J1 and J2, are orthogonal. The wall direction is 200 degrees. The typical joint frequency for the future presplit (extrapolated from the previous wall) is 2j/10m. In one particular area the joint frequency is expected to increase to 6j/10m.

To estimate the quality of the projected future wall the relative angle is required. In this case it is 220 - 200 = 20 degrees. The next step is to draw a vertical line on Figure 5 from 20 degrees until the intersection of the isobars occurs at 2j/10m and at 6j/10m. At each intersection point a horizontal line is drawn and the half-cast factor from the y-axis is estimated. The half-cast factor values correspond to 53% and 30%. These values can then be used by the mining engineer for design modification considerations such as a decrease in the blasthole spacing in the high joint frequency area or an increase in the powder factor. These decisions should be made using all of the engineering information available, as shown in Figure 1. The advantage of a tool such as Figure 5 is that it provides a measure of the likely deterioration of the presplit quality due to jointing. An engineer can then make an informed judgement as to the best approach for the operation's unique set of circumstances.

5 CONCLUSIONS

There are three main properties of joints that can affect presplit quality. There is a theoretical basis for determining the impact on presplit quality of joint frequency, joint interface friction and the relative angle of the joint(s) to the presplit line. These effects are summarised as:
- Increasing frequency will reduce presplit quality;
- Joints not parallel or perpendicular to the presplit line will reduce presplit quality;
- Lower joint friction will reduce presplit quality.

The impact of joint frequency and relative joint orientation to the presplit have been examined in medium strong overburden rock from 6 pits of a large open cut coal mine in Central Queensland, Australia. The results of this study have shown for this rock type that:
- There is a significant decrease in the presplit quality with increased joint frequencies over the range of 2 to 10 joints per 10 metres of face length;
- Presplit quality decreases with an increase of relative orientation from 5 to 30 degrees.

These findings are consistent with theory. However, further data is required to confirm aspects of the

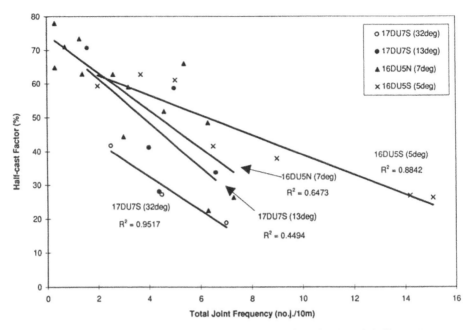

Figure 4. Half-cast factor versus joint frequency for four strips of overburden of similar geology above the D upper seam. The pits are each oriented at different angles to the dominant joint set, J1, as indicated in the Figure. The best fit regression line and the correlation coefficient for each strip are also indicated.

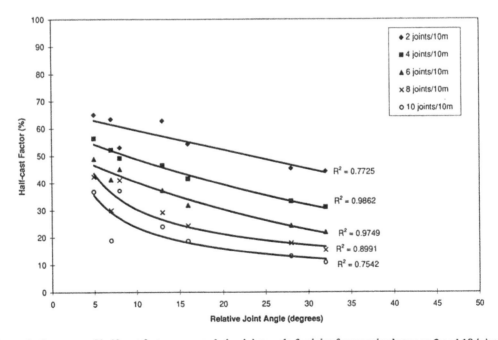

Figure 5. Summary of half-cast factor versus relative joint angle for joint frequencies between 2 and 10 joints per 10 metres. The Figure uses data from all pits at Moura considered in this paper. The best fit curve for each value of joint frequency is indicated, along with the correlation coefficient.

theory that extend beyond the ranges of data considered here.

The results of this study are presented in the form of a set of presplit quality design curves, which can be used as a predictive tool for other presplit walls with similar geology. Standard presplit design calculations can be used initially, with some adjustments according to the nature of jointing. The final design should encompass geological, geotechnical, drilling and economic factors. The design curves developed here demonstrate the potential for improving the engineering calculations for presplitting within this framework.

6 ACKNOWLEDGMENTS

Much of the jointing data and mosaic photographs of the highwall face were collected and prepared by Nigel Godfrey of Garry Leblang and Associates, Toowoomba, Queensland. The assistance of Mr Godfrey and his colleagues in the reported work is gratefully acknowledged. Also, the assistance of staff at BHPAC Moura Mine in the collection of data is also acknowledged. The authors thank BHP Australia Coal Pty Ltd for permission to publish this paper.

7 REFERENCES

Calder, P. 1977. *Pit Slope Manual*. Canmet Report 77-14, Chapter 7: 13-22. CANMET Canada Centre for Mineral & Energy Technology. Minister of Supply & Services Canada.

Chiappetta, R. F. 1991. Pre-splitting and controlled blasting techniques including air decks and dimension stone criteria. *Procs Blast. Technology, Instrumentation and Explosives Applications Seminar.*, Chiappetta, R. F. (ed.), San Diego USA.

Danell, R E. 1989. Estimation of in-situ strength. In A W Khair (ed.) *Rock Mechanics as a Guide for Efficient Utilisation of Natural Resources*, (Proc., 30th US Symposium on Rock Mechanics) p. 571-8. Rotterdam: Balkema.

Dunn, P & Cocker, A. 1995. Pre-splitting - wall control for surface coal mines. In *Proceeding, EXPLO '95 Conference* (Brisbane, 4-7 Sept). p. 307-314. Melbourne: The AusIMM.

Godfrey, N. & Thrift, J. 1995. Analysis of mining conditions by integration of highwall mapping and geological modelling techniques. In *Proceedings, Bowen Basin Geologists Symposium.*

Kutter, H. K. & Kulozik, R. G. 1990. Mechanics of Blasting in a discontinuous rock mass. *Procs. Int. Conf. Mechanics of Jointed and Faulted Rock.*

Rossmanith, H. P. (ed.). Ins. of Mechs., Tech. Uni. of Vienna.

Reinhart, J. S. 1964. Transient stress wave boundary interactions. *Stress Waves in Anelastic Solids*, Kolsky, H. and Prager, W. (eds). Springer-Verlag.

Sanden, B H. 1974. *Presplit Blasting*. Master's Thesis, Mining Engineering Department, Kingston Ontario: Queens University.

Workman J. L. & Calder, P. N. 1991. A method for calculating the weight of charge to use in large hole presplitting for cast blasting operations. In *Proceedings of the 17th Annual Conference on Explosives and Blasting Technique* (Las Vegas, Jan-Feb 1991). Solon, Ohio: Int. Soc. Explos. Eng.

Rock Fragmentation by Blasting, Mohanty (ed.)© 1996 Taylor & Francis. ISBN 90 5410 824 X

Studies on fracture plane control blast with notched boreholes

Dexin Ding & Qiu Ling
Central-South Institute of Technology, Hengyang, Hunan, People's Republic of China

ABSTRACT: The paper deals with the results from theoretical studies and field tests on fracture plane control blasts. A single symmetrically notched borehole was simplified to a pressurized hole with two symmetrical cracks under the state of plane strain. In analyzing the growth and termination of the notch, only the quasi-static pressure was considered. The rock to be blasted is considered to be isotropic. Based on the principles of fracture mechanics, the stress components near the tip of the notch have been calculated, and the criterion for growth of the notch established. The growth direction of the notch is also predicted on the basis of strain energy density factor.

1 INTRODUCTION

In recent years, several investigators have worked on fracture plane control blasting with notched boreholes. Successful application have been reported from several sources (Ding, 1993). So far, this technique has developed to an extent that it can be applied in producing a satisfactory presplit along the excavation contour in rocks.

However, the mechanism of how the notches control the generation and growth of a fracture plane has not been fully understood. Also the amount of explosives for each hole and the spacing of holes are mostly based on trial and error.

As a result, the authors attempt to explain how the notches control the generation and growth of a fracture plane by employing fundamental principles of fracture mechanics and establish an approach in estimating the amount of explosives for each hole and the spacing of holes.

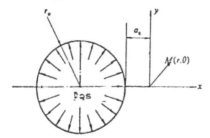

Figure 1: Fracture Mechanics model for fracture plane control blasting with notched boreholes.

2 MECHANISM

Although the fracture plane is achieved in a very short time frame, it is generated at two stages-growth and termination.

2.1 Growth of the Notch

The symmetrically notched boreholes is the most common one to be employed in fracture plane control blasting with notched boreholes. Only such case is considered in this paper.

The boreholes can be simplified to a pressurized hole with two symmetrical cracks under the state of the plane strain. In analyzing the growth and termination of the notch, only the quasi-static pressure is considered, and the rock to be blasted is considered to be isotropic. On these conditions, the fracture mechanics model for notched borehole blasting is represented in Figure 1.

The stress field near the tip of a notch can be given as below (Ouchterlony, 1974; Fourney & Dally, 1977; Holloway et al., 1987; Hong, 1987):

$$\sigma_r = \frac{K_I}{\sqrt{2\pi r}} [\frac{5}{4}\cos\frac{\theta}{2} - \frac{1}{4}\cos\frac{3\theta}{2}] \tag{1}$$

$$\sigma_\theta = \frac{K_I}{\sqrt{2\pi r}} [\frac{3}{4}\cos\frac{\theta}{2} + \frac{1}{4}\cos\frac{3\theta}{2}] \tag{2}$$

$$\tau_{r\theta} = \frac{K_I}{\sqrt{2\pi r}} \, [\, \frac{1}{4}\sin\frac{\theta}{2} + \frac{1}{4}\sin\frac{3\theta}{2}\,] \quad (3)$$

where σ_r, σ_θ and $\tau_{r\theta}$ are the diametrical, tangential and shearing stresses, respectively, r and θ in polar coordinates of a point around the tip of the notch. K_I is defined as stress intensity factor. It can be determined from following formula (Murakami et al, 1987):

$$K_I = F \cdot P_{QS} \cdot [\pi (r_o + a)]^{1/2} \quad (4)$$

where P_{QS} is the quasi-static pressure in the blast hole; r_o, its radius; a, the notch length; and F, the correction coefficient.

When the stress intensity factor is larger than the dynamic fracture toughness of the rock under state of plane strain, the notch will extend, and a fracture plane will develop. The criterion for the growth of the notch can be described as below:

$$K_I = K_{ICD} \quad (5)$$

where K_{ICD} is the dynamic fracture toughness of the rock under state of plane strain. It can be determined from the following empirical formula (Gao, 1993):

$$K_I = 1.6 \, K_{ICS} \quad (6)$$

where K_{ICS} is the static fracture toughness of the rock under plane strain.

2.2 Growth and direction of notch

The growth direction of the notch can be determined with reference to the criterion of strain energy density factor. The criterion states that the notch which has been simplified to a crack will grow further along the direction at which the strain energy density factor achieves its minimum value. The strain energy density factor can be expressed as bellow:

$$S = \frac{K_I}{4G} (3 - 4\mu - \cos\theta)(1 + \cos\theta) \quad (7)$$

where S is the strain energy density function; K_I, the stress intensity factor at the tip of the notch,

(eq. 5); G, the shearing modulus of the rock; μ, the Poisson's ratio; and θ, the angle between the growth direction of the notch and the notch itself.

The conditions under which the stress intensity factor reaches its minimum value can be expressed as follows:

$$\partial S/\partial\theta = 0 \quad (8)$$

$$\partial^2 S/\partial\theta^2 > 0 \quad (9)$$

From condition (8) and eq. 7, the following results can be derived:

$$\theta = 0 \quad (10)$$

$$\theta = \cos^{-1}(1 - 2\mu) \quad (11)$$

Furthermore, the following expression can be obtained from eq. 7:

$$\frac{\partial^2 S}{\partial\theta^2} = \frac{K_I}{2G} [2\cos^2\theta - (1 - 2\mu)\cos\theta - 1]$$

$$(12)$$

so that, only when $\theta = 0$, eq. 9 can be satisfied. This implies that the notch will grow along the same direction.

2.3 Crack termination

With further growth of the notch, the explosion gases gases continuously fill up the fracture. Simultaneously, the quasi-static pressure drops off very rapidly, leading to eventual termination of the crack. The stress state near the blast hole is shown in Figure 2.

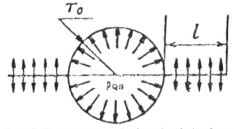

Figure 2: The stress state at the time when further fracturing stops.

The criterion for termination of further extension of the notch can be expressed as below (Murakami et al., 1987):

$$0.2 \, P_{QS} \, [\pi (r_o+1)]^{1/2} \leq K_{ICD} \quad (13)$$

3 ESTIMATION OF AMOUNT OF EXPLOSIVE IN A NOTCHED BOREHOLE

Firstly, the gas pressure can be calculated from the following formula:

$$p = \frac{\rho D^2}{2(K+1)} \quad (14)$$

where p is the gas pressure generated by a detonating charge in Pa; ρ, the density of explosives in kg/m³; D, the detonating velocity in m/s; K, the equivalent entropy coefficient of the explosives, typically 2.

Secondly, the quasi-static pressure can be calculated from the following:

$$P_{QS} = p_k \, (\frac{p}{p_k})^{v/k} \, (\frac{\Delta}{\rho})^v \quad (15)$$

$$\Delta = \frac{4Q}{\pi d^2 H} \quad (16)$$

where P_{QS} is the quasi-static pressure in Pa; v, the gas expansion coefficient without any heat exchange (usually 1.4); p_k, the critical pressure generated by detonating explosives (generally 2×10^9); Δ, the volume density of explosives in a borehole in kg/m³; Q, the weight of explosive loaded in a borehole in kg; d, the hole diameter in m; H, the hole depth in m.

Thirdly, the symmetrically notched borehole is considered equivalent to the plane strain fracture problem under mode I loading. The stress intensity factor can be determined from the following expression:

$$K_I = 0.2 P_{QS} [\pi (d/2+a_o)]^{1/2} \quad (17)$$

Where K_I is the stress intensity factor in N/m³/²; and, a_o, the diametral length of a single notch in m.

Fourthly, in order to make the notches extend and avoid damage to the borehole wall by detonation, the stress intensity factor should be larger than the dynamic fracture toughness of the rock to be blasted, and the quasi-static pressure should be less than the dynamic compressive strength of rock. Therefore, the following requirement can be established:

$$\frac{K_{ICD}}{0.2[\pi(d/2+a_o)]^{1/2}} < P_{QS} < \sigma_{CD} \quad (18)$$

where K_{ICD} is the dynamic fracture toughness of the rock under the state of plane strain; and, σ_{CD}, the dynamic compressive strength of the rock.

Finally, the amount of explosives in a borehole can be estimated from the following expression:

$$\frac{\pi d^2 H \delta}{4} [\frac{K_{ICD}}{0.2 p_k [\pi(d/2+a_o)]^{1/2}} (\frac{2(K+1)g p_k}{\delta D^2})^{v/k}]^{1/v}$$

$$< Q <$$

$$\frac{\pi d^2 H \delta}{4} [\frac{\sigma_{CD}}{p_k} (\frac{2(K+1)g p_k}{\delta D^2})^{v/k}]^{1/v} \quad (19)$$

where Q, the amount of explosives in a notched borehole.

4 ESTIMATION OF HOLE SPACING

Once the amount of explosive in a notched borehole has been estimated, the spacing of holes can be determined (Ding, 1994).

Firstly, since the spacing is much larger than the radius of hole, the stress intensity factor has to be calculated from the following formula:

$$K_I = \frac{P_{QS} \, d}{[(\pi/2) W]^{1/2}} \quad (20)$$

where W is the spacing between centres of the holes.

Then by employing the criterion of fracture mechanics, the following expression for estimating the spacing can be obtained:

$$W = \frac{2}{\pi} \left(\frac{P_{QS}}{K_{ICD}} \right)^2 \qquad (21)$$

5 FIELD TESTS

Five tests were conducted at Dingziwan granite quarry in Hunan province, each consisting of five vertical holes which were 4 cm in diameter and 100 cm in depth. Each of the two symmetrical notches has a diametrical length of 4 mm and a depth of 70 cm from the collar to the bottom.

The properties of the granite are listed in Table 1.

Table 1. Properties of granite.

Dyn. comp. sth.(σ_{CD})	65.7 MP
Dyn. toughness (K_{ICD})	$7.5 \times 10^{1/2} N/m^{3/2}$

The explosive cartridge, made from the No. 2 Explosive for rock blasting was used. Its physical properties are listed in Table 2.

Each explosive cartridge was 25 mm in diameter and 200 mm in length. After loading of the explosive cartridge, the boreholes were stemmed.

Table 2. Properties of explosive No. 2.

Density (ρ)	1.00 g/cm³
Detonating velocity (D)	3000 m/s

The test results are listed in Table 3. The estimated amount of explosives is calculated from eq. 19, and the estimated spacing of holes is calculated from eq. 21.

Table 3. Field test results.

The number of tests	5
Estimated amt. of expl.	49-100 g
Practical amt. of expl.	100 g
Estimated spacing of holes	< 1.02 m
Practical spacing of holes	0.5 m

6 CONCLUSION

It has been shown that both generation and direction of the growth of a fracture plane can be easily accomplished by employing notched borehole blasting.

The fracture mechanics model established in this paper gives a satisfactory explanation of how the notch controls the generation and growth of a fracture plane.

The computation methods, suggested for estimating the amount of explosive in a notched borehole and spacing of holes, are based on the principles of fracture mechanics, and confirmed by a few field tests. Further field trials have to be carried out to validate these findings.

REFERENCES

Ding, D. 1993. Estimating the amount of explosives for fracture plane control blasting with notched boreholes. *Trans. of Nonferrous Metals Soc. of China*, Vol. 3, No. 2.

Ding, D. 1994. Research on mechanical model for presplitting rock. *Journal of Hengyang Inst. of Tech.*, Vol. 8, No. 2.

Fourney, W. L. & Dally, J. W. 1977. Grooved boreholes for plane control in blasting. *NSF/RANN Report*, NSF-RA770216, U.S. National Science Foundation, Washington, DC.

Gao, J. 1993. Studies on the mechanism of split blasting. In: *Selected Papers of China's 4 th National Symp. on Engineering Blasting*, Press of Metallurgical Industry.

Holloway, D.C. et al., 1987. A field study of fracture control technique for smooth wall blasting: Part 2. *Proc. 2nd Int. Symp. on Rock Frag. by Blasting*; Keystone, Colorado, 646-657.

Hong, Q. 1987. Fundamentals of engineering fracture mechanics. *Press of Shanghai Univ. of Transportation*.

Murakami, Y. et al. 1987. *Stress Intensity Factors Handbook*. Pergamon, New York.

Ouchterlony, F. 1974. Fracture mechanics applied to rock blasting. *Proc. 3rd Cong. Int. Soc. Rock Mech.*, U.S. National Acad. of Sci., Washington DC U.S.A..

Fragmentation assessment and image analysis

Rock Fragmentation by Blasting, Mohanty (ed.) © 1996 Taylor & Francis. ISBN 90 5410 824 X

Image analysis of fragment size and shape

Weixing Wang, Fredrik Bergholm & Ove Stephansson
Division of Engineering Geology, Department of Civil and Environmental Engineering, Royal Institute of Technology, Stockholm, Sweden

ABSTRACT: In mining and quarry production, it is well known that the properties of fragmentation, such as size and shape, are very important information for the optimisation of production. For the last ten years, image analysis techniques have been used for fragments measurements, which increases speed and accuracy of analysis. Fragment size measurements on binary images take place after grey scale image segmentation. However, at the size measurement stage, there are several definitions of fragment size applied. When the different size definitions (or methods) are used for the same fragment sample, the resulting size distributions are extremely different, which causes confusion. A good size measurement method should meet at least three criteria. These are rotational invariance, reproducibility and embody overall shape description (elongation / flakiness or angularity / rectangularity). According to these three criteria, this paper analyses and evaluates several existing methods of fragment size measurement, such as Chord sizing, (multiple) Feret diameter, equivalent circle, maximum diameter and equivalent ellipse etc. in image analysis. A program library of aggregate image analysis has been developed for PC computers in C++ programming language. Based on the analyses and evaluations of the existing methods, we propose a new method -- best-fit rectangle for size measurement that satisfactorily meets the criteria of rotational invariance, reproducibility and shape description. This is a blend of previous approaches, combining good properties of least squares with properties of bounding boxes.

1 INTRODUCTION

The size of aggregate is a very important characteristic of the physical properties of aggregate for geology research and aggregate production industry and mining industry. In mining, the size distribution of fragments affects not only rock blasting, but also the whole mining production sequence. In quarry manufacturing, the size of aggregate and its distribution must fit to the requirements of customers, such as highway construction companies, various companies in the building industries etc. In geology, the size and size distribution of gravel, sedimentary are often used for analysing and describing local geological properties in a certain area. So the aggregate size and its distribution are widely applied and studied in the industries and research organisations.

Image analysis techniques have been used for aggregate size measurements for the last ten years. As computers are widely used today, the cost of

image systems is quite low, and aggregate size analysis can be handled quickly and easily.

The main steps for aggregate size analysis in image analysis are (1) grabbing grey-scale images; (2) image enhancement and segmentation and (3) aggregate size, elongation and shape measurement and statistical analysis. The methods can be divided into semi-automatic methods and fully automatic methods (Stephansson, O., Wang, W.X. and Dahlhielm, S., 1992).

When grey-level images are turned into binary ones, based on a semi-automatic method or a fully automatic method, aggregate size, elongation and shape measurements will be carried out. In this working procedure, there is still no single standard. Therefore, different methods for aggregate size and elongation (or shape) measurement are used around the world. When different measurement methods are used on the same sample, the measuring results are extremely different, which causes confusion about aggregate size. In some cases, even using only one of the methods, measured results are different for different images coming from the same aggregate sample, making measurements of poor reproducibility and robustness.

One typical example is shown in Fig. 1. For an aggregate particle with an orthogonal triangle shape, several measuring methods are applied, such as chord sizing, Feret diameter, equivalent circle, maximum diameter and equivalent ellipse (refer to later section). The differences between measured results (sizes and/or elongation) are substantial. This example raises the question of how to establish definitions of size and methods of size measurements applicable to visual views of the particles by using image analysis, which can give out a stable, reasonable measured quantities which well reflect the overall shape.

In order to avoid the above mentioned problems and to contribute to the formulation of a standard of aggregate particle size measurements using image analysis, this paper proposes that four basic requirements on methods for aggregate particle size measurements be obeyed, and we advocate the use of a closed form solution technique defining three robust basic quantities. This new practice would in addition to being rotational invariant and fairly independent of boundary roughness, in a robust way reflect shape characteristics of importance to quarry and mining industry. This technique which we call the best-fit rectangle is a blend of mathematical techniques combining Feret boxes and least squares minimization furthermore, for a standard, it is of importance to use clearly defined software. Most of the methods mentioned in this paper were coded by using C++ programming language for PC computer, in a program library dedicated to aggregate image analysis, which we are developing.

2 BASIC CONDITIONS FOR METHODS FOR AGGREGATE SIZE MEASUREMENTS

For setting up, developing or using a method for aggregate particle size measurements, the following realistic conditions in the field should be considered: (1) aggregate particles are randomly located in an image, in many orientations (2) we cannot expect aggregate particles to be well modelled by such simple regular shapes such as those of circles or rectangles; (3) in quarry or mining industry, the boundaries of aggregate particles are very rough, owing to either their physical properties or to technique of photography or of image digitization

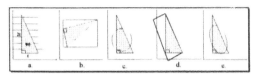

Fig. 1 Various size definitions for an object (shaded). a: Chord sizing in a certain scanning direction; b: Two Feret diameters; c: Equivalent circle (e.g. by area); d: Maximum diameter; and e: Equivalent illipse.

and (4) even though aggregate particles may be of the same size or elongation, they still tend to be of quite different shapes.

Based on considerations like the ones above, a method of aggregate particle size measurement should meet the following four basic conditions:

A. A method should have a size definition where the measured particle size is unique, independently of image scanning direction or particle rotation, which is called rotational-invariance.

B. The method should be reproducible. Rotational invariance promotes reproducibility. If taking images in another laboratory under identical conditions of the same sample just rotating it, or rotating individual particles, a good measurement system should produce, very closely, the same numbers. Hence, reproducibility presupposes at least rotational invariance. (If the particles can be rotated in 3D the only way of ensuring reproducibility would be to capture sufficiently many images from different aspects of the viewed particles).

C. Low boundary roughness sensitivity: if a method measures the size of a particle primarily based on information from the particle boundary, the measured quantities will be dependent on boundary roughness (jaggedness), because perimeter grows substantially with jaggedness of the boundary. This will tend to make such boundary size measurements rather nonrobust and too dependent on image resolution. The discrete sampling of pixels will tend to make such methods non-reproducible. The method of equivalent circle of perimeter (to be discussed below) will produce for jagged boundaries produce circles of sizes that do not reflect the real size.

D. Shape reflection: By this term we mean that the measured sizes should crudely describe the shape (= reflect the shape), which in practice means that overall properties of shape such as elongation and "angularity" be possible to infer from that crude description. It is quite common that aggregate particles are of similar size and elongation but angularity or rectangularity is quite different.

Angularity in a wide sense would mean that the shape deviates from circular shape by having some sharpish ends or be fairly edgy in 3D. Rectangularity is usually defined in particle analysis by circumscribing the particle by a minimum bounding rectangle and check how well the particle fills out the area (Sonka, M.,Hlavac, V., Boyle, R., 1993).

Elongation and angularity (rectangularity) are very important pieces of information in aggregate industry as supplementary attributes of particle size.

Based on these four conditions or requirements on measurement techniques and size definitions, let us analyse and evaluate the existing and widely used methods.

3 REVIEW AND ANALYSIS OF EXISTING METHODS

From a literature review, summarising, the most widely used methods for aggregate size measurement appear to be: (1) Chord sizing (chord size distribution instead of real size distribution); (2) Feret diameter (length L, width W and maximum length $\sqrt{L^2 + W^2}$); (3) Equivalent circle diameter (equivalent perimeter, radius and area of circle); (4) Equivalent ellipse (or rectangle) and (5) Maximum diameter (come from Multiple Feret diameters or radii). Let us analyse and evaluate these widely used size definitions in image analysis, based on the mentioned four basic conditions.

3.1 Chord sizing

The simplest (and earliest introduced) method of applying a size criterion to objects in an image analyser is to scan the binary image in a certain interval distance (i.e. interval lines, cf. Fig. 1(a)). The distances of two adjacent intercept points (chords) are used as the size or size distribution of aggregate particles.

The method is simple and fast, and hence this method is still widely used (Nyberg, L., Carlsson, O., Schmidtbauer, B., 1982). It may be available for the particles with circle shape and the variations of sizes of particles are not too large in one image, and suitable for rectangular particles lying on a same direction that either parallel or vertical to the system scanning direction. Chord sizing does not meet the conditions of rotational-invariance, reproducibility and shape reflection.

The word diameter in this text will mean maximum chord given a certain scanning direction.

Hence, the maximum diameter is then the orientation for which the diameter is maximal (note the double maximisation: maximum over maximal chords). In mathematics, 'diameter' normally means 'maximum diameter'. We define Feret boxes as the bounding box (cf. Fig. 1b) consisting of two orthogonal Feret diameters.

3.2 Feret or Calliper diameter measurement

The Feret diameter, or synonymously the calliper diameter, (see Fig. 1(b)) is the distance between two parallel tangents on opposite sides of the particle. This method was proposed by L.R. Feret in 1931 (Joyce LoeblLtd, 1989) and the measurement is often referred to by his name. In systems employing boundary-coding techniques, a single pass around the boundary of an object noting maximum and minimum x and y co-ordinates will yield vertical and horizontal Feret diameters (length L, width W, and maximum length $\sqrt{L^2 + W^2}$).

It is still used by a lot of image analysis systems. It yields one group of parameters for every single particle, and measures the bounding rectangle circumscribing the particle, but this rectangle is not unique - it varies with the scan direction, and is hence not rotationally invariant and of poor reproducibility. (It is mainly suitable for elongated particles oriented in a specific single direction in an image.)

3.3 Equivalent circle diameter

The method of equivalent circle diameter (see Fig. 1 (c)) has been widely used (Maerz, N.H., Franklin, J.A., Rothenburg, L. and Coursen, D.L., 1987, McDermott, C., Hunter, G.C. and Miles, N.J., 1989, and MacLachlan, R.R. and Singh, A., 1989). The term "Equivalent circle" has two different definitions. They are: (1) the equivalent circle of area and (2) the equivalent circle of perimeter (or mean radius, an approximation of the equivalent circle of perimeter). An equivalent circle is judged to have the same area or perimeter as the object to be measured, and the diameter of the circle taken as a measure of size. These definitions were proposed by Heywood 1964 (Joyce LoeblLtd, 1989).

The "mean radius" has been used by different image system to measure the size of particle, the procedure is to count a number of radii, then average them. The general formula can be presented as

$$\bar{r} = \frac{1}{N} \sum_{i=1}^{N} r_i \qquad (1)$$

where, $\bar{r}$ is mean radius, N is the number of sampled radii (r_i).

When $N \to \infty$, $\bar{r} = \frac{1}{2\pi} \int_{-\pi}^{\pi} r(\theta) \, d\theta$. $\qquad (2)$

From Eq.(2), it can be seen that $\bar{r}$ is the radius of equivalent circle of perimeter.

Let us use the following notation, particle perimeter is P, and area is denoted by A. Then, a dimensionless shape factor is defined as $Sh = \frac{P^2}{4\pi A}$ ($\geq$ 1). The difference of radii (or diameters) between equivalent circle of perimeter r_p and equivalent circle of area r_a can be presented as

$$D_r = \frac{r_p - r_a}{r_a} = \frac{P}{2\sqrt{\pi A}} - 1 = \sqrt{Sh} - 1.$$

So D_r is the function of Sh, it is always greater or equal to zero. It is clear that r_p is always greater than r_a except for particles of the shape of a circle.

From a digitizing technique point of view, the object area calculation is easier than object perimeter calculation, the latter being more sensitive to the boundary roughness of particles, as mentioned. The method of the equivalent circle of area is easy to use. It is rotationally invariant, reproducible and boundary roughness (jaggedness) independent, but has no shape reflection. The mean radius measurement is an approximation of equivalent circle of perimeter with poor boundary roughness independence. to rough boundary of A. The equivalent circle of area is mainly apt for non-elongated particles.

3.4 Multiple Feret diameter and Dot product methods

For elongated objects, length and width are also measured by using Multiple Feret diameters (Fig. 1 (d)), within multiple Feret rectangle boxes, the box with a maximum diameter being selected. This method is based on how many Feret rectangle boxes are taken (Rholl, S.A., Grannes, S.G. and Stagg, M.S., 1987). This kind of measurement includes some shape reflection. The number of the boxes will affect the measurement result, which may cause problems such as poor rotational-invariance and reproducibility.

The dot product can be described as follows:

Let x_i, for i=1,2,..., N be the sampled boundary 2D points. Let u_j for j=1,2,...,D be D so-called reference 2D vectors. Let max and min in Eq. (4) below denote minimization over the index i. In one pass, traversing the points i=1,2,..., N in any order, calculate:

$$x_i^T u_1, x_i^T u_2, ..., x_i^T u_D, \qquad (3)$$

where dot products between the vectors x_i and u_j are denoted $x_i^T u_j$ and save only

$$MAX_j = \max(x_i^T u_j), MIN_j = \min(x_i^T u_j). \qquad (4)$$

for each j=1,2,...,D, as well as those coordinates that give rise to the D maxima and D minima. The multiple Feret method, in the sense that $\max\{L_1, L_2, ..., L_D\}$ is chosen as the maximum diameter, coincides with the dot product method, if new scanning directions are implemented using dot products with new coordinate axes. The only difference is that the number of directions chosen are multiples of 2 with multiple Feret boxes, and multiple Feret is often not implemented as a one-pass algorithm. The importance of the dot product method lies in the fact that there are modifications of it saving more dot products than the ones mentioned above speeding up the calculation of maximum diameter.

3.5 Maximum radius

For maximum radius measurement, after obtaining the centre of gravity of a particle, the orientation is the axis through this centre and lies on the maximum length axis that also depends on how many radii are chosen. After obtaining the orientation of particles by utilizing the procedure, the lenghths and widths of particles can be obtained by using radii in the obtained orientations. With this method, only the length that is approximately close to maximum length can be obtained, the orientation depending on how many radii are used. The requirements of reproducibility and rotational invariance cannot be fully met unless sufficiently many radii are chosen, and many systems seem to use too few radii causing the system to be affected by rotation. Shape reflection only includes elongation, but not angularity or rectangularity.

3.6 Equivalent ellipse or rectangle based on particle's perimeter and area

The method assumes that every particle has a rectangular (or ellipse) shape with length L and width W, the L and W can be calculated based on equivalent perimeter and area. The method is simple, but measuring result is more dependent on particle's boundary rughness, does not give out any shape information. It may be good for the particles of similar shape and slight roughness (jaggedness) of the boundary.

3.7 Equivalent ellipse or rectangle based on moments (see Fig. 1 (e))

This method proceeds as follows: a particle's area, centre of gravity and axis of least inertia (= axis of least 2nd moment) obtained by using zeroth, first and second moments respectively, then a ellipse (or rectangle) that has the same zeroth, first and second moments as the particle has, is made, the length and width are used as the size parameters of the particle. From the least moment concept, the equivalent ellipse and rectangle can be obtained as following description:

Let θ be the direction of the axis of least 2nd moment, and b(x,y) a binary-valued function such that b(x,y) = 1 on the particle and b(x,y) = 0 outside it. Let X = x - mean(x) and Y = y -mean(y). Minimising θ:

$$\min_{\vartheta} = \int\int\limits_{-\infty}^{\infty} \left| X\sin\vartheta - Y\cos\vartheta \right|^2 b(x,y)dxdy$$

Considering the length of major axis as 2α and length of minor axis as 2β, we obtain the minimum and maximum second moments of the ellipse about an axis through its centre: $\pi\alpha\beta^3/4$ and $\pi\beta\alpha^3/4$ respectively. For the equivalent rectangle, the minimum and maximum second moments are $4\alpha\beta^3/3$ and $4\beta\alpha^3/3$ respectively.

This method can produce one group of parameters (length L, width W and ratio L/W) for every individual particle. It is rotationally-invariant because of least second moment theory applied, therefore, it also meet the condition of reproducibility and low boundary roughness sensitivity. As for shape reflection, from the ratio L/W, the particle elongation can be known. The

problem is that it can not distinguish angularity of particles. One example is shown in Fig. 3, three particles have same length L, width W, and area after applying this method, but actually, they are very different. They are rectangle, trapezoid and triangle respectively.

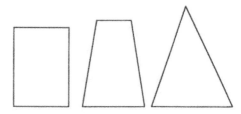

Fig. 3 Three different particles with a same length and width by applying equivalent ellipse

4 A MODEL OF AGGREGATE PARTICLES

Empirically, the following shapes seem to be typical for aggregate particles:

- (a) elliptical-like
- (b) rectangular-like
- (c) trapezoidal-like
- (d) triangular-like

Of course, for the three latter categories, the corners are somewhat rounded, normally. Note also that we use the suffix "like" to stress that this is a crude shape description of the shape of the particle boundary. Fig. 4, below, shows typical shapes. This is an underlying model in the discussion and evaluation of methods in this paper. Trapezoidal shapes ranging from triangles to rectangles form a triangle-rectangle trapezoid family of shapes. A member of that family is uniquely characterised by the area ratio: area of circumscribing (bounding) rectangle to trapezoid area. A triangle occupies half of a circumscribed rectangle (if one rectangle side coincides with one of the sides of the triangle). A trapezoid occupies $(1 + T)/2$ of the circumscribed rectangle where T is the ratio $0 < c/a < 1$, a = base, c = length of side opposite and parallel to a, if aligned to the longest side a. Hence the area ratio is a number in the interval $(1/2,1)$. Crushed rock often belongs to shapes that are members of the triangle-rectangle trapezoid family, or sometimes unions of these shapes, see Fig. 4.

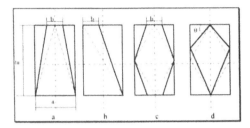

Fig. 4 Typical shapes of crushed aggregate

In order to evaluate the different measurement techniques reviewed in this paper, and the best-fit rectangle approach we designed and created several synthetic pictures, each of them including particles of different shapes, Figure 5 being one of them. In Figure 5, we have created 18 particles in a 512 by 512 synthetic image with different shapes, they are divided in five groups: (1) No. 1 - 4 are rectangles; (2) No. 5 - 8 are ellipses; (3) No. 9 - 12 are orthogonal triangles; (4) No. 13 - 15 are any kinds of triangles and (5) No. 16 - 18 are trapezoids. (The smallest ones are of sizes around 80 pixels.) For every group of the particles, the elongation increases from left to right. The results showed that for example elongation (when choosing 10 different orientations between zero and ninety degrees) had a standard deviation of the same magnitude as the elongation value itself for simple Feret diameters, whereas elongation measured by equivalent ellipse (least 2nd moments) and the best fit rectangle yielded very similar results.

Fig. 5 Aggregate particles with different typical stylized shapes

5 A NEW APPROACH TO COARSE 2D SHAPE AND SIZE MEASUREMENTS

5.1   Area ratio

The word area ratio in this article refers to the ratio between an object and its circumscribed rectangle (see Fig. 6).

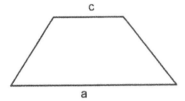

Fig. 6 A trapezoid has area ratio depending on orientation of bounding rectangle, but if the bounding box is chosen so at to be aligned with the (longest) side, the number $(1+c/a)/2$ is always obtained.

5.2 Do coarse 2D-shape descriptions require 3 quantities ?

The concepts of size and shape are rather interrelated. Length and width are size along two different dimensions. Elongation is a shape parameter, but shape is implicit, already, in a length - width representation, for example, by looking at the ratio length to width. Elliptical-like object would have two dimensions - longest and shortest diameter, which happen to be orthogonal to each other. The same for rectangular-like objects or rhomboids. However, although an object that is triangle-like can be described partially by width and length dimensions, a fuller description would involve three measurements, in the simplest case the lengths of the three sides or the lengths of the three bisectors. A good variable for describing the difference between triangles and rectangles are the above-mentioned area ratios. A triangle has area ratio = 1/2, a rectangle has area ratio = 1, and the trapezoid is intermediate (provided the circumscribed rectangle is aligned with the base). It is a semantic question if one wishes to think of area ratio, length and width as global parameters characteristic of the size(s) (dimensional descriptions) or global shape parameters.

In common speech, length and width are referred to as sizes (dimensions), but they could also be called shape descriptors. Discussing triangles, using three numbers for describing them, one could still speak of these numbers as being the dimensions of the triangle (in particular the lengths of the sides or similar entities), or, shape factors. The lengths of the three sides of a triangle describe the shape of triangle. Everyday language does not make a clear distinction between sizes and shape factors. Irrespective of whether these numbers are called sizes or shape factors, they de facto give coarse-scale description of shape.

5.3 A measure of angularity (rectangularity)

As mentioned, if circumscribing a trapezoid or a triangle by a rectangle so that one rectangle side coincides with a side of the triangle or trapezoid, area ratio yields a good measure of angularity.

One way of doing this for particles of more arbitrary shape, approximately, and at the same time in a rotation-invariant fashion, is to choose to circumscribe the object (the particle) by a rectangle, which is chosen to be oriented in the direction of moment of inertia . This is our proposed approach. We call this the best-fit rectangle approach.

The ratio of the particle area to the area of the oriented circumscribed rectangle, is a kind of area ratio, and is called rectangularity if minimal as mentioned by Sonka & Hlavac & Boyle 1993. We will not obtain rectangularity from minimal bounding boxes directly but from the axis of least 2nd moment which yields approximate rectangularity. Note in particular that the maximum diameter approach does not yield even approximate rectangularity, if using bounding boxes (Feret boxes) in that orientation, in particular not for particles that are rather rectangular-like. It should be noted that the best-fit rectangle approach we propose is the only one that satisfies all four requirements we postulated earlier.

5 4   Some details

1. Obtaining orientation of a particle by using least-second moment method (= rotationally invariant) yields a simple closed formula for the orientation. Equivalent ellipse by 2nd moments is of course based on the same formula but there is no shape reflection, since area ratios do not enter.

2. Length and width can be obtained by using a Feret box in the orientation of the least 2nd moment.

Thus, we have rotationally invariant elongation implicitly represented.

3. The area ratio in the orientation of the axis of least 2nd moment, yields approximate rectangularity. For crushed or blasted aggregate materials, the manufacturers want to know not only the elongation of a particle, but also angularity or rectangularity, so the area ratio obtained from the Feret box in the direction of least 2nd moment is of clear practical value. It is less practical to find the rectangularity by a discrete minimisation procedure over possible Feret boxes, since many directions need to be investigated in order to safeguard a rather exact minimum. In Appendix A we prove that for typical shapes of crushed aggregate (the triangle-rectangle trapezoid family) the minimum area Feret box corresponds to alignment to a side of the polygon.

6 AN ILLUSTRATION OF THE BEST-FIT RECTANGLE APPROACH (IMAGE EXAMPLE)

In order to test the best-fit rectangle approach some real images were investigated. First, an aggregate image (see Fig. 7) is captured. In this case, 76 particles were included. A video camera was used Segmentation was done to turn the image into a binary image. Finally size, elongation and 'rectangularity' of the particles were calculated by using Feret boxes in the direction of the axis of least 2nd moment. The results of these calculations (on a PC in the programming language C) are displayed in Fig. 8, Fig. 9 and Fig. 10 respectively.

The original image (Fig. 7) is a grey level image with a resolution 512 x 512 x 8 bits. The image was taken by a CCD camera. The background of image is a piece of conveyor belt. The aggregate particles were taken from a quarry, Vällsta krossanläggning in Upplands Väsby, in a suburb of Stockholm. The colour of aggregate is grey. The particle sizes range from 16 mm to 32 mm. In the following analysis, for convenience, we use pixels as the unit of particle size (3 pixels equal one millimetre, approximately).

The binary image (Fig. 7) was roughly and quickly segmented by a fairly simple local thresholding function. However, some small holes inside particles were not filled, and some noise such as roughness on boundaries, were not smoothed, but these two factors will not affect the size measurements considerably, in this case.

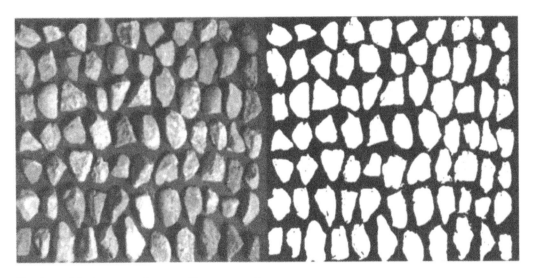

Fig. 7 an original aggregate image and its segmented image

Accumulate percentage (%)

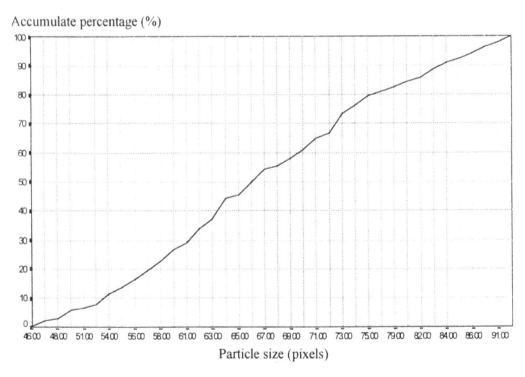

Particle size (pixels)

Fig. 8 Particle size distribution from Fig. 7 using Feret diameter (length) in the direction of the axis of least 2nd moment..

Accumulate percentage (%)

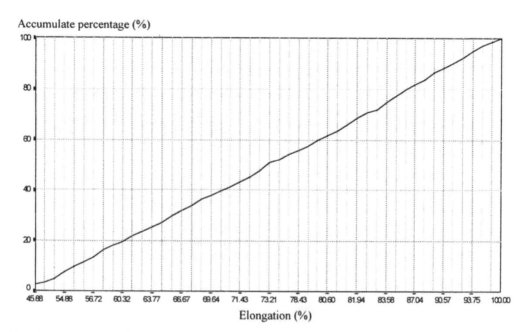

Elongation (%)

Fig. 9 Distribution of particle elongation from least 2nd moment oriented Feret boxes.

Accumulate percentage (%)

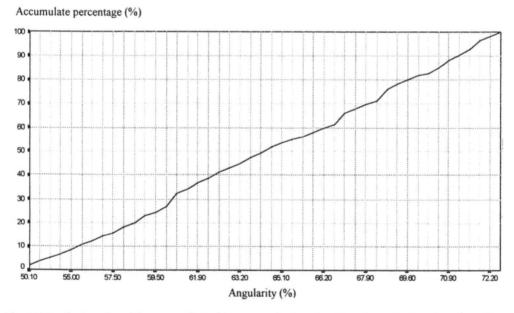

Angularity (%)

Fig. 10 Distribution of particle rectangularity from area ratios based on Feret boxes in the orientation of least 2nd moment. This is a kind of angularity in a wide sense.

In this group of particles (Fig. 7), the particle size ranges from 46 pixels to 94 pixels. the average size is 67 pixels. The elongation, on average, is about 0.73 (73%). The average rectangularity is about 0.64 (64%). From the different distribution curves by applying our new method, more information on size, elongation and rectangularity (angularity) can be obtained.

241

7 FURTHER COMMENTS

In order to analyse how the method meet the conditions of rotationally invariance, reproducibility and shape reflection, we have compared our new method to the existing methods in these aspects. For brevity we do not include these comparisons fully here, which were based on synthetic images. We only make some comments on the size measurements linked to the real image experiment shown in the previous section.

7.1 Size measurements

In Fig. 7, we showed 76 aggregate particles, the size distribution measured by using our new approach (best-fit rectangle), is displayed in Fig. 8. If comparing it with multiple Feret measurement in 9 directions (with angles 0^0, 10^0, .., 80^0), and the equivalent ellipse methods, we obtain the following results in Fig. 11. The curve of best-fit rectangle is a little bit higher than the curve of equivalent ellipse- it locates between the curves of maximum and minimum Feret diameter.

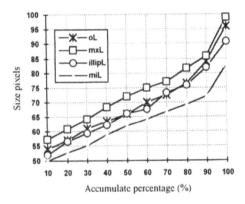

Fig. 11 Particle size measurement with different methods from Fig. 7

Notes: all the following diameters are particle length.
oL: Feret diameter in the direction of particle's orientation;
mxL and miL: maximum and minimum Feret diameters from ten different directions;
illipL: equivalent elliptical length.

In order to understand why the curve of best-fit rectangle is higher than the curve of equivalent ellipse, let us analyse Fig. 5 which presents 18 typical

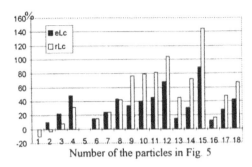

Fig. 12 comparison of eL, rL and cL

Notes:
eL: length of equivalent ellipse;
rL: Length of best-fit rectangle;
cL: diameter of circle of equivalent area;
eLc: (eL-cL)/cL;
rLc: (rL-cL)/cL.

particles. Figure 12 illustrates that the best-fit rectangle is better for representing real particle size of crushed aggregate.

8 DISCUSSION

8.1 Preferred Length-Width Definitions for a Standard

We are interested in setting up a standard for length, width and rectangularity measurements of aggregate. This standard should as mentioned promote reproducibility and therefore be rotation invariant, and have low sensitivity to boundary roughness.
Furthermore, it is good if it is based on an explicit formula, a closed-form solution, since this will facilitate error sensitivity analysis as well as the ease by which it is computed.

If measuring sizes of particles from two-dimensional projections of them, the best candidate for a standard length and width from the Feret box oriented in the direction of the axis of least 2nd moment. This is often a better alternative than the direction of maximum diameter, because although both of them are rotation invariant, the latter orientation estimate is more sensitive to boundary roughness than the orientation of the former (being based on a least squares minimisation). However, there is a yet more important reason for preferring the least 2nd moment oriented Feret box in a standard. The area of that box is of approximately minimal area (the minimisation carried out over bounding rectangles) and that minimal area tells us something about the shape of the particle. It has what we call shape reflection. Approximate rectangularity (a concept described e.g. in Sonka et al. (1993)) is obtained by dividing the area of the 2nd least moment oriented Feret box by the area of the particle. This approximate rectangularity gives a nearly unique number to each member of the triangle-rectangle (trapezoid) family of shapes, which is typical for crushed aggregate. Ellipses and circles are given the single value $\pi/4$. Hence, we have the following observations:

Observation 1: The 2nd least moment Feret box is superior to maximum diameter-based Feret boxes, such as the multiple Feret method, in that the information from length and width of the least 2nd moment oriented Feret box, implicitly, contains at least one more piece of information on crude shape, namely approximate rectangularity.

Observation 2: If crushed aggregate is well modelled by triangular-, trapezoidal- and rectangular shapes, and natural aggregate by ellipsoids, then crushed aggregate is characterised nearly uniquely (depending on a and c of the trapezoid, see Fig. 6) by the approximate rectangularity of least 2nd moment oriented Feret boxes, and natural aggregate by a rectangularity close to 0.78, and mixed aggregate is uniquely characterised except for trapezoidal-like shapes of rectangularity close to 0.78.

Parallelograms with sharp angles will remind of triangles in terms of rectangularity. A hexagon-like shape is like two trapezoids glued together and will have an area ratio that is the average of the individual area ratios (Fig. 4c).

The best-fit rectangle approach only requires calculating three moments about the centre of gravity, and maximum and minimum co-ordinate in a co-ordinate system oriented in the direction of the axis of least 2nd moment, and a simple area ratio. A standard computer program for these measurements should be easy to specify.

REFERENCES

Stephansson, O., Wang, W.X. and Dahlhielm, S., 1992, Automatic image processing of aggregates, ISRM Symposium: EUROCK'92, Chester, UK, 14 - 17 September, 31 - 35.

Joyce LoeblLtd, 1989, IMAGE ANALYSIS: Principle and Practice, Published by Joyce LoeblLtd, UK.

Sonka, M., Hlavac, V., Boyle, R., 1993, Image Processing, Analysis and Machine Vision, Chapman & Hall Computing, London, New York, Tokyo.

Nyberg, L., Carlsson, O., Schmidtbauer, B., 1982, Estimation of the size distribution of fragmented rock in ore mining through automatic image processing, Proc. IMEKO 9th World Congress, May 1982, Vol. V/III, pp. 293 - 301.

Rholl, S.A., Grannes, S.G. and Stagg, M.S., 1987, Photographic assessment of the fragmentation distribution of rock quarry muckpiles, 2nd Int. Symp. Rock Fragmentation by Blasting, Keystone, Colo. 102 - 113 and 501 - 506.

Maerz, N.H., Franklin, J.A., Rothenburg, L. and Coursen, D.L., 1987, Measurement of rock fragmentation by digital photoanalysis, 5th. Int. Cong. Int. Soc. Rock Mech., §1, 687 - 692.

McDermott, C., Hunter, G.C. and Miles, N.J., 1989, The application of image analysis to the measurement of blast fragmentation, Symp. Surface Mining-Future Concepts, Nottingham University, Marylebone Press, Manchester, 103-108.

MacLachlan, R.R. and Singh, A., 1989, Photographic determination of oversize particles of heaps of blasted rock. J.S. Afr. Inst. Min. Metall., 89, 5, 147 - 152.

Rock Fragmentation by Blasting, Mohanty (ed.) © 1996 Taylor & Francis. ISBN 90 5410 824 X

Sampling problems during grain size distribution measurements

R.Chavez, N.Cheïmanoff & J.Schleifer
Paris School of Mines, CGES, France

ABSTRACT : In the past years, numerous developments have been made in the automation of grain size distribution measurements after blasting, especially through image analysis and pattern recognition techniques. The aim of such techniques is to automatize a process which cannot be carried out manually due to the large volumes handled. Efforts were mainly concentrated on the automatic recognition of visible block contours, followed by an individual volumic extrapolation accounting for their shape. However, according to bibliography, little attention was paid to the analysis of the sampling method itself, that is to say to the significance of 2D (or even 1D) sampling used to estimate a 3D physical object. If we define the grain size distribution as volume proportions taken by different size classes within a total volume, then special care must be taken regarding the sampling use. Sampling using images, as well as other more classical methods, do not always respect the equiprobability requirements necessary for an unbiased statistical estimation. In the case of images, important biases can appear due to the insufficient size of the sample and the different probabilities of blocks being visible according to their size. The sampling quality, which depends mainly on these two problems, is a function of the grain size distribution itself, this being what we intend to estimate. The larger the range of sizes to be measured, the worse the quality of sampling. Moreover, the quality also depends on the practical sampling conditions, for example, the practical setting of the image acquisition system. By modifying the location of the camera within the transport sequence of the material, it is possible to significantly improve the sampling quality and make easy corrections. Some attempts for correction will be shown based on probabilistic methods. Special attention is paid to the adopted hypotheses, knowing that they may not be valid for each application. These concern the 3D block modelling and sizing, the probability of each size of being present in the sample, and finally the homogeneity of the whole muckpile.

INTRODUCTION

Measuring the fragmentation of piles after a blast within mining sites reveals difficulties which have lead several operators to look for techniques using images in order to facilitate measuring and to increase the number of blocks taken into account. From what we know from the state-of-the-art, most efforts have been dedicated to problems of image processing itself by trying to use automatic block-recognition algorithms. Substantial progress has been made in this domain. However, the *representativity* of images has also been a constant source of worry, even if the sampling method represented by such a technique has not been, according to our knowledge, analyzed thoroughly into.

On the pile, we define fragmentation as a set of proportions, in volume, upon different classes, characterized by their size, generally the diameter. The problem of estimating this distribution using images may be summarized into two points :

- only a part of the pile is measured, most of it remaining generally inaccessible;

- images contain surface information, i.e. in 2 dimensions, which have to be used to extract a volume-distributed value, i.e. in 3 dimensions.

Is it possible to obtain a representative measurement of the fragmentation of the pile, providing individual proportions upon each size-class, with an acceptable error margin ? To answer this question, it is necessary to analyze errors inherent to the method.

Image acquisition has to be considered as a specific way of sampling in which all visible blocks on the image will belong to the sample. In order to guarantee a representative sampling, in other terms likely to provide an estimation without bias, the grain-size distribution of the sample has to be necessarily representative of the one in the pile. This is not always true because some sampling errors may appear. We do not want to concentrate on estimation errors resulting from insufficient number of samples. In such cases, confidence intervals may easily be computed and it is only up to us to increase the number of images captured to improve estimation. As such, continuous acquisition on dumpers or conveyor belts have a substantial advantage over direct acquisition on the pile. We would rather look for errors likely to generate estimation errors due to the specificity of the sampling method.

Segregating and grouping

The grouping problem may be illustrated in Figure 1. In this extreme case, the sample of visible blocks on surface is completely altering the actual proportions, here represented as numbers. When this type of grouping is achieved on a vertical basis, corresponding to full segregation,

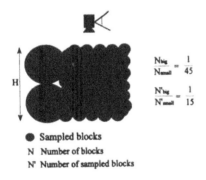

$$\frac{N_{big}}{N_{small}} = \frac{1}{45}$$

$$\frac{N'_{big}}{N'_{small}} = \frac{1}{15}$$

● Sampled blocks
N Number of blocks
N' Number of sampled blocks

Figure 1 Bias due to grouping.

measurement will be completely erroneous because only one class will be visible. These extreme cases may appear currently in real conditions and may concern more specially smaller blocks.

Capturing

If we do not consider segregation and grouping problems and imagine that all sizes are distributed on a homogeneous basis, an other problem will nevertheless occur, related to the geometry and the heterogeneity of the pile. This may be illustrated by a simple experience achieved on a small-scale pile consisting in 3 size-classes of fragments, 6-10, 10-20 and 25-30 millimetres of diameter. The population of each class was respectively 4000, 400 and 40 fragments leading of volume proportions, according to total population, of : 37 %, 43 % and 20 %. When arranging these fragments on one single layer, it is possible to measure them without problem and provide obviously the same proportions, assuming that the image processing technique used is sufficiently accurate. We have then re-arranged the pile in dumping it into recipients of variable size. It is quiet obvious that the number of visible blocks on surface decreases when the base of the recipient decreases and the height increases. What is the effect on the populations visible on surface ? We have repeated the dumping and counting of fragments several times. The means of measured volume proportions are represented in Figure 2. Proportions are given according to the height of overlapping of blocks within the recipient. Actual proportions are presented in the first column , corresponding to the case where all the blocks are arranged in one single layer, thus an overlapping height of H=0. The bias introduced by overlapping may be clearly appreciated. As a paradox, though the volume of the sample decreases, i.e. less fragments are visible on surface, proportions on surface get nearer to actuals as the overlapping increases.

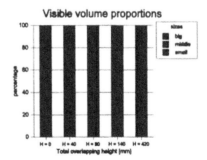

Figure 2 Evolution of volume proportions of fragments on surface according to the total overlapping height (average on several experiments); H=0 corresponds to actual proportions.

The established bias does obviously include some part of error due to segregation. Nevertheless, the convergence towards actuals when increasing the overlapping of blocks indicates the existence of another source of error. This error is related to the "theoretical" positions the blocks may occupy within the volume and is resulting from the non-uniform probability that a block, according to its size, may appear on surface. This point becomes clear if we consider the following example. We have only two classes of fragments of equal volume proportions, and the size of the larger class is twice as large as the size of the smaller class ($d_2 = 2d_1$). We assume that the pile is spread out on a height equal to the size of the larger class : $H = d_2$. We assume that the best sampling consists in taking images of the whole surface of the pile. It is then clear that, even in this case, error is substantial because, due to the difference of diameter, probability of large blocks to be seen, and thus captured, is close to 1 whereas probability of small blocks would be, in theory, close to 0.5.

The more the sizes of blocks are heterogeneous and the more this error is important. The difference of probability may be seen as an "error of capturing". This error is likely to appear when only the surface of overlapping blocks are considered during measuring. The two above-mentioned errors are related to the heterogeneous nature of the pile and its geometric layout. They both lead to an over-estimation of larger blocks. It is the control over these errors that will determine the representativity of the method.

SAMPLING ERRORS ACCORDING TO GY

Pierre Gy has developed a complete theory about sampling split material. It is especially concentrated on the estimation of the grade of ore, but it may also be applied to the estimation of humidity or fragmentation. Gy states that heterogeneity is the specificity of split material. This aspect is present in two ways : the heterogeneity in constitution, because the fragments may have different characteristics, and the heterogeneity in distribution, due to segregation and regional variability of the fragments. This is why a good sampling is difficult and why some rules must be respected. According to Gy, to get a *representative* sampling, it has to be *correct* and provide a *uniform sampling (access) probability for all the fragments*. To achieve this, a correct integration law, a correct cutting and a correct capturing are required. These three elements are the basic operations of a sampling process such as ours.

The integration defines the places on the pile where the sampling takes place. For integration to be approximately right, it is necessary and sufficient that the sampling probability density function $g(x)$ is uniform within the domain D_L

delimiting the portion and equal to null outside. In order to achieve this, Gy defines integration laws, i.e. sampling procedures, enabling random access to the samples at any point of the portion.

The cutting represents the way to sample and defines the domain to be taken around the point previously selected by integration. A cutting error occurs when the geometry of the volume to be extracted differs from volume-models defined according to the dimension of the portion. These volume-models must be :
- for 3 dimensions : a cubic volume
- for 2 dimensions : a square-section column containing the whole thickness of the portion
- for 1 dimension : a slice of constant thickness containing the whole section of the portion

The capturing is the physical process of sampling, which Gy has studied generally for mechanical means. Some devices do not give the same capturing probability to the individual fragments because of their different sizes or density. Gy defines a correct capturing when the probability for any fragment to be captured is equal to 1.

Not respecting these rules will involve estimation errors of the considered variable (grade or fragmentation). Sampling will not be representative. According to the practical conditions of sampling, Gy concludes that three-dimensional portions can not be sampled correctly, two-dimensional portions generate some difficulty and that only one-dimensional portion can be sampled correctly, provided some rules are respected.

Interpreting according to Gy the sampling using images

Current methods for image acquisition are the following :
- directly on the surface of the muck-pile of blasted blocks;
- on the surface of the trucks while loading and transporting the blocks;
- on the conveyor belt.

Taking images on the pile

It is clear that taking the image directly on the pile is not a correct integration, because the sampling points are only determined by their access facility, that is on surface. Correct sampling should enable access to any point of the pile and to take out constant volumes.

Some authors, using then image processing, justify their approach in stating that the surface is representative of the whole pile. According to Gy, a pile of fragmented material could not be qualified as homogeneous, at least because of segregation due to gravity. Moreover, in order to check the accuracy of such a statement, it should be necessary to take into account the cinematic of the blast and study the resulting degree of homogeneity of the fragments. Some authors state that an opencast blasting operation will generate an internal structure for the fragment distribution within the pile according to their origin on the face. As an example, it is common that the portion corresponding to the face-bottom will end in the lower part of the pile, thus completely concealed. On the other hand, the upper part of the pile is originated from the upper part of the face. This indicates that it is not always secure to speculate on a homogenizing action from the blast and that a pile partially measured may not be representative. To know of this method allows to simply compare two different piles is a question without answer.

Some operators carry out image acquisition on the pile as loading goes along. If carried out on a regular basis, so that an image is acquired after a constant volume has been removed, such a method could be associated with a systematic random integration, which could guarantee some representativity. However, taking only images, which are surfaces, is a *non-correct cutting* and therefore involves a non-correct sampling. Thus this method is not representative, according to the definition of Gy, because a correct procedure should collect volumes, and more precisely cubic volumes.

Taking images on the trucks

Carrying out loading operations is the first step in handling the pile. The problem becomes, in theory at least, less critical than taking images on three-dimensional portions. We may already state that this sampling is not correct, if we do consider the problems presented at the beginning of this article. Interpreting the resulting error is rather ambiguous, because it may be seen according to two different points.

The first one would be to say that each individual truck-load forms a two-dimensional portion. Correct sampling of such portions requires to collect constant-section-columns along the whole thickness of the portion. Taking only surfaces generate an important *cutting* error.

The second point is to consider the loading

operation as the actual collecting. Integration is correct, because any point within the pile may be accessed. Cutting is also correct because each load is a volume, even if it is not exactly a cubic volume. But, while dumping the material into the truck, a selective action consists in taking the blocks remaining on surface. Due to this random selection, individual blocks will be allocated with a probability Pi to appear on surface, in other terms to be part of the sample. A *non-correct capturing* has to be associated to this action, because the probability of a block to be on surface will not be equal to 1 and, moreover, this probability varies from one block to another.

This illustrates the ambiguous interpretation of the sampling error. The problem remains however the same. Aside from the category in which this error is classified, the crucial parameter for using images is the individual probability for a fragment to appear on surface during the loading operation. For this method, the error is originated in a *non-uniform probability for fragments to remain on surface*. This aspect is due to the identified problems which are the segregation and grouping and the capturing error.

Taking images on the conveyor belt

The pile on a conveyor belt can be considered as one-dimensional portion which has to be sampled by collecting transverse sections of constant thickness according to a uniform distribution along the length. Images will display sections of constant thickness, but they are not sections. Only blocks located on surface are collected, thus leading to the same error as for the trucks.

These considerations, as well as small-scale experiments, tend to show that our way of sampling is not correct and that a bias is likely to occur. Three questions must be answered :

- how important is this error and how does it vary according to the different ways of acquiring images ?
- if necessary, is it possible to correct at the end of the line a measure in order to make it representative ?
- are results able, despite their errors, to compare different piles ?

We would like to answer these questions by concentrating on the image acquisition on the trucks. The reason why is that a shovel loading fragments into the truck may be seen as a random operation during which fragments have a probability Pi to remain on surface depending, among other causes, on their size. It is then possible to model the *capturing probability* related to the sole geometric conditions of fragment overlapping. This probability is not equal to the probability for a fragment to appear on surface, because it is necessary to consider as well segregation and grouping problems. However, knowing the capturing error will result into knowing the importance of the two sampling error components.

Modelling the error of capturing

Relation between sample and probability of appearing on surface

We consider the event "a fragment F_i appears on surface" as an independent random variable which may take only two values corresponding to the two natural states : "the fragment is on surface" or "the fragment is concealed". We allocate to each fragment F_i the probability P_i to be on surface and thus $1-P_i$ to be concealed. The random variable obeys a Bernouilli law with parameter P_i. If we consider that probability P_i within a given pile depends on the fragment size, then all fragments of size k will have the same probability P_k to appear on surface. The number X_k of blocks, within the size-class k, on surface is the sum of N_k independent Bernouilli variables of same statistic law, with N_k the number of blocks of size-class k present within the pile. Mean and standard deviation for binomial variable X_k are :
$$E(X_a) = N_a P_a \; ; \quad V(X_a) = N_a P_a (1-P_a).$$
The mean of number X_k of fragments of size-class k to appear on surface is directly dependent on the individual probability for a fragment to appear. If we know this probability, it is then theoretically possible to estimate the actual number N_k out the observed X_k.

Modelling the probability of capturing

The probability for a fragment F_i of size D_i to appear on surface depends on the number of positions it may take within the height H of the pile. When fragments have the same size, the number of positions it may take is H/D_i. The probability it may appear on surface after dumping is then $P_i = D_i/H$. When fragments have different sizes, the number of positions depend on :
- its own size;
- the sizes represented within the pile;
- the number of fragments for each of these sizes.

The probability to appear on surface depends of the pile fragmentation. If we know the average size D_m,

average according to the number, of the grain-size distribution, the number of positions fragment F_i may take within the height H, may expressed, on a simplified basis, as :
$$N_H = 1 + (H - D_i)/D_m$$
with D_i as the fragment diameter. Its probability to appear on surface will be :
$$P_i = \frac{1}{1 + (H - D_i)/D_m}$$

The component $(H-D_i)/D_m$ represents the average number of blocks which may take place within the remaining height $H-D_i$, if considering that other blocks have a size equal to average size in numbers D_m.

We have compared these theoretic probabilities to the small-scale experiment observed frequencies, using the pile consisting in three classes of known proportions, which was dumped into recipients of different shapes. The number X_k, of fragments of size-class k visible on surface, divided by the number N_k, of fragments present within the pile, represents the empiric frequency F_k of the fragments to appear. When repeating this experiment several times, using the same recipient, the mean of empiric frequencies provides an estimate of the actual probability for fragments of size-class k to appear on surface. The results of comparing computed probabilities and observed frequencies are shown in figure 3. It is possible to see that for the two upper classes, the two probabilities are rather close once the height is above H=90. It is necessary to remain careful when interpreting deviations for H=40. This height is too small for the average size used in the simple model may really represent the number of positions in which a fragment may located. For the lower class, computed probabilities are substantially larger, except for the last height H=420. These deviations reflect segregation and grouping problems penalizing smaller sizes.

From these experiments, it is possible to make the following conclusions :

As established for classes 2 and 3, the theoretical model for capturing probability illustrates quite well the ability of fragments to reach surface. This ability, quantified as a probability factor, depends on the number of positions blocks, according to their size, may take within the height H. Difference between these probabilities describes the method's *capturing error*. Using this simple model, it is possible to explain convergence of the sample towards actual proportions when the overlapping height increases : for a given fragmentation, the capturing error decreases when the ratio H/D_{max}

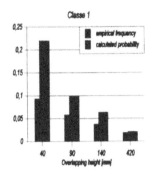 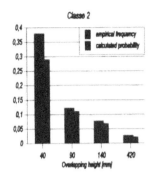 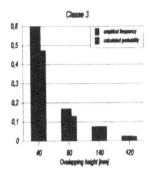

Figure 3 :comparing computed capturing probabilities and observed frequencies

(with D_{max} being the size of the upper class) increases. In our case, it becomes nearly irrelevant when this ratio reaches 14 (H/D_{max} = 420/30 = 14). This is because the capturing probabilities deviation between classes decrease when H increases. It is quite clear that, for a given height H, the capturing error is reduced as much as the maximal size is small and the size distribution is homogeneous.

Important deviations for the smaller size-class between observed frequencies and computed probabilities, indicate that smaller block positions may not be modelled by simple geometric stacking. This is especially true when much larger blocks are present within the pile. This leads to segregation and grouping error. As seen from the experiments, this error is always present and inhibits any possibility to correct the sample with the sole capturing probabilities. The evidence of these errors and the possibility to compute a capturing error offer new conclusions for a field-trial.

Sampling errors on the field

From what has been said, the quality of our sampling is directly related to the level of fragmentation heterogeneity within the pile. In mines and quarries, fragmentation is usually very heterogeneous, thus forfeiting any speculation upon representative sampling. It has been shown that the height of the dumped material is the relevant parameter for the capturing error.

In order to check how it varies and to get closer to real conditions, we have built a grain-size distribution consisting in several classes as shown in Table 1. Computed on a theoretical basis, deviations, due to capturing error according to different pile heights, are shown in Figure 4.

In this example, capturing error has nearly no impact when the height is above 1500mm. On the contrary, it is very important when the height is close to the maximal size (H = 650 and 800mm). These deviations represent the minimum theoretical error. In practice, it has to be added to segregation and grouping errors. They may never mutually compensate because these two sources tend to under-estimate smaller blocks to the benefit of larger ones.

The established deviations may discriminate the different ways to acquire images in the field. According to the production environment and the transport process, the volumes we have to deal with have geometry with varying overlapping heights. Methods having the possibility to compute the capturing error, in other terms including a loading operation considered as random operation with a constant overlapping height, are to be considered as much better cases.

Table 1 : example for grain-size distribution

Size [mm]	50	150	250	350	450	550	650
Volume [%]	21	19	17	14	12	10	7

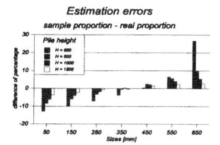

Figure 4 Capturing error progression according to pile height for a given size-class distribution.

As for conveyor belts, it will be difficult to estimate this error because the pile on the belt does not have a constant geometry with a constant overlapping height, even with the dumping process. Knowing this capturing error will allow to tackle later on errors due to segregation. Projects are currently carried out to estimate on an indirect basis proportions of smaller classes subject to segregation and grouping.

The last question we would like to deal with is to know if comparisons between piles is nevertheless possible even with these errors.

First of all, it must be stated that comparing fragmentation using images must be done with extreme care due to the influence of geometric conditions on the sample on surface. Some studies aim to compare fragmentation of piles on trucks to fragmentation after dumping on the dump, both being assessed with images. However, according to the variation of the overlapping height in the dump, visible fragmentation on surface may vary substantially versus the one visible on the trucks, without any change of global fragmentation. It would be hasty to assign these observed deviations to external factors, such as destruction of fragments during transport, and not take into account they may be due to the fluctuating of samples. A good rule to stick to would be to compare only what should be compared, in other terms samples collected in the same conditions and likely to present the same errors.

When comparing different piles, respecting the above-mentioned rules, it is nevertheless important to ask if surface fragmentations reflect actual differences, thus enabling to distinguish a larger fragmentation from a smaller one. Such a comparative method requires that the bias always moves in the same direction when the average size for the piles changes.

Concerning the capturing error, it has been demonstrated that the bias results from the non-uniform theoretical probability for a sample, according to its size, to appear on surface, leading to a non-equal probability to be collected. Figure 5 illustrates the evolution of the capturing probabilities for three classes versus the average size of the pile. It can be seen that deviations between probabilities increase slightly on a linear basis with the average size. It means that with a larger fragmentation a larger fragment will be more likely to appear than a smaller one. Capturing errors vary in the same direction than the fragmentation variation. Fragmentation differences between piles are even emphasized.

Concerning the segregation and grouping error, decreasing the average size should reduce empty spaces between larger blocks, in which smaller blocks will tend to group or fall. This segregating phenomenon should be attenuated, bringing smaller blocks to surface. As a result, this error also varies in the right direction and emphasize differences between piles.

According to these considerations, we think it is realistic to compare different piles using the surface fragmentation.

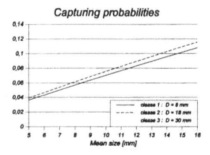

Figure 5 Evolution of capturing probabilities versus average size for three size-classes and H=140mm

CONCLUSIONS

Correct sampling requires a uniform probability to collect any fragment within the pile. This uniformity is basically not respected when using images, or any other method dealing only with the pile surface, to acquire surface information. Acquiring images directly on the heap lead, at the very begin, to a poor probability for some fragments to be collected because it would be wrong to speculate on a homogenizing action of the blast on the pile, thus giving equal chances to all fragments to appear on surface. When acquiring images on the trucks, the probability for a fragment to appear on surface is not uniform. Disparities are due to the heterogeneity of sizes and depend, on one hand, on the block overlapping height (capturing error) and, on the other hand, on segregation and grouping problems. These problems may lead to substantial errors in some cases. It is possible to correct the capturing error using a simple model validated on a small-scale sample. However, we do not know more details about the influence of segregation. Comparing piles should not cause major problems. The sampling method basically respects relative deviations between different fragmentations.

It is possible to conclude that measuring using image processing is able to characterize fragmentation after blast. However, they are not able to provide an estimate without bias for the volumic proportions of different size-classes within the pile. Research is carried out to improve this point.

REFERENCES

Farmer I.W., & al., 1991, "Analysis of rock fragmentation in bench blasting using digital image processing", Proceedings of the 7th International Congress on Rock Mechanics, Aachen, Germany, 1991, pp. 1037-1042.

Gy, P., 1966, "L'échantillonnage des minerais", Revue de l'Industrie Minérale, Juin 1966.

Gy, P., 1967, "L'échantillonnage des minerais en vrac", Mémoires du B.R.G.M. N°56, Tome 1, 1957.

Gy, P., 1975, "Théorie et pratique de l'échantillonnage des matières morcelées", Editions PG, 1975.

Hunter, G.C., & al., 1990, "Review of images analysis techniques for measuring blast fragmentation", Mining Science & Technology , Vol. 11, N°1, pp. 19-36.

Kemeny J.M., 1994, "Practical technique for determining the size distribution of blast benches, waste and heap leach sites" Mining Engineering, Nov. 1994, pp. 1281-1284.

Rock Fragmentation by Blasting, Mohanty (ed.) © 1996 Taylor & Francis. ISBN 90 5410 824 X

SIROJOINT and SIROFRAG: New techniques for joint mapping and rock fragment size distribution measurement

L.C.C.Cheung – *CSIRO Division of Exploration and Mining, Melbourne, Vic., Australia*

J.M. Poniewierski – *ICI Explosives, Kurri Kurri, N.S.W., Australia*

B.Ward – *CSIRO Division of Applied Physics, Sydney, N.S.W., Australia*

D. LeBlanc – *CSIRO Division of Exploration and Mining, Perth, W.A., Australia*

M.J.Thurley & A.P.Maconochie – *CSIRO Division of Exploration and Mining, Brisbane, Qld, Australia*

ABSTRACT: SIROJOINT and SIROFRAG are two aspects of a computer based, active stereo image analysis system that uses a scanning light stripe and a pair of CCD cameras to determine the 3D geometry of rock walls and blast muckpiles. This 3D information is used to determine the joint structure in the rock face and the fragment size distribution in the blast muckpile.

1 INTRODUCTION

The effect of rock structure, or jointing, has long been recognised as a major factor in determining blast fragmentation distributions. Yet only a minor portion of blasting research has attempted to characterise and account for the rock mass structure and its control over blasting (Doucet *et al.* 1994; Kleine, Cocker & Cameron, 1990; Kondos, 1983). A major reason for the paucity of such research has been the difficulty with using current techniques to quickly and reliably obtain the data (pre-blast structural data and post-blast fragmentation size distribution data) required to develop blast models that account for rock structure.

The development of quick and objective methods to remotely measure the 3D structure of rock surfaces and to measure the post-blast rock fragment size distribution in a muckpile in a mine environment has been the subject of a collaborative research project between CSIRO[1], BHP Australia Coal Pty Ltd (BHPAC) and ICI Explosives, since the beginning of 1993. SIROJOINT and SIROFRAG, the concepts of which were first described in 1991 (Ord and Cheung 1991; Ord *et al.* 1991), are two related and complementary computer-based active stereo image analysis systems. CSIRO is the prime research contractor for the project and is utilising resources from several divisions including: Exploration and Mining, Applied Physics and Mathematics and Statistics.

The imaging system being developed is based on a novel prototype scanner that projects a light stripe

that can illuminate an object in full daylight in a mine environment and a pair of video cameras that captures the image, allowing computation of the 3D geometry of the scanned objects. For SIROJOINT that object is a rock face and for SIROFRAG the object is a blast muckpile.

SIROFRAG developments were discussed in detail at the EXPLO-95 conference (Poniewierski *et al.* 1995). This paper will concentrate mainly on SIROJOINT developments.

By knowing the pre-blast 3D structure of rock surfaces derived by SIROJOINT it will be possible to calculate pre-blast in situ block size distributions. Using the post-blast rock fragment size distribution derived by SIROFRAG, it will then be possible to obtain a quantitative measure of blast efficiency.

2 HARDWARE DESCRIPTION

The main hardware components are the scanner; consisting of a flash lamp and a parabolic mirror both mounted on a motorised rotational stage; a scanner controller to position the scanner; a power supply unit to power the flash lamp; a pair of CCD cameras with electronic shuttering; a frame grabber to capture the digital images; and a personal computer.

A flash lamp system is used in preference to a laser system as it is cheaper, reasonably robust, easily replaced and is less restrictive in use when considering eye safety. The system uses Xenon filled flash tubes with a 2 mm bore and a 5 inch arc length, operating at a maximum pulse energy of 60 Joules.

A cylindrical elliptical reflector focuses the light from the flashlamp onto the wall. The reflector is

---

[1] Commonwealth Scientific and Industrial Research Organisation

Figure 1. Scanner (reflector, flashlamp and rotational stage), power supply unit and one of the CCD cameras in the laboratory.

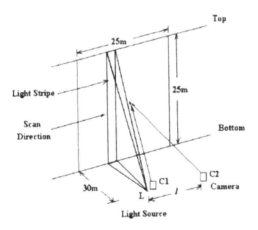

Figure 2. Measurement configuration for SIROJOINT.

made from electro-polished aluminium sheet and the desired shape is obtained by compression. The vertical plane of the reflector is tilted to match the angle of the wall face (this ensures a vertical stripe across the whole field of view). A *Galil* motion controller and a *Meade* motorised rotational stage are used to position the reflector to an accuracy better than one thousandth of a degree. A photo of the reflector, flashlamp and base is shown in Figure 1.

To enable capture of images under full sunlight conditions it is necessary to use fast shuttering

(100μs) of the CCD cameras, coincident with the pulsed light source generation. Additionally, to capture both odd and even interlacing of the CCD image of the stripe it is necessary to flash the lamp twice and synchronise the framegrabber capture with the respective interlacing signals from the cameras.

3   MEASUREMENT CONFIGURATION

A typical measurement configuration is shown in Figure 2. The light source L generates a light 'stripe' which extends over the entire height of the wall to be measured. The light stripe is approximately 100 mm wide when the distance between the light source and the wall is 30 m and is scanned across the wall from left to right. Images are captured by two cameras, $C1$ and $C2$. Camera $C1$ is a reference camera and is positioned adjacent to the scanner. Camera $C2$ is offset from the scanner by a distance l. The value of l is chosen to optimise relief sensitivity versus shadowing. The light stripe is scanned across at a rate of 0.1 m/s and is pulsed on at 1 sec intervals for a duration of 100 μs. For each stripe the (x, y, z) positions of pixels at the centre of the stripe image relative to one of the cameras (the reference camera) are then calculated. The registered reflectivity value of these pixels at the stripe centre are also recorded as a 'grey level' between 0 and 255. Higher walls may be scanned by using additional cameras or by repositioning cameras $C1$ and $C2$ (Ward *et al.* 1995).

4   SOFTWARE DEVELOPMENT

The software is Microsoft Windows™ based and uses advanced visualisation methods to control the hardware, link video data, and interpret the data as either joint structure information or fragmentation size distribution information.

A software feature called *Linked Views* has been developed for SIROFRAG and SIROJOINT. It has the capabilities of displaying a number of 'sub-windows' that are linked to the same data set. Actions in one sub-window, such as the selection of a feature, automatically modify/update the other linked sub-windows. The windows that are linked include displays of the captured digital image of a rock face or muckpile; a three-dimensional reconstruction image; for SIROJOINT, the equal area/angle projections as pole plots or contour plots, spacing and persistence graphs and joint area distribution graphs; and for SIROFRAG, various size distribution graphs.

For example, if a region of the equal area plot is selected, the rock structures giving rise to the poles within the selected region are highlit in the digital image and in the wireframe image, as shown in

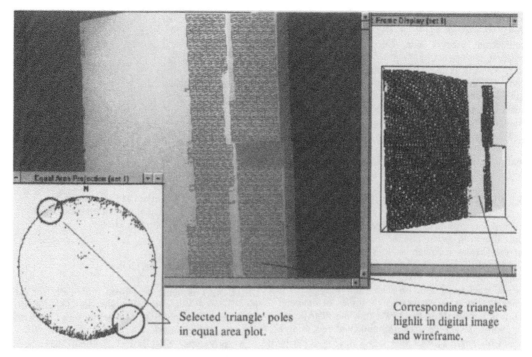

Figure 3.   A screen capture example of Linked Views with scanned laboratory 'face'.

Figure 4.   Screen capture image of 3D reconstruction of scanned laboratory 'face', showing four 'joint' surfaces.

Figure 3. (Figure 4 shows the 3D reconstruction of the same scanned laboratory 'face' shown in Figure 3.) Conversely, users can select a region within the digital camera image and the poles corresponding to that region will be highlit in the equal area plot.

With advanced visualisation features such as Linked Views, users can easily relate an area or a feature in the captured digital image to the corresponding results in the equal area plot, persistence, spacing or size distribution graph.

5   CALIBRATION

To determine the 3D positions of stripe centres on the object being scanned, it is necessary to know the orientations and positions of the two cameras used. This is found using a calibration technique called Relative Camera Calibration (Cheung *et al.* 1989).

The two cameras are positioned as indicated in Figure 2, to view the same section of object space. The reference camera ($C1$) is set as the origin with the x, y, z co-ordinates (0,0,0) and its orientation angles around the X, Y, Z axes as (0,0,0). By identifying any fifteen distinct matching points (control points) in the two camera images of the common scene (selected on the PC screen using the mouse) and by knowing the distance between two of those points or between the two cameras, the calibration process can calculate the accurate position and orientation angles of the second camera (*C2*) as well as the 3D co-ordinates of the fifteen control points. With the two cameras calibrated, the 3D co-ordinates of stripe centres on the scanned surface can be readily calculated by a simple triangulation process.

The system calibration process and data collection process have been rigorously tested in the laboratory and successfully proven to work.

6 VERIFICATION OF CALIBRATION

One of the tests to check and verify the validity of the calibration system and the measurement accuracy of the system hardware is to use the SIROFRAG functionality and scan a set of geometric shapes, for which the resulting distribution of chord lengths (stripe segments falling on each individual shape) is known. For a set of mono-size circles of diameter *D*, it is known (Hornby, 1994) that the distribution of chords is such that the probability of recording a chord of length *L*, such that $l_1 < L < l_2$ is given by equation 1:

$$p(l_1 < L < l_2) = \left\{1 - \sqrt{\left(1 - \left[\frac{l_2}{D}\right]^2\right)}\right\} \tag{1}$$
$$- \left\{1 - \sqrt{\left(1 - \left[\frac{l_1}{D}\right]^2\right)}\right\}$$

The distribution this equation gives is known as the "unit chord distribution" for circles.

Figure 5 shows a set of circles, all of 200 mm diameter, which was scanned using the developed system. Figure 6 shows the measured unit chord distribution against the theoretical distribution for an "infinitely thin" linear probe.

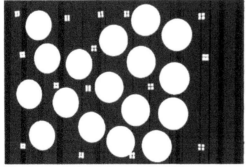

Figure 5. A set of mono-size (200 mm diameter) circles scanned to test system calibration accuracy (NB: Markers used in calibration process, but removed before scan).

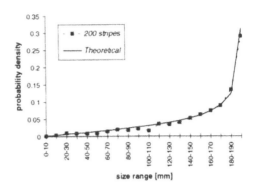

Figure 6. Comparison of experimentally determined and theoretical unit chord distributions for a set of 200 mm diameter circles.

7 SIROJOINT OPERATIONS

SIROJOINT is being developed principally for use in structural mapping of highwalls and for the determination of pre-blast *in situ* block size from the measured joint pattern.

A prototype SIROJOINT system has been extensively tested in the laboratory and has been field trialled at two quarries and a coal mine. A rock face from one of the quarries and an example of a projected light stripe on this face are shown in Figure 7. The cameras and projector were placed approximately 9 metres from the rock face with the two cameras approximately 7 metres apart. The scan this stripe was taken from was conducted during the afternoon with full sunlight on the rock face.

In normal operation a scan of 100 stripes is taken and a 3D map of the rock face is produced. Small triangles, as shown in Figure 3 for a laboratory 'face', are formed between adjacent 3D data points and the plane equations of each of these triangles is calculated. Joint plane identification is then done by merging adjacent triangles with similar orientations (with appropriate filtering), as shown in Figure 8, where three of the parallel 'joint' surfaces of the laboratory 'face' have been identified and are outlined (planes *B*, *C* and *D* of Figure 4).

With the joint planes identified, the equal area/angle pole plot (Figure 9), persistence and spacing information can then be obtained.

8 FUTURE

For SIROJOINT further laboratory and field testing will develop the automatic recognition of geological structures and extraction of numerical data including statistical distributions of joint length and spacing. In particular the reflectivity information from the stripe projection, and passive image analysis of the

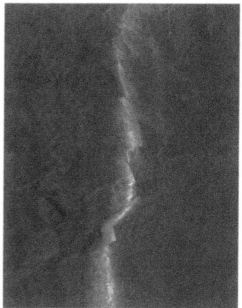

Figure 7. Quarry rock face and projected light stripe on the rock face.

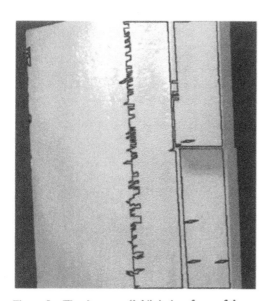

Figure 8. The three parallel 'joint' surfaces of the laboratory 'face' identified by triangle merging.

CCD camera images of the rock face need to be used in conjunction with the triangle merging in order to further refine the joint identification process (real faces are still subject to significant noise in the plane identification process).

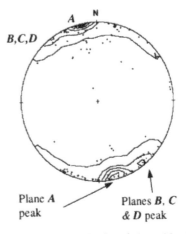

Figure 9. Equal area pole plot of planes identified by triangle merging for laboratory 'face'.

For SIROFRAG, further field and laboratory work will examine the issue of fines (the fragments below the system resolution), the effects of mixed resolution loss (distant fragments are lost to system resolution while similar size but nearer fragments are still resolvable) and validation of the statistical equations used to determine the three dimensional size distribution (i.e. system interpretation accuracy).

ACKNOWLEDGMENTS

The authors wish to thank BHP Australia Coal Pty Ltd, ICI Explosives and CSIRO for granting permission to publish this paper and for financial support of the project. The other team members who have laboured over several years on this project and whose work is encapsulated in this short paper are also acknowledged: M. Boland, P. Cusack, P. Fairman, M. Ghaffari, Z. Hegedus, D. Martin, A. Ord, B. Oreb, Z. Quinn, and C. Walsh

REFERENCES

Cheung, C.C., B.F. Alexander, W.A. Brown & K.C. Ng 1989. Relative camera calibration, *ICIP'89 IEEE International Conference on Image Processing*, IEEE Singapore Section, Singapore, September, pp 461-465.

Doucet, C., M. Paventi, Y. Lizotte & M. Scoble 1994. Structural control over fragmentation: characterisation and case studies, *Proc. 10th Annual Symp. On Explosives and Blasting Research (ISEE)*, Jan. 30- Feb. 4, Austin TX, USA, pp. 121-133.

Hornby, P. 1994, Some comments on SIROVISION stereology, unpublished CSIRO internal report.

Kleine, T., A. Cocker & A. Cameron 1990. The development and implementation of a three dimensional model of blast fragmentation and damage, *Proc. Third International Symposium on Rock Fragmentation by Blasting (FRAGBLAST '90)*, Brisbane Australia, 26-31 August., pp 181-187.

Kondos, P.D. 1983. *Rock Structure: An Important Factor in Forecasting Blast Fragmentation*, M.Eng. Thesis, unpublished, McGill University, Montreal, Canada.

Ord, A. & L.C. Cheung 1991. Image analysis techniques for determining the fractal dimensions of rock joint and fragment size distributions. *Proc. 7th Intl. Conf. on Computer Methods and Advances in Geomechanics*, Cairns, Australia, pp. 87-91.

Ord, A., L.C. Cheung, B.E. Hobbs & D. Le Blanc 1991. Automatic mapping of rock exposures for geotechnical purposes. *2nd Aust. Conf. on Computer Applications in the Mineral Industry*, Univ. of Wollongong, Australia, pp. 205-210.

Poniewierski, J., L.C.C. Cheung, A. P. Maconochie 1995, SIROFRAG - a new technique for post-blast rock fragmentation size distribution measurement, *Proc. EXPLO-95 - A Conference Exploring the Role of Rock Breakage in Mining and Quarrying*, Brisbane, Australia, 4-7 September, pp. 263-268.

Ward, B.K., B.F. Oreb, M. Ghaffari, A.P. Maconochie, L. Cheung & J. Poniewierski 1995. The design and performance of an optical system used to generate a light stripe for relief measurement on 60 metre highwalls in open-cut mines, *Proc. Ann. Conf. Australian Optical Society*, Brisbane, 4-7 July, 12 pp.

Rock Fragmentation by Blasting, Mohanty (ed.)© 1996 Taylor & Francis. ISBN 90 5410 824 X

Two dimensional polyomino simulation of muckpile swell

A.T.Spathis
ICI Explosives Asia-Pacific, Kurri Kurri, N.S.W., Australia

ABSTRACT: The swell of a muckpile is a complex function of the rock fragmentation and the disposition of the rock within the muckpile. The digability of a muckpile appears to be a function of this swell. Polyominoes are used to represent rocks of non–uniform shape. Numerical simulations using such particles indicate that particle size is a more significant influence on the muckpile swell than is the detailed shape of the particles. A limitation of the model is the restriction on particle movement after it lands on the muckpile.

1 INTRODUCTION

The primary outputs of a blast include: fragmentation, heave, damage and environmental effects. A mine or quarry must deal with the fragmented rock using ground engaging equipment such as draglines, shovels or front end loaders. The efficiency of this equipment is strongly influenced by some measure of the digability of the muckpile. The digability is a complex function of the fragmentation distribution, the heave delivered by the explosive and the design of the excavator. Furthermore, the digability appears to be affected by the swell of the muckpile. Here, the swell (SW) is defined as the final post–blast muckpile volume (Vol (post)) divided by the pre–blast, intact volume (Vol (pre)):

$$SW = \frac{Vol\ (post)}{Vol\ (pre)} \qquad (1.1)$$

or,

$$SW = \frac{S + V}{S} \qquad (1.2)$$

where S is the solids volume and V is the voids volume. Clearly, the swell will have a global value for the total blast volume, but it will vary on a local value throughout the muckpile.

The preferred method for assessing the post blast digging, haulage and crushing processes is to measure these over a period of time at the mine or quarry (see, for example, Hawkes *et al*, 1995). However, models of blasting offer the opportunity to perform idealised experiments which may lead to improved blast designs and hence, better mine equipment productivity.

The present paper looks at two dimensional simulation of muckpile swell. Most blast models (see, for example, Preece and Taylor, 1990) rely on spherical (three dimensional) or disc (two dimensional) particles to represent the rock fragments; alternatively, they develop a global muckpile profile from which the global and local swell may be determined (Yang and Kavetsky, 1988; Favreau, 1993).

The model employed in the present work uses a finite set of irregular two dimensional particles called polyominoes (Golomb, 1994). Particles are selected from this set and deposited at arbitrary locations in a bin of fixed dimension. Each particle selection and placement is done randomly. No rotations of the particles is permitted after placement in the muckpile.

2 POLYOMINO MODEL OF SWELL

Polyominoes are clusters of contiguous squares which meet along edges. No connection is allowed at vertices alone. The most well known polyomino is the domino. Polyominoes have been used to consider the so-called tiling problem. Penrose (1995) describes this as: " given a set of polygonal shapes, decide whether these shapes will tile the plane; that is, is it possible to cover

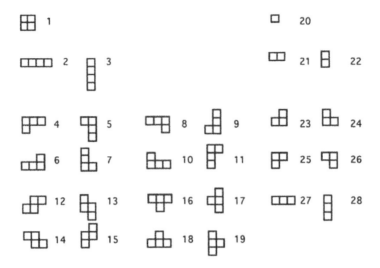

Figure 1. Low order polyominoes. Number 20 is a monomino, number 21 is a domino, numbers 23 and 27 are trominoes and numbers 1, 2, 4, 12 and 16 are tetrominoes. The other particles are formed by rotations and reflections of those in the base set of polyominoes. The numbers shown here are used to identify the particles throughout the paper.

the entire Euclidean plane using only these particular shapes, without gaps or overlaps ? "

The present application of polyominoes is a stark contrast to the tiling problem. Here, I am concerned with the relative amount of voids created by tiling the plane in a random manner. It would appear that the solution to either problem is non–trivial.

Figure 1 shows the low order polyominoes, including rotated polyominoes and reflected asymmetric polyominoes. Here, the order of a polyomino is defined as the number of unit squares used to construct the polyomino. An explicit formula for the number of polyominoes of order n has not been found (Golomb, 1994). If P(n) is the number of polyominoes of order n, then for large n, it has been shown that,

$$\frac{P(n+1)}{P(n)} \to K^n \qquad (2.1)$$

where $3.9 < K < 4.65$, and is known as Klarner's Konstant.

Table 1 lists the number of polyominoes of order n for n up to 15. Also shown in the table is the corresponding number of polyominoes which contain holes. For my purpose such polyominoes are not desirable, as rock fragments are not often created containing such voids! In any case, it is not until n=7 that we find such a particle.

The polyomino model of muckpile swell uses the base set of polyominoes including those formed by rotation and reflection of the base set of polyominoes. A Monte Carlo procedure is used in the model. Polyominoes are chosen randomly from a given set and then deposited at a random location at the top of a square bin. The particle settles in a vertical direction until it strikes the bottom of the bin or another particle which has been previously deposited. The process continues until the bin is full. The swell, as defined by

Table 1. Number of polyominoes, P(n), of order n Q(n) is the number of polyominoes with holes.

n	P(n)	Q(n)
1	1	0
2	1	0
3	2	0
4	5	0
5	12	0
6	35	0
7	108	1
8	369	6
9	1 285	37
10	4 655	195
11	17 073	979
12	63 600	4 663
13	238 591	21 474
14	901 971	96 496
15	3 426 576	425 365

Equation (1.1), is calculated for a number of simulations using the same particle set. The mean and standard deviation of the swell is reported for each set of simulations.

The particle width is defined as its greatest horizontal dimension. For example, a horizontal domino has width two while a vertical domino has width one. Figure 2 shows the particle width (or size) distributions for two sets of particular polyominoes. The first set is for the 19 tetrominoes and the second is for these plus the extra 9 polyominoes of lower order. Obviously, reflected particles and rotated particles are included. The sizes in the latter set are skewed slightly towards the smaller widths.

It is worth noting that polyominoes of a given order have the same area but a different shape to each other. This contrasts with the more common usage of single sized circles for modelling swell. The global swell for circles placed on a square lattice is 1.2732 and for a hexagonal lattice it is 1.1027. The latter swell represents the minimum swell or maximum packing density for such a particle set.

3 SIMULATIONS USING THE POLYOMINO MODEL

A typical output obtained from a simulation is shown in Figure 3. The dark areas represent the solid material and the white represents the voids. In this example, the square tetromino (identified as particle 1 in Figure 1) has been used exclusively for the simulation. The tetromino has dimensions 2 x 2 units. The bin has dimensions of 100 x 100 units.

(a)

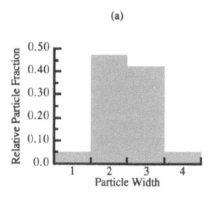

(b)

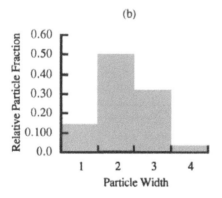

Figure 2. Polyomino width distributions for
(a) tetrominoes (particles 1-19 in Figure 1)
(b) 1-, 2-, 3-, 4 - ominoes (particles 1-28 in Figure 1)

Figure 3. Typical output from the polyomino model (upper figure). Here, the square tetromino is used exclusively. The dark areas represent solid material and the white areas represent the voids. The enlargement (lower figure) illustrates a statically unstable staircase of tetrominoes.

A significant, and in terms of actual muckpiles, an unrealistic feature of the simulation output is the statically unstable arrangement of some of the structures generated by the model. This is illustrated in the enlargement shown in Figure 3. The model also produces large void spaces due to these structures and subsequent bridging. In effect, the model produces an overestimate of realistic muckpile swell.

The present study investigates the influence of particle shape on the muckpile swell while acknowledging the stated model limitations.

Convergence tests for the predicted swell were performed for a number of model arrangements. It has been found that a bin size of 200 by 200 units yields adequate results. Figure 4 shows the predicted swell for such a bin size. Results are included for all individual polyominoes up to order four. Single unit width particles are excluded for obvious reasons.

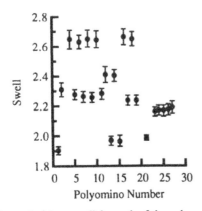

Figure 4. Mean swell for each of the polyominoes in Figure 1 apart from particles 3, 20, 22 and 28. The error bars are three standard deviations above and below the mean. Calculations are based on a bin size of 200 x 200 units using 100 simulations.

Each result in Figure 4 was generated for 100 individual simulations. The mean and three standard deviations above and below the mean swell are plotted. The swell ranges from a minimum of approximately 1.9 to a maximum of approximately 2.65.

There is some obvious clustering of the predicted swell for different polyominoes. Similar swells are obtained for polyominoes which are symmetric or antisymmetric reflections of a given polyomino. For example, particle numbers 4, 6, 8,

10 form one group where particles 8 and 10 are symmetric reflections of particle 4 and particle 6 is an antisymmetric reflection of particle 4.

The width of the particles has a significant influence on the swell: in general, larger swells arise for larger particles. This result is particularly clear when we compare polyominoes with different width to height ratios. The polyominoes 4, 6, 8, 10 have a swell of approximately 2.65, while the polyominoes 5, 7, 9, 11 have a swell of approximately 2.27. These two sets of polyominoes have identical shapes but are simply rotated by 90 degrees and hence change their width from 2 units to 3 units. An extreme example of this phenomenon occurs for the straight polyominoes: 21, 27 and 2. When these particles are rotated by 90 degrees, they fill the bin completely.

Skew polyominoes (13, 15, 12, 14) are quite efficient in filling a bin. Although polyominoes 12 and 14 have a width of 3 units, their swell is not too much greater than polyominoes with a width of 2 units. Straight polyominoes are more efficient than might be expected from their relative size.

A simple summary of the simulation results of Figure 4 is given in Table 2. The polyominoes are identified by their basic shape; L tetrominoes are particles 4 to 11 and L (or right) trominoes are particles 23 to 26, and so on.

It is possible to model a large number of combinations of the various polyominoes including those of higher order than those depicted in Figure 1. Two particular combinations are considered: combinations of polyominoes which are simply rotations of a given polyomino; and, combinations of polyominoes which have similar swell values when each single polyomino is used exclusively in the model.

Table 2. Summary of single particle simulation results in descending order of swell.

Polyomino	Width
L, T tetromino	3
Skew tetromino	3
Straight tetromino	4
L, T tetromino	2
Straight tromino	3
L tromino	2
Straight domino	2
Skew tetromino	2
Square tetromino	2

Figure 5 shows the first of these groupings of polyominoes. Again, we see that similar swells are obtained for groups of polyominoes which are reflections of another group (for example, particles 4-7 and particles 8-11). The general trend is for groups of polyominoes which have larger widths to produce larger swells. The largest swell arises from the straight tetromino and its rotated companion.

A simulation using all the tetrominoes (1-19) results in a swell of approximately 2.48. A smaller swell of approximately 2.1 arises when just the 1-, 2- and 3-ominoes are used. Combining all the polyominoes (1-28) yields a swell of approximately 2.39. The introduction of the smaller particle set produces a small but significant decrease in swell.

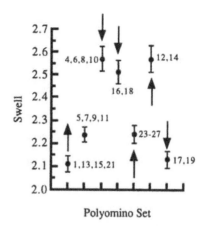

Polyomino Set

Figure 6. Mean swell for polyomino sets where each set consists of polyominoes which have a similar swell when used alone (see figure 4). The error bars are three standard deviations above and below the mean. The arrows indicate the direction of the change in swell compared with the swell calculated for the individual particles. Calculations are based on a bin size of 200 x 200 units using 100 simulations.

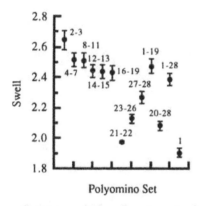

Polyomino Set

Figure 5. Mean swell for polyomino sets where each set consists of a polyomino and its rotations. Also shown are the results for all the tetrominoes (1-19), all the 1-, 2-, 3-ominoes (20-28) and a combination of these two sets (1-28). The error bars are three standard deviations above and below the mean. Calculations are based on a bin size of 200 x 200 units using 100 simulations.

It is natural to consider what happens when we combine polyominoes which, when used alone, yield approximately the same swell. The results of these simulations are shown in Figure 6.

The interaction between the particles in each set modifies the net swell. Both increases and decreases in swell are observed for the polyomino sets. The arrows in Figure 6 indicate the direction of change in swell for the combinations of polyominoes compared to the individual particle swells. The sets containing small particles and/or skew polyominoes generally result in an increase in swell. The sets containing large particles and/or L or T tetrominoes generally result in a decrease in swell.

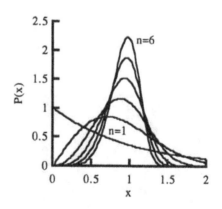

Figure 7. Normalised Rosin-Rammler distributions for exponents n=1 to n=6.

4 SIMULATION OF ROSIN-RAMMLER DISTRIBUTIONS OF POLYOMINOES

As a final application of the polyomino model of swell, I shall consider the case of a distribution of particles which follow the so-called Rosin-Rammler curve. This has the following normalised functional form,

$$P(x) = n(x)^{n-1} \exp(-x^n) \qquad (4.1)$$

where x is the size dimension of interest and n is the uniformity index. This distribution is known more widely as the Weibull distribution. Normalised continuous Rosin-Rammler distributions are shown in Figure 7. The distributions are based on Equation (4.1).

Field measurements of rock fragmentation are often described by the Rosin-Rammler distribution (Cunningham, 1983). The number of size categories is often restricted to no more than five or six. The following modelling uses ten size categories.

The basic Rosin-Rammler distribution curve has been discretised by fitting a histogram to ten different size categories. The polyomino shapes used in the model are square polyominoes with edge dimensions ranging from one unit to ten units. The sizes have been scaled so that the ten unit wide polyomino corresponds to two units in the normalised distribution curve. Obviously, the five unit polyomino corresponds to one unit in the normalised distribution curve.

The results of a typical simulation using a uniform distribution of the ten square polyominoes is shown in Figure 8.

Two further simulation outputs are shown in Figure 9. These are for Rosin-Rammler distributions of the ten square polyominoes. The upper figure is for n=2 and the lower figure is for n=4 in Equation (4.1). It is evident that the lower figure has a more narrow distribution of particle sizes; that is, there are fewer large and fewer small particles. Both these outputs were generated using an identical set of random numbers for the particle selection phase of the simulation.

This means that at each stage of the simulation the same random number was used to select the appropriate particle, based on the relevant Rosin-Rammler distribution. This explains the gross similarity between the two simulations shown in Figure 9. Indeed, much greater similarity occurs for distributions with consecutive exponent values.

5 DISCUSSION AND CONCLUSIONS

The remarkable results obtained from the simulations for Rosin-Rammler distributions of square polyominoes are shown in Figure 10. The

Figure 8. Section of a typical output from the polyomino model for a uniform distribution of square polyominoes of size one to ten units. The dark areas represent solid material and the white areas represent the voids. The bin has dimensions of 100 x 100 units.

Figure 9. Outputs generated by the polyomino model for a Rosin-Rammler distribution of square polyominoes with n=2 (upper figure) and for n=4 (lower figure). The square polyominoes range in size from one to ten units. The bin has dimensions of 100 x 100 units.

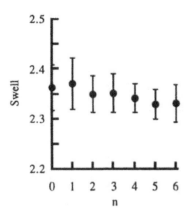

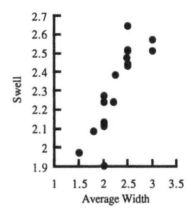

Figure 10. Swell predicted from the simulation of Rosin-Rammler distributions of square polyominoes. The exponent n is the uniformity index in the Rosin Rammler distribution. The n=0 point corresponds to a uniform distribution of square polyominoes. The polyominoes range in size from one to ten units. Calculations are based on a bin size of 500 x 500 units using 20 simulations. Similar results were obtained for a bin size of 200 x 200 units using 100 simulations.

Figure 11. Swell versus average width of polyomino groups shown in Figures 5 and 6.

model predicts approximately equal global swell for all the distributions considered. Indeed, even a uniform distribution of square polyominoes yields a similar swell.

The initial modelling used a bin size of 200 x 200 units with 100 simulations. The results in Figure 10 used a bin size of 500 x 500 units with 20 simulations. Both these model parameters gave the same result - namely, that the global swell is almost constant irrespective of the value of the Rosin-Rammler uniformity index. There was a slight difference in the predicted absolute value, with the larger bin size yielding a slightly higher swell.

The question to be answered is " Why do we obtain such a result ? ". The results from the polyomino model of swell for the low order polyominoes (Figures 4-6) suggest that the particle widths have the most significant influence on the global swell.

For a continuous Rosin-Rammler distribution (or Weibull distribution) given in Equation (4.1), the mean particle size of the distribution is given by,

$$\mu = \Gamma\left(1 + \frac{1}{n}\right) \qquad (5.1)$$

where Γ is the gamma function. The mean of the normalised Rosin-Rammler distributions is 1.0 for n=1 and it has a minimum value of 0.886 for n=2. The mean lies between these two values for all other values of n considered here. The present implementation of a discrete Rosin-Rammler distribution will yield similar mean values.

It appears that the mean value of the size distribution of a set of polyominoes correlates with the global swell in the model (Figure 11).

However, when we introduce the data obtained from the Rosin-Rammler distribution simulations (a swell of approximately 2.35 and an average width of approximately 5 units), it is clear that the simple (say almost linear) correlation between global swell and average width breaks down.

The following conclusions may be drawn from the present study:

• The results from the Rosin-Rammler distribution of square polyominoes do not confirm the almost linear correlation between swell and average width.
• The influence of the particular shape of the polyominoes on the global swell appears to be a second order effect. (Attempts to look at the influence of particle sphericity - particle perimeter to particle area in this case - are not useful as all the tetrominoes have the same value for such measures apart from the square tetromino).
• For the 1-, 2-, 3- and 4-ominoes there is a good correlation between the global swell predicted by the model and the average width of the polyomino groups considered;

265

• For the Rosin-Rammler distributions of square polyominoes, there is no difference between the global swell predicted for distributions with n=1 to n=6. A uniform distribution of such particles also exhibits the same swell.

• Interaction between different polyominoes does influence the resultant swell, but again, it appears to be a second order effect.

• Further studies need to investigate the 'tiling' properties of polyominoes in the case where complete covering of the plane is not the primary requirement.

• Notwithstanding the above conclusions, it is necessary to move to models which remove the statically unstable structures exhibited by the present model. A simple starting point would be to use a 'maximally avoiding' random number generator for the particle deposition part of the model.

6 ACKNOWLEDGMENTS

The author would like to acknowledge discussions with his colleagues at ICI Explosives. In particular, Michael Noy provided some valuable information on the assessment of rock fragmentation.

7 REFERENCES

Cunningham, C. 1983. The Kuz-Ram model for prediction of fragmentation from blasting. *Proc. 1st Intnl. Symp. Rock Fragmentation by Blasting*, August, 439-453, Lulea, Sweden

Favreau, R F 1993. The prediction of blast–induced swell by means of computer simulations. February, *CIM Bulletin.*, 69-72

Golomb, S W 1994. Polyominoes: puzzles, patterns, problems and packings. Princeton University Press, Princeton.

Hawkes, P J, A T Spathis and G Sengstock 1995. Monitoring mine equipment productivity improvements in coal mines. *AusIMM Explo 95*, 127-132, 4-7 September, Brisbane.

Penrose, R 1995. Shadows of the mind: A search for the missing science of consciousness. Vintage Press, Sydney.

Preece, D S and L M Taylor 1990. Spherical element bulking mechanisms for modelling blasting induced rock motion. *Proc. 3rd Intnl. Symp. Rock Fragmentation by Blasting*, 189-194, 26–31 August, Brisbane.

Yang, R L and A Kavetsky 1988. A three dimensional kinematic model of muckpile formation in bench blasting. *AusIMM Explo 88*, 45-49, November, Melbourne.

Explosive energy and fragmentation

Rock Fragmentation by Blasting, Mohanty (ed.) © 1996 Taylor & Francis. ISBN 90 5410 824 X

Blasting-crushing-grinding: Optimisation of an integrated comminution system

Kai Nielsen
Department of Geology and Mineral Resources Engineering, Norwegian University of Science and Technology, Trondheim, Norway

Jan Kristiansen
Dyno Nobel, Lierstranda, Norway

ABSTRACT: The paper describes and presents the results of several industrial and laboratory blasting, crushing, and grinding tests and experiments investigating how blasting can influence the subsequent crushing and grinding operations. The paper will also discuss how these results can aid in evaluation of the whole comminution system.

1 INTRODUCTION

The primary purpose of drilling and blasting is to fracture solid ores and rocks and prepare the material for excavation and subsequent transport.

The most common practise today is therefore to evaluate the blasting results only in connection with the loading of the material and the secondary blasting which may be needed. The fragmentation must be "fine enough" for the loader, the muckpile must be "loose enough" to facilitate digging, the bottom must not be "too hard" in order to keep the designed level, and there must not be "too many" boulders.

These are all qualitative criteria, and even if "good fragmentation" and effcient loading is economically very important, such an evaluation cannot give much information about whether the blast is anywhere near an optimal design or not.

Blasting, however, plays a wider role than just fragmenting the rock. It is the first step of an integrated comminution process leading from solid ore to a marketable product, and industrial experience and observations show that blasting will influence the processing steps following after loading, hauling and primary crushing in the mine.

Blasting operations in the mining industry should therefore be designed and optimised in order to obtain the lowest overall product costs or the highest operating profit.

The questions are then: How will the primary drilling and blasting influence the costs of the subsequent steps in the production process, and how far down the flowsheet will blasting have a significant influence on the economic results?

In order to get at least some answers to these questions, several full scale industrial and laboratory tests have been undertaken, investigating how blasting can influence the subsequent crushing and grinding operations.

2 INDUSTRIAL TESTS

The industrial blasting and crushing tests were done as part of a research programme investigating how the choice of drillhole diameter and the detonation velocity of the explosive would influence the product quality and the generation of fines in crushed aggregate production. Two separate series of tests were done in two different quarries.

2.1 VOD and the generation of fines

The first series consisted of six blasts, two with ANFO and four with two different types of slurry. (Kristiansen, 1994a; Kristiansen, 1995a).

The blasts were drilled with 76 mm bits and consisted of 22 to 30 drillholes with a pattern of 2.0 x 3.5 meters. The bench height varied between 17 and 19 meters.

Table 1 shows the characteristic theoretical parameters for ANFO and the average values for the two slurries which were very similar.

Because the drilling pattern was the same for all the blasts, the powder factor for the slurry blasts were 0.8 kg/m³ and 0.5 kg/m³ for the ANFO, due to the different densities of the two explosives.

The blasted rock was first crushed in a jaw crusher and then in a gyratory crusher to minus 100

TABLE 1. Explosives parameters.

TYPE OF EXPLO-SIVE	Density g/cc	Energy MJ/kg	Gas volume liter/kg	VOD m/s
ANFO	0.85	4.0	970	2 200
Slurry	1.25	3.1	910	4 500

TABLE 2. Percentage of fines minus 12 mm after blasting and two stage crushing.

TYPE OF EXPLOSIVE	Percentage - 12 mm	Average percentage
ANFO	21.9	21.8
ANFO	21.6	
Slurry	26.6	24.1
Slurry	23.9	
Slurry	23.1	
Slurry	22.9	

TABLE 3. Drilling parameters.

Dia-meter mm	Burden m	Spac-ing m	Area per hole m^2	Stem-ming m
76	2.2	4.5	9.9	2.7
89	2.5	4.5	11.3	3.0
102	3.0	4.5	13.5	3.5
114	3.3	4.5	14.9	3.8

TABLE 4. Blasting parameters.

Diameter mm	Planned powder factor kg/m^3	Realised powder factor kg/m^3	Explo-sive per surface area gram/cm^2
76	0.51	0.56	2.4
89	0.56	0.53	2.8
102	0.61	0.56	3.2
114	0.66	0.60	3.6

mm before the material passed over a 12 mm screen.

The two belt conveyors transporting the 12-100 mm and the minus 12 mm material from the screen, have been fitted with belt scales. This made it easy to register the proportion of fines minus 12 mm after blasting and two stage crushing.

Table 2 shows the proportion of fines minus 12 mm for the six blasts.

Considering that the slurry blasts had a much higher powder factor than the two ANFO blasts, and that the VOD of the slurries were nominally twice that of ANFO, there is an unexpected small difference in the amount of fines between the blasting done with slurry and the two ANFO blasts.

This can at least be partially explained by the fact that the ANFO blasts had a coarser primary fragmentation than the slurry blasts. This means that the crushers had to do more crushing work on the ANFO material, generating more fines in the two crushing stages compared with the more finely fragmented slurry material.

2.2 Drillhole diameter and the generation of fines

The second series of full scale testing was done using drillhole diameters varying between 76 mm and 114 mm. (Kristiansen, 1995b).

The blasts consisted of 116 to 125 holes in 6 and 7 rows. The bench height was 11 meters for all the blasts.

Table 3 shows the drilling parameters for four blasts drilled with the various hole diameters. The explosive was a slurry type with the same properties as for the previous full scale test series.

It was planned to increase the powder factor with 0.1 kg/m^3 per 25 mm increase of the drillhole diameter in order to obtain a similar fragmentation for the different blasts. This was, however, difficult to realise in practise.

Table 4 shows the planned and real powder factors for the various drillhole diameters and the amount of explosive per surface area of the wall of the hole.

The amount of explosive per surface area of the drillhole wall is a relative measure of the amount of detonation energy which is transferred from the explosive to the surrounding rock.

After blasting, the muck was crushed in a jaw crusher and a gyratory crusher to minus 70 mm, and then screened at 32 mm. The proportion of fines minus 32 mm for the different drillhole diameters was found to be:

76 mm diameter: 26.2 per cent
89 mm diameter: 31.1 per cent
102 mm diameter: 32.7 per cent
114 mm diameter: 35.6 per cent

Figure 1 shows the percentage of fines plotted as a function of the drillhole diameter. Seen the fact that this was a series of full scale blasts with all the variations which may occur both in explosives performance and rock quality, the results are good. There seems to be a linear correlation between the drillhole diameter (and the amount of explosive per wall area) and the proportion of fines.

This has earlier been found to be the case in a series of laboratory blasting experiments with small diameter holes. (Kristiansen, 1994b).

3 LABORATORY EXPERIMENTS

Two series of laboratory grinding tests have been done with seven different ore types in order to investigate whether the amount of blasting energy can influence the grindability of these rocks. (Nielsen, 1995; Nielsen and Kristiansen, 1995a; 1995b).

3.1 VOD and grindability

In the first series of experiments, grinding tests were done on four types of igneous rocks: A fine grained granite, a coarse grained monzonite, a very fine grained gabbro, and a very fine grained quartz diorite. These rocks are used for crushed aggregate production.

Solid boulders were picked from each quarry and cut by diamond saw to form cubes of 500x500x500 mm size. The cubes were quite homogenous without visible cracks. Three cubes were used for each blasting series, and were blasted one at a time inside a cylindrical steel chamber which contained the fragments.

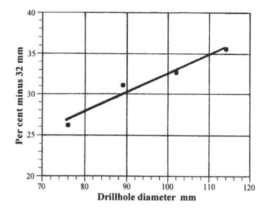

Figure 1. Percentage of fines minus 32 mm after two stage crushing.

TABLE 5. Fragmentation of blocks blasted with high VOD extra dynamite and low VOD dynamite.

ROCK TYPE AND EXPLOSIVE	k_{50}, mm	k_{80}, mm
Granite, dynamite	72	> 100
Granite, ex. dyn.	48	86
Monzonite, dynamite	73	> 100
Monzonite, ex. dyn.	36	66
Diorite, dynamite	100	>> 100
Diorite, ex. dyn.	65	99

All four rock types were blasted with a high velocity dynamite, Extra dynamite, with a measured VOD of around 6000 m/s.
In addition was three cubes each of granite, monzonite and quartz diorite blasted with regular dynamite with a VOD of about 3500 m/s.

The fragments from each block was collected and sieved after blasting, and it was found that all the blocks shot with extra dynamite had a finer fragmentation than the corresponding blocks shot with dynamite.

Table 5 shows the average sizes, k_{50}, and the k_{80} (80 per cent passing) sizes of the primary fragmentation for the three rock types which had been shot with both explosives.

Two samples of each rock types blasted with each explosive were prepared for the grinding tests. One set of samples were taken from the fragmented material minus 8 mm, and the other set was taken from the coarse fraction plus 20 mm. Each sample represents an average of the three blocks in each of the blasting series.

From the coarse fraction, all the fragments with a visible part of the sawn surface were picked out. This was done to assure that the coarse grinding samples came from the outer parts of the cubes furthest away from the centrally placed hole with the charge. It is reasonable to assume that these fragments had been exposed to the lowest level of blasting energy.

It is also reasonable to assume that much of the fine fraction minus 8 mm would originate from the inner part of the cubes, closer to the explosive, thus having been exposed to a higher level of explosive energy.

The coarse samples were crushed to minus 8 mm, and both the coarse and the fine samples were sieved into several fractions, and discarding the minus 2 mm material, they were remixed to make grinding samples with the same weight and size distribution.

All the samples were then dry ground for 10 minutes under identical conditions in a batch type ball mill.

After grinding, each batch was split and sieved with the finest sieve being 0.104 mm (150 mesh Tyler).

Figure 2 shows typical size distribution curves for the two coarse and fine samples of granite blasted with the high VOD Extra dynamite. The

curves show that the fine sample which was assumed to have been exposed to a higher level of explosive energy, were indeed easier to grind than the coarse sample.

The same holds true for all the other samples.

The results of each grinding test can be further evaluated and compared with the others by using Bond's Third Law of Comminution which states:

$$W = \frac{10 \times W_i}{(P_{80})^{0.5}} - \frac{10 \times W_i}{(F_{80})^{0.5}}$$

where:

W = Energy consumption, kWh/tonne
W_i = Bond's Work index, kWh/tonne
P_{80} = Product size, 80 per cent passing, micron
F_{80} = Feed size, 80 per cent passing, micron

Since all the fine/coarse sample pairs were ground for 10 minutes, it is assumed that the energy input, W, was the same.

The realtive Work index, W_i, can then be calculated from the sieve data using Bond's formula. The feed size, F_{80}, was 7460 microns for all the batches.

Table 6 shows the results of these calculations and also the ratio W_{if}/W_{ic} for the work indices of each corresponding pair of fine and coarse samples. The table further shows the percentage minus 0.104 mm, and the per cent increase of minus 0.104 mm material for the fine samples compared with the corresponding coarse ones.

It can be seen that all the fine samples had a lower Work index and were easier to grind than the corresponding coarse ones.

The results also show that the Work indices for the coarse samples are lower when the blocks had been shot with extra dynamite, compared with the blocks that had been blasted with the lower VOD dynamite.

The same is also true for the fine samples of granite and monzonite, but not for the quartz diorite.

The ratio W_{if}/W_{ic} is higher for all the sample pairs blasted with extra dyanmite compared with those shot with dynamite. This means that the difference in grinding resistance is smaller between the fine and coarse sample shot with the high VOD explosive.

The reason is most likely that dynamite with its lower VOD and lower accoustic impedence, will generate relatively   fewer micro cracks in the coarse material from the outer parts of the cube.

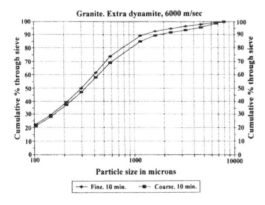

Figure 2.   Size distributions after grinding of granite blasted with extra dynamite.

TABLE 6. Results of grinding tests done on different rock samples blasted with low VOD dynamite and high VOD extra dynamite.

ROCK TYPE AND EXPLOSIVE	P_{80}	W_i	Per cent - 0.104	W_{if}/W_{ic}	Increase - 0.104
Granite, Fine, DYN	820	4.3	23.2		
Granite, Coarse, DYN	1120	5.5	21.2	0.78	9.4
Granite, Fine, XDYN	790	4.2	22.4		
Granite, Coarse, XDYN	960	4.8	21.3	0.88	5.2
Monzonite, Fine, DYN	1720	8.0	20.7		
Monzonite, Coarse, DYN	2940	14.6	19.0	0.55	8.9
Monzonite, Fine, XDYN	1560	7.3	21.3		
Monzonite, Coarse, XDYN	2600	12.5	19.0	0.58	12.1
Diorite, Fine, DYN	5150	42.8	15.7		
Diorite, Coarse, DYN	5970	74.5	12.7	0.57	23.6
Diorite, Fine, XDYN	5190	43.8	14.4		
Diorite, Coarse, XDYN	5900	70.5	12.5	0.62	15.2
Gabbro, Fine, XDYN	3710	20.8	21.5		
Gabbro, Coarse, XDYN	4710	33.7	19.3	0.62	11.4

TABLE 7. Distribution of the 8-6 and 6-4 mm fractions after crushing of coarse samples.

ROCK TYPE AND EXPLOSIVE	8-6 mm	6-4 mm	W_i
Granite, ex. dyn.	47.0%	53.0%	4.8
Granite, dynamite	49.4%	50.6%	5.5
Monzonite, ex. dyn.	52.1%	47.9%	12.5
Monzonite, dynamite	53.8%	46.2%	14.6
Diorite, ex. dyn.	54.5%	45.5%	70.5
Diorite, dynamite	55.2%	44.8%	74.5

This effect can also be seen in connection with the crushing and sieving of the coarse samples as shown in Table 7. There was a higher proportion of the 8-6 mm fraction for all the coarse samples which had been shot with dynamite. The grindability of the three rock types is also reflected by the increase of the 8-6 mm fraction percentages from granite to diorite.

3.2 Powder factor and grindability

The second series of tests were done on three different ore types using diamond drill cores of 63 mm diameter. The ore types were a low grade quartz banded iron ore (taconite), a nepheline syenite ore, and an ilmenite ore.

Two sets of cores, each of three pieces, were blasted in the blasting chamber using 10 gram PETN/m detonating cord fastened along the length of the cores. One set of cores was blasted with one length of PETN cord, and the other with two lenghths fastened diametrically along the cores.

After the cores had been blasted, each set was crushed to minus 8 mm, and grinding samples were prepared the same way as for the first test series. Grinding samples were also made from reference cores which had not been blasted.

The grinding samples of the heavier taconite and ilmenite ores were prepared in order to give approximately the same volumetric filling of the ball mill as for nepheline syenite. The feed size was 7460 micron for all the samples.

Each sample was then dry ground for 10 minutes in the batch ball mill.

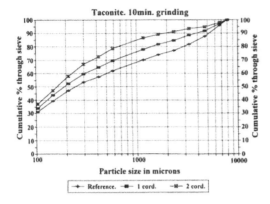

Figure 3. Size distributions after grinding of
taconite diamond drill cores blasted
with detonating cord.

After grinding, each batch was split and sieved
as in the first test series.

The curves on Figure 3 show the typical size
distributions of one of the grinding series consisting of taconite cores which had been blasted with
one or two cords respectively, and the reference
core.

The size distribution curves for nepheline syenite
and ilmenite are very similar.

The curves show clearly that the core samples
which had been blasted, were easier to grind than
the reference sample. Also, the core which had
been blasted using two detonating cords had a
better grindability than the sample which had been
shot with one cord.

The curves further show that the amount of fines
minus 0.104 mm will increase as the amount of
applied explosive energy is increased.

Again using Bond's formula, the results can be
summarised as shown in Table 8.

4 SUMMARY OF TEST RESULTS

The results of the industrial and laboratory tests
can be summarised as follows:

* Increasing the drillhole diameter and using the
  same explosive, will generate more fines after
  blasting and crushing
* Using an explosive with a high VOD will lead
  to more fines after blasting and crushing compared with an explosive with a lower VOD,
  keeping the same powder factor.
* Using an explosive with a high VOD will
  increase the grindability of the ore after blasting and crushing.
* Increasing the powder factor and using the
  same explosive, will increase both the crushability and the grindability of the ore.

5 OPTIMUM BLASTING

The results of the tests and experiments presented
above show that the amount of explosive energy
and how this energy is applied, will influence the
crushability and grindability of hard and competent
rock materials.

TABLE 8. Results of blasting and grinding tests on diamond drill cores.

ORE TYPE AND EXPLOSIVE	P_{80}	W_i	Per cent - 0.104	W_{bla}/W_{ref}	Increase - 0.104
Nepheline Reference	1090	5.4	22.8	---	---
Nepheline 1 cord	900	4.6	23.3	0.85	2.2
Nepheline 2 cords	730	3.9	25.5	0.72	11.8
Taconite Reference	2910	14.4	31.4	---	---
Taconite 1 cord	1420	6.7	34.0	0.47	8.3
Taconite 2 cords	730	3.9	37.2	0.27	18.5
Ilmenite Reference	5030	40.0	15.2	---	---
Ilmenite 1 cord	4630	32.2	15.5	0.81	2.0
Ilmenite 2 cords	3500	18.8	17.1	0.47	12.5

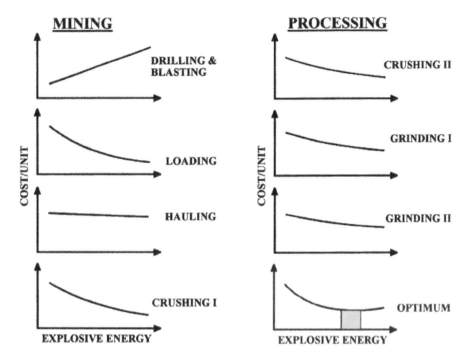

Figure 4.   Optimum blasting leading to cost minimisation.

The results are also consistent with industrial experience in Australian quarrying (Kojovic & al. 1995; LeJuge and Cox 1995), and mining of quartzite in Norway (Nielsen 1987).

Blast design should therefore aim at an economic optimum by balancing the use of chemical energy applied to the rock by the explosive, and the electrical energy applied in crushing and grinding.

5.1 Crushing and grinding

Looking at optimisation as a question of minimising costs, the concept of optimised blasting for a mining and processing operation can be illustrated as in Figure 4.

An increased level of applied explosive energy will lead to finer fragmentation and a larger number of micro cracks in the primary fragments. (Revnivtsev 1988).

Using the same drillhole diameter and explosive, an increased amount of explosive energy will necessitate higher costs for drilling and blasting.

Finer fragmentation will, however, result in higher productivity, less wear, and less maintenance for loading, hauling and primary crushing.

(MacKenzie 1966; Nielsen 1983; Nielsen 1986).

An increased amount of explosive energy will also lead to reduced costs for the the subsequent crushing and grinding as demonstrated by the investigations presented here.

The optimum blast design can be based on either minimisation of product costs or maximisation of operating profits. This design may, however, also lead to an increased capacity for the whole production system.

If this capacity can be utilised and the product(s) sold, the increases in revenues and profits may prove substantial, and this may in turn shift the optimum blast design towards even higher energy levels.

The industrial tests and laboratory experiments showed that crushability and grindability can be enhanced by adapting one of the following practices, or by a combination of them:

* Increase the powder factor.
* Increase the drillhole diameter.
* Use an explosive with a higher VOD.

5.2 Crushing and screening

Optimum blast design is not always a matter of minimising the costs for mining and subsequent processing by enhancement of the crushability and grindability of the ore.

A number of minerals will just be crushed and screened in several stages after blasting, and are then sold in sizes varying from 1 mm to several centimeters. Among such products are crushed aggregate, granulated materials for the production of refractories, and lump ores such as iron, manganese and quartz for smelting.

Too much fines often mean a loss of revenue for these producers, as for example in Norwegian hard rock quarries where the proportion of minus 4 mm material will typically be around 20% of the total production. Such an amount of fine material can often not be absorbed by the local market, and what can be sold, will fetch a price which is only half of that of the coarser fractions.

The generation of fines both during blasting and subsequent crushing is influenced by the blast design. A finely fragmented blast will yield a higher proportion of fines after crushing. But just reducing the powder factor in order to obtain a coarser fragmentation may not be the right solution.

Experience shows that a reduced powder factor which results in a coarser primary fragmentation, may also change the crushing results both in the secondary and tertiary stage.

Figure 5 shows how both the capacity and throughput fluctuations of the secondary crusher can change for two different blast designs in the same quarry. (Bearman 1995).

Diagram (a) shows the capacity of the secondary crusher with a powder factor of 0.65 kg/m^3, and (b) shows the capacity when the powder factor was reduced to 0.52 kg/m^3.

The crushing results were also different after the tertiary stage with 90% passing 20 mm using the higher powder factor, and only 78% passing 20 mm after the powder factor was reduced.

The industrial tests and laboratory experiments showed that the generation of fines during blasting and crushing can best be reduced by adapting one ore more of the following measures:

* Reduce the drillhole diameter.
* Better distribution of the explosive in the rock mass.
* Use an explosive with a lower VOD.
* Decoupling of the explosive in the drillhole.
* Dilute the explosive in the upper part of the drillhole.

5.3 Practical optimisation

Trying to pin-point an optimal blast design from an economic point of view, will be very difficult in actual operation because practical blasting will seldom give consistent results from one blast to the next.

This is partially due to varying rock conditions which will always influence the results, but more variation is probably caused by actual implemetation of the blast design.

The most common problem experienced in the field, involves the very real differences between blasting operations "as designed" and "as built". It is estimated that 60 per cent of current blasting problems are caused by such differences. (Scott 1992).

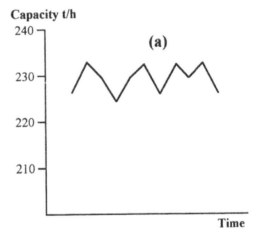

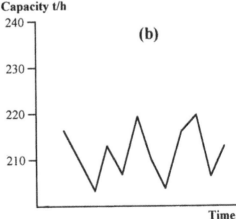

Figure 5. Changes of capacity and throughput in secondary crushing with two different powder factors: (a) = 0.65 kg/m^3, and (b) = 0.52 kg/m^3 (After Bearman).

276

Implementation differences should therefore be corrected as far as practically and economically possible before any attempt is made trying to optimise blasting. This can be achieved by using common Quality Assurance methods and techniques. (Nielsen 1993).

6 CONCLUSIONS

The potential gains which can be realised by total blast optimisation from solid rock to a marketable product, is most often overlooked. The reason is simply the way most mineral operations are organised. The excavation, transport, and primary crushing of the ore is the responsibility of the mining department, whereas the responsibility for secondary/tertiary crushing, grinding, and further treatment lies with the processing department. The two departments will try to make their own operations as cost effective as possible, a practice which may easily lead to sub-optimisation when looking at the whole production process.

Looking at blasting as the first step of an integrated comminution process should therefore constitute an important part of the overall analysis of the mineral production system.

REFERENCES

Bearman, R.A. 1995. Crushing plant performance - A function of blast fragmentation? Keynote presentation: 4th Nordic Aggregate Research Conference. Helsinki, Finland. November 1995.

Kojovic, T. & al. 1995. Impact of blast fragmentation on crushing and screening operations in quarrying. Proc: EXPLO 95 Conference: 427-436. AusIMM/SEE. Brisbane. September 1995.

Kristiansen, J. 1994a. Blastability of rock. Full scale tests at Åndalen quarry. Report No. 548095-3, Norwegian Geotechnical Institute. Oslo. June 1994. (In Norwegian).

Kristiansen, J. 1994b. Blastability of rock. Small scale experiments with rock cubes. Report No. 548095-4, Norwegian Geotechnical Institute. Oslo. December 1994. (In Norwegian).

Kristiansen, J. 1995a. A study of how the velocity of detonation affects fragmentation and the quality of fragments in a muck pile. Proc: EXPLO 95 Conference: 437-444. AusIMM/SEE. Brisbane. September 1995.

Kristiansen, J. 1995b. Blastability of rocks. Full scale blasting tests with different drillhole diameters. Report No. 548095-5, Norwegian Geotechnical Institute. Oslo. March 1995. (In Norwegian).

LeJuge, G.E. and Cox, N. 1995. The impact of explosive performance on quarry fragmentation. Proc: EXPLO 95 Conference: 445-452. AusIMM/SEE. Brisbane. September 1995.

MacKenzie, A. 1966. Cost of explosives - Do you evaluate it properly? Mining Congress J. Vol 54: 32-41. May 1966.

Nielsen, K. 1983. Optimization of open pit bench blasting. Proc: 1st International Symposium on Fragmentation by Blasting: 653-664. Luleå, Sweden. August 1983.

Nielsen, K. 1986. Optimum fragmentation in underground mining. Proc: 19th International APCOM Symposium: 746-753. SME. University Park, Pennsylvania. April 1986.

Nielsen, K. 1987. Recent Norwegian experience with polystyrene diluted ANFO (ISANOL). Proc: 2nd International Symposium on Fragmentation by Blasting: 231-238. Society of Experimental Mechanics. Keystone, Colorado. August 1987.

Nielsen, K. 1993. Safety versus production? Quality Assurance in blasting. Proc: FRAG-BLAST 4, Rock Fragmentation by Blasting: 487-494. Vienna. July 1993.

Nielsen, K. 1995. Laboratory blasting and grinding expriments on diamond drill cores. Internal report. SINTEF Rock and Mineral Engineering. Trondheim. May 1995. (In Norwegian).

Nielsen, K. & Kristiansen, J. 1995a. Can blasting enhance the grindability of ores? IMM Transactions, Vol. 104: A144-148. London. Sept.-Dec. 1995.

Nielsen, K. & Kristiansen, J. 1995b. Blasting and grinding - An integrated comminution system. Proc: EXPLO 95 Conference: 113-117. AusIMM/SEE. Brisbane. September 1995.

Revnivtsev, V.I. 1988. We really need revolution in comminution. Proc: XVI International Mineral Processing Congress: 93-114. Stockholm. June 1988.

Scott, A. 1992. A technical and operational approach to the optimization of blasting operations. Proc: MASSMIN 92: 247-252. SaIMMM. Johannesburg, 1992.

Rock Fragmentation by Blasting, Mohanty (ed.) © 1996 Taylor & Francis. ISBN 90 5410 824 X

On the characteristics of rock and energy under high strain rate

Yalun Yu
Research Institute of Mines, University of Science and Technology, Beijing, People's Republic of China
Jindo Qi
College of Resources Engineering, University of Science and Technology, Beijing, People's Republic of China
Mingte Zhao
Guangxi Institute of Geology, People's Republic of China

ABSTRACT: This paper describes the experimental procedures used to determine results and present the relationship between impact energy and specific surface area of the crushed rock. Furthermore the relationship between grain size and void ratio is formulated and the compressive strength of rock is discussed in connection with its grain size. The apparatus used in this study consists of the split Hopkinson bar, a pressure vessel and a hydraulic actuator. Six kinds of rocks were selected for the test. The experiments were made at the strain rate 10^3 sec^{-1} and under a confining pressure varying 0.1 to 100 Mpa.

1 INTRODUCTOION

As rock or the rock mass is subject to high — speed impact loading, such as drilling, blasting, crushing and earthquakes it fails within only 10^{-4} to 10^{-2} seconds. In this dynimc loading condition, prorerties of rock strength, deformation and failure will be considerably different from those in a static condition. These changes are not only due to the strain rate and applied energy, but also caused by the type and the structure of the rock. Therefore, it is of significance to both the theory of rock mechanics and the practice of rock engineering to study fracture properties of rock under high speed impact loading.

The paper describes the experimental procedures used to determine results and present the relationship between ipmact energy and specific surface area of the crushed rock.

Furthermore the relationship between grain size and void ratio is formulated and the compressive strength of rock is discussed in connection with its grain size.

2 EXPERIMENTAL PROCEDURES

2.1 Experimental apparatus

Fig. 1 illustrates the split Hopkinson bar (SHB) testing apparatus used in this study. Semiconductor eletric — resistance strain slices are mounted on both the in put bar and output bar to detect stress waves. These are recorded on the instantaneous wave storage unit and processed by a computer to determine the stress — strain curve under high speed dynamic loading.

2.2 Energy of the stress wave

The energy of incident stress wave W_I, the energy of reflected stress waves R_R, the energy of transmitted stress wave W_T and the absorbed energy W_L can be calculated by the folliwing equations:

$$W_I = \frac{A_b \cdot C_b}{E_b} \int_o^t \sigma_I^2(t) \cdot dt \qquad (2.1)$$

$$W_R = \frac{A_b \cdot C_b}{E_b} \int_o^t \sigma_R^2(t) \cdot dt \qquad (2.2)$$

$$W_T = \frac{A_b \cdot C_b}{E_b} \int_0^t \sigma_1^2(t) \cdot dt \qquad (2.3)$$

$$W_L = W_I - W_R - W_T \qquad (2.4)$$

where σ_1, σ_R, σ_T = the stress amplitudes of incident, reflected and transmitted waves respectively; A_b = the cross-sectional area of the elastic bar; C_b = the bar wave velocity; E_b = the elastic modulus of the bar; t = loading time.

2.3 Rock specimens

The rock used in this study are sandstones A and B, and Akiyoshi limestones. The latter are classified on the basis of grain size as fine, medium, coarse and coarser grained limestones. Their physical and mechanical properties are presented in Table 1.

3 TEST RESULTS

3.1 The relationship betwwen specific surface area and incident ecergy

The characteristics of crushed rock can be shown in terms of size distribution. In this investigation the size distribution reduced to the specific surface area of rock fragments. The distribution approxi-

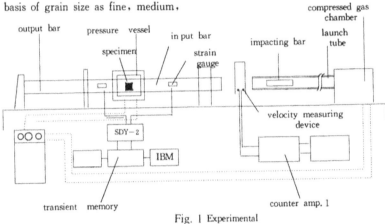

Fig. 1 Experimental apparatus

Table 1. Physical and mechanical properties of rock

rock	apparent specitic gravity (g/cm³)	true specific gravity (g/cm³)	porosity (%)	P—wave velocity (km/sec)	uniaxial compressive strength (kg/cm²)
SANDSTONE(A)	2.58	2.70	4.30	3.25	1493
SANDSTONE(B)	2.54	2.69	5.60	3.05	1235
AKIYOSHI LIMESTONE					
fine grained	2.64	2.72	2.94	2.92	630
medium grained	2.64	2.72	2.94	3.31	620
coarse grained	2.65	2.72	2.57	3.92	450
coarser grained	2.65	2.72	2.21	4.11	460

mately follows a Rosin — Rammler distribution curve. Therefore, the specific surface area can be calculated by the following equation:

$$S_V = \frac{\Phi}{D_e} 1.065e \frac{1.795}{n^2} \qquad (3.1)$$

where S_v = specific surface area; n = index of uniformity; D_e = characteristic szie factor; Φ = shape factor of specific surface area. D_e, n can be obtained from the Rosin—Rammler size distribution curve.

Fig. 2 shows the relationship between specific surface area (S_V) and incident energy (W_1). Form fig. 2 both the limestone and sandstone relationships are fitted by straight lines. Therefore, specific surface area increases linearly with the increase of incident energy, i. e., dW_1/dW_V = constant.

Since the absorbed energy of rock (W_L) is directly proportional to incident energy (Fig. 3), there is a similar relationship between the specific surface area (S_V) and the absorbed energy of rock.

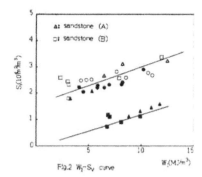

Fig.2 W_1-S_V curve

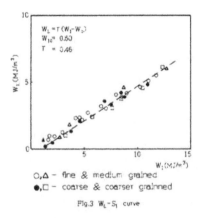

O,△ — fine & medium grained
●,□ — coarse & coarser grainned

Fig.3 W_L-S_1 curve

3. 2 The relationship between the mechanical property of rock and grain size

In order to examine the effect of grain size on the mechanical behavior of rock, the grain size of the limestone was determined by means of an optical microscope. Specimens where polished using 1000 # corrundum, and surfaces were regularly divided into square meshes measuring form 90 to 640 mm wide, according to size. Then, the grains that are contained in a square mesh are counted with an optical microscope, and the square area that is occupied by a grain is calculated. Finally the grain size is determined form the square root of the average area of a grain. Table 2 presents the results obtained form these measurements. Optical microscope photographs are shown in Fig. 4 and Fig. 5.

From Tables 1 and 2, the relationship between grain size and void ratio seems to be given by a negatively sloped function. Repressenting the particle size by x in mm and the void ratio by y in %, we obtain the following equation:

$$Y = 3.76 - 0.45x \qquad (3.2)$$

In general, the compressive strength of limestone is inverseiy proportional to grain size, i. e., the smaller the grain size, the higher the compressive strength.

This is mainly due to the fact that limestone is composed of calcite and that dislocation of calcite is arrested at the grain boundary.

Therefore the finer the grain size, the more crystal boundaries there are, and the larger the action to block dislocation movement. This results in increased compressive strength.

Talbe 2 Mean grain size

AKIYOSHI LIMESTONE	measured section area (mm²)	number grains	section area of particle (mm²)	equivalent size of grain (mm)	mean grain size (mm)
fine grained	2. 2 2. 2	280	0. 047	0. 31	0. 28
medium graind	20. 0 20. 0	472	0. 847	1. 04	0. 92
coarse grained	50. 0 50. 0	410	6. 098	2. 76	2. 76
coarsser grained	80. 0 80. 0	330	19. 394	4. 97	4. 97

Fig. 4 Fine grained limestone

fig. 5 Medium grained limestone

4 CONCLUSION

This paper describes the characteristics of rock under dynamic loading that is obtained by the SHB method. It is concluded that the specific surface area of limestone rock fragments of different grain sizes is proportional to the incident of absorbed energy. Furthermore, the effect of grain size on the compressive strength of limestone is discussed. The characteristics of rock under high−speed impact loading is of significace for the development of rock dynamic and engineering design of drilling, blasting,crushing and earthquake eangineering. In the near future, the experiment described in this paper will be perfomed in various other rocks.

REFERENCES

Lundberg, B. A. 1976. Split Hopkinson bar study of energy absortion in dynamic rock fragmentation.

   Int . J. Rock Mech. Min. Sci. &. Geomech. 13,187—197.

Rock Fragmentation by Blasting, Mohanty (ed.)© 1996 Taylor & Francis. ISBN 90 5410 824 X

Blasting as a comminution process: A useful tool for prediction of fragment size and explosive energy

U. Prasad & A. R. Laplante
Mining & Metallurgical Engineering, McGill University, Montreal, Que., Canada

B. Mohanty
Mining & Metallurgical Engineering, McGill University, Montreal & ICI Explosives Canada, McMasterville, Que., Canada

ABSTRACT: Both grindability and blastability may be seen as the energy spent in the process to get a desired product size from a specified sized rock mass. Four different rock types and a well documented case study of two blasts were selected for this study. The operating work index (calculated from the explosive energy spent per blast hole for the specified blast geometry) corresponding to each blast has been used to predict the product size, P_{80}, of the other. A method has been proposed to estimate the energy required in blasting. Feed size (defined by the blast geometry or the joint pattern, whichever is lower), F_{80}, desired product size, P_{80}, and the work index, determined in the laboratory by standard Bond rod mill test, are used as inputs in Bond's law of comminution. The explosive energy utilized has been correlated with the blast geometry, the desired products size and the rock type. The choice of feed size, F_{80}, is discussed in detail, and a case is made for the use of much a smaller value, the average of effective burden and spacing. An attempt has also been made to correlate laboratory determined work index with elastic and strength properties.

1 INTRODUCTION

In blasting the usual strength and elastic properties of the rock to be blasted are not always adequately measured. Furthermore, even if these properties were known, their direct applicability to the blast design, prediction of explosives consumption, and prediction of blast results, is open to question. One alternative approach would be to treat the blasting process from an energy balance point of view, as in comminution investigation. Explosive energy and blast fragment size have been estimated, albeit empirically, using comminution size distribution models by several authors (Kuznetsov 1973, Cunningham 1983, Da Gama 1971 and Just 1973). The present work deals with the estimation of the fragment size and the explosive energy using Bond's law (1952) of comminution. The concept of work index and its potential applications has been explored. Further, a method has been proposed to estimate the explosive energy and the fragment size by simulating blast geometry (feed size), fragment size (product size), explosive energy (total work input) and characteristic properties of rocks (laboratory determined work index or field determined operating work index).

1.1 Work index

The work index (W_i) originates from the Bond's theory of comminution:

$$W = 10 \ W_i \ (\frac{1}{\sqrt{X_p}} - \frac{1}{\sqrt{X_f}}) \quad (1)$$

where W is the total energy input, X_f is the feed size, and X_p the product size (both 80% passing). W_i is the work index (an intrinsic property of a material, relating energy input in kWh/st (1 kWh=3.6 MJ), required to break a given material from a theoretically infinite size to 80 % passing 100 micrometers, P_{80}. It is the proportionality constant and takes into account of material characteristics, method of size reduction and efficiency of the grinding operation. Bond used this index to model grinding circuits by assuming that an almost negligible change occurs in work index during grinding. Thus, the energy requirement for a material in a standard grinding mill (2.44 meter inner diameter overflow mill operating under a given set of standard conditions) can be predicted for a specified feed and product size.

Bond made use of separate bench-scale laboratory tests both for rod and ball mills. He determined the laboratory scale work index (W_i) by equating the work applied in the 2.44 meter mill to the number of revolutions to obtain the same size reduction. In contrast, the work index calculated from the mill based on the power drawn from the motor and the feed and product size, is known as operating work index (W_{op}). The operating work index during blasting is calculated by the explosive energy (kWh/t) spent in getting a desired product size with the assumption of feed size to be infinity (Bond 1954, Bond & Whitney 1959) and the 80 % passing size to be converted into micrometre. The ratio of operating work index, W_{op} to lab-scale work index, W_i is called the efficiency factor of the size reduction process (Rowland 1973).

1.2 Operating work index

The operating work index, W_{op} (kWh/t) of a blast can be calculated by knowing total energy spent, W (kWh/t), the feed size, F_{80} (μm), and the product size, P_{80} (μm). Equation 1 can thus be re-written as,

$$W_{op} = \frac{W}{10 \ (\frac{1}{\sqrt{P_{80}}} - \frac{1}{\sqrt{F_{80}}})} \quad (2)$$

The work index (W_i) is also used to represent the relative resistance of breakage of different materials. In the present work the work index (W_i) is used in this form, i.e. as a relative breakage behaviour. Its measured value takes into account, at least implicitly, strain energy (elastic and plastic energy), surface energy, kinetic energy (some of the kinetic energy is translated into heat, material and machine vibration, and sound generation), and finally material-material and material-machine friction.

1.3 Laboratory work index

The work index can be determined in different types of equipment such as ball mill, rod mill etc. depending upon scale of operation. The ball mill

test is used to measure the work index in finer sizes, 100 percent -3.36 mm (6 mesh), which is roughly the product of the rod mill. This size is too small to be well correlated with blasting energy requirements. The rod mill, however, with a feed size of 100 % passing 13.2 mm, is more suitable, and was our choice for the study.

This test is conducted in a standard laboratory batch mill of 30.5 cm diameter and 61 cm long with a wave type internal lining. It can be tilted 5° from horizontal on both sides to reduce preferential isolation of coarse material. The mill is charged with two 4.4 cm diameter iron rod weighing 6.5 kg each, and six 3.2 cm diameter rod weighing 3.5 kg each. The rod length used is 53.3 cm and the total weight of the system is 33.4 kg. The mill is rotated at a fixed speed of 46 revolutions per minute. The material used for the test should be less than 12.7 mm size and the volume of the sample to be 1250 cm^3 (Bond 1952). Tests can be conducted at all sizes ranging from 4.75 mm to 208 μm (4 to 65 mesh). The product from the mill is screened and oversize material is mixed with unsegregated original sample to make up the initial volume of 1250 cm^3. This new mixed sample is placed in the mill and is ground for the estimated number of revolution so as to achieve the 100 percent circulating load. The grindability of the material is calculated keeping the same volume of material in the mill. The average grindability of the last three cycles at steady state is used to calculate the work-index using a standard formula (Bond 1952).

2 PROPERTIES OF ROCK SAMPLES

The properties analyzed in this study were bulk density, unconfined compressive and tensile strength (Brazilian), and dynamic moduli of elasticity. The tensile and compressive strengths were measured by a servo-controlled stiff testing machine according to ISRM standards. The dynamic moduli were calculated indirectly using bulk density, and the measured longitudinal (P) and shear (S) wave velocities.

3 RESULTS

The physico-mechanical properties of the four rock types (Prasad 1994) are shown in Table 1 along with the respective work indices; where, W_i is the work index, σ_c the unconfined compressive strength, σ_t the tensile strength

Table 1: Properties of test rocks.

Item	Unit	Granite	Gneiss	Limestone	Marble
W_i	kWh/t	7.8 ± 0.1	10.8 ± 0.1	17.0 ± 0.2	19.2 ± 0.2
σ_c	MPa	129.9 ± 4.6	50.6 ± 24.8	167.4 ± 11.7	147.2 ± 104.8
σ_t	MPa	8.2 ± 1.4	5.6 ± 4.3	13.7 ± 1.5	15.5 ± 5.3
ρ	kg/m³	2643 ± 12	2789 ± 24	2606 ± 27	2864 ± 28
C_p	m/s	4170 ± 100	4770 ± 290	5300 ± 318	5430 ± 170
C_s	m/s	2670 ± 90	3080 ± 110	3250 ± 195	2830 ± 85
Y	GPa	43.44 ± 1.80	60.45 ± 4.45	65.98 ± 5.74	60.25 ± 3.06
K	GPa	20.84 ± 2.74	28.18 ± 8.12	36.50 ± 9.83	53.86 ± 5.62
μ	GPa	18.84 ± 1.27	26.46 ± 1.90	27.52 ± 3.32	22.94 ± 1.40
ν	-	0.15 ± 0.05	0.14 ± 0.09	0.20 ± 0.08	0.31 ± 0.02

(Brazilian), ρ the density, C_p the P-wave velocity, C_s the S-wave velocity, Y, K, and μ the Young's, bulk and shear modulus, respectively, and ν the Poisson's ratio. The attempted correlation of work index with normalized strength properties, seismic velocities and Young's, bulk and shear modulus are shown in Figures 1, 2 and 3, respectively. Altogether nine different rock properties were compared to the work index. The density and the compressive strength were found to be independent of work index, whereas, the tensile strength was found to be only slightly correlated. The W_i is found to have an approximately linear trend with P-wave velocity, but virtually no correlation with S-wave velocity. The moduli of elasticity calculated from P and S wave velocities, however, show greater correlation with work index. Overall the best correlation is obtained with measured bulk modulus as shown in Fig. 3. Although the reason for this is not known at present, it appears that the compression forces govern the fracture process inside a rod mill, and hence the improved correlation between work index and compressibility (or bulk modulus).

4 CASE STUDY

To compare the fragment size and specific grinding energy (calculated from Bond's law) with respect to blast results and explosive energy, a blast conducted in a limestone quarry was selected as a case study (Mohanty & Chung 1990). This is the same quarry from which the sample of the limestone was selected for the determination of the work index and related properties.

The quarry employs a single bench and each blast consisted of 36 holes, initiated with short period detonators in a staggered 3-row pattern. The burden, spacing, collar length and the bench height were 2.4 m, 2.7 m, 1.2 m and 6.3 m, respectively. Each 75 mm diameter borehole with no subgrade contained on average 22 kg of 65 mm diameter detonator-sensitive plastic-wrapped explosive. The coupling ratio achieved was about 92 %. The explosive types used were emulsions with varying aluminum contents. The calculated and measured detonation properties of the explosives are shown in Table 2.

The ratio of the total measured energy to the calculated 'ideal' energy (calculated from thermodynamic equations) denotes the 'efficiency' of the explosive energy release. This is in fact the 'actual' energy released by the explosive in that diameter. The blast results are summarized in Table 3.

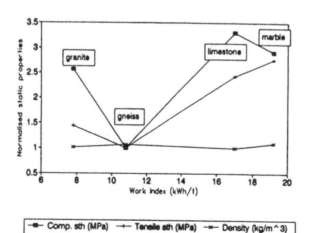

Figure 1: Work index and static strength properties for four rock types, plotted with those of gneiss as the reference.

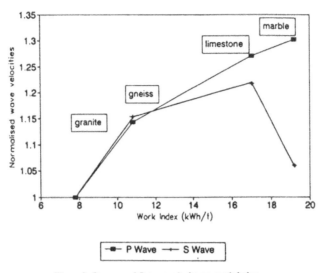

Figure 2: P-wave and S-wave velocity vs. work index.

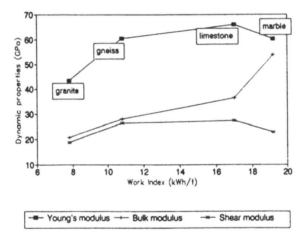

Figure 3: Work index and dynamic moduli for four rock types.

Table 2. Measured detonation properties of explosives.

Explosive	Emulsion #1 (0% Al)	Emulsion #2 (5% Al)
Density (kg/m³)	1120	1150
Bubble Energy (MJ/kg)	1.65	1.85
Shock Energy (MJ/kg)	1.11	1.24
Total measured energy(MJ/kg)	2.76	3.09
Calculated ideal energy (MJ/kg)	2.88	3.65
Energy efficiency (%)	96	85

Table 3. Blast results from case study.

Explosive	Emulsion #1	Emulsion #2
Maximum throw (m)	21	32
Height of muck pile (m)	5.0	3.8
Backbreak (m)	0.5	1.5
Face velocity (m/s)	8.0	11.3
Fragment size, P_{80} (m)	0.92	0.66

These represents the average of two blasts, with each explosive.

The size distributions from the resulting blasts were measured by photographic analysis by means of a 1.83 m x 1.83 m plastic grid (each grid measuring 0.61 m x 0.61 m), placed at random at numerous sites on the muck pile.

5 ENERGY UTILISATION

The energy breakdown in the test blasts in question is analyzed in terms of the amount of explosive per borehole (kg), powder factor (kg/t), theoretical ideal energy (MJ/t) and actual measured energy (MJ/t). These parameters are shown in Table 4. Since the blast geometry is the same for both the explosives (106.3 t of rock blasted per borehole for both), the explosive content per hole differs very little as the two explosives have virtually the same density. As a result the powder factor (kilogram of explosive spent in blasting per unit weight of rock) is almost identical (0.20 and 0.21 kg/t). However, the actual
energy density (calculated from the measured energy in MJ/t) comes out to be very different (energy densities of 0.55 vs. 0.64 MJ/t). It has to be noted that the explosive energy is responsible for the 'overall' blast results. These include, as Table 3 shows, fragmentation, and heave. Although, in a properly designed blast, a major fraction of the explosive energy is utilized in producing fragmentation, it is by no means the only usage of the explosive energy. In fact, a significant fraction of the energy is 'lost' through elasto-plastic deformation of rock, ground vibrations and air shock (from high velocity gases escaping through cracks and stemming ejection), and heating the surrounding rock.

5.1 Blast operating work index

In this section the concept of work index is applied to blasting for prediction of fragment size distribution for a given energy density. The operating work index of a blast can be calculated by knowing the total energy spent, W, the feed size, F_{80} (μm), and the product size, P_{80} (μm). The operating work index calculated for either blast can be used to predict the fragment size (80 % passing) for the other blast by assuming a simple global distributions. Similarly for the

Table 4. Energetics of the test blasts.

Explosive	Emulsion #1	Emulsion #2
Explosive (kg/hole)	21.36	21.93
Rock blasted (t/hole)	106.3	106.3
Powder factor (kg/t)	0.20	0.21
Ideal energy (MJ/t)	0.58	0.76
Actual energy (MJ/t)	0.55	0.64
" " (kWh/t)	0.15	0.18

known operating work index of the blast the explosive energy requirement can be estimated for different product size and blast geometries. Furthermore, the ratio of operating work index to the laboratory determined work index can also be used in evaluating the efficiency of a blast.

The analysis, however, depends entirely on an accurate measure or estimate of feed size. The assumption of feed size, F_{80}, as infinity as well as that guided by drilling and firing pattern have been examined as the two possible approaches. In the first approach the operating work index for the two blasts is estimated at 13.4 kWh/t and 13.1 kWh/t. This is approximately 25 % lower than the measures laboratory work index of 17 kWh/t for the same rock. Although, the results obtained with an infinite feed size are in good agreement with some previous work, (Bond 1954, and Bond and Whitney 1959), we know it to be incorrect as energy utilisation (and to some extent, fragmentation) would then be largely independent of actual blasting pattern and rock structure.

In the second approach the F_{80} is assumed to be the arithmetic average of the effective burden and effective spacing. For the case studies in question, these are 0.9 m and 7.1 m, respectively. The effective burden is taken to be 10 % larger than the actual distance between successive firing lines, perpendicular to the direction of throw. The additional amount is intended to incorporate the extent of backbreak behind the row of blast holes. The operating W_{op} comes out to be 25.7 and 22 kWh/t for the two blasts, respectively, which is 50 % and 30 % higher than the laboratory rod mill work index of 17 kWh/t. At this stage of the investigation, it is not possible to state categorically that one approach is better than the other. Both approaches ('infinite' feed size vs. average of 'effective burden and spacing') show significant discrepancy between laboratory W_i and blast operating work index, W_{op}.

Clearly, the proper choice probably lies somewhere in between. However, what is most encouraging and obvious is that both approaches are within reasonable range of energy use.

5.2 Estimation of explosive energy

In this section, a method is proposed to predict explosive energy requirements (in terms of MJ/t of rock blasted or kg of explosive per borehole) as a function of the operating work index of the blast. The explosive energy is calculated with reference to emulsion #2 by using its density and actual energy density factor as shown in Table 2. A finite feed size (the same exercise could be performed with an infinite feed size) equal to the average of the effective burden and spacing, 4.05 m, and an operating work index of 22 kWh/t calculated for second blast with emulsion #2 are used as inputs in Equation 1. Energy requirements in terms of kWh/t, MJ/t and kg/hole of emulsion #2 required for different product sizes, P_{80}, of the limestone rock type are is shown in Table 5. The required emulsion (kg/hole) shown in column 4 of the above table implies that the same blasting pattern (borehole diameter and burden-spacing) can not be employed for all desired product sizes (or, at constant product sizes, with rock types of different work index). It may be physically impossible to increase the charge weight in a borehole in a major way, as it is constrained by the borehole diameter and the bench height. In this case, one would have to recommend a larger borehole diameter or reduced burden and spacing, or a more energetic explosive. Of course, any change in explosive type or diameter may require modification to the energy balance shown in the Table 2. Increasing borehole diameter leads to more ideal behaviour from the explosive, which would affect the energy-size reduction relation.

Similar to the exercise shown in Table 5, the explosive energy (in terms of MJ/t or kg of emulsion per borehole) for a specific rock type (limestone in this case) and for a specified product size (P_{80}=0.66 m) can also be calculated for different blast geometries (assuming the same firing sequence). The average of the effective burden and spacing (feed size, F_{80}) for blast

geometries of 2 m X 2.5 m, 2.4 m X 2.7 m, 4 m X 4 m, 6 m X 6 m comes out to be 3.4 m, 4.1 m, 6.6 m and 9.9 m, respectively. The total energy required in terms of kWh/t, MJ/t and kg of explosives (emulsion #2 in the present case) is shown in Table 6. The variation in explosive energy per borehole for the above two cases, i.e. with respect to product size, P_{80} and feed size, F_{80}, are shown in Figure 4. The two curves represent the postulated variation of explosive consumption per hole for either product size of feed size, respectively, when the other is fixed. For example, for a product size of 0.66 m (P_{80}), if the effective feed size is increased by 50 % (from 4 m to 6 m), the weight of explosive in each hole must be increased by 15 % (i.e. from 22 to 25 kg). Similarly, if a 25 % finer fragment size is desired (0.8 m to 0.6 m), the explosive weight per hole would have to be increased by 25 % (i.e. from 19 to 24 kg). This applies to the test limestone in question; similar curves for other blast sites could be generated provided properly documented and monitored case studies of blasts are available.

Since the laboratory W_i differs significantly from the operating work index, the laboratory W_i alone cannot be used in predicting explosive energy requirement. Assuming the ratio of the operating work index of the second blast (calculated with F_{80}=4.05 m) to the laboratory work index, 1.3, to be the same for all rock

Table 6. Explosive energy (MJ/t or kg of expl./hole) for different feed sizes for P_{80} of 0.66 m.

F_{80}(m)	W(kWh/t)	W(MJ/t)	Emulsion (kg/hole)
3.4	0.129	0.600	20.6
4.1	0.137	0.640	21.9
6.6	0.158	0.733	25.1
9.9	0.171	0.795	27.3

Table 7. Explosive energy requirements (kg/hole) for different rock types (feed size, F_{80}=4.06 m; product size, P_{80} =0.66 m).

Rock Type	W_i(kWh/t)	W(kWh/t)	W(MJ/t)	Exp.(kg/hole)
Granite	7.8	0.082	0.294	10.2
Gneiss	10.8	0.113	0.407	14.9
Limestone*	17.0	0.178	0.640	21.9
Marble	19.2	0.201	0.723	27.2

(*: used as the reference)

Table 5. Explosive energy requirement for different fragment sizes for test limestone.

P_{80}(m)	W(kWh/t)	W(MJ/t)	Emulsion #2 (kg/hole)
0.80	0.150	0.542	18.6
0.70	0.169	0.609	20.9
0.60	0.192	0.692	23.7
0.50	0.222	0.800	27.4

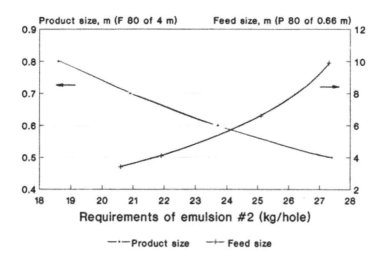

Figure 4: Estimate of explosive energy vs. feed and product size.

types, the energy requirement with reference to emulsion #2 can be calculated. Table 7 shows the required blasting energy for fragment size, P_{80}, of 0.66 m, for the other three rock types.

CONCLUSION

It has been shown that comminution properties established through laboratory studies may have significant application in predicting rock fragmentation by blasting. Despite the generalization in equating available explosive energy in a borehole with degree of fragmentation of the surrounding rock, the estimates of energy requirement appear within reasonable range for a well documented case study. Although a feed size based on the blast geometry has been proposed in this study. the issue remains unresolved until additional data becomes available. However, the common practice of assuming infinite feed size would be unacceptable for fragment size prediction in blasting.

There is a significant difference between work index determined in the laboratory for the test limestone and that deduced from the blasting case study. It raises the question of whether the blasting operating index is primarily a function of rock properties or blast geometries and rock structures. Furthermore, for the comminution approach to be useful, accurate information on fragment size distribution in a muckpile is critical.

This would require much higher precision and reproducibility than currently available with the various photographic techniques in use for assessing fragmentation.

ACKNOWLEDGEMENT

This work was supported by a research grant from the Natural Science and Engineering Research Council of Canada.

REFERENCES

Bond, F. C. 1952. Third theory of comminution. *Trans. SME/AIME*. 193: 484-494.

Bond, F. C. 1954. Which is more efficient Rock Breaker? *Eng. Min. Journal*, 155: No. 1, pp. 82.

Bond, F. C. & Whitney, B. B. 1959. The work index in blasting. *Quart. of Colorado Sch. of Mines*. Vol. 9, No. 3.

Cunningham, C. 1983. The Kuz-Ram model for prediction of fragmentation from blasting. *1st Int. Symp. on Rock Fragmentation by Blasting*, Lulea, 439-453.

Da Gama, C. D. 1971. Size distribution general law of fragments resulting from rock Blasting. *Trans. SME/AIME*. Dec. 314-316.

Just, G. D. 1973. The application of size distribution equation to breakage by explosives.

National Symp. on Rock Fragmentation.
Adelaide, 18-23.

Kuznetsov, V. M. 1973. The mean diameter of
fragments formed by blasting rock. *Soviet Min.
Sci.*, 9: 144-148.

Mohanty, B. & Chung, S. 1990. An integrated
approach in evaluation of blast-a case study.
Proc. *3rd Int. Symp. on Rock Frag. by Blasting*,
353-360, Brisbane, Aust. Inst. of Mines & Met
pp 353-360.

Prasad U. 1994. Energy utilisation in
comminution and its application to rock blasting.
M. Eng. Thesis, McGill University, Montreal,
Canada.

Rowland, C. A. 1973. Comparison of work
indices calculated from operation data with those
from laboratory test data. *10th IMPC*. London.

Controlling the patterns of fragmentation in blasting and mechanical crushing operations

V. P. Tarasenko
Moscow State Mining University, Russia

ABSTRACT: This paper offers a new approach to calculating a reproducible /guaranteed/ granulometric composition of the end products of rock fragmentation. This approach resulted in finding an original solution to the problem of determining specific energy needed for rock blasting and mechanilcal crushing operations.

The observable similarities in the curves describing the distribution of the fragments according to their sizes for different methods of rock disintegration may infer that there exists a certain logic for the formation of the fragments in these operations. The author believes that he found a mathematical approximation for that logic, based on the value of the parameter ξ, which determines sizes of the fragments:

$$\xi = \sqrt{\frac{W_o}{R_*}} = \left(\frac{\Delta}{q}\right)^{0,25} \sqrt{\frac{2n}{\lambda}} \qquad /1/$$

where:

W_o - an optimal line of the least resistance;

R_* - a radius of the zone of multiple disintegration; $R_* = 0,5 d\lambda$

d - a diameter of the drillhole;

Δ - density of the charge; $10^{-3} Kg/m^3$

q - a specific per unit charge; Kg/m^3

n - an abstact parameter, indicative of the linear scale of the fragmentation;

λ - an indicator of relative rock ductility;
$$\lambda = 6,5 - 0,5\left(\delta/\tau - 3\right)$$
δ, τ - respectively, rock strength limits for compression and for pushing.

Equation /1/ allows to give a material substance to and futher develop M.A.Sadovskiy's hypothesis to the effect that nature seems to give preference to certain sizes of fragments over the others when

rock is crushed. These preferred pieces have one more than trivial distinction: the increment of their sizes follows a geometric progression /referred hereafter as a preferred sizes sequence/ with the denominator of ξ.

Ordinary, this sequence may be presented in the following way:

$$d_*, d_*\xi^{1/2}, d_*\xi, d_*\xi^{3/2} \ldots d_*\xi^i \qquad /2/$$

It is worth noting that the value of parameter ξ, calculated for blasting disintegration, remains constant for all other stages of rock fragmentation. The sequence /2/ forming piece of d_* size changes its dimentions at each different stage of rock fragmentation. In blasting operations $d_* = 0,17d$, while in mechanical fragmentation and comminution $d_* = d_n/3\lambda$ where d_n is a nominal diameter of the piece at every stage of disintegration, corresponding to sieve perforation which passes 95 percent of the broken rock.

The average geometric d_{av} size /diameter/ of a blasted rock piece constitutes:

$$d_{av} = d_*\xi^{3/2} = 0,17d\xi^{3/2} \qquad /3/$$

In mechanical crushing and comminution its value will look like this:

$$d_{av} = d_*\xi^{3/2} = \frac{d_n}{3\lambda}\xi^{3/2} \qquad /4/$$

The equations below permit to determine, that is the probability of obtaining the total number of pieces with sizes less than d_i:

$$d_i = d_*\xi^{i/2} \left(i = 2n, \; i = 2n-1, \; i = 2n-2 \ldots i = 0\right)$$

where: $n \leq \lg (d_m / d_*) / \lg \xi^{\frac{1}{2}}$

d_m - the largest fragment of rock at a given stage of disintegration.

Thus the likelyhood of obtaining fragments of the sizes less than $d_{i=2n}$ is:

$$P_{-d_{i=2n}} = 1 - 1 / \left(\xi^{0,5} \right)^{2n}$$

For $d_i = 2n - 1$

$$P_{-d_{i=2n-1}} = P_{-d_{i=2n}} - 1 / \left(\xi^{0,5} \right)^{2n-1}$$

In general form, from $i = 2n$ to $i = 3$

$$P_{-d_i} = P_{-d_{i+1}} - 1 / \left(\xi^{0,5} \right)^{i} \qquad /5/$$

With $0 \leq i < 3$

$$P_{-d_i} = P_{-d_{i+1}} / \xi^{0,5} \qquad /6/$$

This evaluation of the probability P_{-d_i} of obtaining fractions based on the value of ξ , that is the sequence of the preferred fragments,is valid when the method of disintegration is best suited to the conditions of given rock.Therefore the granulometric composition,calculated on the basis of equations /2/,/5/ and /6/ should be treated as reproducible and guaranteed.

Only through an analysis and a comparison between compositions a reliable benchmark for choosing the most effective methods of disintegration can be established.

The correlation of a specific charge and energy released in the blast may be described as follows:

$$A = qU \qquad /7/$$

where: A - total specific energy needed for blasting;
U - specific energy of the charge.

The equation /7/ shows that the degree of precision in evaluating the energy required for a blasting operation largely depends upon the adequacy of the specific charge to the rock conditions.As for other types of disintegration,such as mechanical crushing and comminution,their energy needs are determined on the basis of the empirical criteria,proposed by Kirpichev-Kick,Rittenger and by Bond.These equations represent two-parameter functions, accounting in

different ways for the inpact of the initial and final fragment sizes upon specific energy consumed in the process of disintegration.

In contrast to mechanical crushing and comminution, we know much more about the interdependence in blasting operations between energy needs and rock conditions.The application of parameter ξ offers new ways for analyzing and predicting energy requirements for disintegration purposes.

Indeed,by combining equation /3/ and equation /1/ and by taking into consideration the value of /7/, we will get a new equation:

$$\frac{d_{av}}{d_*} = \left[\left(\frac{\Delta U}{A} \right)^{0,25} \left(\frac{2n}{\lambda} \right)^{0,5} \right]^{3/2} \qquad /8/$$

From equation /8/ it follows that:

$$A = \frac{4 \Delta U n^2}{\lambda^2} \left(\frac{d_*}{d_{av}} \right)^{8/3} \qquad /9/$$

Since $d_* / d_{av} = \xi^{1,5}$,then equation /9/ may be transformed into:

$$A = \frac{4 \Delta U n^2}{\lambda^2 \xi^4} \qquad /10/$$

In case of a given pair "charge-rock", number /10/ would look more convinient in this form:

$$A = K_d / \xi^4 \qquad /11/$$

where $K_d = 4 \Delta U n^2 / \lambda^2$ and represents a factor of proportionality.

Let us look now at the range of values for K_d starting with easy to blast, dry rock, where igdanite can be effectively used. It has on average a calculated volume concentration of energy of $3,4 \, MJ/m^3$

$$K_d = \frac{4 \cdot 3,4 \cdot 28,6^2}{6,5^2} = 263 \, MJ/m^3$$

With the average density of fragmented rock $\gamma = 2,8 \, t/r$ and taking into consideration that $1 kwh \doteq 3,6 \, MJ$ we will get

$$K_d = \frac{263}{2,8 \cdot 3,6} = 26 \, \frac{kwh}{t}$$

Similarly,for hard to blast rock with very high ductility / $\lambda = 2,7$/ and for powerful high explosives of $6,5 \, MJ/m^3$

$$K_d = \frac{4 \cdot 6,5 \cdot 28,6^2}{2,7^2} = 2917 \, MJ/m^3$$

provided that $\gamma = 2,8 \, t/m^3$

$$K_d = \frac{2917}{2,8 \cdot 3,6} = 289 \, \frac{kwh}{t}$$

The specialists know and use widely for the evaluation of the amount of energy required for rock mechanical disintegration Bond's index of crushing effort W_i. This index defines the volume of energy needed to turn a unit of rock, consisting of pieces of indefinite sizes $d_f = \infty$, into a condition when 80 percent of fragments have a diameter $d_p = 100 \, mic$

$$W_i = K_g \left(\frac{1}{\sqrt{d_p}} - \frac{1}{\sqrt{d_f}}\right) = K_g \left(\frac{1}{\sqrt{100}} - \frac{1}{\sqrt{\infty}}\right) = 0,1 K_g \quad /12/$$

where is a factor of proportionality, introduced by Bond.

Now, let us compare the values calculated above and the range of values for K_d with the published absolute values and the range of changes for coefficient K_b in rocks of various hardness.

Table 1 contains the relative data:

Category of hardness	Very hard	Hard	Mean	Soft
Coefficient of hardness	15	8-10	3-4	1-2
Bond's Index W_i (kwh/t)	19-29	12-17	7-10	< 5
$K_g = 10 W_i$	190-290	120-170	70-100	< 50

The comparison proves meaningful proximity both in the absolute values and in the range of changes for K_d and K_g. It should be noted however, that Index used as the basis for calculating K_g, is a convinient but rather an artificial and abstract criterium. The fragmentation /comminution/ of indefinite mass into pieces, 80 percent of which have sizes of 100 mcm, has nothing to do with the real technological processes of the rock mechanical disintegration. Neither Index W_i can be verified through lab tests using a double pendulous hammer.

Bond's superior engineering competence and his vast knowledge of experimental data on mechanical disintegration allowed him to offer an elegant empiric correlation between W_i and K_g. In contrast to K_g, indicator K_d was established analytically and equation /11/ permits to give a clear interpretation of its physical import.

Indeed, $K_d = A$ when $\xi = 1$, but $\xi = \sqrt{W_0 / R_*}$, that is $\xi = 1$, with $W_0 = R_*$ or when radii of the general and local breaking efforts have a same value. That means that indicator K_d represents the value of specific energy spent on rock disintegration in the zone of local application of the breaking effort and reflects the pattern of dissipative losses in that zone.

The energy released in an industrial blast impacts upon the rock in a mechanical way and for that reason the dissipation of the energy in the zone of local application of the breaking effort is identical both for blasting and mechanical operations. The indicator K_d can be easily calculated and in view of the analysis, referred to earlier in the paper, may be used instead of the empirical coefficient K_g in all its applications.

Index W_i retains its usefulness since it is a part of many empiric equations in the trade handbooks which are used as practical references for technical and economic evaluations in mechanical rock crushing operations.

The significance of indicator K_d lies in its function to be a basis for evaluating energy requirements both for blasting and mechanical rock disintegration operations. In this sense K_d may be treated as a universal indicator of rock pulverizability, as a technological parameter characteristic of rock resistance to forced fragmentation.

The introduction of K_d removes the existing uncertainty both in the meaning and in the quantative value of the pulverizability factor. Besides, both K_d and ξ are equally applicable to mechanical methods of disintegration which allows to use uniform approach towards analyzing and calculating energy needs and granulometric composition for blasting and crushing operations.

In view of the fact that energy requirements were determined using as a criterium of fragmentation median-sized pieces of the broken rock, the previously advanced statement will be true if we include into the calculation respective values for the said pieces. Correspondingly the basic Bond's equation /12/ should be augmented by the following: the median size of the rock entering the crusher is $d_f = d_{*(6)} \xi^{15}$, while on leaving the crusher it should be $d_p = d_{*(2)} \xi^{15}$. Here $d_{*(6)}$ and $d_{*(2)}$ are the first members in the sequence of the preferred sizes in case of blasting and initial stage of crushing respectively.

In keeping with equation /12/ and the comments above, let us establish the value of E_2 - energy requirements for initial rough crushers:

$$E_2 = K_d \left(\frac{1}{\sqrt{d_{*(2)}} \frac{1}{\xi} \cdot 10^3} - \frac{1}{\sqrt{d_{*(6)}} \frac{1}{\xi} \cdot 10^3} \right) \quad /13/$$

Transformations in the brackets will lead us to:

$$\left(\frac{1}{\sqrt{d_{*(2)}} \frac{1}{\xi} \cdot 10^3} - \frac{1}{\sqrt{d_{*(6)}} \frac{1}{\xi} \cdot 10^3} \right) = \frac{\sqrt{\frac{d_{*(6)}}{d_{*(2)}}} - 1}{31,62 \, \xi^{0,75} \sqrt{d_{*(6)}}} \quad /14/$$

Taking into consideration that $d_{*(6)} = 0,7 d$ while $d_{*(2)} = d_{n(2)}/3\lambda$ as seen in equations /13/ and /14/, we will have:

$$E_2 = K_d \frac{\sqrt{\frac{0,5 \, d\lambda}{d_{n(2)}}} - 1}{31,62 \, \xi^{0,75} \sqrt{0,17 \, d}} \quad /15/$$

The energy needs E_s for a second stage crusher when it deals with median sized pieces d_s on the input, equal to median fragments exiting the first stage of a rough crusher, that is when $d_s = d_{*(6)} \xi^{1,5}$ will amount to:

$$E_s = K_d \left(\frac{1}{\sqrt{d_{*(s)}} \frac{1}{\xi} \cdot 10^3} - \frac{1}{\sqrt{d_{*(2)}} \frac{1}{\xi} \cdot 10^3} \right) /16/$$

After substituting values for $d_{*(s)} = d_{n(s)}/3\lambda$; $d_{*(2)}$) and further transformations we will get:

$$E_s = K_d \frac{\sqrt{\frac{d_{n(2)}}{d_{n(s)}}} - 1}{31,62 \, \xi^{0,75} \sqrt{\frac{d_{n(2)}}{3\lambda}}} \quad /17/$$

Performing the same procedures, we can determine E_f energy requirements for final stage crushers:

$$E_f = K_d \frac{\sqrt{\frac{d_{n(s)}}{d_{n(s)}}} - 1}{31,62 \, \xi^{0,75} \sqrt{\frac{d_{n(s)}}{3\lambda}}} \quad /18/$$

CONCLUSION:

The meaning of this paper is in that it offers a new approach to calculating a reproducible /guaranteed/ granulometric composition of the end products of rock fragmentation. This approach resulted in finding an original solution to the problem of determining specific energy needed for rock blasting and mechanilcal crushing operations.

It was found that so called Bond's Third Law of rock fragmentation, established empirically, represents a singular case / $\xi = 1$; $K_g = K_d$ / within the proposed general approach. But since that singular case is supported by the analysis of the vast empirical data and is widely recognized and used, its very concurrence with the new general approach, described in this paper, proves the validity of this approach. The structure and physical meaning of new parameter K_d give grounds to treat it as a universal indicator of rock pulverizability. As such it helps to overcome the drawbacks inherited in index W_i, namely the uncertainties and limitations of its application for the evaluation of rock resistance to crushing and comminution. In contrast to $K_g = 10 W_i$, parameter K_d takes into account the granulometric composition of the end products and kind of the machinery used. The ideas about the preferred sequences in the fragments of disintegration /for different methods/ and the determination of the median geometric sizes /diameters/ of the fragments, based on $d_{av} = d_* \xi^{1,5}$ permits to drop parameter d_{80}, the truth of which has been long suspected. The instruments for calculating the preferred sizes of the fragments in disintegration processes as well as the sums of the granulometric compositions and the specific energy requirements open new and promising perspectives for improving efficiency in rock blasting and crushing operations.

REFERENCE:

1. Tarasenko V.P. Physic-technical base of charge calculations in open pits. MSMU, 1985.
2. Sadovsky M.A. Applied seismology of century last years. Physic of Earth, N2, 1992
3. Tarasenko V.P. Cuanting of piece dimention and granulomethry in blasting. Reports of Academy of Science, 1990, Vol. 315, N2, p. 323-326.

Rock Fragmentation by Blasting, Mohanty (ed.) © 1996 Taylor & Francis. ISBN 90 5410 824 X

The importance of velocity of detonation in rock blasting – A full scale study

Magnus Gynnemo

Department of Geology, Chalmers University of Technology, Göteborg, Sweden

ABSTRACT: There is a need to identify which parameter that governing the amount of fines (0-8 mm) produced in the bench blasting process. The objective of the study was to investigate how velocity of detonation effects the rock fragmentation in different types of crystalline bedrock. This paper describes a series of full-scale bench blasting experiments in two crystalline bedrocks, granitic gneiss and dolerite. The two rock types, widely used as aggregates, where chosen because they differ in petrology and mechanical properties. In the test, we use an explosive with two different kinds of sensitizers, glass microspheres (Emulan 7500, VOD=5300 m/s) and expanded polystyrene spheres (Emulan 7500, S VOD=4300 m/s). It was found that a lower velocity of detonation may not result in coarser fragmentation because the geology seems to be the dominant factor in the fragmentation process.

1 INTRODUCTION

The main problems today with bench blasting operations are the excess of fines (0-8 mm) produced in the fragmentation process. One percent reduction of fines will save a million SEK/year for a major producer of aggregate.

SveBeFo (Swedish Rock Engineering Research Foundation) started a research project, called SveBeFo 440, to find new methods to control the fragmentation process and improve the quality of the crushed aggregates.

One of the main objectives of the SveBeFo project "Fragmentation of Aggregate Rock", planned by a steering group during 1992 (Andersson, Ronge and Svensson, 1992), is to identify the parameters that govern the amount of fines (0-8 mm) produced during normal bench blasting operations.

The first part of the project is to investigated how velocity of detonation of the explosive effects the fragmentation results in two crystalline rock types. We decreased the velocity of detonation (VOD) value without changing any other explosive properties. To accomplish that, we chose an AN-based bulk emulsion explosive Emulan 7500 and Emulan 7500 S. Emulan 7500 uses glass microspheres, as sensitizer, and Emulan 7500 S uses expanded polystyrene spheres (Nitro Nobel AB, 1993). The

results from these tests give us better understanding and data for the second part of the project now in preparation.

2 FIELD EXPERIMENTS

The project is a full scale study, and so far 12 test blasts in two normal sized quarries in western Sweden have been investigated (Table 1).

We started in the Precambrian granitic gneiss with

Table 1 Test blast design.

Data	Site 1	Site 2
Rock type	Precambrian granitic gneiss	Permian dolerite
Bench height (m)	10	18
Holes per round	18	24
Rows per round	3	3
Hole inclination,deg	15	12
Hole diameter, mm	76	76
Burden, m	2.8	2.5
Spacing, m	3.3	3.0
Stemming, mm	4-8	4-8
Detonators	Nonel Unidet	Nonel U.

6 test blasts during 1993. The project continued during 1994 with six field tests in a Permian dolerite. The test blasts were monitored concerning:

1. geology: structural and geological mapping.
2. bench geometry; topographical mapping and bench height.
3. drilling: bit wear, hole position, hole depth and in-hole deviations.
4. rock mass permeability by water loss measurements in blast holes.
5. charges: type of explosive, density, specific charge, stemming material.
6. breakage: VOD measurements and high speed filming of bench face.
7. post blast: profile and swelling of muck pile, fragment size distribution (screened off before primary crusher).

Rock mechanic parameters (Brook, 1993) were measured with test equipment called Bemek Rocktester (Röshoff, 1992) on core samples taken from representative blocks in the muckpile.

Strength and shape properties were measured on crushed core material according to a project "Rock Material" conducted at the Department of Geology at Chalmers University of Technology (Högström, 1994).

3 SIZE DISTRIBUTION

3.1 *Muckpile*

Figure 1a-b shows the comparison of the average distribution curve of 34-36 sieving results (0-30 mm) received from three test blasts using the explosive with high VOD (Emulan 7500) with the average distributions curve of 34-36 sieving results from three test blasts using the low VOD explosive (Emulan 7500 S) in the two quarries. Results from the granitic gneiss and from the dolerite quarry show that velocity of detonation seems to have little or no effect on the fragment size distribution. The low VOD explosive resulted in less than 1% reduction of the 0-30 fraction after blasting.

3.2 *Rockmass*

To evaluate quantitatively the size distribution curve in-situ of a rockmass before blasting, we used a simple and cheap photographic method. The

method is based on photographs taken perpendicular to an important joint system in the rockface. The block fragments in the picture are digitised as polygons and a PC computed the polygon areas. By converting the polygon areas to a volumetric distribution, it will be possible to compare the size distribution in the rockmass with the average size distribution curves from the muckpiles. The average size distribution curves (figure 2a-b) for the two test sites included three test blasts using Emulan 7500 and three with Emulan 7500 S.

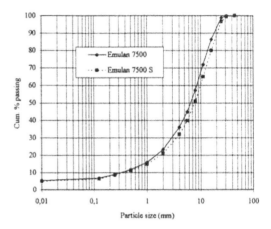

Figure 1a. Comparison of the Velocity of detonation (VOD) effect on 0-30 mm fraction from muckpiles in the granitic gneiss quarry.

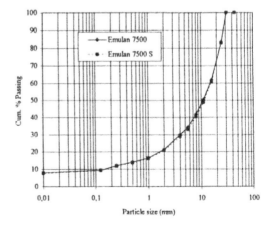

Figure 1b. Comparison of the velocity of detonation (VOD) effect on 0-30 mm fraction from muckpiles in the dolerite quarry.

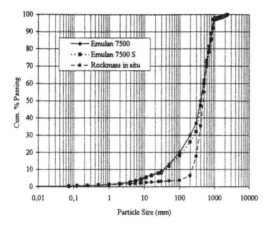

Figure 2a. Comparison between size distributions curves from the rockmass in-situ and from sieved rounds in the granitic gneiss quarry.

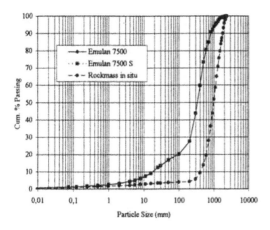

Figure 2b. Comparison between size distribution curves from the rockmass in-situ and from sieved rounds in the dolerite quarry.

Figure 2a-b shows how the rockmass reacted differently in the two quarries. The difference in the fragmentation done by the explosives in the two quarries may be a result of the higher blast hole confinement in the dolerite, or by geological differences between the two test sites, such as fracture density and mechanical properties.

4 FIELD TEST PROBLEMS

A full scale test conducted in a heterogeneous crystalline rockmass is a task with interesting and difficult problems to elucidate. During the field tests we identified a list of problems either of sientific or practical nature:

4.1 Hole measurements

It is difficult to
1. drill the blast holes according to theoretical drill hole pattern
2. avoid blast hole deviations.
3. repeat the test blast design in more than one quarry.

4.2 Charging

It is difficult to
1. measure the amount of explosive in each blast hole due to leakage in the bore hole.
2. know if the water content in the bore hole effects the VOD in the explosive.

4.3 VOD measurements

It is difficult to
1. obtain velocity of detonation values because the rockmass contained many open joints and explosion gases disturbed (dead-pressed) measurements in the surrounding holes (explained by Hanasak et al., 1993).
2. get reliable values. The explosive with low velocity of detonation showed a detonating behaviour with both high and very low velocities, 400-5000 m/s.

4.4 Screening

It is difficult to
1. screen a muckpile of 12000 metric ton in different fractions.
2. measure with accuracy the weight of the different fractions in the muckpile during a long time period.

4.5 Laboratory testing

It is difficult to find a representative sampling method.

4.6 *Evaluation*

It is difficult to

1. control and explain the scatter in the field test results.
2. control the explosive performance in the fragmentation process.

5 CONCLUSIONS

The results from extensive surveys in two quarries in western Sweden have been used to assess the influence of velocity of detonation on the fragmentation process. The main conclusions of this work (Gynnemo, 1995) are: 1. The velocity of detonation (VOD) has insignificant influence on the fragment size distribution in the two tested rock types.

2. The reduction of VOD has negative quality effects on the granitic gneiss but the quality ameliorated in the dolerite (Gynnemo, 1995).

3. Explosives can differ in efficiency depending on

the geology (Hagan, 1993). The fragmentation work done by the explosive is more pronounced in the dolerite compared with the granitic gneiss.

4. Geology and fracture density appears to be two important factors in the fragmentation process.

6 ACKNOWLEDGEMENTS

The author gratefully acknowledges my colleagues at Department of Geology, Chalmers University of Technology, Göteborg. Supervisors have been Professor Gunnar Gustafson and senior lecturer Bo Ronge .

I also wish to thank Mathias Jern and Petter Tyrenius, geology students, who performed the in-situ rockmass curves.

REFERENCES

Andersson, P., Ronge, B & Svensson, U., 1992. project plan "BeFo 440 Fragmentation of aggre gate rock" version 5. Unpublished document (in Swedish).

Brook, N., (1993). The Measurement and Estima tion of Basic Rock Strength. Comprehensive Rock Engineering, Ed. J. A. Hudson, Pergamon Press, Oxford, V. 3, 1st ed., 41-81.

Gynnemo, M., 1995. "The importance of velocity of detonation in rock blasting" Blast News (Nitro Nobel: Sweden), No 3, Oct. 1995, pp. 32-35. (in Swedish).

Hagan, T. N., 1993. The importance of some per formance properties of bulk explosives in rock blasting. Paper in Proc. of the Fourth Symposium on Rock Fragmentation by Blasting, Rossmanith (ed). Balkema, Rotterdam.

Hanasaki, K., Terada, M., Sakuma, N., Yoshida, E., and Matsuda, K., 1993. Studies on the sensitivity of dead pressed slurry explosives in delay blasting: Paper in Proc. of the Fourth Symposium on Rock Fragmentation by Blasting, Rossmanith (ed). Balkema, Rotterdam.

Högström, K., 1994. A study on strength parameters for aggregates from south-western Swedish rocks. Department of Geology, Chalmers university of technology and University of Göteborg, Publication. A 76. Göteborg. 74 p.

Nitro Nobel AB, 1993. Marketing Handbook (in Swedish).

Röshoff, K., 1992. Rock Tester Version 1.1 - User Manual.

Rock Fragmentation by Blasting, Mohanty (ed.)© 1996 Taylor & Francis. ISBN 90 5410 824 X

Changes in fragmentation processes with blasting conditions

S. Bhandari

Mining Engineering Department, J.N.V. University, Jodhpur, India

ABSTRACT: Simultaneous study of fragmentation, displacement and ground vibrations from a series of blasts on a reduced scale in a granite bench 1.5 m at burdens of 0.5 m, 0.75 m and 1.0 m and variable spacings showed that the utilsation of explosive energy changes with the blasting parameters. Small scale blasting tests also indicate with the changes in orientation of joints, fragmentation changes substantially with development of cracks according to joints.

1. INTRODUCTION

When an explosive charge, confined within a blasthole is initiated, reactions take place resulting in production of large amount of gases at very high temperature and pressure in a very short time. An important characteristic of high explosives is the production of very large amount of energy per unit of time. The gas pressure acts on the walls of the hole and thus subjects the media beyond the hole to vast stresses and strains.

More than 300 years ago Vauban indicated that rock was broken due to lifting action of gases produced in blasting. His views with some modifications were believed to be true till the later part of 1950's. Hino (1956a and b) and Duvall and Atchison (1957) indicated that rock is broken mainly by reflection of shock wave that travels outward spherically from the charge.

However, later workers indicated that it was the joint action of gaseous pressure and shock waves which was responsible for rock fracture and fragmentation under the action of explosives (Saluja, 1962; Kutter 1967; Persson et al., 1970; Kutter and Fairthurst, 1971; Bhandari, 1975; Bhandari, 1979). A widely accepted process of energy utilisation and fragmentation process is still lacking. Further, with the change in blasting conditions changes in rock fragmentation processes occur. In the investigations presented here it is shown that with the change in blast conditions like burden and spacing, orientation of joints some changes in the relative roles of different mechanisms involved in rock fragmentation processes occur.

2. PRESENT UNDERSTANDING OF ENERGY UTILISATION & FRAGMENTATION PROCESS

As a result of detonation of explosive energy is released some of which is useful and much of it is wasted. Useful part of the energy is capable of doing the work of fragmentation and displacement and wasteful part of energy can be put to no work, which causes many harmful effects.

Any one mechanism, as to how the useful energy is utilised is not adequate to explain the whole process of fragmentation involved in various conditions and material types. The energy evolved on detonation of explosives is utilised in the fragmentation process by two groups of mechanisms (Fig. 1). First, a stress wave of extremely short duration results from the detonation of the explosive; it is followed by quasi-static gas pressure generated by the gas products of explosion. A small zone of crushed rock is created immediately surrounding the hole, on detonation of explosive. Intensity of crushing and fracturing decreases as the distance from the hole walls increases till it reaches the transition zone beyond which other effects occur. The stress pulse propagates as cylindrical or spherical wave into the surrounding rock and induces besides the radial compressive stress, a circumferential tensile stress

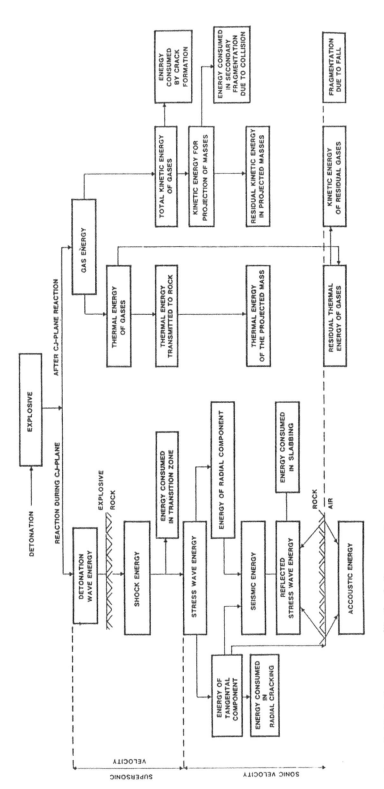

Figure 1. Explosive energy utilisation in fragmentation processes.

around the borehole. As this stress exceeds the tensile strength of rocks a pattern of radial fractures is created. As the stress wave travels outwards from the borehole, its amplitude is rapidly attenuated, so that after some distance no further crack initiation and eventually no crack propagation can occur. If, however, this stress pulse reaches a free surface, it is reflected from there and its originally compressive radial component is reflected as tensile stress. This newly generated tensile stress may be of sufficient magnitude to exceed the tensile strength of the rock, and this results in surface parallel scabbing or spalling of the rock. Multiple reflections of outgoing and reflected waves occur while fracturing takes place, dictating flaw initiation sites.

As a result of quasi-static gas under high pressure acting in the widened borehole and on the surfaces of the radial fractures, it causes further propagation of the cracks. The gases also find their way into the stress induced radial fractures. In addition, flextural failure may occur at the surface, when the layers between cavity and free surface are bent outwards by the expanding gases. The resulting rock fragments are finally pushed outwards and ejected. During ejection process there is some consumption of energy in the collision of fragments and further fragmentation takes place.

The explosive detonation also produces energy which does not in itself, lead to fragmentation and does no useful work during blasting operations. This energy can be called as waste energy which finally yields accoustic energy, thermal energy in the fragmented mass and released gases, light energy and seismic energy.

How much is the role of each of these mechanisms is not yet certain. Partly this uncertainity is because roles of different mechanisms are affected with the changing blasting conditions. In the practical blasting, conditions varied are the blast parameters, rock parameters and explosive parameters. Some important parameters which influence blasting results are known to be the burden, spacing, orientation of joints in the rock. Investigations were carried out to examine the influence of the variation of these parameters. In the literatature simultaneous study of fragmentation, displacement (throw) and seismic energy has been lacking, therefore, reduced scale tests were carried out in which size of fragments in terms of boulders, their throw, quantity of rock broken and the peak particle velocity generated were determined.

3. CHANGES IN THE PROCESS WITH BLAST PARAMETERS

A series of blasting experiments (Bhandari, 1990) on a 1.0 m high bench of joint free granite rock were carried out. In each test four holes in single row were drilled with jack hammer at a constant burden with variable spacing to burden ratios of 1, 2 and 3. The burdens used were 0.5 m, 0.75 m and 1.0 m. Explosive used was cartridged slurry of 25 mm diameter with a constant quantity at 0.5 kg in each hole. All the holes were simultaneously fired using detonating cord. Before each blast the face was straightened by carrying out smooth blasting.

Measurements were made for the volume of rock broken by taking profile of the face before and after blasts. The volume of ten largest boulders were determined while taking into account their shape and also their throw distance was measured. Vibration measurements were made 20 m behind the blast and in each case peak particle velocity was measured. Maximum distance to which any of the ten largest boulders was thrown was determined and sum of throw of the ten largest boulders was determined.

The results of bench blasts (Table 1) showed that the smaller burden gave smaller volume of rock in small sized boulders but maximum throw distance was very large with comparatively less ground vibration levels (Fig. 2a). At intermediate burdens volume of rock broken was large only at larger spacing, the size of large boulders was small, the sum of throw distances of ten largest boulders was fairly large but vibration levels were less (Fig. 2b). At the largest burdens volume of rock broken was very large, with large size boulders thrown to a small distance with large vibration levels (Fig. 2c).

These tests showed that with respect to fragmentation, the results were similar to as that for small scale tests. It can be deduced that the utilisation of explosive energy changes for the same rock and the same explosive with the change in blasting parameters. This is on account of quantitative changes in the role of different mechanisms involved in blasting. At smaller burdens stress wave energy is better utilised in causing scabbing fractures as well as in increasing microfractures in the rock. At small burdens gas energy does not get enough time to cause radial cracks and their bifurcation, before it is vented. Further, stress wave transfer as seismic energy is reduced thereby causing reduced amount of

Figure 2 Reduced scale test showing results (a) with small (0.5 m) burden (b) with   intermediate (0.75 m)
burden (c) with large (1.0 m) burden

Table 1: RESULTS OF BLASTING TESTS WITH VARIABLE BURDEN/SPACING

Blast No.	Burden m	Spacing m	Total Breakage Volume m³	Vol.of Ten Largest Boulders m³	Max. Throw m	Σ Throw Distance m	Σ Throw Index	Peak Particle Velocity mm/s
AF 3	0.50	0.50	2.34	1.31	200	164.5	12.90	42.5
BF 2	0.50	1.00	2.78	0.84	75	156.0	14.50	47.5
AF 10	0.50	1.50	3.80	0.63	40	149.0	12.54	60.0
AF 4	0.75	0.75	2.81	1.75	100	257.0	40.86	75.0
BF 1	0.75	1.50	5.10	3.61	35	136.5	65.50	62.5
AF 6	0.75	2.25	6.75	1.64	17	98.5	22.62	47.5
AF 5	1.00	1.00	4.40	3.74	20	89.0	34.01	70.0
BF 3	1.00	2.00	8.80	7.74	3	27.0	22.08	87.5
AF 1	1.00	3.00	12.50	10.99	3	12.5	22.67	100.0

vibrations. The gas energy is utilised in causing greater throw.

At larger burdens the role of stress waves is considerably decreased in fragmentation and the waves are transferred to the surrounding rock as seismic wave and thereby causing increased vibrations. Rock broken is coarser because the role of stress waves is decreased. Further, the throw distance of the large boulders is reduced indicating that the gas energy is mainly utilised in causing radiating cracks.

At the intermediate burdens the boulder size is not very large though the quantity of rock broken is high. The sum of throw distances of ten largest boulders is very high indicating that fragments have been thrown to greater distances, thus the gas energy is not being wasted but utilised in displacement. Vibration levels are lesser. This suggests greater utilisation of stress wave energy in fragmentation.

The effect of increased spacing is that each hole is breaking separately and the closer spacing results in enhancement of seismic energy resulting in increased vibration and airblast.

4. FRAGMENTATION PROCESSES IN JOINTED MEDIA

In practice blasting has to be accomplished in rock mass having fractures, beddings and/or joints. Rock formations as they occur are not homogeneous and

isotropic and even on small scale the homogeneity varies. The structural control has a considerable influence on the geomechanical and dynamic properties of the rock formations. The strength of rock mass decreases with the increase in frequency of joints and the deformability of rocks depend on their orientation. It is the interaction between the rock mass and stresses generated due to explosive detonation which may produce favourable or harmful blasting results. Sometimes the joint planes add to the performance of explosive induced fragmentation mechanism.

In the presence of joints, interaction between joints and related mechanisms for rock breakage take place and rock mass loosening also occur. In some situations the role of stress waves in jointed rocks dominates the fragmentation phenomena. In some other situations, the role of quasi-static gas pressure is to aid the fragmentation of weak jointed formations. The orientation of joints change the growth of fractures and hence fragmentation.

The relationship between orientation of joints and some blast parameters have been studied on laboratory scale by Bhandari (1983), Bhandari and Badal (1990) and Badal (1990). Unlike in field blasting situations where variations of rock and its discontinuities are not always apparent either on the top or at the face of the bench, in the laboratory scale tests joints were placed in known directions. Figure 3a shows a block with horizontal joints which has been blasted and Figure 3b show top view of a

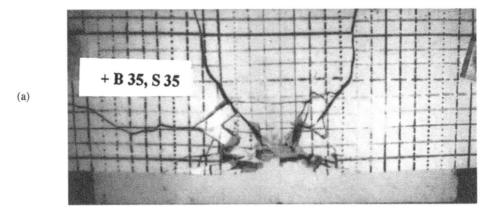

(a)

(b)

Figure 3 (a) Front view of blasted block of horizontal joints
         (b) Top view of reassembled fragments for horizontal joints (showing fracture pattern)

blocks with horizontal joints where fragments have been reassembled. Based on the fractures present, the fragmentation process can be explained as : the detonation of explosive inside the hole, formed a crushed zone at the hole periphery and formation of radial cracks resulted from the blasthole walls. The propagating stress waves, were responsible for the extension of radial fractures to greater length, which also developed in the block behind the hole. In all slabs forming bench, the development of fracture pattern was not identical. It is believed that the propagation of the stress waves arrived at the free face after crossing the burden distance from where these waves got reflected and travelled back towards the borehole axis. During this interval the newly reflected waves have sufficient strength for further fracturing and extension of fractures which were formed in the earlier stages of stress wave motion resulting in fine fragmentation between blasthole and the free face. The fracture network

was so dense that the reassembly of the broken fragments became very difficult for this study as can be seen in Figure 3b. The stress waves which travelled towards the side and in the back of the block, attenuated and weakened gradually on reaching the ends of the block, so that the reflected waves do not carry sufficient energy for further fracturing of rock on the bench, only large pie shaped fragments could be created in the adjoining area of the blasthole.

Figure 4a & b show front view of a block with vertical joints parallel to face and the top view of reassembled fragments. Based on the presence of fractures the fragmentation process can be visualised: after detonation of the charge, the stress waves propagated in all directions. First formation of crushed zone occurred, then fracture development started from the blast holes, immediately the stress waves propagating towards

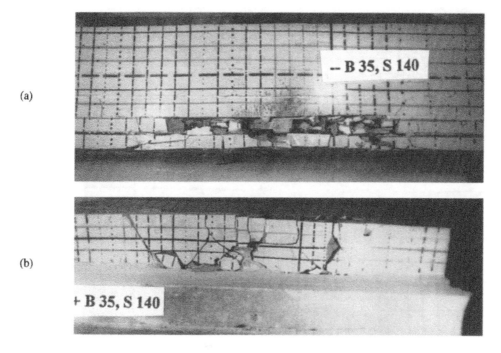

(a)

-- B 35, S 140

(b)

+ B 35, S 140

Figure 4 (a) Front view of the blasted block with vertical joints parallel to the face
(b) Top view of reassembled fragments for vertical joints parallel to the face

the free face, encountered a joint plane. At this plane, considerable attenuation of stress waves took place, some part of it reflected and travelled back to the blasthole, and remaining part of the waves which propagated towards the free face, was weakened and were unable to create hardly any fracture. Thus the stress waves caused reduced fractures in the slab at free face and reflected back. Since these waves were too weak to create any new fractures, they could only aid the extension of fractures. Thus causing a highly fractured zone in the slab containing explosive and between the holes. No significant fracture could grow from the blastholes. The slab containing explosive was subjected to the repeated loading of stress waves, which had enough strength in the reflected tensile stress waves to propagate the fractures and further growth of the fractures resulted in a dense fracture network.

The propagation of stress waves towards the side ends of the block created parallel fractures in the direction of propagation. The reflected waves which travelled back from the side end of the block were too weak to create any useful fragmentation except the possibility of increase in fracture width.

Figure 5a illustrates top of blasted block containing vertical joints perpendicular to the face. In this orientation of joints the vertical blastholes were not crossing any of the joints and the slab containing hole and explosive charge extended up to the free face. On detonation of explosive inside the hole stress waves propagated in all the directions towards free face. The reflected waves from free surface intensify the fracture network. The breakage extent is reduced and terminated at joint planes.

Figure 5b and c show two views of a block with joints perpendicular to the face which are not vertical but are steeply dipping. In this fractures develop in a direction following the joint planes.

The filler material of the joints also have influence on the fragmentation. Bhandari (1975) showed that the fine fragmentation increased in case of cemented joints compared to joints which were filled with weaker material or were open joints. Thus indicating that participation of stress waves was better in case of strong joints.

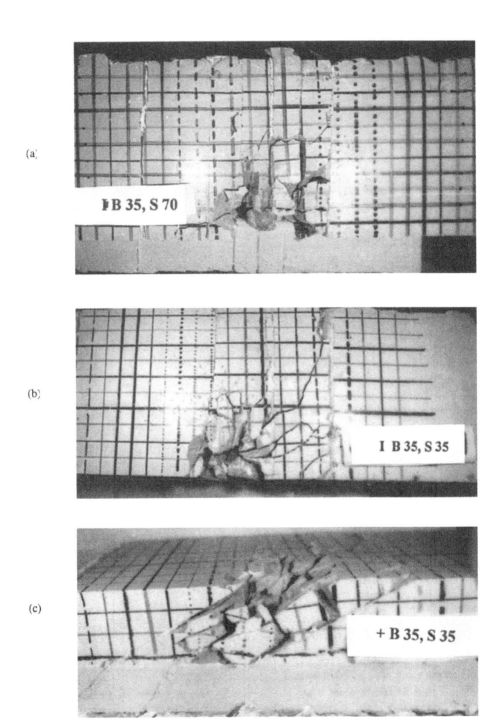

(a)

(b)

(c)

Figure 5(a) Top view of reassembled fragments for vertical joints perpendicular to the face
(b) Top view of reassembled fragments for steeply dipping joints perpendicular to the face
(c) front view of reassembled fragments for steeply dipping joints perpendicular to the face

5. CONCLUSIONS

For the same rock and explosive with the change of blasting parameters the quantity of rock broken, large sized fragmentation, throw and ground vibrations changed indicating that the way energy is utilised changes. Also as yet due to involvement of numerous factors complete understanding of the process of fragmentation by blasting in rocks with joints has not been achieved.

One must utilise that part of energy which is needed for the type of fragmentation and other blasting results by altering blast designs. The concept could be further extended to the selection of explosives and initiation pattern.

REFERENCES

Badal R., Studies on Fragmentation by Blasting of Rock with Discontinuities. Ph.D. Thesis, University of Jodhpur, 1990, 215 p.

Badal, R. and Bhandari, S., Fragmentation Mechanism in Rock Joints, International Society of Rock Mechanics Regional Symposium on Rock Slopes, New Delhi, 1992, pp 337 -385.

Bhandari, S., Studies on Rock Fragmentation by Blasting, Ph.D.Thesis, University of New South Wales, 1975a, 201 p.

Bhandari, S., Burden and Spacing Relationship in the Design of Blast Patterns, 16th Symp. on Rock Mech. University of Minnesota, USA, 1975b, pp 210-220.

Bhandari, S., On the Role of Stress Waves and Quasistatic Gas Pressure in Fragmentation by Blasting, Acta Astranoutica, Vol 6, 1979, pp 365-381.

Bhandari, S., Influence of Joint Directions in Blasting, Proc. 9th Conf. Explosive and Blasting Technique, Dallas, 1983, p. 839-369.

Bhandari, S., Changes in Energy Utilisation with Blast Parameter Variations, Int. Symp. on Explosives and Blasting Tech., Inst. of Engineers (I), New Delhi, 1990.

Bhandari, S. and Badal, R., Post Blast Study in Jointed Rocks, Int. J. Engg. Fracture Mechanics, 1990a, pp 439-445.

Duvall, W.I. and Atchison, T.C.; Rock Breakage by Explosives, U.S. Bureau of Mines, Report of Investigation 5356; 1957, 52 p.

Hino, K., Fragmentation of Rock Through Blasting and Shock Wave Theory of Blasting, Symp. on Rock Mech., Quart. Colorado School of Mines, Vol. 51, No. 3., 1956a, pp 191-209.

Hino, K., Fragmentation of Rock Through Blasting, Quart. Colo. Sch. Mines, 51, 1956b, 189 p.

Kutter, H. K., The Interaction between Stress Waves and Gas Pressure in the Fracture Process of an Underground Explosion in Rock with Particular Application to Pre-splitting, Ph. D. Thesis, Univ. of Minnesota 1967, 171 p.

Kutter, H.K. and Fairhurst, C., On the Fracture Process in Blasting, Int. J. Rock Mech. Min. Sci., Vol. 8, 1971, pp 181-202.

Persson, P.A., Lundborg, N. and Johansson, C.H., The Basic Mechanism of Rock Blasting, Proc. 2nd Cong. Int. Soc. Rock Mechanics, Belgrade, Vol. III, 1970

Saluja, S.S., Study of the Mechanism of Rock Failure Under the Action of Explosives, Ph.D. Thesis, University of Wisconsin, 1962, 175 p.

Specialized blast design

Rock Fragmentation by Blasting, Mohanty (ed.) © 1996 Taylor & Francis. ISBN 90 5410 824 X

A study of free toe-space explosive loading and its application in open pit blasts

G.J. Zhang

Department of Mining Engineering, Anshan Institute of Iron and Steel Technology, Liaoning, People's Republic of China

ABSTRACT: In this paper, problems associated with the continuous explosive loading in deep boreholes of open-pit blasts are analyze. A new loading practice (toe-space loading) and its blasting mechanism are described. It was found that several benefits could be obtained using the toe-space loading practice. These include more rational explosive energy distribution, decreased borehole pressure, prolonged gas pressure effects and improved blast performance.

Crater and small bench tests were carried out for different toe-space coefficients ϕ (ratio of the toe-space length to charge length). During the test, the effect of the toe-space loading on rock fragmentation was studied. It was found that the toe-space loading was superior to continuous loading structure. The optional toe-space coefficient was found to be in a range of 20% to 25%.

The toe-space loading technique was applied to open-pit iron mines. It was shown that placing one spacer of 1.0m to 1.5 m long at the bottom of the borehole was sufficient for rock fragmentation without leaving tight bottoms or unbroken toes. By measuring several parameters of the blast performance, it was shown that the blast effectiveness was improved significantly.

1 INTRODUCTION

Varying of the explosive loading structure in a borehole is an effective way to control the distribution of the explosive energy and to improve blast effectiveness. The continuous explosive loading practice has been used for a long time. In this loading configuration, the distribution of explosive along the borehole length is not even. The major part of the explosive column is located at the lower-half part of the bench. I particular, this is the case when large-diameter boreholes and high-density explosives are used.

Therefore, in some cases, the upper part of the bench can not be fragmented properly. Large boulder rate from the upper part of the bench may occur. In addition, with such loading configuration, overbreak or significant damage ti the new pit floor can be induced by the blast. This may reduce the drill hole efficiency. Since the explosive charge is more confined at the lower part of the bench, excessive vibration may be produced from the continuous loading structure. Furthermore, if there is water at the bottom of the hole, the continuous loading

structure becomes unsuitable for non-water resistant explosives, such as ANFO.

Improved blast performance has been obtained by some mines with air decks at the bottom of the hole. However, it was found that the air-decking is not easy to implement in deep boreholes in open pit blasts.

In this paper, a free toe-space loading configuration and its blasting mechanism are described. A specifically designed bottom spacer (refer to Figure 1) was used to obtain air or water space at the bottom of the hole. crater and small bench tests were carried out for different toe-space coefficients ϕ (ratio of the toe-space charge length). The loading technique was applied to open-pit iron mines. By measuring several parameters of the blast performance was improved significantly with the toe-space loading.

2 TOE-SPACE LOADING CONFIGURATION

The toe-space loading structure is one in which an unloaded column is left at the bot-

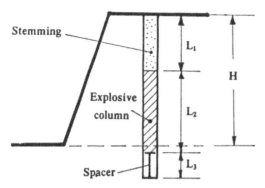

Figure 1. Bottom-space loading structure for open pit blast.

tom of the hole and on the top of it there is a continuous explosive column(as shown in Figure 1). A specifically-designed spacer was placed at the bottom of the blast hole to obtain the bottom space. The spacer is easy to use in practice.

The explosive loading is high-speed dynamic. The strength of rock mass is strain-rate dependent. It increases with increase of loading rate. This means that the slower the loading rate, the better the rock is fragmented for the same amount of explosive energy. In addition, some investigators have shown that the higher the initial borehole pressure, the more fines will be generated around the blast hole. Therefore, in such case, the energy from the blast is wasted in producing undesirable fines.

Therefore, in order to increase the blasting effectiveness, it is important to be able to control the blast loading rate and as well as the amplitude of the initial borehole pressure. By using the bottom-space loading structure, it is possible to control the borehole pressure duration and the pressure amplitude.

Air or water can be the medium occupying the bottom space of the borehole. However, their working mechanisms are different during blasting. They are described as follows.

2.1 Working mechanism for the bottom air-space structure

Existence of proper air space results in decrease of the peak value of initial borehole pressure since a part of the explosive energy is stored in the air. Therefore, premature escaping of explosion gas due to local breaking can be reduced. The duration of the borehole pressure can be prolonged. As results, the distribution of explosive energy along the borehole is towards uniform.

Consequently, the finely broken zone can be reduced and the fragmented zone enlarged. The utilization of explosive energy can be increased and the blast effectiveness can be improved. Intensity of blast vibration can also be reduced.

2.2 Working mechanism of bottom water-space structure

Different from air, water features isotropic uniform pressure transferring and less compressibility. Under the high pressure, the compressibility of water is larger than that of rock. During the blasting, the water can act as a buffer layer between explosion gas surface and rock. With such a buffer layer, explosion energy can be transferred uniformly to the wall of borehole. In addition, the high-pressure "water wedge" action is favorable to the further development of blast-induced rock fracture. Therefore, the proper water space can improve the explosion pressure parameters and prolong the pressure duration.

3 FIELD COMPARISON TEST WITH CRATER BLASTS

In order to examine the effectiveness of the bottom-space loading structure, crater blasts were tested. A total of 25 blasts were fired. During the tests, effects of different lengths of the bottom space on the blast effectiveness were compared with that of the continuous loading structure. The tests were carried out in both dry holes(the air bottom-space)and water holes(the water bottom-space). The tests were conduced in five different rock types, which include an iron ore, serpentinite, dolomite, skarnization marble and skarn. The test sites were all competent rock. Table 1 lists tensile and compressive strengths of the rocks. As seen in the table, the compressive strength of rocks varied from 92.5MPa to 154.6MPa. The rock type varied from relative soft to hard rock.

Table 1. Tensile and compressive strength of rocks at the test sites.

Rock type	Tensile strength (MPa)	Compressive strength (MPa)
Iron ore	7.6	154.6
Serpentinite	4.5	92.5
Dolomite	4.6	94.8
Skarnization marble	10.5	214.3
Skarn	6.5	133.7

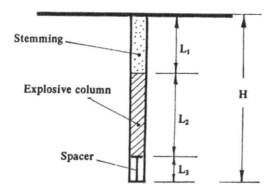

Figure 2. Bottom-space loading structure for crater blast.

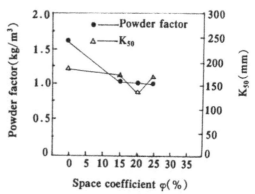

Figure 3. K_{50} and powder factor vs. space coefficient φ for blast in the Iron ore.

In order to ensure the results to be comparable for the same rock type, the blasts were conducted at the same site with the blast holes drilled more than 3m apart. For the water bottom-space tests, permeability tests were conducted before loading explosive in order to ensure that the space was filled with water during blasting.

For the crater tests, vertical holes were drilled 1m deep and 45mm in diameter. Cartridges of Rock explosive #2 were used and initiated with blasting cap #8. The blast holes were stemmed with 0.4m of clay. Figure 2 shows the loading structure for the crater test.

A bottom-space coefficient φ is defined as ratio of the space length (L_3) to the explosive column length (L_2), as shown in Figure 2. After each blast, rock fragments were screened and weighed manually. The dimension of the crater was measured. The fragment size of 50% passing (K_{50}) and the powder factor were used to measure the blast effectiveness. The test results are shown in Table 2. As seen in Table 2, for each type of rock, test #1 is for the continuous loading structure ($\varphi = 0$). Figures 3～7 display K_{50} and the powder factor against the space coefficient φ for the five types of the rocks, respectively. From the figures, we can seen that there is the optimum value of the space coefficient φ for each rock type. Under the test condition described, the optimum φ values for the air bottom-space (dry holes) loading structure are 20%, 30% and 25% for iron ore, serpentinite and dolomite, respectively. Whereas, the optimum φ values for the water bottom-space (wet holes) loading structure are 20% and 25% for skarnization marble and skarn, respectively. As seen in Figures 3～7, at the optimum value of the space coefficient, the powder factor is lower than that of the continuous loading structure ($\varphi = 0$). This means that with the optimum space

length the blast effectiveness is improved and the amount of explosive can also be saved. The results show that the bottom-space loading structure of air or water is superior to the continuous loading structure ($\varphi = 0$). The test results also show that the optimum φ value appears to increase with the decrease of the rock compressive strength.

Table 2. The results of crater tests.

Ore or Rock	Test No.	L_2 (m)	L_3 (m)	φ (%)	Powder factor (kg/m^3)	K_{50} (mm)
Iron ore (Dry hole)	1	0.60	0	0	1.648	192.99
	2	0.52	0.08	15	1.032	168.23
	3	0.50	0.10	20	0.984	142.31
	4	0.48	0.12	25	0.974	161.62
Serpentinite (Dry hole)	1	0.60	0	0	1.086	234.69
	2	0.52	0.08	15	0.937	152.41
	3	0.50	0.10	20	0.753	136.69
	4	0.48	0.12	25	0.709	51.75
	5	0.46	0.14	30	0.657	96.08
	6	0.44	0.16	35	0.675	93.40
Dolomite (Dry hole)	1	0.60	0	0	1.100	167.20
	2	0.52	0.08	15	0.962	157.35
	3	0.50	0.10	20	0.780	157.09
	4	0.48	0.12	25	0.726	99.52
	5	0.46	0.14	30	0.714	103.84
Skarnization marble (Water hole)	1	0.60	0	0	1.316	187.10
	2	0.52	0.08	15	1.026	143.26
	3	0.50	0.10	20	0.929	104.25
	4	0.48	0.12	25	0.923	151.63
	5	0.46	0.14	30	0.910	150.07
Skarn (Water hole)	1	0.60	0	0	1.061	139.29
	2	0.52	0.08	15	0.965	211.70
	3	0.50	0.10	20	0.796	261.36
	4	0.48	0.12	25	0.741	112.61
	5	0.46	0.14	30	0.716	183.08

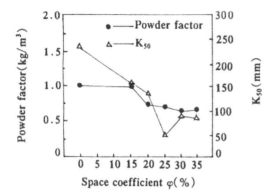

Figure 4. K_{50} and powder factor vs. space coefficient φ for blast in Serpentinite.

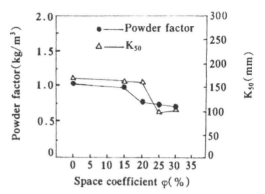

Figure 5. K_{50} and powder factor vs. space coefficient φ for blast in Dolomite.

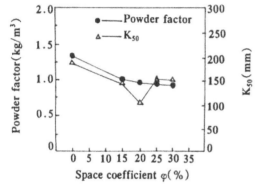

Figure 6. K_{50} and powder factor vs. space coefficient φ for blast in Skarnization marble (water bottom-space hole).

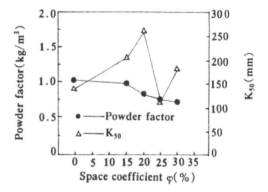

Figure 7. K_{50} and powder factor vs. space coefficient φ for blast in Skarn (water bottom-space hole).

4 SMALL BENCH TEST

In order to test the loading technique in bench blasting, four small bench blasts were conducted, each of which includes a pair of holes simultaneously initiated. The borehole diameter was 41mm. Rock-explosive # 2 was used. The test was conducted in quartzite. The design parame-

ters of the small bench blasts are listed in Table 3. After each blast, the fragments was screened and weighed manually. The measured results are shown in Table 4. The data in Table 4 were analyzed by regression. Values of K_{50} were obtained for each blast, as listed in the table. The optimum value of the space coefficient φ is also about 25%.

Table 3. Design parameters of small bench blast.

Parameters	φ							
	0%	0%	25%	25%	30%	30%	35%	35%
Height of bench(m)	1.00	0.95	1.00	0.95	1.00	1.00	1.00	1.00
Toe burden(m)	0.75	0.76	0.75	0.75	0.80	0.80	0.80	0.80
Hole spacing(m)	0.75	0.70	0.70	0.70	0.70	0.70	0.70	0.70
Hole depth(m)	1.20	1.20	1.20	1.15	1.30	1.30	1.30	1.30
Overdepth(m)	0.20	0.25	0.20	0.20	0.30	0.30	0.30	0.30
Explosive quantity(kg)	0.24	0.24	0.24	0.24	0.25	0.25	0.25	0.25
Explosive column length(m)	0.21	0.21	0.21	0.21	0.22	0.22	0.22	0.22

Table 4. Results of small bench tests (Powder factor: 0.50kg/m³).

φ (%)	Blasted quantity (kg)	Grade size (mm)									K_{50} (mm)
		≤30	≤80	≤130	≤180	≤230	≤280	≤330	≤380	>380	
0	2544	105	110	157	130	130	155	599	510	648	320.48
25	2596	100	227	331	302	245	335	340	430	286	243.88
30	2611	149	151	295	318	311	204	220	409	554	274.49
35	2653	72	220	320	200	228	325	334	330	524	311.65

5 FIELD TRIALS IN AN OPEN PIT MINE

In order to apply the loading technique to production blasts, field trials were carried out at Dong'anshan iron mine of Anshan Iron and Steel Company. The mining is a large open pit operation. The bench height of the blasts was about 12m. The free face angle was 65 to 70 degrees to the horizontal. Blast holes were vertical and 270mm in diameter. An emulsion explosive was used for water holes and ANFO for dry holes. A non-electric milliseconds delay system was used to initiate the blasts. Two or three rows of blast holes were blasted for each blast. The number of blast holes per blast varied from tens to a couple of hundreds. Plastic and bamboo spacers were designed with diameter of 220mm and varying length from 1m to 1.5m (1m, 1.2m, 1.4m and 1.5m). The spacer was placed at the bottom of the hole before loading explosives. Crushed stone was used as stemming material. By using the bottom-space loading structure, the quantity of the explosive loaded in the hole was reduced. For an emulsion hole, with a spacer at the bottom, the explosive in the hole was reduced by an amount of 35kg ~ 65kg (depending on the length of the spacer) compared with continuous loading structure. For an ANFO hole, with a spacer at the bottom of the hole, the explosive loaded was reduced by an amount of 40kg~65kg.

The trials were carried out at 18 blast zones in the mine with blasted quantity of 1.9 million cubic tons. The blast effectiveness with and without spacer were compared. Several parameters were used to compare the blast effectiveness. These include the width of the muckpile (the front shock), the volume of tight bottom and large boulder rate. Table 5 shows the blast design parameters along with the parameters measuring the blast effectiveness. Furthermore, the material consumption for the secondary blast of boulders, road cleaning rate and truck loading efficiency were measured. Table 6 shows the material consumption for the secondary blast, road cleaning rate and truck loading efficiency.

As seen in Table 5, the blast effectiveness was improved significantly by using the bottom-space loading structure compared with the continuous loading structure ($\varphi = 0$). The large boulder rate was decreased by 30% ~ 50% with the bottom-space loading structure. Other improvements can be seen from Tables 5 and 6 as follows. There is no tight bottom for all the tested blasts. The width of the muckpile (front shock) was reduced. The explosive powder factor was decreased by 17% ~ 24%. The net truck loading efficiency was increased by 15% ~ 20%. The material consumption for the secondary blast was reduced by 50%.

At present, the bottom-space loading structure has been applied in Dong'anshan Iron Mine of Anshan Iron and Steel Company, Zhijiazhuang Iron Mine of Hebei Province and Xigou Limestone Mine of Jiuquan Iron and Steel Company, etc. All these mines have obtained good economic benefits.

Table 5. Design parameters of production blast and results.

No. of blast Zone	Rock kind	φ (%)	Height of bench (m)	Toe burden (m)	Hole spacing (m)	Row spacing (m)	Over depth (m)	Explosive Height (m)	Powder factor (kg/m³)	Front shock (m)	Tight bottom (m³)	Boulder rate (%)
3	Phyllite	18	14.2	10.8	7.3	6.2	2.3	7.9	0.53	25.7	0	0.04
		0	12.8	11.8	7.2	6.5	4.0	9.1	0.64	29.1	146	0.28
7	Hematite	24	15.0	9.8	7.7	6.3	1.9	6.9	0.40	18.5	0	0
		0	14.6	10.3	7.4	7.0	2.5	9.6	0.56	24.5	0	0.09
11	Quartzite	23.6	15.1	9.2	7.1	6.2	4.1	6.4	0.44	13.0	0	0.09
		0	15.1	9.7	6.2	5.8	2.7	7.8	0.55	22.0	40	0.48
17	Mixture	24.0	12.1	7.6	7.2	5.8	2.5	4.3	0.83	20.5	0	0.05
		0	10.2	9.1	7.5	5.9	5.8	6.1	0.96	23.6	0	0.15

Table 6. Some measured parameters of blast performance.

No. of blast zone	φ (%)	Blast material consumption for blasting boulders					Excavator efficiency	
		Explosive (kg)	Non-eleo-tric cap (piece)	Detonation tube (piece)	Fire detonator (piece)	Blast fuse (piece)	Road cleaning (m/h)	Net loading efficiency (min/truck)
3	18	0.6	4	20	2	3	54	4.00
	0	96.6	401	2370	18	27	32	4.87
7	24	0	0	0	0	0	86	3.80
	0	7.5	60	600	2	3	54	4.33
11	23.6	2	9	90	2	3	63	3.66
	0	20.6	152	1465	6	5	41	4.83
17	24.0	2	20	200	2	3	45	3.83
	0	4.5	32	320	2	3	30	4.33

6 CONCLUSIONS

According to the comprehensive tests of the crater blasts, the small bench blasts and the open pit production blasts, conclusions can be drawn as follows.

6.1 The bottom-space loading structure is superior to the continuous loading structure. The former can distribute the borehole pressure more evenly along the borehole and input the explosive energy more uniformly within the overburden than the continuous loading method.

6.2 The effect of different lengths of the bottom spacer on blast performance was studied with the crater and small bench tests. The optimum value of the space coefficient was found is about 20% ~25%. Although the value may vary for different blast design, the finding here can serve as a reference for selection of the spacer size in practice.

6.3 Trials of production blasts show that the bottom-space loading structure was practical and that the blast performance was significantly improved with this technique. The technique is easy to implement and costs very little. It is reliable in practice.

6.4 Further tests are required to find the optimum of the space coefficients for different blast designs, such as, for open pit coal mines and very high benches(about 50m).

REFERENCES

Baranov, Y. G. 1986. Raising the utilization effectiveness of deephole charges blasting energy in water-bearing rocks. *Refractory*. 2:28—31.

Lin, D. Y. 1989. Blasting action and effectiveness analysis of the bottom air-space loading structure. *Non-ferrous Metal*. 4:1—6.

Padukv, V. A. 1987. Raising the fragmentation effectiveness of rock blasting. *Comprehensive Utilization of Mineral Raw Materials*. 5:7—10.

Zhang, G. J. 1991. Research on the application of deephole bottom air-space explosive-loading blasting. *Gold*. 11:27—31.

Zhang, G. J. 1991. Test research on the incline deephole bottom-space explosive-loading blasting. *Metal Mines*. 6:22—24.

Rock Fragmentation by Blasting, Mohanty (ed.) © 1996 Taylor & Francis. ISBN 90 5410 824 X

Numerical modelling of the effects of air decking/decoupling in production and controlled blasting

Liqing Liu & P.D. Katsabanis
Department of Mining Engineering, Queen's University, Kingston, Ont., Canada

ABSTRACT A series of numerical simulations were performed to investigate the physical processes and effects in rock blasting with air decked/decoupled explosive charges. A newly developed continuum damage model for blasting analysis was successfully used to model the movement of detonation products in the borehole chamber with an air space. The model evaluates rock damage according to constitutive relationships and the concepts in statistical fracture mechanics, and predicts fragment size distribution considering the equilibrium between kinetic energy and fracture surface energy.

It is shown that a secondary loading wave is generated due to the repeated movement of the detonation products within the borehole chamber when blasting with an air deck. As a result, the potential energy retained in the explosion gas is released and a significant part of it is transmitted into the rock burden. It is found that there exists a minimum beneficial air deck length which is determined by the balance between the energy loss from primary loading to stemming and the energy gain from secondary loading. Blasting results are modified with an air deck longer than the minimum beneficial length. The best blasting results are obtained with the longest air deck provided that the shortened stemming can confine the explosion gas in the borehole chamber. It is concluded that with the use of air decking, the volume of broken rock can be increased or the explosive usage can be reduced. The simulation results have been verified by experiments in model scale. It is also revealed that some variations to top air decking such as bottom and mid-column air decks do not benefit production blasting in the axisymmetric case of the simulation conditions. Three charging strategies, i.e., stemming decking, decoupling and air decking which are used in controlled blasting have been modelled. It is shown that air decking has the maximum damage potential to rock mass and it is recommended for both production and controlled blasting. Simulation results also suggest that the use of air decking for non-ideal explosives like ANFO is much more important than it is for some more ideal explosives like emulsion in maximizing energy utilization.

1 INTRODUCTION

Rock blasting with air decked explosive charges has been a practice of long history. The earliest document regarding the advantages of air decking in production blasts may date as far back as one century ago to the work by Knox (1893) who had an invention patented in German in 1893. A copy of the original drawing as used by Foster (1894) to introduce this invention is shown in Fig. 1, where **A** is the powder, **B** the air space, and **C** the stemming material. In the days of Knox, the modern technique of blasting and explosive technology was still in its early development. Knox might have judged the results by intuition and was not possibly able to provide a theoretical proof for what he observed. This may partly explain why his invention had not received much attention during the long period of over one hundred years. In the 1940's and 1950's, Mel'nikov and Marchenko (1971) conducted some experiments with the purpose to verify the advantages of blasting with air decking and to investigate the mechanisms. Their work seems to confirm the findings

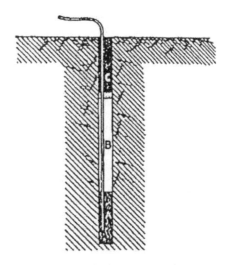

Fig. 1 Original drawing used by Foster (1894) to introduce the invention by Knox (1893)

by Knox. In the early 1980's, Fourney et al (1981) conducted experiments in Lucite models and found some interesting crack initiation and propagation phenomena in the portion of model blastholes that contained an air deck. Chiappetta et al (1987) conducted high speed photography studies of the surface effects of air decking. As a result, a new technique of pre-splitting with air decking (Bussey and Borg, 1988) was developed to replace conventional methods such as stemming decking and decoupling.

Since then air decking has been applied in a variety of situations. However there is evidence that since the mechanisms are not well understood, the use of air decking does not always modify the blasting results. To fully exploit the advantages of air decking and make it an established technology in engineering blasting, the study of the mechanisms seems to be of critical importance. To serve this purpose, the present effort is devoted to a numerical investigation into the mechanisms and effects in rock fragmentation with air decked explosive charges in production as well as in controlled blasting.

2 THE NUMERICAL MODEL AND THE MODELLING TECHNIQUES

A constitutive damage model for blasting analysis developed by the authors was used in this study. The model is based on continuum mechanics and statistical fracture mechanics. In the model, the activation and growth of micro cracks are calculated by integrating a crack density function over time. The crack density function is based on the well established experimental facts that rock materials do not fail under the action of a stress not exceeding the static strength, a certain time duration is required for a dynamic failure to occur and the dynamic fracture stress is approximately cube root dependent on the loading rate. In the model, the damage coefficient is defined as the probability of fracture at a certain crack density. The minimum damage value is 0.632, which means that in order to form fragments, each unit volume has at least one crack. Fragment size is calculated considering the equilibrium between kinetic energy and fracture surface energy, with the changes in loading rates, material density and stiffness taken into account. The model has been calibrated by field tests in granite and has been found to be in conformity with the theory of explosive energy partitioning in rock mass. The details of the model work are described in the publications by the authors (Liu, L. and Katsabanis, P.D., 1995a, 1995b).

Before the effects of air decking in blasting are studied, the role of air in air decking should be clarified. It has been suggested by some researchers that in blasting with air decking, the presence of air in the air deck plays a critical role in obtaining the advantages. According to Chiappetta et al (1987), the air acts as an "energy accumulator", it "stores" and then "releases" energy into the rock medium. However, the density of air (1.29 kg/m^3, at 0°C and 760 mm Hg) is only about one thousandth of that of a typical solid explosive. During a blast, the air is initially static and at room temperature and pressure. The air material under such physical conditions offers virtually no resistance to the expansion of the detonation products which have a temperature and pressure about 3-4 orders higher in magnitude. The detonation products transform some energy to the air by compressing and heating it. However, even if the air is to be as energetic as the detonation products, the fraction transfered is at most in the order of one thousandth of the energy retained in the latter, which is obviously negligible. In order to concentrate on the most important aspects in blasting with air decking, the presence of air in the air deck was neglected in this study. Only the detonation of explosives and the movement of detonation products within the air chamber are considered, the air deck being regarded as a "vacuum deck".

In this study, only the rock damage induced by shock/stress wave is modelled and possible penetration by the detonation products into cracks in later stages of the rock fragmentation process is not considered. The movement of detonation products is modelled using the surface contact capabilities of ABAQUS/Explicit (version 5.4) (Hibbitt, Karlsson & Sorensen, Inc., 1994) assuming that detonation products can move freely in the borehole chamber upon their formation. The damage model was coded as user subroutines in this program.

3 THE GEOMETRICAL MODEL, ROCK AND EXPLOSIVE PROPERTIES

In the present work, blasting with a borehole drilled perpendicular to a surface is studied. In numerical analysis, it is simplified as an axisymmetric problem. Actual blasting results (Katsabanis, 1994) and numerical simulations(Liu & Katsabanis, 1995b) show that with a borehole 50mm in diameter and an explosive column (emulsion) 1.6 m in length, the critical stemming length is approximately 1.6 m at which the two damage zones (one at the surface and the other next to

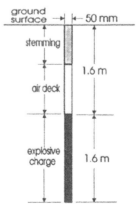

Fig.2 Geometry of the finite element model

the explosive charge) are separated but are about to coalesce. For an easy demonstration of any changes in simulation results, the above values are used as a starting point for a numerical observation of the effects and possible benefits of air decking. The geometry of the finite element model is shown in Fig. 2. The rock material simulated was the granite gneiss of the test site of

Table 1: Parameters of granite used in this study

$E, (GP)$	51.8
ν	0.33
$\rho, (t/m^3)$	2.55
$\sigma_c, (MP)$	215
$\alpha , (x10^{10})$	7
β	2
$\theta_c , (x10^{-3})$	0.1411
$K_{IC}, (N/M^{2/3}, x10^6)$	3.1

Table 2: JWL equation of state parameters of the explosives used

Constants	Emulsion	ANFO
$A, (GPa)$	214.36	229.23
$B, (GPa)$	0.182	0.54969
R_1	4.2	6.5
R_2	0.9	1.0
ω	0.15	0.35
$E_0 , (J/Kg, x10^6)$	3.2	3.6962
$VOD, (m/s)$	5500	3900
$\rho_0, (g/cm3)$	1.31	0.95

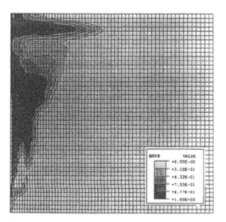

Fig. 3 Damage contours of the model with full stemming, showing two damage zones at critical state

Queen's University. Its material constants are listed in Table 1. The explosives simulated were an emulsion and crushed ANFO. The JWL equation of state parameters are listed in Table 2. Fig. 3 shows the predicted damage contours with full stemming (no air deck) in the finite element model. The damage zones are at a critical state between coalescence and separation. In the simulations described in this study, the definition of an air deck on top of the explosive column was performed by removing the stemming material in that position. The air deck was placed on top of the explosive column and its length changed from zero, in the case of full stemming to the collar length. To simplify the problem, it was assumed that the stemming material has the same properties as the rock medium.

4 MOVEMENT OF DETONATION PRODUCTS IN THE BOREHOLE CHAMBER

In a blasting process, initially the only difference between a fully stemmed borehole and that with an air deck is the space available for the expansion of the detonation products. Therefore, the characteristics of the movement of detonation products in the borehole chamber should be studied. It is found that the differences in the movement of the detonation products are in fact responsible for the variations in blasting results.

When there is an air deck on top of the explosive charge, the detonation products move rapidly through the air deck and collide with the stemming at its bottom. Upon the initial collision, stemming is compressed to a certain degree, and a reflected compressional wave impacts the borehole gases. This results in compression of the gases followed by a later expansion, forming a reverberating movement until an equilibrium state in pressure within the borehole chamber has been reached. Fig. 4 shows the initial movement of detona

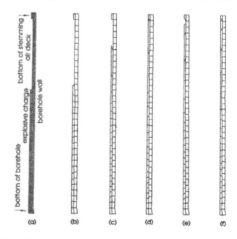

Fig. 4 Initial movement and the first cycle of reverberation of the detonation products (a)the borehole chamber; (b)at 350μs; (c)at 750μs; (d)at 1.05ms (e)at 1.3ms; (f)at 1.45ms

321

tion products and the first cycle of reverberation in a borehole which has an air deck with a length of 0.96 m. The explosive modelled was a typical emulsion.

Fig. 4(a) shows the element set which defines the borehole chamber and explosive charge. The reaction time of the explosive column is 291µs. Before this time, the top part of the explosive column remains undisturbed. Fig. 4(b) shows the state at 350µs after initiation when the explosive column has fully transformed to detonation products and the top of it begins to move upward. Fig. 4(c) shows the state at 750µs at which detonation products are moving towards the stemming. Fig. 4(d) shows the initial collision between the gas front and stemming bottom at 1.05ms. It has taken 759µs for the gas front to travel through the air deck (0.96 m), so the average velocity of the gas front before the initial collision is about 1300 (m/sec) while the density of the gas front at this time is about 0.22(g/cm^3). The increase in gas volume and the decrease in gas pressure and density are demonstrated by an increase in the length of the top explosive elements. Fig. 4(e) shows the state after the initial collision. At this time, the stemming is deformed as a result of the collision, the gas front is compressed and the direction of its movement is reversed. At the time the gas front impacts the stemming, its pressure is relatively low. After that, there is a sudden rise in gas pressure because of a reflection from the stemming upon collision and the compression by the gas behind the front. When the gas front is compressed to its minimum volume, it begins to expand again and move upward toward stemming. Fig. 4(f) shows the second collision between the gas front and stemming at 1.45ms. The same mechanisms are observed in different air deck lengths. However, the time needed to reach equilibrium in pressure tends to increase with air deck length.

During the detonation process, only a fraction of the total energy is transmitted into the rock medium as shock energy while a significant amount of it is retained in the detonation products. The existence of an air deck on top of the explosive charge allows this part of energy in the detonation products to be released when it expands in the air deck. It was found that the energy relationships are totally changed by the introduction of an air deck. Firstly, the energy retained is transformed into kinetic energy, driving the detonation products into rapid movement; then the kinetic energy is imparted to the rock mass upon collisions in the form of strain energy and elastic dissipation. The extra strain energy is responsible for the enhanced rock breakage. A more detailed description of the energy transformation relationships has been presented by the authors in another publication (Liu & Katsabanis, 1995c).

5 EVIDENCE AND EFFECTS OF SECONDARY LOADING

During the movement of the detonation products in the borehole chamber, there are repeated increases and decreases in the pressure of the detonation products

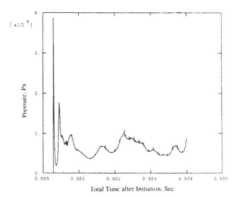

Fig. 5 Pressure history of the top most explosive element in a model with full stemming

until an equilibrium state is reached. Each pressure change acting on the borehole wall stimulates a stress wave propagating in the rock mass and transmits some energy into it.

With a fully stemmed borehole, the movement of the detonation products is strictly confined in a limited space. Fig. 5 shows the pressure history of the top most explosive element in a model with full stemming. The pressure peak on the left corresponds to a sudden rise in pressure when this part of explosive is detonated. After the detonation, the pressure rapidly reaches a stable and relatively high value (about 0.7×10^9 Pa). At an equilibrium state, the pressure level is high but changes very little with time. The stress field around the borehole chamber is approximately quasi-static and decays with distance very rapidly. The high pressure is balanced mainly by the plastic strain in the vicinity of the borehole chamber. Detonation products at this state have a significant amount of potential energy which can be released only when there is a sudden drop in the pressure. Such a pressure change is made possible by the use of an air deck. Fig. 6 shows the pressure history

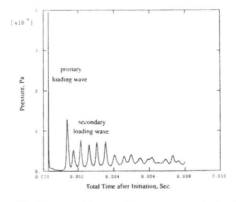

Fig.6 Pressure history of the top most explosive element in a model having an air deck 0.96m long, showing the appearance of a secondary loading wave

of the top most explosive element when the air deck is 0.96 m long. It is shown that there is a sudden rise in the pressure when this explosive element is detonated. Then the pressure drops rapidly when the detonation products are travelling upward through the air deck. At the time the gas front impacts the stemming, a series of pressure waves is stimulated as the detonation products reverberate in the borehole chamber. This process is accompanied by a rapid release and transformation of the energy carried by the detonation products. As compared to the primary loading wave due to the detonation of explosives, this series of pressure waves may be termed secondary loading wave. The secondary loading wave is quite independent of the primary one. The time lag between the secondary loading and the primary loading is determined by the air deck length and the travelling velocity of the gas front. As shown by Fig. 6, for the secondary loading wave, although the pressure level is not as high as in primary loading, there are numerous loading pressure pulses compared to only one in primary loading. Furthermore the time duration for each pressure pulse is significantly larger than that in primary loading.

The additional stress waves stimulated by the secondary loading wave act on the rock medium after the passage of the primary loading wave and cause the damage of the rock mass to accumulate on the basis of the damage resulted from primary loading. This point can be shown by the differences in the damage growth history of the rock elements in a finite element model with different air deck lengths. Fig. 7 shows the damage history

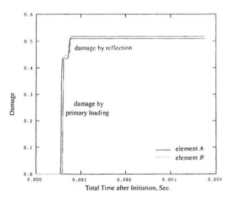

Fig. 7 Damage growth history of elements *A* and *B* in a model with full stemming

of two selected elements from a model with full stemming, in which element *A* is about 0.47 m away from the borehole center and 0.64 m below the ground surface, while element *B* is at the same level as element *A* but 0.9 m away from the borehole center. Both elements are in the area which separates the upper and lower damage zones. The damage for elements *A* and *B* is driven by primary loading and the reflection wave from the ground surface and the value remains unchanged after it is slightly larger than 0.5 (0.9 ms after initiation). As a comparison, Fig. 8 shows the

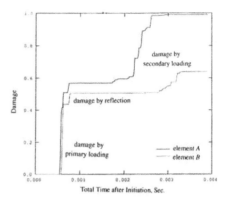

Fig. 8 Damage growth history of elements *A* and *B* in a model having an air deck 0.96m long

damage history of the same two elements in a model which has an air deck 0.96 m long. Here element *B* has a damage value of about 0.5 after the primary loading and the loading by the reflection wave, which is not large enough to from fragments. When the secondary loading wave arrives at 2.5 ms, the damage value of the element is increased to a value near 0.632, the minimum damage value to form fragments. Similarly, as shown in the same figure, the damage value of element *A* increases from below the minimum to a value at which nearly full fragmentation can be expected. Damage growth driven by the reverberation of detonation products is evidenced by the small steps in the curves. The secondary loading waves shown here seem to conform the experimental work by Marchenko (1954), who showed that to produce maximum fracturing, numerous extra loading waves tended to increase the volume of material removed. Also, according to the experimental work by Fourney et al (1981), it is concluded that the collision between the detonation products and stemming reinforces the pressure wave. The reinforced pressure has larger magnitude and longer duration, which appears to be very effective in initiating and propagating fractures.

6 EXISTENCE OF A MINIMUM BENEFICIAL LENGTH

Although the use of an air deck facilitates the release of the energy retained in the detonation products, it also weakens primary loading in the position of stemming. A significant amount of energy is transmitted into the rock medium for rock breakage through stemming during the loading process. Part of it will later be reflected at the free face and be consumed in forming slabs near the free face. According to Brinkman (1987), shock energy appears to be most effective in fracturing near the blasthole where it produces the greatest rock loading. Since the interface between the explosive and stemming is the position where the explosive column is closest to the free face, any weakening in loading in this position will reduce the amount of energy transmitted into the rock medium and the blasting results. When there is an air deck on top of the explosive column, the stemming is separated from the explosive charge. As a result, primary loading due to the detonation of the explosive charge cannot impact the stemming and impart the energy to it. Instead, the stemming is only loaded by the explosion gas which has to travel through the air deck before its initial collision with it. The energy loss from primary loading to stemming is compensated by secondary loading. However, if the air deck is not long enough, only a small portion of potential energy in the explosion gas can be released. This portion is too small to compensate the said energy loss. As a result, the use of an air deck does not modify but slightly deteriorates the blasting results. With the increase in air deck length, the energy transmitted into the rock medium by secondary loading increases rapidly. The increased amount of energy can fully compensate the energy loss from primary loading to stemming and enhance the rock fracturing and fragmentation processes. There exists a break-even point in air deck length at which energy loss from primary loading to stemming is just compensated by energy gain from secondary loading by detonation products. The length of air deck at this point may be termed minimum beneficial length. The blasting results at the minimum beneficial length are equivalent to those of a fully stemmed borehole. The blasting results with an air deck shorter than the minimum beneficial length are slightly inferior than those of a fully stemmed borehole. Since the minimum beneficial length is resulted from the uncompensated energy loss from primary loading, its existence and actual value depend on how important primary loading is, i.e., on the explosive properties as well as rock types.

7 BLASTING RESULTS AS A FUNCTION OF AIR DECK LENGTH

In the present study, the simulated final blasting results are evaluated in four aspects: total volume of fragments predicted, change in specific charge, fragment size distribution and the diggability of fragments.

The total volume of fragments is the sum of the volume of all the elements in the model which have a damage value larger than or equal to 0.632, the minimum damage value to form fragments. The total volume of fragments at different air deck lengths is shown in Fig. 9. In the figure, air deck length is represented by the percentage of original stemming length, i.e., 0% represents full stemming while 100% represents an open hole (no stemming). As shown, when the borehole is fully stemmed, the volume of fragments predicted is 5.25 (m^3). As the length of air deck gradually increases, the predicted volume slightly decreases. This trend continues until the air deck length is 30% (0.48m), when the smallest amount of volume is predicted. The decrease in volume is attributed to the energy loss from pri

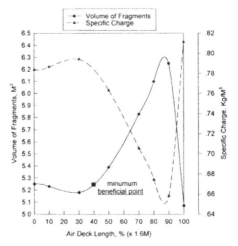

Fig. 9 Volume of fragmnets and specific charge as a function of air deck length

mary loading to stemming, and that the energy gain from secondary loading is still not enough to compensate the said energy loss because of a too short air deck length. At an air deck length of 0.48 m, the difference between the energy loss and energy gain is the greatest. As the air deck length further increases, more potential energy in the explosion gas is released. Energy gain from secondary loading slowly overtakes the energy loss, and the predicted volume begins to increase. At an air deck length of about 40% (0.64 m), when the energy loss is offset by the energy gain, the predicted blasting results are approximately equivalent to those when full stemming is used. This point is the minimum beneficial point, and the air deck length at this point is termed the minimum beneficial length for the rock-explosive(granite-emulsion) combination modelled.

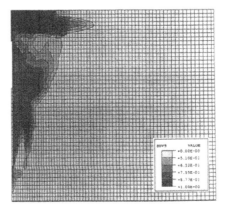

Fig. 10 Damage contours at an air deck length 0.96m

After the minimum beneficial point, volume increases very rapidly with air deck length and seems to be proportional to it. This phenomenon is due to the more complete conversion of potential energy to strain energy. Interestingly, this increase in the volume of fragments seems to occur mainly near the position of the air deck, or above the lower damage zone, see Fig. 10 and 11 for damage contours when air decks are respectively 0.96m, 1.44m in length. As compared to the damage contours for full stemming (see Fig. 3), the two damage zones have partly coalesced at an air deck length of 0.96m while they have completely coalesced when the air deck is 1.44m long. The predicted volume reaches its maximum (6.25 m^3) at an air deck length of 90% of the original stemming length (1.44m). This corresponds to about 19% increase as compared to the fully stemmed. The simulated fragment volume change with air deck length agrees with the conclusion drawn by Chiappetta et al (1987) from their high speed photography study of the surface effects of air decking, that the maximum utilization of energy is obtained with a combination of the longest air deck and the shortest stemming deck that will provide adequate confinement of borehole gases and control of air blast.

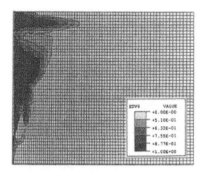

Fig. 12 Damage contours along a model without stemming (at 2ms after initiation)

similarity occurs because there is virtually no damage induced by secondary loading. This prediction seems to be in good accordance with the experimental or computational findings by Hommert et al (1987) and Simha et al (1987) that the rock burden can be fractured without stemming and the blasting results are just as good as fully stemmed. However, gas penetration of cracks is not modelled in this study, which may be of some importance in enhancing rock fragmentation in later stages.

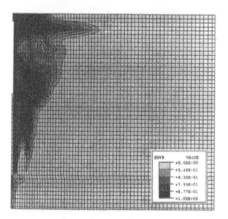

Fig. 11 Damage contours at an air deck length 1.44m

As a result of the contribution made by secondary loading to the fracturing of the rock burden, a better utilization of explosive energy is achieved. When the air deck is 1.44 m long, an optimum specific charge (0.658kg/m^3) is obtained, which is 16% lower than the specific charge in the case of full stemming (0.784kg/m^3). The change of specific charge with air deck length is also shown in Fig. 9. A decrease in specific charge represents a reduction in explosive consumption for the same amount of rock material blasted and a better utilization of explosive energy. In addition to a reduction in the specific charge, as was described, the increase of air deck length also leads to a modified fragment size distribution and improved diggability of fragments.

What is of theoretical interest is the simulation results of a model without stemming. The predicted damage zone and the volume of fragments (5.1m^3) are very similar to that obtained with full stemming. Fig. 12 shows the damage contours formed when the gas front escapes from the borehole (at 2ms after initiation). The

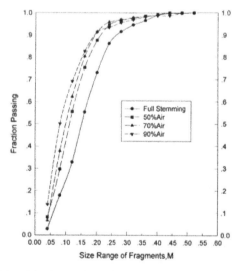

Fig. 13 Statistical fragment size distribution curves of the models with full stemming, 50%, 70% and 90% of air deck

The average fragment size tends to get smaller as the air deck length increases. This is not surprising because more energy is consumed in the comminution of the rock burden when an air deck is used. Fig. 13 shows the statistical fragment size distribution curves of the models with full stemming, 50%, 70% and 90% air deck lengths. As shown in this figure, with the increase in air deck length, not only the average fragment size decreases, but also the size range that the fragments take gets narrower indicating that the distribution of the fragment sizes is more uniform. The characteristics of the fragment size distribution as predicted here is in good accordance with the experimental work by Mel'nikov and Marchenko (1971), who reported an increased degree and modified uniformity of fragmentation with the use of an air deck.

8 EXPERIMENTAL VERIFICATION IN MODEL SCALE

The effects of air decking on the blasting results have been verified in model scale. The charge configuration of a borehole with a top air deck is shown in Fig. 14.

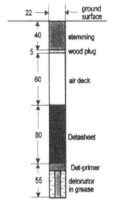

Fig. 14 Charging configuration of the testing boreholes, unit in mm

The borehole was 22mm in diameter and 240mm in length and it was drilled perpendicular to the ground surface in granite. An electric detonator with a small primer of 1 gram PETN was placed at the bottom of the borehole to initiate a small column of high explosive (Detasheet C) on top of it. Since the diameter of the detonator (6mm) was smaller than the borehole diameter, an empty space existed at the bottom of the borehole. To eliminate the influence of this space so that it did not form a bottom air deck in the borehole, it was filled by grease. Detasheet is a plastic explosive and its shape can be changed as required for charging easiness. A wood plug 5mm in thickness was used to retain the stemming material at the borehole collar. The stemming material was gravel about 1mm to 5mm in size. To ensure that the stemming material work effectively in preventing the explosion gas from early venting, a plastic glue was used to mix the gravel before filling the borehole collar. The plastic glue can provide a strong bond between the stemming material and the borehole wall after it cures. The space between the top of the explosive charge and the bottom of the wood plug was the air deck used in the shot. For a fully stemmed borehole, the wood plug was not used and all the space

above the explosive charge was stemmed with plastic glue mixed gravel.

To observe the effects of air decking, the amount of explosive used was properly controlled so that there were apparent surface effects but not a complete fracture of the rock burden. In the tests, 30 grams of Detasheet was used. With an air decked borehole, a shallow surface crater was obtained with a radius of 330 mm. To reveal the damage below the surface crater, the fragments were removed and the fines cleaned. It was found that there were numerous radial cracks radiating from the borehole with the longest reaching a distance of 210 mm. A close-up view of the crack pattern is shown in Fig. 15. As a comparison, with a fully

Fig. 15 Radial Cracks surrounding a borehole shot with air decking

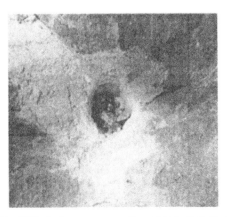

Fig. 16 Rock damage surrounding a borehole shot with full stemming

stemmed borehole (Fig. 16), the surface crater found after the shot had a smaller radius (264mm). A part of the surface rock near the borehole collar was not fragmented. The most impressive difference is that in the tests with full stemming, the radial cracks did not appear. Fig. 16 shows the rock damage near the borehole collar.

The differences between the results may be attributed to the fact that in an air decked borehole, secondary loading by repeated movement of the explosion gas aided in the fracture of the rock burden. However, in a fully stemmed borehole, secondary loading is not formed.

9 VARIATIONS TO TOP AIR DECKING

There are a number of variations to top air decking in practical applications, such as bottom and mid-column air decking. Their influences on blasting results are modelled and analyzed in this section.

In an axisymmetric environment, when the air deck is placed at the bottom of the explosive column, it is far away from the only free face. The gas front travels downward and impacts the bottom of the borehole. The secondary loading by the reverberating movement of the detonation products is mainly applied to the position near the borehole bottom where the resistance of the rock burden to fracture is the maximum. Thus extra energy transmitted by secondary loading to the rock mass is mainly dissipated in the rock mass without fracturing the burden. Therefore, the use of an air deck at the bottom of the explosive charge does not modify blasting results. Fig. 17 shows the predicted damage contours resulted from the use of bottom air deck,

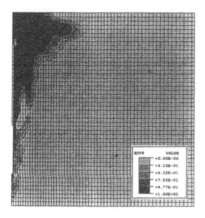

Fig. 17 Damage contours with a bottom air deck, explosive bottom initiated

explosive charge bottom initiated. The explosive column modelled was an emulsion, the charge length is 1.6 m long as before and the stemming is 0.64 m long. It is predicted that a small diggable crater can be obtained with a volume of only 4.1 m³. When the explosive column is top initiated, only sporadic fragments are predicted in the vicinity of the borehole mouth and no diggable crater is formed, as demonstrated by Fig. 18.

The use of mid column air decking is originated from the benefits of top air decking and the presumed advantages of detonation wave collision in a continuous explosive column. It is claimed by some researchers

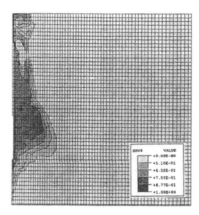

Fig. 18 Damage contours with a bottom air deck, explosive top initiated

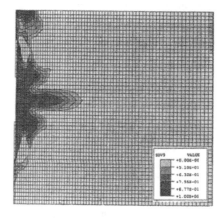

Fig. 19 Damage contours with a mid-column air deck, explosive columns initiated simultaneously from opposite ends

(Chiappetta, R.F., 1987, Mel'nikov et al, 1979) that the rock breaking capabilities can be significantly enhanced by creating a wave collision within a mid-column air deck by using precise delay electronic detonators (Chiappetta, 1992) or detonating cords (Mel'nikov et al, 1979) (which results in an unbalanced collision). Fig. 19 shows the predicted damage contours along a model when the continuous explosive column is divided into two small columns each 0.8 m in length. Between them there is an air deck 0.96m long, while the stemming above the top explosive column is 0.64 m in length. The top explosive column is top initiated while the bottom one is bottom initiated so that the gas fronts travel against each other and collide in the center of the air deck. According to the simulation, damage to the rock burden concentrates in the center of the air deck (and is somehow symmetrical to it) but there is very little damage near the ground surface. The blasting effects of detonation wave collision with a continuous explosive column in crater blasting was examined by Katsabanis (1994). It was concluded that such a collision did not modify but deteriorated the blasting results. In the experimental work by Katsabanis, a wave collision of a continuous explosive column below the ground surface resulted only in a small number of visible cracks radiating from the top of the borehole, which otherwise would produce a large excavatable crater. In an axisymmetric environment, given the fact that wave collision in a continuous explosive column does not facilitate rock fragmentation, the so-called advantages with mid column air decking (wave collision) can be negated logically. However, there might be some advantages to use mid-column air decking in other environments or with different rock-explosive combinations, which will be investigated in future studies.

10 DAMAGE POTENTIAL OF THREE CHARGING STRATEGIES

In controlled blasting such as presplitting, the purpose of blasting is to create a fracture plane through the axis of boreholes in a line. The spacing between the boreholes and the explosive usage depend largely on the damage potential of a single borehole.

There are mainly three charging strategies for presplitting which have been developed over the years, i.e., stemming decking, decoupling and air decking. The damage potential to the rock mass of each charging strategy will be modelled in this section in view of the effects of secondary loading. Only one borehole is analyzed, therefore, it should be noted that the damage pattern predicted is different from shooting more than two boreholes simultaneously, in which case the damage is directed along the line of boreholes by the interaction between them. However, from the simulated damage potential for a single borehole, important information regarding explosive consumption and justifiable borehole spacing can be obtained. In the simulations, the borehole was assumed to have a diameter of 100mm and a length of 4m and it was drilled perpendicular to the ground surface in granite mass. To use such a borehole for presplitting with air decking, the amount of ANFO usage is in the order of 3.6kg, determined according to the guidelines provided by McGill (1991). Fig. 20 shows the charging configurations corresponding to three strategies, assuming the same stemming length of 0.8m is used and the rest 3.2m of the borehole is for charging variations. In Fig. 20(a), there are 5 small ANFO charges, each 0.72kg in weight. The space between the charges is filled by stemming material. In Fig.20(b), the ANFO charge of 3.6 Kg is shaped into a continuous cylinder 39mm in diameter and 3.2 m in length. The decoupling ratio is 1.28. In Fig. 20(c), the same amount of ANFO has the shape of a short cylinder 100mm in diameter and 480mm in height and it is simply placed at the bottom of the borehole, leaving an air deck 2.72m long on top of it. In all three configurations, the charge or the series of charges is bottom initiated. Fig. 21 shows the predicted damage contours. In controlled blasting, normally low-strength non-ideal

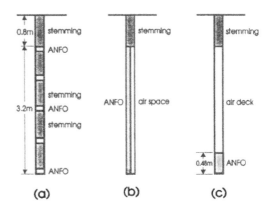

Fig. 20 Three charging strategies (a)stemming decking; (b) decoupling; (c) air decking

explosives are used, in which shock energy occupies only a small portion of the total energy and rock damage due to primary loading is not as important as with some more ideal explosives like emulsion. However, with a non-ideal explosive like ANFO, more energy is retained in the explosion gas and available for release if there is enough space for gas expansion. In stemming decking, only very limited damage is produced by primary loading. The movement of explosion gas is strictly restricted by the stemming material and the borehole wall, therefore, secondary loading due to gas movement does not appear. Consequently the damage range of stemming decking is the smallest as shown in Fig. 21 (a). Bussey and Borg (1988) reported that with stemming decking, even when the ANFO usage is nearly two times as high as in air decking, the spacing that can be used is still 25% smaller than the latter.

With a decoupled charge, although damage due to primary loading is negligible due to the low strength of ANFO and the presence of an air space between the charge and the borehole wall, there is ample space for gas movement. The damage is induced by secondary loading resulted from the gas movement in the borehole chamber. Since the explosive charge is distributed uniformly in the borehole and the VOD of ANFO is much higher than the velocity of the gas front, the explosion gas reaches an equilibrium state in pressure quickly, hampering further reverberation and ending secondary loading. The time at which the explosion gas reaches equilibrium in pressure is about 2ms after initiation, compared to about 6ms in air decking. The final damage contours are shown in Fig. 21(b). Obviously, damage is larger than stemming decking but still quite small.

Fig. 21 (c) shows the damage contours with a concentrated ANFO charge at the bottom of the borehole and an air deck on top of it. It is shown that blasting with air decking has the largest damage potential to the rock mass. The air space in Fig. 21(c) has the same volume as in Fig. 21(b) for the decoupled charge. However, the expansion history of the explosion gas is quite different from that with a decoupled charge. With a concentrated charge, the explosion gas travels upward over a distance of 2.72m, releasing energy to the rock mass until an equilibrium state is reached. While with a decoupled charge, the explosion gas travels radially a distance of only 30.5mm before it hits the borehole wall. With a concentrated charge, the reaction time of the charge is 123 μs, but there is no significant rock damage until 1ms after initiation. This point along with the fact that little damage is induced in the vicinity of the concentrated charge suggests that secondary loading is responsible for rock damage, see Fig. 21 (c). The damage potential shown above for three charging strategies is in fact an indicator of explosive energy utilization. For the rock-

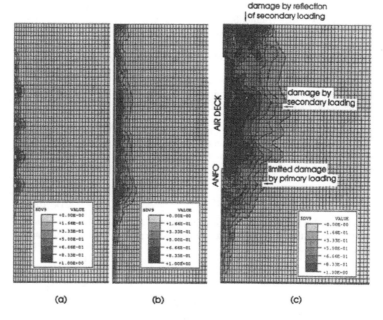

Fig. 21 Illustration of the damage potential of three charging strategies (a)stemming decking; (b)decoupling; (c)air decking

explosive (granite-ANFO) combination studied, stemming decking has the poorest energy utilization, decoupling has the intermediate and air decking has the maximum. Considering other advantages associated with blasting with air decking such as simplicity in operation and that it is less labor intensive and time saving (Bussey & Borg, 1988, McGill, 1991), the strategy of charging with air decking is recommended not only for controlled blasting, but for production blasting as well.

The properties of explosives have significant influence on the air decking effects. Of the two explosives modelled in this study, the effects of air decking for emulsion seem to be only auxiliary in rock breakage since emulsion is nearly an ideal explosive and most of its chemical energy is released in primary loading. However, rock damage induced by primary loading for ANFO, a typical non-ideal explosive, is very limited and almost negligible as compared to that due to secondary loading, see Fig. 21(c). If the detonation products are strictly confined, most of the explosive energy for ANFO would have been wasted without blasting effects, see Fig. 21(a). Therefore, it seems that the use of air decking in maximizing energy utilization for non-ideal explosives is much more important than it is for ideal explosives.

11 CONCLUSIONS

The present study has shown that a series of unique physical processes are associated with rock blasting with air decked explosive charges, and such physical processes are responsible for the potential benefits. Under the simulation conditions used in this study, the following conclusions can be drawn:

(1) When there is an air deck on top of the explosive column, it is rapidly occupied by the detonation products. The detonation products move repeatedly in the chamber, impacting the borehole wall until an equilibrium state in pres
sure is reached.

(2) Due to the repeated movement and impact, a significant amount of the potential energy retained in the detonation products is released and imparted to the rock mass, which forms a secondary loading wave in addition to the primary loading wave generated by the detonation of the explosive charge.

(3) There exists a minimum beneficial air deck length which is determined by the equilibrium of the energy loss from primary loading to stemming and the energy gain from secondary loading. Only when the air deck is longer than the minimum beneficial length can the blasting results be modified as compared to that obtained in the case of full stemming.

(4) Under the simulation conditions, as the air deck length increases beyond the minimum beneficial length, the volume of the fragmented rock mass increases, the average fragment size gets smaller, the size distribution becomes more uniform and the diggability of the damage zone is improved.

(5) In the axisymmetric case studied, only the air deck on top of the explosive column benefits rock fragmentation. Variations including bottom air decking, decoupling and mid-column air decking (wave collision) do not modify blasting results according to the conducted simulations.

(6) In the three charging strategies used in controlled blasting, stemming decking has the poorest damage potential to the rock mass, decoupling the intermediate and air decking the maximum.

(7) In maximizing explosive energy utilization, the use of air decking seems to be much more important to non-ideal explosives like ANFO than it is to some more ideal explosives like emulsion.

ACKNOWLEDGMENT

The authors wish to extend their appreciation to Mr. R. Heater for his participation and assistance in the experimental work.

REFERENCES

Brinkman, J.R. (1987) Separating Shock Wave and Gas Expansion Breakage Mechanisms, Second International Symposium on Rock Fragmentation by Blasting, Keystone, Colorado, August 23 -26, 1987, pp6 - 15

Bussey J. and Borg D.G. (1988) Presplitting with the New AIRDEK Technique Proceedings of the 14th Conference on Explosives and Blasting Technique, Society of Explosives Engineers Annual Meeting, Jan. 31 - Feb. 5, 1988, Anaheim, California, pp197 - 217

Chiappetta, R.F. (1992) Precision Detonators and Their Application in Improving Fragmentation, Reducing Ground Vibrations and Increasing Reliability -- A Look into the Future, Fourth High-Tech Seminar, Nashville, Tennessee, USA, June 20 - 25, 1992

Chiappetta, R.F., Mammele, M.E. (1987) Analytical High-Speed photography to Evaluate Air Decks, Stemming Retention and Gas Confinement in Presplitting, Reclamation and Gross Motion Applications, Second International Symposium on Rock Fragmentation by Blasting, Keystone, Colorado, August 23 -26, 1987, pp257 - 301

Foster, C. Le Neve (1894) A Text-Book of Ore and Stone Mining, London,

Fourney, W.L., Barker, D.B., Holloway, D.C. (1981) Model Studies of Explosive Well Stimulation Techniques, International Journal of Rock Mechanics, Min. Sci., and Geo. Mech., Volume 18, pp.113 - 127

Hibbitt, Karlsson & Sorensen, Inc. (1994) ABAQUS/ Explicit (version 5.4), User's Manual, 1994

Hommert, P.J., Kuszmaul, J.S., Parrish, R.L. (1987) Computational and Experimental Studies of the Role of Stemming in Cratering, Second International Symposium on Rock Fragmentation by Blasting, Keystone, Colorado, Aug. 23-26, 1987

Katsabanis, P.D. (1994) Development of a Model to Predict the Effects of Delay Blasting on Rock Fragmentation and Throw, Phase I: Progress Report, prepared for Thiokol Corporation, Elkton, Maryland, Sept. 3, 1994

Knox (1893) German Patent Specification, No. 67,793, 1893

Liu, L., Katsabanis, P.D. (1995a) A Constitutive Model for Predicting Rock Fragmentation by Blasting, 1995 APS Topical Conference Shock Compression of Condensed Matter, Seattle, USA, Aug. 13-18, 1995

Liu, L., Katsabanis, P.D. (1995b) Development of a Continuum Damage Model for Blasting Analysis (under review by Int. J. Rock Mech. Sci. & Geomech. Abstr.)

Liu, L., Katsabanis, P.D. (1995c) A Numerical Study of the Physical Processes in Rock Fragmentation with Air Decked Explosive Charges: Axisymmetric Case (under review by International Journal of Mining and Geological Engineering)

McGill, M.D. (1991) Air-Deck: An Update, Proceedings of the 17th Conference on Explosives and Blasting Technique, Las Vegas, Nevada, USA, Feb. 3-7, 1991

Melnikov, N.V. (1940) Utilization of Energy of Explosives and Fragment Size of Rock in Blasting operations, Gorn. Zh. No. 5

Mel'nikov, N.V., Marchenko, L.N., Seinov, N. P., N.P., Zharokov, I. F. (1979) A Method of Enhanced Rock Blasting by Blasting, IPKON AN SSSR, Moscow, Translated from Fiziko - Techniskie Problemy Razrabotki Poleznykh Isko-Paemykh, No. 6, pp 32-42, November -December

Melnikov, N.V. and Marchenko, L.N. (1971) Effective Methods of Application of Explosion Energy in Mining and Construction, 12th Symposium Dynamic Rock Mechanics (New York, AIME), Chap. 18, pp. 350 - pp 378

Simha, K.R.Y., Fourney, W.L., Dick, R.D. (1987) An Investigation of the Usefulness of Stemming in Crater Blasting, Second International Symposium on Rock Fragmentation by Blasting, Keystone, Colorado, Aug. 23-26, 1987

Rock Fragmentation by Blasting, Mohanty (ed.) © 1996 Taylor & Francis. ISBN 90 5410 824 X

The development and trialling of a cement grout blasthole stemming enhancement cone

T.N. Little
ICI Explosives, Kurri Kurri, N.S.W., Australia

C.E. Murray
Newcrest Mining Ltd, Telfer Gold Mine, W.A., Australia

ABSTRACT: This paper discusses the development and field testing of a Stemming Enhancement Cone (SEC) for its potential introduction into the Eastern Goldfields of Western Australian surface mining operations. The function of the SEC is to improve blasting efficiency by enhancing the effect of the stemming in the blasthole thereby reducing the associated mining costs. For these trials the SEC was constructed of a high early-strength cable bolt grout. The SEC unit was a solid cone placed just above the explosive column within the stemming. The SEC was trialled at the bottom of an abandoned section of an open pit operation in transition sulphide ground conditions. Only crater blast configurations were used and all test holes were nominally 3 m deep and 76 mm in diameter. Test measurements included motion analysis using video techniques, crater or mound profile using manual methods and full waveform airblast monitoring. The SEC has the effect of increasing the surface area affected by the blasthole and locking the stemming into the blasthole for a longer period of time when compared to normally stemmed blastholes. Furthermore, the holes containing SEC generally had lower airblast levels and low ejection heights. The major conclusion is that the introduction of a SEC for routine open pit gold mine blasting is not recommended.

1 INTRODUCTION

Surface mining operations are moving closer to built up areas and the mining of lower grade ore is becoming more common. It has therefore become necessary to examine ways of breaking the rock more efficiently, with less impact on the surrounding environment, while maintaining the required fragmentation. The objective of blasting is to transfer the explosive energy into energy used to fragment and move the rock to the desired muckpile profile as efficiently as possible. Any other effects waste energy and therefore are costly, in excess explosives used, in the damage caused by airblast and flyrock, or in increased wear and tear on excavating equipment caused by poor fragmentation.

Airblast, flyrock and fragmentation can all be improved by confining the explosive gases in the borehole longer to prevent premature venting. Premature venting of explosive gases directly produces airblast and flyrock, in the form of stemming ejection, while degrading the fragmentation of the material.

Ground vibrations are partially reduced with the correct use of stemming but are more effectively controlled with the accurate timing of the initiation devices and therefore will not be included in this report.

The above adverse effects can be controlled or reduced to acceptable levels with the correct use of stemming in most cases.

There are three methods of ensuring that the stemming will remain in the borehole longer.

(i)   By improving the quality of the stemming, with respect to its retention in the borehole, under high pressure.

(ii)   By increasing the length of stemming in the borehole so that the expanding gases must overcome a larger resistance before ejection can occur.

(iii)   By incorporating a mechanical device into the borehole to enhance the "locking" ability of the stemming and thereby increasing the stemming retention time in the hole.

Where the desired results cannot be obtained with the use of stemming alone, due to logistic or economic barriers, other means of explosive confinement have to be used. Method three, using a Stemming Enhancement Cone, is the subject of this paper.

1.1 Background enhancement cone

A blast control plug was designed by Paul Worsey, funded by the U.S. Department of Energy. It is designed to operate with drill cuttings and enhance the "locking" effect of the stemming, which in turn reduces mining costs. It has been claimed Worsey(1990) that substantial reductions in stemming ejection velocity or ejection elimination were achieved whilst reducing stemming lengths by up to 35% over conventional practice. Peak airblast pressures were also reduced by between 8 and 25 dBL. Our cement grout SEC is shown in its operating position in Figure 1.

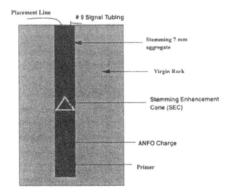

Figure 1 Typical loaded blasthole with a SEC in the stemming.

1.2 Placement into blasthole

Two hole diameters of stemming are placed into the hole before the SEC is installed. This protects the SEC from the destructive forces of the detonating explosive column. The placement line is tied to a length of surveyors flagging, then using a piece of 20 mm PVC tubing the plug is placed in the hole to the correct orientation. The PVC is then removed from the hole leaving the flagging behind. The hole is then back filled as usual using drill cuttings. Ensure that the flagging doesn't fall down the hole on top of the SEC as this will adversely affect its performance.

1.3 Working principle

The reaction mechanism which enables the SEC to enhance stemming characteristics is as follows. On detonation of the hole the rapid expansion of the explosive gases attempts to eject the stemming and thereby forces the SEC up into the stemming. Due to the shape of the SEC this action forces the stemming against the borehole wall which effectively locks the stemming column in place with a greater than normal lateral force. As soon as this stemming material fails in shear, the plug moves further up the hole and repeats the process by producing new shear components on new stemming material. Because the shear component in the stemming is greater and achieved at a quicker rate when compared to standard stemming, the stemming retention time in the hole is increased. The locking mechanism for the stemming is not dominated by the characteristics of the stemming itself, but more by the influence of the SEC; drill cuttings can therefore be used instead of importing special crushed aggregates.

The SEC is only suited to dry holes. According to Worsey (1990) this is attributed to fluidization as a result of the relative incompressibility of water and its lubricating effect under pressure. Also the benefits of the SEC are minimal in oxide ground due to the dilation of the borehole under high pressures and therefore not allowing the full potential of the SEC to be recognised.

2 PRODUCTION OF THE SEC

The moulds were made from 125 mm diameter solid nylon rod. A 90° cone was lathed into the nylon which enables plug diameters ranging up to 110 mm in diameter to be produced.

A 4 mm hole was drilled in the apex of the cone, (base of the mould), to allow a piece of 3 mm diameter braided cord to be fixed to the apex of the cone. The cord will allow the SEC to be lowered and placed correctly in the hole.

The cones were manufactured oversize so that its base can be rounded off to prevent damage to leads going down the hole while maintaining the diameter of the plug at 90% of the hole diameter. After the cones were taken out of the moulds they were placed in water for a minimum of fourteen (14) days to cure. They were then sized to 68 mm.

The only difference between the Paul Worsey's cone design and the one used in the trials is that the placement rod on the initial design has been placed with the cord/flagging arrangement. This was to enable easy manufacture of the cones and also enables marginally more stemming to be placed into the hole.

2.1 Cast material

Foseco Technik's Lokset CB Cable Bolt Grout was used for the 68 mm diameter plugs. This was available and exhibited a high compressive strength.

Lokset CB is a Type C grout supplied as a ready to use powder requiring only the addition of water to produce a plastic free flowing non-shrink grout for gap widths of 10 - 125 mm. The material is a blend of cements, graded fillers and additives which impart controlled expansion, while minimizing water demand. The low water requirement ensures high early strength and long term durability. The filler grading is designed to aid uniform mixing and minimize segregation and bleeding.

2.2 Grout strength

The compressive strength of the grout reaches 44 MPa in 7 days and 60 MPa in 14 days with a maximum strength of 94 MPa in 28 days. The early strength is important due to the restricted time between manufacture and use.

2.3 Site selection

An area of competent "hard rock" had to be found for the trials of the Stemming Enhancement Cone. Two sites were considered, an abandoned aggregate quarry and the bottom of an open pit. The bottom of a pit being ideal because ground conditions matched conditions in normal blasting operations.

The area chosen for the trails in determining the effect of the Stemming Enhancement Cone was Kalgoorlie Consolidated Gold Mines, Mt Percy Operations Mystery Pit. An abandoned section of the pit was selected with suitable ground conditions for the field trials. This area of the Mystery Pit was mainly a sulphide section which was required for the consistent testing of the SEC.

Ausdrill Pty Ltd was approached and agreed to donate the drill rig and drillers' time to drill the 52 holes required for the trials. The selected area of the pit was drilled on a 5 m x 5 m square pattern to a depth of three metres. The spacing of five metres was considered necessary to eliminate any interaction between holes. This spacing proved to be adequate.

2.4 Initiation system

KCGM agreed to supply the explosives and initiation system required for the blasts. The explosive selected was ANFO to be initiated with "K" booster and a Nonel system. Number 12 Nonels were used as a surface delay (400 ms) with number 9 Nonels being used down the holes (250 ms). The surface detonators were covered with 150 mm of sand to prevent any interference in the airblast monitoring.

A minimum of 400 ms delay between each hole was required to give a clear result on the airblast records for each hole. If the delay is shorter the signals may merge and give false readings of the peak pressures generated by each hole.

All the holes were dry and ANFO was used on all shots. The density of free pour ANFO has been taken as 0.80 g/cm^2. This gives a mass per unit length of 3.53 kg/m in a 76 mm hole. The holes drilled for the trials were all 76 mm in diameter. Figure 2 shows the

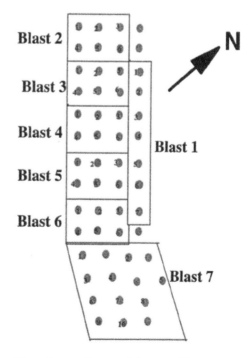

Figure 2   Layout of the seven blasts

layout of the holes and the blasts that they were fired in. The airblast monitoring location was approximately 200 m in the south-east direction.

3 MONITORING SYSTEMS

3.1 *Video recording*

The video camera used to monitor the blasts was a SONY H18 CCD-V900 operating at 25 frames per second. This gives a frame interval of 40 ms. The camera was mounted on a tripod in the same position for all the blasts enabling easy comparison of the series of blast undertaken.

From the video we can determine the ejection velocities and height. A rough estimate of the stemming retention time can also be quantified along with the surface area affected by each hole.

3.2 *Playback equipment*

The video was replayed using a SHARP VC-486X video cassette recorder and NEC ST-2102 PAL television.

When viewing the blasts the ejection was classified into three categories:

- Stemming ejection.
- Cratering.
- Combination of the above.

The relative magnitude of the combination is reported as the percentage of stem ejection in the blast dynamics results whereas cratering details are listed in the post-blast results for each trial.

3.3 *Airblast monitoring*

The system used to measure the airblast is shown in Figure 3.

The individual components of the measurement system consisted of the following B&K components: #4147 condenser microphones, #2639T preamplifier and a #2804 two channel power supply. The microphone unit was covered with the recommended windscreen.

The preamplified airblast signal was a.c. coupled and fed into a computer-basted data acquisition system which consisted of a Toshiba 3200 Laptop with a

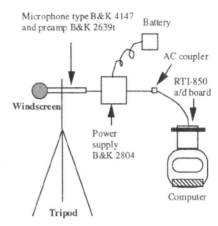

Figure 3 Airblast Monitoring Equipment

plug-in RTI-850 High Resolution Data Acquisition Card. The Data Acquisition Operating System was installed on the Toshiba computer and enabled the data to be acquired in real-time. (Blair and Little, 1991)

A duplicate system was used to enable a comparison of the signals and ensure the accuracy of the results.

3.4 *Post blast visual inspection*

A visual inspection of each hole after the blast gives added information that the monitoring equipment cannot pick up. These areas include:

- The depth of the crater or the height of the mound created by the blast.

- The diameter of the effected area due to the crater or mound created.

- Degree of fragmentation.

4 BLASTING TRIALS

4.1 *Blast number one*

There were seven holes in the blast. All holes were charged with ANFO with various charge lengths using two different types of stemming. The layout of the trials is seen in Figure 2.

The holes used were mostly in an oxide section of the blast pattern. Holes five, six and seven did show some sulphide chips in the drill cuttings indicating that the ground was in a transition zone. Hole seven was mainly sulphide cuttings. These holes were used so that the remaining sulphide holes could be used for the trials. The oxide nature of these holes while not fatal for this initial trial did create an unwanted variable.

Two types of stemming were used to determine which type would give the best stemming ejection results on the video tape and airblast monitor. For this blast the monitoring equipment failed to trigger and as a result no data was obtained. Photographs were taken and a visual inspection of the craters was carried out. However, the findings were generally inconclusive.

Table 1. Hole by hole blast design parameters for Blast No 1

Hole No	Charge Length (m)	Charge Mass (kg)	Stemming Length (m)	Stemming Material
1	1.6	5.65	1.0	Cutting
2	1.8	6.35	1.2	Cutting
3	1.6	5.65	1.4	Cutting
4	1.8	6.35	1.0	7 mm agg
5	1.7	6.00	1.2	7 mm agg
6	1.6	5.65	1.4	7 mm agg
7	2.2	7.77	0.8	7 mm agg

Table 2. Hole by hole blast design parameters for Blast No 2

Hole No	Charge Length (m)	Charge Mass (kg)	Stemming Length (m)	Stemming Material
1	1.0	3.53	2.0	Cuttings
2	1.6	5.65	1.4	Cuttings
3	2.0	7.06	1.0	Cutting
4	2.0	7.06	1.0	7 mm agg
5	1.6	5.65	1.4	7 mm agg
6	1.0	3.53	2.0	7 mm agg

Table 3 Hole by hole blast dynamics results for Blast No 2

Hole No	Stemming Ejection Mode	Peak Rock Motion Height (m)	Peak Rock Motion Velocity (m/s)	Airblast Level (dB Lin)
1	0%	4.00	11.75	119.3
2	70%	16.11	103.75	133.5
3	60%	11.65	100.00	126.9
4	80%	19.90	182.25	143.7
5	65%	9.06	63.25	125.7
6	0%	2.23	8.88	119.3

4.2 Blast number two

Six holes were trialled, as shown in Figure 2, with the blast being recorded with the video camera and airblast monitor. The airblast recording showed the individual holes clearly with pressures ranging from 119 dB to 144 dB. The wind conditions were still and had little or no effect on the airblast. The video gave frame by frame results at 40 ms intervals.

The drill cuttings indicated that the holes were mainly all in sulphide materials with about 0.3 m of damage oxide material on the surface.

Holes No 1 and No 6 did not appear to have any stemming ejection and hence the peak rock motion velocity refers to the face velocity. The degree of stemming ejection is a subjective assessment by the researcher. It represents the percentage of the rock motion attributed to the stemming ejection and implies that the remaining percentage can be attributed to the face or crater ejection mode.

The results of Blast No 2 indicate that for the Mystery Pit blasting environment and the design used there is no observable difference between the performance of the drill cuttings and imported 7 mm crushed aggregate. It may be that the 7 mm aggregate was too fine, although it complies with the general rule of thumb for stemming particle size.

Holes No 2 and No 5 with a charge length to stemming length ratio of close to 1:1 gave measurable stemming ejection levels that were considered suitable for subsequent trial design.

4.3 Blast number three

Based on the results from Blast No 2 the following hole charge/stemming configurations were used. Holes No 1 and No 6 were loaded differently because they were short holes to start with. These

Table 4 Hole by hole post-blast results for Blast No 2

Hole No	Crater Depth (m)	Mound Height (m)	Feature Diameter (m)	Fragments
1	1	None	2.5-3.0	Large
2	None	1	4.0	Medium
3	0.75-1.0	None	2.5-3.0	Small
4	None	0.5	2.0	Medium
5	None	0.5	1.5-20	Medium
6	None	0.5	1.5-2.0	None

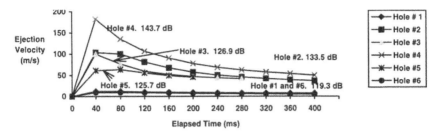

Figure 4   Motion Velocity Versus Elapsed Time for Blast No 2

Table 5 Hole by hole blast design parameters for Blast No 3

Hole No	Charge Length (m)	Charge Mass (kg)	Stemming Length (m)	Stemming Material
1	1.3	4.59	1.2	7 mm agg
2	1.5	5.30	1.5	SEC
3	1.5	5.30	1.5	7 mm agg
4	1.5	5.30	1.5	SEC
5	1.6	5.65	1.4	7 mm agg
6	1.3	4.59	1.2	SEC

Table 6 Hole by hole blast dynamics results for Blast No 3

Hole No	Stem Ejection Mode	Peak Motion Height (m)	Peak Motion Velocity (m/s)	Peak Airblast Level (dB Lin)
1	Minor	14.00	33.25	NR
2	70%	20.44	66.75	NR
3	None	8.00	33.25	NR
4	50%	16.22	55.58	NR
5	40%	12.00	61.00	NR
6	None	8.67	39.00	NR

Table 7 Hole by hole post-blast results for Blast No 3

Hole No	Crater Depth (m)	Mound Height (m)	Feature Diameter (m)	Fragments
1	None	1.0	3.0	Med-Large
2	0.5	None	3.0	Large
3	0.5 - 1.0	None	3.0	Small
4	1.0	None	3.0	Large
5	1.5	None	3.0	Very Large
6	1.5	None	3.0	Large

holes also gave an opportunity to look at the effects of different stemming/charge lengths. This method of dealing with any short holes was carried out for the remaining trials.

The initiation system and sequence was the same as that used for Blast No 2. The airblast recording of this trial was triggered early and therefore failed to pick up the blast.

The results of Blast No 3, see Figure 5, show a strong correlation between degree of stemming ejection and the motion velocity. This was a surprise but the introduction of the SEC did not markedly affect the results.

4.4   Blast number four

This trial follows the principles outlined in the previous blast, the pattern layout also being the same. The blast was recorded on both the video and the airblast monitor.

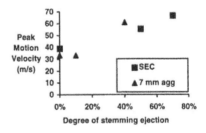

Figure 5   Peak Motion Velocity

Table 8. Hole by hole blast design parameters for Blast No 4

Hole No	Charge Length (m)	Charge Mass (kg)	Stemming Length (m)	Stemming Material
1	1.5	5.30	1.5	SEC
2	1.5	5.30	1.5	7 mm agg
3	1.5	5.30	1.5	7 mm agg
4	1.5	5.30	1.5	SEC
5	1.5	5.30	1.5	7 mm agg
6	1.5	5.30	1.5	SEC

Table 9. Hole by hole blast dynamics results for Blast No 4

Hole No	Stem Ejection	Peak Motion Height (m)	Peak Motion Velocity (m/s)	Airblast Level (dB Lin)
1	None	9.18	22.75	125.6
2	50%	9.82	29.50	121.1
3	None	13.45	30.33	122.9
4	50%	12.73	47.75	123.8
5	50%	7.73	33.00	119.4
6	50%	14.18	56.75	125.8

Table 10. Hole by hole post-blast results for Blast No 4

Hole No	Crater Depth (m)	Mound Height (m)	Feature Diam (m)	Fragments
1	None	1.0	3.0	Med to Large
2	None	0.2	2.0	Small
3	0.5	None	1.5- 2.0	Small to Med
4	None	0.2	2.0	Small
5	1.0	None	2.5	Small oxide
6	1.5-2.0	None	4.0	Large

Table 11. Hole by hole blast design parameters for Blast No 5

Hole No	Charge Length (m)	Charge Mass (kg)	Stemming Length (m)	Stemming Material
1	1.7	6.0	1.3	SEC
2	1.7	6.0	1.3	7 mm agg
3	1.7	6.0	1.3	7 mm agg
4	1.7	6.0	1.3	7 mm agg
5	1.7	6.0	1.3	SEC
6	1.7	6.0	1.3	SEC

Table 12. Hole by hole blast dynamics results for Blast No 5

Hole No	Stem Ejection	Peak Motion Height (m)	Peak Motion Velocity (m/s)	Airblast Level (dB Lin)
1	None	8.50	13.92	122.2
2	70%	14.67	31.25	124.0
3	30%	12.25	29.25	122.9
4	30%	12.08	29.25	121.2
5	None	11.67	54.25	127.4
6	None	13.00	27.75	124.7

4.5 Blast number five

There was a slight delay difference on this blast compared to Blast No 2. The down hole delays for hole three and four were 200 ms (No 8 Nonels) instead of the normal 250 ms (No 9 Nonels). This

Table 13. Hole by hole post-blast results for Blast No 5

Hole No	Crater Depth (m)	Mound Height (m)	Feature Diameter (m)	Fragments
1	None	1.0	3.0-4.0	Small
2	1.0	None	2.5-3.0	Small
3	1.0	None	2.5-3.0	Medium
4	1.0-1.5	None	3.0-4.0	Large
5	1.5	None	3.0	Small
6	2.0	None	3.0	Large

Table 14. Hole by hole blast design parameters for Blast No 6

Hole No	Charge Length (m)	Charge Mass (kg)	Stemming Length (m)	Stemming Material
1	1.4	4.94	1.5	7 mm agg
2	1.4	4.94	1.5	7 mm agg
3	1.3	4.57	1.5	7 mm agg
4	1.5	5.30	1.5	SEC
5	1.4	4.94	1.5	SEC
6	1.5	5.30	1.5	SEC

Table 15. Hole by hole blast dynamics results for Blast No 6

Hole No	Stem Ejection	Peak Motion Height (m)	Peak Motion Velocity (m/s)	Airblast Level (dB Lin)
1	None	3.56	13.88	116.5
2	50%	8.89	18.06	119.2
3	60%	11.33	27.75	118.0
4	None	5.78	22.25	118.2
5	None	6.33	13.00	118.2
6	30%	5.33	12.00	113.9

will have no effect on the stemming ejection characteristics, just the timing of the ejection. The charge length was increased slightly in this trial to try and obtain better stemming ejection definition. The blast was again recorded on the video camera and the airblast monitor.

4.6 Blast number six

This trial followed the same guidelines as set down for trials three and four. Unfortunately, there were a number of short holes in this blast. All holes were given 1.5 m of stemming, with and without the SEC. The length of the stemming was deemed to have more influence on the blast results than the length of the explosive column. In general the holes were only 100 mm short. The blast was again recorded with airblast monitor and video camera. The airblast signal is shown in Figure 7 with the individual hole signals clearly visible.

Table 16. Hole by hole post-blast results for Blast No 6

Hole No	Crater Depth (m)	Mound Height (m)	Feature Diam (m)	Fragments
1	None	1.0	2.0	Medium
2	1.5	None	2.0-2.5	Small-Med
3	2.0	None	2.0-2.5	Small
4	None	1.5-2.0	2.0-2.5	Medium
5	None	1.5	1.5-2.0	Medium
6	None	1.0	1.5	Med-Large

Table 17. Hole by hole blast design parameters for Blast No 7

Hole No	Charge Length (m)	Charge Mass (kg)	Stemming Length (m)	Stemming Material
1	1.5	5.30	1.5	SEC
2	1.7	6.00	1.1	SEC
3	1.4	4.94	1.5	SEC
4	1.5	5.30	1.5	SEC
5	1.35	4.77	1.5	SEC
6	1.9	6.71	1.1	SEC
7	1.45	5.12	1.5	SEC
8	1.5	5.30	1.5	SEC
9	1.5	5.30	1.5	SEC
10	1.5	5.30	1.5	SEC

4.7   Blast number seven

The remaining 14 holes were used as a control to determine the variability due to geology.

4.8   Further Analyses

A preliminary analysis using all the data exhibited a high degree of scatter so it was decided to concentrate on a reduced data set. Also, in order to

Table 18. Hole by hole blast dynamics results for Blast No 7

Hole No	Stem Ejection Mode	Peak Motion Height (m)	Peak Motion Velocity (m/s)	Airblast Level (dB Lin)
1	60%	16.00	40.50	119.9
2	70%	24.31	58.63	127.6
3	None	11.38	23.13	123.1
4	None	1.60	19.25	120.4
5	None	1.54	15.38	122.8
6	None	7.08	23.06	144.2
7	60%	21.24	78.88	123.0
8	None	6.15	13.50	124.0
9	None	11.31	30.75	116.1
10	None	9.00	30.75	125.8

Table 19. Hole by hole post-blast results for Blast No 7

Hole No	Crater Depth (m)	Mound Height (m)	Feature Diam (m)	Fragments
1	None	1.0	2.5	Small
2	1.0	None	4.0	Medium
3	None	0.5	1.5	None
4	None	Radial	Cracks	None
5	None	1.5	2.5	Small to Med
6	None	1.75	3.5-4.0	Medium
7	0.75	None	2.0	Small
8	None	1.5	3.0	Medium
9	None	1.5	3.0	Medium
10	0.5	None	2.3-3.0	Medium

focus on the major question at hand, that is, "Is this type of SEC worth using?" we restricted all further analysis to only those blastholes with the following design:

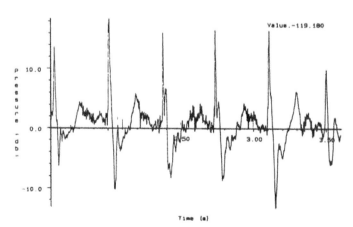

Figure 7   Airblast Waveform Record for Blast No 6

338

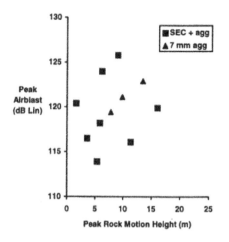

Figure 8 Plot of peak airblast versus peak ejection height for the reduced data set.

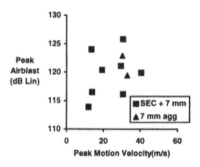

Figure 9 Plot of peak airblast versus peak rock motion velocity for the reduced data set.

- charge length 1.5 m of ANFO
- stemming length 1.5 m
- charge weight 5.3 kg

Eleven such blastholes contained the SEC and four blastholes with this design had only the 7 mm aggregate stemming and a compilation of these results are presented as Tables 20 and 21 respectively .

Figure 8 illustrates the lack of difference between those blastholes that had the SEC and those that only had 7 mm aggregate stemming. One obvious limitation of this data is that the peak motion height data contains information on two distinct types of events, they are stemming ejection and free face motion.

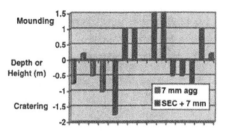

Figure 10 Cratering and mounding features for the reduced data set.

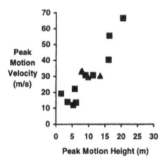

Figure 11 Peak motion velocity versus peak motion height.

This plot illustrates the lack of difference between those blastholes that had the SEC and those that only had 7 mm aggregate stemming. One obvious limitation of this data is that the peak motion velocity data contains information on two distinct types of events, they are stemming ejection and free face motion.

The results illustrated in Figure 10 show a large range of post-blast features from 1.75 m deep craters to 1.5 m mounds. No apparent difference is obvious between those blastholes that contained SEC and those that contained only the 7 mm aggregate.

A strong correlation between the peak motion velocity and peak motion height is evident in Figure 11. However, the data for the 7 mm aggregate blasthole is sitting centrally in the middle the SEC data. This strong correlation can be explained in terms of the stemming ejection motion having high motion velocities and peak heights. On the other hand, the free face motion (horizontal in crater blasts) has significantly lower

Table 20　　　　Listing of Identical Blasthole with SEC

Blast No /Hole No	Stem Ejection Mode	Peak Motion Height (m)	Peak Motion Velocity (m/s)	Peak Airblast Level (dB Lin)	Crater Depth (m)	Mound Height (m)	Feature Diameter (m)	Fragments
B3H2	70%	20.44	66.75	-	0.5	None	3.0	Large
B3H4	50%	16.22	55.58	-	0.5 - 1.0	None	3.0	Small
B4H1	None	3.56	13.88	116.5	None	1.0	3.0	Med to Large
B4H4	50%	9.82	29.50	121.1	None	0.2	2.0	Small
B4H6	50%	5.33	12.00	113.9	1.5 - 2.0	None	4.0	Large
B6H6	30%	5.33	12.00	113.9	None	1.0	1.5	Med to Large
B7H1	60%	16.00	40.50	119.9	None	1.0	2.5	Small
B7H4	None	1.60	19.25	120.4	None	Radial	0.0	None
B7H8	None	6.15	13.50	124.0	None	1.5	3.0	Medium
B7H9	None	11.31	30.75	116.1	None	1.5	3.0	Medium
B7H10	None	9.00	30.75	125.8	0.5	None	2.3 - 3.0	Medium
Sum		100.72	317.21	1068.70			27.65	
Sample size		11	11	9			11	
Average		9.16	28.84	118.74			2.51	
Std Dev		5.77	17.73	3.95			1.00	
Coeff Var		0.62	0.61	0.03			0.40	

Table 21　　　　Listing of Identically Blasthole with only 7 mm aggregate

Blast No /Hole No	Stem Ejection Mode	Peak Motion Height (m)	Peak Motion Velocity (m/s)	Peak Airblast Level (dB Lin)	Crater Depth (m)	Mound Height (m)	Feature Diameter (m)	Fragments
B3H3	None	8.00	33.25	-	0.5 - 1.0	None	3.0	Large
B4H2	50%	9.82	29.50	121.1	None	0.2	2.0	Small
B4H3	None	13.45	30.33	122.9	0.5	None	1.5 - 2.0	Small to Med
B4H5	50%	7.73	33.00	119.4	1.0	None	2.5	Small
Sum		39.00	126.08	363.40			9.25	
Sample size		4	4	3			4	
Mean		9.75	31.52	121.13			2.31	
Std Dev		2.28	1.63	1.43			0.48	
Coeff Var		0.23	0.05	0.01			0.21	

motion velocities up to about 30 m/s and peak motion height up to approximately 11 m.

5　　DISCUSSION

A comparison between the results in Table 20 and Table 21 indicate that the mean peak motion height, mean motion velocity and mean peak airblast level is less for the SEC than the blastholes that contained only the 7 mm aggregate stemming. Thus it can be claimed that the SEC is working to some extent. Because of the high variability in the motion data the only reliable data is thought to be the airblast data. The statistics indicate that 66% of the airblast results using the SEC will be between 114.79 dBL and 122.69 dBL. For the 7 mm aggregate stemming 66% of the airblast result would be expected to be between 119.70 dBL and 122.56 dBL. However it should be remembered that the later limits are based on a very limited data set.

A number of limitations with the trials and subsequent analyses have been identified. These are as follow:

• The frame by frame replay of 40 ms on the video camera was too insensitive to obtain accurate initiation times and elapsed times of the events.

- The scale of the image played back on the television produced errors in the order of 12.5 m/sec. Where the image was not as sharp this error increased.

- Problems introduced by overlapping blast motion events.

- The cut-off point for the peak ejection height was sometimes impossible to pick due to dust cloud blurring.

- The geology variation in the trial site and the blast damage zone was not known.

- The motion data contains information on two distinct type of events, they are stemming ejection and free face motion.

6   CONCLUSION

(1) The results of blast #2 indicate that for the Mystery Pit blasting environment and the design used there is no observable difference between the performance of the drill cuttings and imported 7 mm crushed aggregate.

(2) The construction and use of the cement grout stemming enhancement cone is a fairly straight forward process.

(3) In the particular environment where these trials were conducted it was found that our SEC did not significantly improve crater blast performance. It may be that to be fully effective an undamaged hard rock mass is required.

(4) The high degree of variability in the performance from blastholes with the same design indicates the paramount importance of local geology (and blast damage pre-conditioning) on blasting performance.

(5) If further trials are undertaken it is suggested that they should involve:

- determining the interval between hole initiation to the first sign of venting gases.

- full scale bench blasting in a hard rock environment.

- comparisons between the commercially available plugs and a home-made Stemming Enhancement Cone, similar to the one described in this paper.

- clearly identifying the differences between purely stemming ejection events, purely free face motion and hybrid combination events.

(6) The major conclusion is that the introduction of a SEC for routine open pit gold mine blasting is not recommended.

ACKNOWLEDGEMENTS

The authors would like to thank the Western Australian School of Mines, Kalgoorlie Consolidated Gold Mines, Ausdrill Pty Ltd and Total Energy Systems for providing the infrastructure, trial site, drilling and explosives respectively. Dr J Jiang is also thanked for his assistance with the monitoring. ICI Explosives are commended for sponsoring the publishing of this paper. Dr P Warburton and Dr D Blair both of ICI Explosives also provided valuable comment and review.

REFERENCES

Armstrong, L.W. (1994). The quality of stemming in assessing blasting efficiency. ME Thesis Univ. of New South Wales, Australia

Little, T.N. (1994). Practical approaches to airblast monitoring and management. Proc. of Open Pit Blasting Workshop 1994 (OPBW94). Curtin University, Perth, Sept. 10-12.

Murray, C.E., (1991). The development and testing of a stemming enhancement cone designed to improve explosive performance. Western Australian School of Mines, BEng. Thesis.

Worsey, P. (1990). Stemming ejection comparison of conventional stemming and stemming incorporating blast control plugs for increased explosion energy use. Proc. of 3rd Int. Fragblast Conf., Brisbane, Ausstralia, August 26-31.

Rock Fragmentation by Blasting, Mohanty (ed.)© 1996 Taylor & Francis. ISBN 90 5410 824 X

Feasibility of air-deck blasting in various rock mass conditions – A case study

A.K.Chakraborty & J.L.Jethwa

Central Mining Research Institute, Regional Centre, Shankar Nagar, Nagpur, India

ABSTRACT: A feasibility study on the use of air-deck blasting was conducted in the footwall (F/W) and hangwall (H/W) sides of the overburden benches at Dongri Buzurg Mine of Manganese Ore (I) Ltd. in Maharashtra, India. Bieniawski's (1973) RMR ratings of the rock mass of each blast site was determined. Blast performance like fragmentation, powder factor, throw, overbreak, underbreak and ground vibration were monitored both with and without air-deck. Relations between the average fragment size and specific surface area with powder factor and RMR were determined on the basis of trial blast results. Analysis of the blast results indicated a 13 per cent saving in explosive consumption, reduction in ground vibration by 40 per cent, in overbreak by 50 per cent and in throw by 45-50 per cent in air-deck blasting compared to those in conventional blasts. On the basis of trial blast results, specific blast patterns with air-decking were recommended in different sets of rock mass units defined by RMR ranges of 25-35, 35-50 and 50-65.

1 INTRODUCTION

In blasting, much of explosive energy is wasted in super fragmentation or production of under-size fragments. Further, the efficiency of the loading equipments is reduced due to generation of oversize boulders. Therefore, a method should be evolved for more efficient utilisation of the explosive energy to obtain uniform fragmentation. By introducing one or more air-gaps in the explosive column, it is possible to utilise the shock energy in a more purposeful manner, though the intensity of the gas energy is, thereby, reduced (Melnikov et al. 1972). Melnikov et al. (1972) also opine that shock energy from either ends of the explosive column collides in the air gap provided by air-decking and generates a new source of high pressure. The shock waves, after collision, move in the opposite directions and are reflected from the hard obstacles at the hole bottom and the stemming column. These again collide in the air column. The process is repeated for a number of times and restricts attenuation of shock waves in the surrounding media and increases the duration and utilisation of the shock waves in rock fragmentation. It is claimed that the energy utilisation with charges having air-gaps is as high as 50 per cent (Marchenko, 1982). Whereas that in case of blasting was found as 15 per cent only (Singh, 1988). Moxon et al. (1993) found that mid-column air-decks produce better fragmentation than those obtained by bottom-column or top air-deck and that the optimum size of the deck depends on the geo-mechanical properties of rocks. They also suggested that the total length of the air-deck should not be more than 15-35

per cent of the explosive column, the lower value being more suitable for harder formations. It was possible in the former USSR to reduce the explosive consumption up to 30 per cent in surface blasting by using air-decks (Moxon et al., 1993). A feasibility study on the use of air-deck blasting in Dongri Buzurg opencast mine of Manganese Ore (I) Ltd. (MOIL) was carried out with a view to improve fragmentation with reduced explosive consumption. The present report includes the field investigations and analysis of data to assess the feasibility of air-deck blasting in different rock mass conditions.

2 GEO-MINING DETAILS OF THE MINE

2.1 Location

Dongri Buzurg Mine is situated near the village Balapur Hamesha in the Bhandara district of Maharashtra State. This place is about 6 kms west of Goberwahi on Tumsar-Katangi State Highway, which is about 25 kms away from Tumsar town and approximately 125 kms away from Nagpur. A location map of the mine is given in Fig. 1.

2.2 Geology

2.2.1 Rock succession

The manganese ore horizon occurs in the lower part of a sequence of metasedimentary rocks of Sausar group of Pre-cambrian age. The footwall and the hangwall

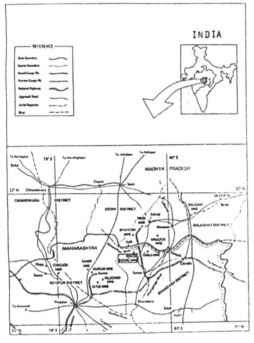

Fig.1 Location map of Dongri Buzurg Mine

Table 1. Rock succession in Dongri Buzurg mine

Rock type	Sausar series	
Mica schist	Munsar stage	
Manganese ore/	Lohangi Stage	Pre-Cambrian
Quartz muscovite/ schist	Sitasaongi stage	
Muscovite gneiss		

rocks belong to Sausar series and comprise of Tirodi, Sitasaongi and Munsar formations. The Munsar formation forms the footwall of the ore horizon. The Sitasaongi formation followed by Tirodi gneisses form the haning wall. The rock succession in the area is as given Table 1.

2.2.2 Geo-technical Observations

Bieniawski's (1973) rock mass rating (RMR) was determined in the F/W and the H/W side of the opencast mine (Table 2) to assess the nature of the rock mass in different locations. The RMR varied from 25 to 65.

2.3 Mining Methods

The central portion of the ore deposit is being worked by opencast mining. Underground mining is being carried out in the eastern and the western flanks. In opencast workings, 110mm large diameter drilling and shovel-dumper combination are deployed for development only. Jack hammer drilling and manual loading are being used for selective ore mining. Details of mechanisation used in developments are listed in Table 3.

Table.2
Rock mass rating (RMR) observed in different locations of Dongri Buzurg mine

Sl no	Location rock mass	Type of rock mass	Ratings						RMR
			UCS	RQD	Spacing of joints	Joint condition	Ground water	Joint orientation	
1	H/W, 10 ft levels, Ch. 3000 ft	Quartzite-muscovite gneisses	7	13	15	20	10	0	65
2	H/W, 10 ft level, Ch. 2900 ft	same as 1	7	13	8	20	10	0	58
3	F/W, 160 ft level, Ch. 2800 ft	Muscovite-schist	2	8	10	20	10	-25	25
4	F/W, 160 ft level, Ch. 3000 ft	same as 3	2	8	20	20	10	-25	35
5	F/W, 160 ft level, Ch. 2900 ft.	same as 3	2	8	20	12	10	-25	27
6	H/W, 35 ft level, 5000 ft	Micaceous schist	4	13	8	12	10	0	47

* UCS - Uniaxial compressive strength
  RQD - Rock quality designation

Table 3. Details of mechanisation used in Dongri Buzurg mine

Sl. no.	Unit operation	Mechanisations
1	Drilling	Type: ICM, Make: Ingersol Rand, Size: 110 mm
2	Mucking	Poclain, Type: CK170, Make: L&T, Capacity: 1.7 m³
3	Hauling	Tippers - Make: M/S Hindustan Motors; Capacity:8m³ - Make: Ashok Leyland Capacity: 6m³

Slurry explosive cartridges of 83 mm diameter, 0.45m length and 2.78kg weight were used in blasting of 110m holes. The powder factor varied from 0.3 to 0.4 kg/m 3, the higher value used for harder formations.

2.4 Major Blasting Problems

The following major blasting problems were identified in the developmental activities :

1. Generation of oversize boulders, specially in the F/W side.

2. Formation of underbreak or toe in the bench foot.

3. Overbreak in the bench top causing drilling problems in the subsequent blasting rounds.

3 TRIAL BLASTS

Six trial blasts were conducted out of which five blasts numbering 1,2,3,4 and 5 were conducted with air-decking at different locations of the F/W and the H/W sides of the mine with various blast geometry like hole depth, burden and spacing. The air column

was provided with wooden stick placed in between the cartridges in the explosive column. Blast no. 6 was conducted in the hangwall side without any air deck. The blast input parameters and the blast results are listed in Tables 4 and 5.

4 ANALYSIS

Fragmentation assessment was made on the basis of measurements of about 200 rock fragments and visual observations of the undisturbed muck pile, muck in the dumpers and in the dump site and two dimensional analysis of the photographs of the muck pile. Curves between the average fragment size and the cumulative percentage are drawn in Fig. 2.
The mean fragment size (MFS, m) in the present study was estimated using the following relation (Moxon et al., 1993):

$$MFS = \frac{\sum Mass \% \times Mean\ sieve\ size}{100} \qquad (1)$$

The specific surface area (SSA, m^2/t) was estimated using the relation given below (McCabe et al., 1967):

$$SSA = \frac{6\lambda}{\rho} \cdot \frac{\phi_n}{D_{n'}} \qquad (2)$$

where, λ = shape factor = 1.5
ρ = density (T/m³) = 2.766 T/m³
ϕ_n = mass fraction of the total sample that is retained by the screen `n'
$D_{n'}$ = arithmatic average of the particle diameter of Dn and Dn-1

The values of MFS and SSA of the muck produced in different blast rounds are given in Table 6.

Table.4 Blast input parameters

Blast no.	Location and formation	RMR	Bench height (m)	Hole depth (m)	No.of holes	Burden x spacing (mxm)	Charge /hole /delay (kg)	Maximum charge/ length	Air-deck number & length
1.	H/W, Quartzite-muscovite-gneisses	65	6.0	6.5	11	2.5 x 3	11.5	62.95	One of 0.9m
2.	H/W, same as 1	58	8.0	8.5	26	2.5 x 3.5	22.24	266.88	One of 0.9m
3.	F/W, Muscovite-schist	21	9.5	10.5	11	2 x 3.5	26.41	132.08	Two each of 0.9m
4.	F/W, same as 3	35	9.5	4.0	26	2 x 2.5	8.34	98.69	One of 0.9m
5.	F/W, same as 3	27	10.0	10.0	9	2 x 2.5	22.24	100.08	Two each of 1.5m
6.	H/W Micaceous schists	47	7.5	8.0	11	2.5 x 3	16.68	66.72	No decking

Table.5 Result of trial blasts

Blast no.	Powder Factor (kg/m 3)	Fragmentation	Overbreak (m)	Throw (m)	Toe	Vibration (mm/s) at distance of (m)
1.	0.278	Good	Almost nil (last line of breakage along the last row of holes)	Tight muck pile, 10m throw in one side	Nil	Negligible
2.	0.31	Good	0.3 - 0.7	15.5	Nil	--
3.	0.39	Poor	0.1 - 0.2	13.5	Few	42.434 at 60 m
4.	0.33	Good	0.1 - 0.2	9.0	Nil	30.02 at 62 m
5.	0.44	Fair	0.5	6.0	Few	20.87 at 88 m
6.	0.3	Fair	1.5	14.5	--	17.685 at 110m

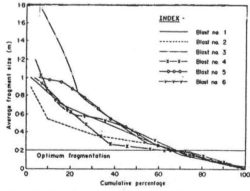

Fig.2 Cumulative size analysis of the blasted muck

Table 6. Mean fragment size (MFS) and specific surface area (SSA) of the muck produced in different blast rounds

Blast no.	MFS (m)	SSA (m²/T)
1	0.322	52.95
2	0.277	40.5
3	0.5	34
4	0.32	41.5
5	0.42	36.8
6	0.39	39.5

Considering the capacity of loading bucket as 1.7 m3 the optimum fragment size works out to be 0.20 m (Rzhevsky, 1985). The mean fragment size of 0.277m in case of blast no. 2 is the closest to the optimum value of 0.2m where a low powder factor of 0.31 kg/m³ was used though the RMR in this region is 58. generally, in conventional blasting i.e. without air-decking, a powder factor of 0.35-0.4 kg/m³ is used in similar rock masses.

The slope of the curve in blast no. 2 is minimum, indicating a more uniform fragmentation. The slope of the curves (Fig. 2) in case of F/W blasting (no. 3, 4 and 5) are higher than those in case of H/W blasting (no.1, 2 and 6) as the over size boulders in F/W side are relatively more than those in the H/W side. This was due to the presence of frequent open and weathered joints in the F/W rock masses. The mean fragment size in blast no. 1 and 4 are almost the same though the former is at the H/W side having RMR of 65 and the later is in the F/W side having RMR of 35. In the H/W rock, the mean fragment size in case of air-decking (0.322m in blast no. 1 and 0.277m in blast no. 2) are lower than that without air-decking (0.39m in blast no.6).

The spacing to burden ratio was kept as 1.75 in blast no. 3 which was the maximum of all the trial blasts. Despite this, the mean fragment size was as high as 0.5 m. Such a large mean fragment size attributes to the presence of intra-hole joints and fractures.

It can be seen that the SSA value is the highest in blast no. 1 (conducted in H/W side) where air-decking was used. Its value in blast no. 6 (conducted in H/W side), where no air-deck was used, is lower than that in blasts no. 1, 2 (in H/W side) and 4 (in smaller benches of the F/W side) where air-deck was used with nearly similar charge density. Further, the SSA value in blast no. 6 is higher than those in blasts no. 3 and 5 where higher powder factor compared to blast no. 6 was used. This was mainly due to the unfavourable joint orientation and closely spaced joints in the F/W side. As, the SSA value indicates the degree of fragmentation, it can be said that, a better fragmentation can be obtained in air-deck blasting compared to the conventional blasting in similar rock mass conditions.

From the above analyses it appears that both the MFS and SSA values are influenced by the rock mass onditions and the charge parameters. Olofsson (1988) comments that the rock will be broken more if the specific charge is increased. Thus, it is obvious that the mean fragment size (MFS) will reduce but the specific surface area (SSA) will increase with the charge amount. Moxon et al. (1993) also found that the specific surface area increased with the explosive quantity. But the effects of explosive quantity on the degree of fragmentation, measured in terms of specific surface area and mean fragment size, in different rock mass conditions can not be assessed from the available literatures. Based on the results of the trial blasts conducted the relations between the SSA or the MFS with powder factor (pf, kg/m³) and the rock mass rating (RMR) were determined (Figs. 3 and 4) as given below:

$$SSA/pf = 8.99(RMR^{0.7}) \qquad (3)$$

$$MFS*pf = 2.55(RMR^{-0.82}) \qquad (4)$$

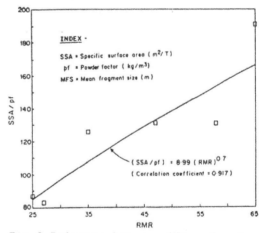

Fig.3 Relation between SSA, pf and RMR

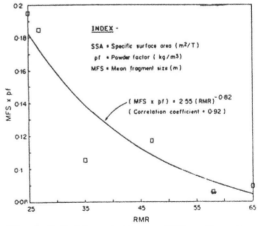

Fig.4 Relation between MFS, pf and RMR

The rock beam stiffness is less in the higher benches compared to those in the lower benches. As a result, the process of separation along the joints due to flexural rupture is much easier in the higher benches. The fragmentation was much better in 4m benches (blast no.4) than that in 10m benches (blasts no. 3 and 5) though the rock mass conditions were nearly the same. Further, benches of 6-8m height are the most suitable in Dongri Buzurg mine considering the boom length of the shovels and the size of the blast holes. Thus, in-case of the F/W side, where the rock mass is highly jointed and foliated, 6m benches seem to be most feasible for obtaining good fragmentation.

The powder factor in blast no 1 was as low as 0.278 kg/m3. The maximum powder factor used for trials in the H/W side was 0.31 kg/m 3 in the formation where conventionally 0.35-0.4 kg/m 3 was being used earlier. Thus, a saving of about 13 per cent in the explosive consumption in the H/W side was obtained. In case of the F/W blasting, it is better to compare the powder factor in air-deck blasting with the ideal one as desired fragmentation was not obtained with the practiced one. Considering the jointed formation in the F/W, a closer spacing of the blast holes are required for a good fragmentation. As per the classification by Ghose (1988), a powder factor of 0.5 kg/m 3 should be optimum in such condition. Considering this, the maximum charge factor used in H/W side i.e. 0.44 kg/m^3 in blast no. 5 was 13.6 per cent less than the required.

The overbreak decreased considerably with air-decking. There was almost no overbreak in blast no.1. Further, the maximum overbreak of 0.7m observed in air-decking was considerably less than an average overbreak of 1.5m, obtained in the conventional blasting without any air-deck.

The occurrence of underbreak or toe was common both in the F/W and the H/W blasting. The blasted muck was found to be locked inside the toe in a number of blasts all along the F/W side. This was mainly due to the joints dipping away from the bench face. No toe was obtained in blast no. 1 though the muck pile was a bit tight. Some underbreak was, however, found in the F/W side in the blast no. 3 and 5 which was due to misfires and unfavourable joint orientation i.e. joints dipping away from the free face.

From the limited data on ground vibration, the following ground attenuation equation was obtained by regression analysis for air-deck blasting:

$$V = 363.6 \ (D/Q^{1/2})\text{-}1.326 \qquad (5)$$

where, V = peak particle velocity, mm/s

Table.7 Predicted vs observed throw in trial blasts

Blast No.	Type of blast	Average throw (m)		Per cent reduction	Remarks
		Predicted	Observed		
1.	Air-deck	11	6.0	45	
2.	- do -	15.4	15.5	Nil	High throw observed due to low burden in 1st row of holes
3.	- do -	26.6	13.5	50	
4.	- do -	18.2	9.0	50	
5.	- do -	33.6	6.0	82	Less throw observed due to misfires
6.	Without any deck	14	14.5	Nil	

$D/Q^{1/2}$ = scaled distance, m/ûkg

$\quad$ D = distance of the measuring point from the blast site, m

$\quad$ Q = maximum charge per delay, kg.

In blast no 6 where no air decking was used, the peak particle velocity at the scaled distance of 14.446 was observed as 7.685 mm/sec. But, as per equation 5, the peak particle velocity in air-deck blasting at the scaled distance of 14.446 m/(kg)$^{1/2}$ works out to be 10.53 mm/sec. The observed ground vibration in case of conventional blasting (blast no. 6) was higher than the predicted one as the shock energy in air-decking was poorly utilised in fragmentation and a higher percentage of it was wasted in the surrounding media to create higher ground vibration.

Except in case of blast no. 1 where the muck pile was found to be somewhat tight, the throw in all the blasts using air-deck was satisfactory. The research performed by Swedish Detonic Research Foundation (SVEDEFO) shows that the forward movement of muck pile in conventional charging should be 140c-28 m where `c' is the specific charge in kg/m^3 (Olofsson, 1988). A comparison is made below in Table 7 between the predicted throw after SVEDEFO and the observed throw.

Thus, it can be seen, that in most of the cases the muck throw is lower than the predicted value in air-deck blasting. This means that, less explosive energy is used in rock throw in air-deck blasting compared to the conventional blasting (blast no. 6) where the observed value is almost the same as the predicted one.

5 SCOPE OF IMPROVEMENTS IN AIR-DECK BLASTING

The following advantages (Table 8) of the air-deck blasting compared to the conventional blasting were surfaced from the trial blast results.

Table 8. Benefits of air-deck blasting over conventional blasting as per trial blast results

Sl. no.	Parameter	Benefits
1.	Fragmentation	- Mean fragment size close to optimum fragment size in H/W side
2.	Powder factor (Kg/m3)	- Powder factor improved by 13 per cent
3.	Ground vibration	- Reduced by 40 per cent
4.	Overbreak	- Reduced by 50 per cent
5.	Toe	- Significant improvement in H/W side and no secondary blasting required, nominal improvement in F/W side
6.	Throw	- 45-50 per cent less than generally observed in conventional blasting

6 RECOMMENDED BLAST DESIGN WITH AIR-DECKING

On the basis of trial blast observations, the following blast patterns are suggested (Table 9) for different rock mass conditions with 6m high benches and assuming that other site conditions remain unchanged.

Table.9 Recommended blast patterns with air decking in different rock masses

Sl. no.	Rockmass rating (RMR)	Burden (m)	Spacing (m)	Charge /hole (kg)	Powder factor (Kg/m)	Charge distri- bution
1	25-35	2	2.5	12.5	0.41	1P+1C# 1P+1.5C
2	35-50	2.5	3.0	13.9	0.3	1P+1.5C# 1P+1.5C
3	50-65	2.5	3.5	15.3	0.29	1P+2C # 1P+1.5C

where, P is the cartridge for primary charge, C is the cartridge for secondary charge and # denotes 0.9m long air-deck column.

It was estimated from equation 4 that the mean fragment size with the recommended powder factor should lie between 0.3-0.35m, which can be considered as good.

7 CONCLUSIONS

To improve fragmentation at reduced explosive consumption, the feasibility of air-deck blasting was studied in the footwall and hangwall sides of the Dongri-Buzurg mine of Manganese Ore (I) Ltd. with various blast patterns. The fragment size analysis indicated that the specific surface area is directly related to the powder factor and rock mass rating (RMR). Similarly the mean fragment size is inversely proportional to the powder factor and rock mass rating. Further, in air-deck blasting, a saving of 13 per cent in explosive consumption and 40 per cent reduction in ground vibration, 50 per cent reduction in overbreak and 50 per cent reduction in throw were visible compared to those in conventional blasting. The improvements were more significant in the hangwall side compared to those in the footwall side because of favourable joint patterns in the former case. On the basis of trial blast results blast patterns with air-deck were suggested for different rock mass conditions.

REFERENCES

Bieniawski Z. T. (1973), Engineering classification of jointed rock masses, Trans. of South African Inst. of Civil Engineers, Vol. 15, No. 12, pp. 335-344.

Chakraborty A. K., Goel R. K., Murthy VMSR, Kumar P, Jethwa J.L. and Dhar B. B. (1995), Project report on A feasibility study on the use of air-deck blasting in Dongri Buzurg mine of Managanese Ore India Ltd., April.

Ghose A.K. (1988), Design of drilling and blasting sub systems - A rock classification approach, Mine Balkema, Rotterdam, 335-340.

Hagan T. N. (1992), Charges with lower effective densities: A means of reducing costs when blasting rocks of low to moderate strength

John M. (1971), Properties and classification of rock with reference to tunnelling, CSIR Report, MEG, 1020, Pretoria, South Africa.

MeCabe W. L. and Smith J. (1967), Unit operations of chemical engineering, McGraw-Hill Book Company, New York.

Melnikov N. V. and Marchenko L. N. (1971), Effective methods of application of explosive energy in mining and construction, Proc. 12 th. Symposium Dynamic Rock Mechanics, New York, 359-378.

Marchenko L. N. (1982), Raising the efficiency of a blast in rock crushing, Soviet Mining Science, Vol. 18,#5, Sept.-Oct., pp. 46-51.

Moxon N.T., Mead D. and Richardson M. (1993), Air-decked blasting techniques: some collaborative experiments, Trans. Inst. Min. Metall. (Sec. A: Min. industry), 102, Jan-Apr., pp. A25-A-30.

Olofsson O. S. (1988), Applied explosives technology for construction and mining, Sweden.

Pal Roy P., Singh R. B. and Mandal S. K. (1995), Air deck blasting in opencast mining using low cost wooden spacers for efficient utilisation of explosive energy, Accepted for publication in Jr. of Mines, Metals and Fuels.

Rzhevsky V.V. (1985), Opencast mining Unit operations, Mir Publishers, Moscow.

Singh D. P. and Sastry V. R. (1988), Effect of controllable blast design factors on rock fragmentation, Jr. of Mines, Metals and Fuesl, December, pp. 539-548.

ACKNOWLEDGEMENTS

The authors are thankful to Director, Central Mining Research Institute for permission to publish the paper.Thanks are due to the staff and management of Manganese Ore (I) Ltd. for the cooperation provided for the study. The views expressed here are of the authors and not necessarily of the institute they belong.

Blasting and productivity

Rock Fragmentation by Blasting, Mohanty (ed.) © 1996 Taylor & Francis. ISBN 90 5410 824 X

Measuring the effect of blasting fragmentation on hard rock quarrying operations

Leven Moodley & Claude Cunningham
AECI Explosives Limited, Modderfontein, South Africa

Hennie Lourens
Hippo Quarries Eikenhof, Limited, Johannesburg, South Africa

ABSTRACT

While it is well accepted that fragmentation has a pivotal effect on downstream quarrying costs, little quantitative information has been accumulated upon which to formulate reasoned blasting strategies. In a study at a producing quarry, loads from benches producing different fragmentation were monitored during loading and crushing. Dig time, truck load mass, fragmentation sizing, time through the crusher and power consumption were monitored for each load taken in three different blasts. Fragmentation was evaluated using ICI's Powersieve™ optical system. At the crusher, power consumption was more sensitive to rock dimensions than to tonnage. The study was effective in exploring the relationship between rock fragmentation and quarry profitability. The conscientious use of good optical sizing technology will be important for guiding blast design in optimizing quarrying operations.

1 INTRODUCTION

The quality of rockbreaking greatly influences the profitability of any mining venture. As control is improved over each aspect of the blasting operation,

- production becomes more predictable,
- working costs fall,
- income increases and
- the ability to plan accurately is enhanced.

Advances in blasting technology such as Computer Aided Blasting (CAB), electronic timing, and specialized explosives formulations are giving engineers an increasing ability to control blasting results, but at a price. The problem is knowing if this price can be justified by improved cash-flow for the mine.

The shear volume of broken rock from blasting normally defies serious attempts to define the muckpile characteristics during routine production, and the results of most investigations are highly debatable owing to variability in:

- pit layout,
- deployment and condition of equipment, and
- personnel parameters.

It is difficult to motivate resources under these conditions, but *failure to implement effective technology forfeits real cash-flow*, so it is extremely important to find ways to make the connection between rock-breaking results and their impact on the whole operation. It is not surprising therefore that so much is being done to elevate the state of monitoring technology in these areas.

2 QUARRYING OPERATIONS

The investigation reported in this paper was undertaken with the objective of evaluating the impact of fragmentation on productivity and costs at the Eikenhof quarry near Johannesburg, South Africa. No attempt was made to influence the blasting process: the starting point of the investigation was rock from the available muckpiles. The quarry was a good venue for the following reasons:

- The rock type is an exceptionally hard, abrasive, andesitic felsite, typically of UCS exceeding 400 MPa, and well known for the heavy duty it imposes on the whole sequence of mining equipment, from drill to crusher. The effect of fragmentation was therefore expected to be marked.

• Single shift operation enabled consistent manning and recording of the data.

• The electrical control room was positioned conveniently for access during monitoring.

The planned production rate of 600 tons per hour is loaded at the bench faces using a wheeled Komatsu WA600 front-end loader with a bucket capacity of 5.5m^3, and a track driven DEMAG H85 face shovel with a bucket capacity of 5.0 m^3. These load into a fleet of four Cat 760C 35t trucks, which tip into the 1067mm gyratory crusher set to 125mm. The crushed rock passes from there to a secondary crusher set to 38mm with a capacity of 400 tph, and thence to four tertiary crushers set to 13mm and with a capacity of 200 tph. The final sales products are:

Sand,
6.7 mm
9.5 mm
13.2 mm
19.2 mm
26 mm and
37 mm crusher run.

3 MONITORING

The key principle in this exercise was to follow each load of rock from bench to primary crusher, monitoring the expenditure of effort on that particular load. This required patience on the part of production personnel, as trucks had to queue at the crusher and wait for each load to be processed before being allowed to tip. The obvious worth of the exercise provided management with the motivation to permit this.

For each load the following data were gathered:

• Dig time
• Load mass
• Fragmentation in the mouth of the crusher
• Crush time
• Crusher power

Unfortunately, the first half of the trial was conducted without load mass information, as the weigh-bridge was not initially available.

3.1 Loading

In order to assess the digability of the muckpile and obtain comparative digging effectiveness of the loaders, the parameter monitored was the time for which the loader bucket was in contact with the muckpile. Four buckets made up a full truck load, and dig time was taken as the sum of these times, which were adjusted to take into account the different bucket sizes of the loaders. The time of day and truck number was also recorded with the first bucket of each load, thus giving cycle times.

When the portable weigh-bridge became available, the mass of each truck, with a full fuel tank, was calibrated at the weigh-bridge computer: the identity code for the truck was input by the operator when weighing each load.

3.2 Primary Crusher

The current drawn by the primary crusher, and the time to process each load, were captured using a digital multimeter with an RS232 interface, connected to a dedicated PC located in the motor control centre. Duration time of the load in the crusher mouth was recorded on a stop-watch by an observer at the tip, who also recorded time of day to ensure correlation with the recorded peaks on the PC. The power drawn by each load was then converted to kWh of energy.

3.3 Fragmentation Measurement

It is relatively simple to determine, for each loading cycle, the cycle time, load mass and crusher power consumption. However, the dimensions of the rock fragments being handled must radically influence the relationship between these parameters. Really useful output from the exercise can therefore only be extracted if a measure of fragmentation is also obtained for each load.

Elsewhere, we have discussed the difficulty of achieving an accurate evaluation of fragmentation in blast muckpiles (Cunningham, 1995), and concluded that while insurmountable difficulties present themselves in obtaining an accurate size versus mass analysis of normal mine production, useful output can be obtained from imaging systems when the characteristics and limitations of such systems are properly understood.

Conveniently, ICI released its Powersieve™ system during the project, and the opportunity was taken to evaluate it. The system uses the threshold binary imaging and erosion/reconstruction technique to identify areas of particles of given minimum dimension. Steps in processing are as follows:

a) For each load dumped into the crusher, place a scale object amongst the fragments in the area of interest and capture on film or digital imaging system.

b) Import the image into a PC image-processing package.

c) Optimize the visual quality of the image using available tools to adjust contrast, brightness, sharpness, etc.

d) Read the enhanced image into Powersieve™ software, and use editing tools to achieve a binary image separation of adequate quality.

e) When satisfied, commission fragmentation analysis by erosion. In this exercise, the "bin sizes" selected for analysis of the rock incremented from 0.1m upwards in 0.1m steps.

f) Print out the images obtained at each step and compare with original photograph to ensure a good match, then take the reported areas for each size range and store these in the database against load number.

The time to process each image depended a great deal on the quality of the image, the speed of the computer and the skill of the operator. For a quality result, one image could take anything from 15 to 60 minutes: Figure 1 represents output from processing one image. The following points arose during adoption of the technology:

• The method is powerful provided no attempt is made to extrapolate from particle area to particle mass. Close examination proved that *in many cases and for many reasons, the particle areas presented for processing would correlate with, but not be proportional to mass.*

• An important feature of the software is the editing capability, especially that which enables the operator to define areas of fines. Patently, attempting to define individual particles below a certain size is both impossible and pointless, *yet areas of fines must be differentiated from areas of voidage.* The images suffered from adverse lighting conditions in the mouth of the crusher, requiring extensive work to modify the effects of airborne dust, excessive contrast when sunlight penetrated into the mouth, and insufficient resolution and contrast on dull days.

• Overlapping fragments and pools of finer particles on large rocks demanded alertness and discernment from the operator during editing. Since a large rock would greatly outweigh any smaller rocks on its surface, these had to be eliminated prior to thresholding.

• The method categorizes area of particle for a particular second dimension - i.e., a particle seen to measure 0.5m by 0.2m would be registered as an 0.2m particle having area 0.1m^2.

• No attempt was made to force a fit to any artificial fragmentation curve: output was observed area, m^2 versus second particle dimension, m.

The Achilles Heel of the system, and any optical system, is dependence on the patience and commitment of the operator: paying less attention to quality in editing could have greatly reduced the time required to produce an answer. A real strength of the Powersieve™ technology is the separation of fragments into images for each bin size, enabling sensible comparison between the original image and what was detected. Based on this comparison, analysis was occasionally repeated so as to obtain better results. 43 images were processed, of which five needed re-processing. Many of the images were very poor, and required protracted editing, but a useful fragmentation analysis was obtained in every case.

4 RESULTS

The relationships we report on in this paper are,

Digging versus:
- loader type,
- crusher power,

and

Crusher power versus:
- load mass
- fragmentation

4.1 Dig Times

4.1.1 Type of Loader

Interesting results were obtained when examining the effect of loader type on dig times. These are reflected in Table 1.

The results show that digging was very different at the two sites. The front-end loader found digging easier at pile B, while the hydraulic face shovel found easier digging at pile A.

Intuitively, taking the known characteristics of the loader types, this suggests that Pile B was more scattered than Pile A and therefore better suited to the mobility of the wheeled loader. The same blast layout is used throughout the quarry, and the characteristic dig times for these sites indicate that

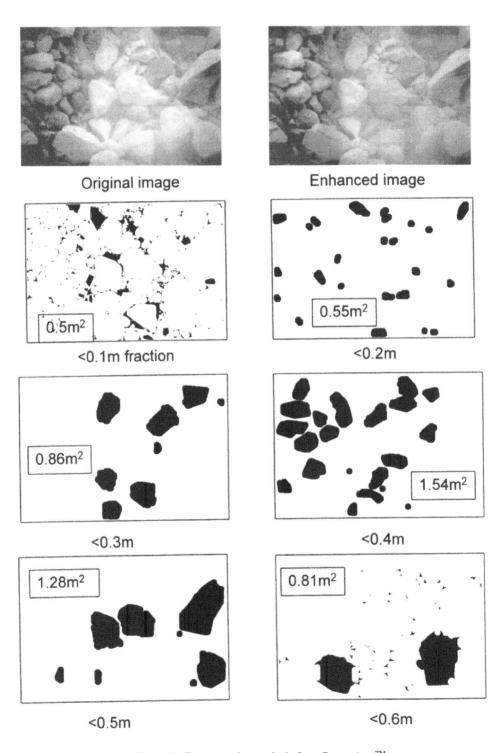

Original image

Enhanced image

0.5m²

<0.1m fraction

0.55m²

<0.2m

0.86m²

<0.3m

1.54m²

<0.4m

1.28m²

<0.5m

0.81m²

<0.6m

Figure 1: Fragmentation analysis from Powersieve™

Loader:	Front-end Loader		Face shovel	
Muckpile:	A	B	A	B
Mean, seconds	51	44	61	77
S.D, seconds	8	8	7	7
Comment		Easiest	Easiest	

- the consistency of the discrepancy in digging should be confirmed by further work, and
- if warranted, the blasting should be tailored to accommodate the bench conditions and loader type.

The front-end loader delivered quicker dig times than the face shovel, because the latter was used only for tough digging at the back of the blast. At pile A the face shovel dig-times were 20% longer, and at pile B 75% longer than what was achieved by the front-end loader, allowing for different bucket sizes.

4.1.2 Dig time versus crusher power

Intuitively, it is to be expected that high dig times would be matched by higher residence times in the crusher. As Figure 2 indicates, no relationship was evident. For a range of dig times ranging from 25 to 120 seconds, crush times of between 82 and 193 seconds were obtained, with equal scatter across the range. If there was no correlation with crusher time, it might at least be expected that there would be a link between dig time and crusher power. As shown in Figure 3, however, no link was found: if anything, there is a slight inverse relationship, but not enough for drawing conclusions. In order to test that there is at least a link between the time of

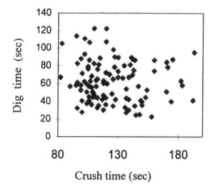

Figure 2: Dig time versus
Crush time

crushing and the power consumed, Figure 3 also plots these parameters and shows a strong correlation.

These findings are interesting and deeper investigation is in progress. The missing dimension so far is that of fragmentation, which is still to be investigated using the present data.

4.2 Crusher Power

The available data for analysis in this arena was the mass of each load, the fragmentation analysis for half of the loads, and the power versus time data for the crusher.

4.2.1 Load mass

It seemed reasonable to assume that there would be a direct correlation between load mass and crusher power consumption (the correct term for this is electrical *energy* to the crusher, but common usage of the word '*power*' instead of 'energy' is being observed in this paper). Figure 4 shows that load masses varied from 25t to 34t, with power off-take per load ranging between 0.4 and 6.9 kWh, but without discernible trends linking these parameters. Clearly, at this quarry anyway, the power consumption is not primarily related to the mass of rock being crushed. The only remaining parameter is fragmentation, which was then examined.

4.2.2 Fragmentation

The fragmentation analysis had identified, for each photograph, the combined visible areas of fragments having dimensions within the specified ranges. Since not all the rock could be seen, it was proposed to plot area of observed sizes against crusher power. It seemed likely that larger rocks would cause the greatest power drain on the crusher. It was therefore assumed that if the data was filtered to produce plots of fragment area against power for increasing fragment dimensions, then there would be a steep relationship between

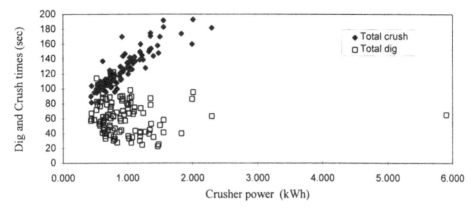

Figure 3 Dig time and crush time versus crusher power

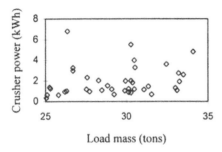

Figure 4 Crusher power versus Load mass

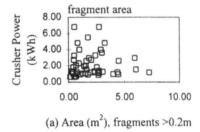

(a) Area (m^2), fragments >0.2m

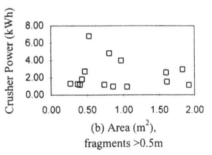

(b) Area (m^2), fragments >0.5m

Figure 5: Crusher power vs

area and power for the bigger fragment sizes, and a relatively flat gradient for the whole range of fragment sizes. Accordingly, for each image, kWh of crusher power per load was plotted against fragment areas for the following:

All sizes greater than 0.2m
All sizes greater than 0.3m
All sizes greater than 0.4m
All sizes greater than 0.5m

Rather surprisingly there was no discernible trend of this kind (figure 5).

We were sure that the image sizing process was giving reasonable estimates of what could be seen. The only conclusion that could be drawn, was that not enough of the rock was visible to the imaging, and known rock mass had to be combined with the indicated fragmentation. It was thus decided to include load mass and apply indicated proportional

sizing from the fragmentation imaging. Taking this step was testing a significant assumption, that the range of fragmentation that could be seen was representative of the total tonnage known to be in the crusher. Since it was quite obvious when doing the fragmentation exercise that significant omissions and unavoidable errors were included in the analyses, there did not seem to be much hope that a better correlation would be achieved. The percentage size passing for observed area was

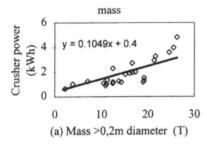

(a) Mass >0,2m diameter (T)

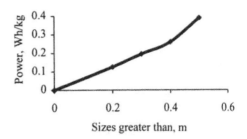

Figure 7: Effect of fragmentation on
crusher power.

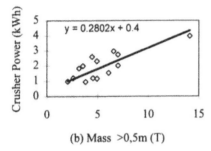

(b) Mass >0,5m (T)

Figure 6 Crusher power vs load

therefore multiplied by the known tonnage of each load, and new graphs drawn. This time a strong trend appeared, with power consumption ramping at an increasing gradient as the calculated tonnage of larger fragments increased. Although there was significant scatter, straight-line relationships could be derived for each size range, the gradients of the lines representing the crusher power per kg of rock for that size range. These are shown in figure 6. If the gradients are plotted against fragment size, then a curve is obtained indicating the electrical energy to the crusher against tons of rock of a particular size range. The relationship obtained is shown in Figure 7.

This is a significant finding, because it enables one of the direct costs of processing large fragments to be determined. It is a reasonable assumption that the greater is the power being consumed by the crusher, the greater is the wear and tear on the crusher components, and a serious attempt should be made to establish this relationship. In addition, however, the unit cost of power consumed is less important than the peak power demand created by tough crushing situations. The exercise captured power peaks and troughs, and it is hoped to work through the data and determine whether peak power is also affected by fragment size.

5 CONCLUSION

Much useful learning has developed from this exercise:

a) With the interest of quarry management, relatively important and reliable data can be gathered for full scale rock breaking operations. The monitoring of bucket-muck intimacy time, weighing of haul trucks and capturing of crusher power are not really difficult operations, but yield valuable information about working costs and productivity.

b) Fragmentation analysis by imaging has an indispensable role to play, as sizing is one of the key parameters in determining working costs, and certainly appears to be more important than mass throughput. Since working costs are normally calculated on throughput, there is real potential to gain better control by intelligent monitoring of fragmentation.

c) Intuition has never been a good guide to rockbreaking. This exercise demonstrates that it should always be extremely tentative, as most of the assumptions that were tested proved to be false.

6 ACKNOWLEDGMENTS

The authors are grateful to the management of Hippo Quarries Ltd, and AECI Explosives Limited for their sponsorship of this programme and permission to present this paper.

7 REFERENCES

Cunningham, C.V.G., 1995, Methods of evaluating and predicting fragmentation. Proc. 6th High Tech. Blasting Seminar, 317-334, BAI Inc., Allentown

Rock Fragmentation by Blasting, Mohanty (ed.) © 1996 Taylor & Francis. ISBN 90 5410 824 X

Drill and blast data analysis procedure for the purpose of blast performance improvement in underground drifting

P. Moser, T. Oberndorfer & M. Siefert
Department of Mining Engineering and Mineral Economics, Montanuniversity Leoben, Austria

ABSTRACT: Within the EU-research project "Blasting Control", jointly operated with the Ecole des Mines de Paris, the Politecnico Torino, Italy, the French explosives manufacturer Nitrobickford and the Greek mining company Bauxite Parnasse, the Department of Mining Engineering of the Montanuniversity Leoben, Austria developed a method for the quantification and evaluation of drilling and blasting parameters. This method incorporates monitoring of geometric, explosive, vibration and granulometry data. All data are compiled in a database and are provided for direct retrieval by demand of blasting experts in a graphical environment. This paper focuses on the advantages of such a graphical system both for research tasks and applications in real mining operations. It describes the procedure of data monitoring and presents a selection of graphical outputs generated during data processing.

1 INTRODUCTION

In July 1995 a three years research project called "Blasting Control", funded by the European Union under the Research Program BRITE EURAM was completed (The Austrian part of the project was funded by the Austrian Foundation for research promotion FFF). The aim of the project was to contribute to an improved blasting in underground drifts. This was intended to be achieved through the development of a methodology for the evaluation of a blasting operation. This comprised the development of practical methods for the evaluation and quantification of the parameters involved in blasting i.e. the quantification of the situation before blasting, the blasting process and the blasting results.

The main project achievement was the development of a diagnosis system applicable in underground drifting, in order to identify possible reasons for blast failures or poor blast performance. Such a diagnosis system consists of a set of measurement devices to measure relevant blasting parameter and a standard procedure to

process the parameters measured and to bring them into a standardized form. This diagnosis system enables blasting experts to analyze a blast, to quantify the results achieved (deviation between planned and actual) and to make suggestions for an improved blast design.

During the project emphasis was put on the fact, that the tools developed and tested within the project should be applicable in mining operations under real mining conditions. A further important aspect was, that the parameters measured should be available as quantified figures as soon as possible after the blast. This should help to use the quantified information about the blasting result for the design of the next rounds.

The research project comprised two major phases. The first phase focused on the development and adaptation of appropriate monitoring devices and measurement techniques. This included 29 full scale blasts (12 m², 3m pull) at the research site "VEBSTER" of the Department of Mining in Austria. All these blasts were completely monitored in terms of geometry, drilling performance, rock mass conditions, vibrations, muck pile shape and granulometry.

During the second phase the procedures developed were tested in a mine in production in Greece, the mining company being a partner of the project. During this phase the monitoring results were interpreted by a group of blasting experts and the resulting suggestions of improvements were then implemented and tested in the mine again. This final phase proved the advantage of having quantitative figures instead of qualitative descriptions. However, it also illustrated the importance of convenient presentation of the results - long lists of numbers are only of little help.

2 DATA MONITORING

The entire procedure of monitoring a blast round comprised the following steps:

1. Drilling of blastholes: The drilling procedure was completely monitored in terms of online measurement of the drill speed and the hammer performance. These data serve as a basis for the future development of a system to determine rock parameters from drilling parameters for blastability prediction.
2. Surveying of the boreholes (hole position, length and direction). Surveying of the boreholes through monitoring the position of the jumbo boom would have been possible in principle, but required equipment and time was not available during the tests performed. However for future practical applications this way of borehole surveying should be considered.
3. Charge control (number and position of cartridges, delay sequence)
4. Blast performance control through vibration measurements followed by muck pile surveying.
5. Mucking, accompanied by fragmentation measurement (by photographic analysis of the blasted rock in the LHD shovel).
6. Surveying of profiles of the blasted area to measure the actually achieved void (volume blasted, advance).

Such detailed monitoring resulted in a huge number of data, which had to be re-processed in order to provide comprehensive and meaningful information for blast performance evaluation through an expert. In total 42 blast rounds with an advance of around 3 m each were monitored in three different types of rock.

3 DATA PROCESSING AND ANALYSIS

3.1 General

All data monitored were compiled in a database (MS-ACCESS). For data analysis the software SURPAC was chosen. SURPAC is a general purpose mine planning program developed by Surpac Software International, Perth, Australia. Typical applications of this software are in all fields of mining, in particular for any geometric problems. SURPAC incorporates all necessary mechanisms required for related problem solving, including database structures (geological information, surveying, block modelling) and a wide range of modelling and processing techniques for any geometric shapes, either in surface, underground or deposit context. The routines are - however - not linked to a specific mine planning procedure in its strict sense, which makes the program extremely versatile for any geometric problems related with complex geometries.

For the specific task of "Blasting Control" the software "SURPAC" was used both for visualization and computation tasks:

1. Visualization: Visualization proved to be an extremely important part in the project, as this is a convenient way to organize the mass of monitored data, yet not loosing the relation between distinct data. Appropriately organized plots help the blasting experts to judge on position of boreholes, distribution of explosives, shape of resulting excavation, etc. Visualization can be done both in 3D and 2D, whatever the task requires. Furthermore due to the interactive design of SURPAC visualization is not restricted to predefined views, but can be adapted to whatever the blasting experts think is helpful in a specific case.

2. Computation: SURPAC was also used to calculate volumes of any 3D shapes relevant for judging blasting procedure and results. SURPAC was chosen for this task as it can handle volume calculations of any complexity. In particular the feature of intersecting different 3D shapes is of high interest in this respect, because in many situations 2D plots are more easy to read and interpret. Certain tasks can only be solved with this procedure, e.g. following the course of over- and underbreak along the drift axis.

Despite the generally interactive design of data manipulation, the macro facility allows fast calculations and output preparations on standardized, pre-defined topics without any manual interference of the user. Therefore it is not necessary to have a skilled SURPAC user for data processing. Still further investigations on specific issues of interest may be completed by specially trained personnel.

3.2 Design parameter processing and analysis

3.2.1 Drillholes

Drillhole data processing consisted of the creation of surfaces related to the drilled volume. Above all these are the faces built by the collar and the end of the boreholes respectively, and an "envelope surface" constructed by the connection of the contour holes.

The calculation steps (see Figure 1) comprised:
1. Coordinates of collar and end of drillholes are extracted by SURPAC routines from the drillhole database.
2. Collar and endpoints were transformed into "relative coordinates".
3. Surface creation for collar and faces.
4. Creation of the "envelope surface" between the two faces, created by linear interpolation between the two contour profiles.
5. Connection of the created surfaces (3,4) to one closed solid and calculation of the volume (volumes are defined as

voids, i.e. the volume is negative).
6. Retransformation of the created surfaces into real world coordinates, for eventual needs (e.g. viewing of a series of blastrounds).

No explicit calculations reflecting the comparison between designed and actual borehole positions were executed, e.g. calculation of the minimum, maximum and average hole displacement. In a first step the merely visual interpretation seemed to be sufficient (Figure 2). The same applied for comparison in a 3D sense, i.e. calculating the volume drilled outside the designed profile due to incorrect positioning or drilling deviation. The possibility to display the designed volume simultaneously with the actually drilled usually gives a good impression of the reasons for a poor blast performance.

3.2.2 Blast zones

The steps enumerated above were executed almost in the same way for the blast zone calculation (Figure 3). The volumes given in the summary table represent the volume excluding the pervious, inner shape (Total volume = cut volume + production volume + contour volume).

A slightly different process is applied for the volume calculation assigned to each blast hole. This is possible, because the assigned shape is always in principle a body with two connected triangles, which simplifies the solid-creation.

These files are compiled and formatted within "EXCEL".

3.2.3 Charged volume

Calculation of the charged volume is identical with the calculation of the drilled volume. The only difference is, that the physical collar and endpoints of the borehole are replaced by the coordinates, which describe the beginning and the end of the explosives column within the borehole (end of borehole and end of charged borehole are usually identical). These coordinates are extracted from the borehole database by standard SURPAC commands.

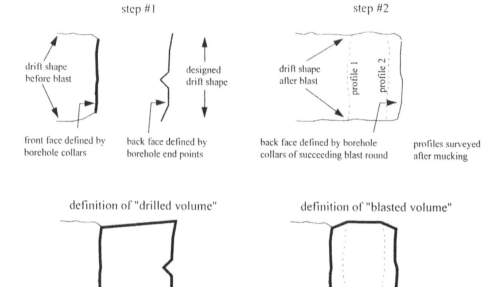

Figure 1:   Sketch of the geometric data analysis

The charged volume is also separated into cut, production and contour-zone.

3.3 Result parameter processing and analysis

3.3.1 Blasted volume

The calculation of the actually blasted volume (Figure 1) comprised:

1. retrieving front and back face, which were already calculated according to the previous chapter. The back face is in fact the front face (collar face) of the next blast round. Hence volume calculation can only commence after these data are available (next blast round drilled and collar positions surveyed). For the final blastround in a series of measurements the resulted back face must be surveyed additionally.
2. Creation of the envelope surface between the faces including the profiles surveyed in between. The envelope is defined as the connection between the contour boundary of the front face, the 1st and 2nd profile and the contour boundary of the backface.
3. Connection of the surfaces and creation of a closed solid and calculation of the volume.

No explicit calculations were done with respect to the drilled volume - blasted volume comparison in the sense of more detailed investigations on "where" overbreaks occur. The viewing capabilities of SURPAC seemed to be sufficient for interpretation (at current stage).

3.3.2 Advance

The actually achieved advance is defined as the average distance from front to bottom face (as defined in blasted volume) in direction of the center hole (representing the required direction of drifting).

The calculation is based on intersecting both surfaces with a series of lines in direction of the center hole and on a regular grid, where the distance between the

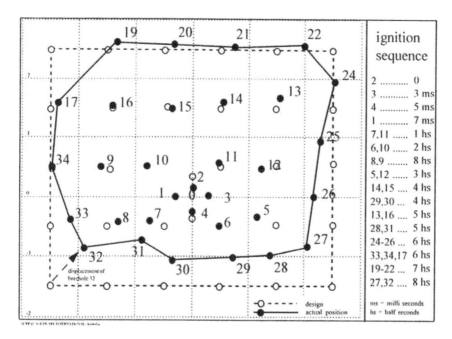

ignition sequence	
2	0
3	3 ms
4	5 ms
1	7 ms
7,11	1 hs
6,10	2 hs
8,9	8 hs
5,12	3 hs
14,15	4 hs
29,30	4 hs
13,16	5 hs
28,31	5 hs
24-26	6 hs
33,34,17	6 hs
19-22	7 hs
27,32	8 hs

○ ‑ ‑ ‑ ‑ design
● ———— actual position

ms = milli seconds
hs = half seconds

Figure 2:   Comparison of designed and actual borehole collars of a blast round. Additional information such as ignition sequence, charged explosives or stemming can be displayed

intersection points gives the advance at this position.

3.3.3 Overbreak

Through a comparison of the actually blasted profile with the planned profile the overbreak was calculated. Summarizing is done in EXCEL (Figure 4).

4 OUTPUT PRESENTATION

There is a huge variety of possible outputs. In this paper just a small selection can be presented, in particular those which are included in a standard processing run. Additional information is only processed on request, if indications are that these information can contribute to the solution.

Besides the calculation in SURPAC (where all the results are stored and thus is the central reference for presentation, in particular for interactive processing) the results are provided in form of 3D-view (usually in color), 2D-plots, and report files (note-files), the latter usually imported and re-formatted in EXCEL to tables and diagrams.

3D views are extremely helpful for interactive investigations, as they give a good overview over the situation. However, due to the versatile viewing features (view angle, scale, elements to be displayed, shading, etc.) they are less appropriated be included into standard outputs for each blastround. An example of a graphical output is shown in Figure 5.

The main 2D plots prepared are:
1. Comparison of planned and real position of collar of drillhole, inc. display of the contour boundary. Each drillhole is labeled with drillhole identification number (ID) and the blast delay stage.
2. Section views of drift excavated. These plots incorporate the drift profiles, the planned profile (deduced from design drift, i.e. the designed profile along the designed drift axis), the drillhole path within the section limits labeled by

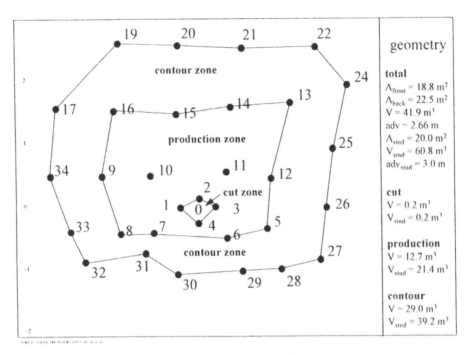

Figure 3: Volume and advance for cut, production and contour zone

drillhole ID and explosive status at the section (stemming, explosive, unloaded, cap). The plots are identified in the title block by the section number which correlates with the section number displayed in the overview map.

3. Overview map shows the general situation of the mine. This includes the planned drift and the actually achieved tunnel. In addition the drift faces of the rounds blasted and the positions and labels of sections of the slices are plotted (Figure 6).

5 SUMMARY AND OUTLOOK

The described way of drill and blast data measurement and analysis turned out to be quite useful for the analysis of poor blast performance and blast failures. The quantitative blast data analysis procedure enabled blasting experts to determine the reasons for a poor blast performance in a mine in operation. Based upon the data visualized, the jumbo operators and the blasting foremen were trained and

considerable improvement in the drifting work could be achieved.

As the standard output features described are executed by a series of SURPAC macros, the processing time usually does not exceed for all outputs 4 hours (drilled, charged and blasted volume, slicing, plot preparation, plotting time). But it has to be mentioned, that the run of a macro is highly sensitive to any import errors according to the rules of the general input procedure. Incorrect number or odd position of profile measurement points, etc. cause problems. These problems can only be overcome by interactive interference of the user.

What still is rather time consuming and what interferes the drifting work is the large amount of surveying work necessary for the measurement of relevant data (collar position surveying 1 hour, borehole deviation surveying 1 hour, surveying profiles (3 profiles per blast) 1½ hours, importing raw-data to ACCESS and exporting to SURPAC, including checking 2 hours).

Therefore future development work should concentrate on the development of measuring equipment for online hole position

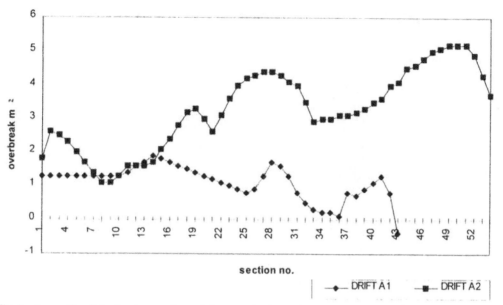

OVERBREAK ERZBERG

Figure 4:   Graphical presentation of the trend of overbreak for the VEBSTER test series

determination (e.g. through boom position monitoring) and hole deviation determination (e.g. through measuring the forces resulting from the bent drill steel).

6 REFERENCES

Moser, P. and H. Wagner: Blasting control for underground mining with special reference to in-situ measurement and tests. Consolidated Project Report.   Department of Mining Engineering and Mineral Economics. Montanuniversity Leoben, Austria. 1995, 10 pp.

Oberndorfer, T. and M. Siefert: Drill- and Blastdata Analysis and Visualization with SURPAC. Research Report. Department of Mining Engineering and Mineral Economics. Montanuniversity Leoben, Austria. 1995, 37 pp.

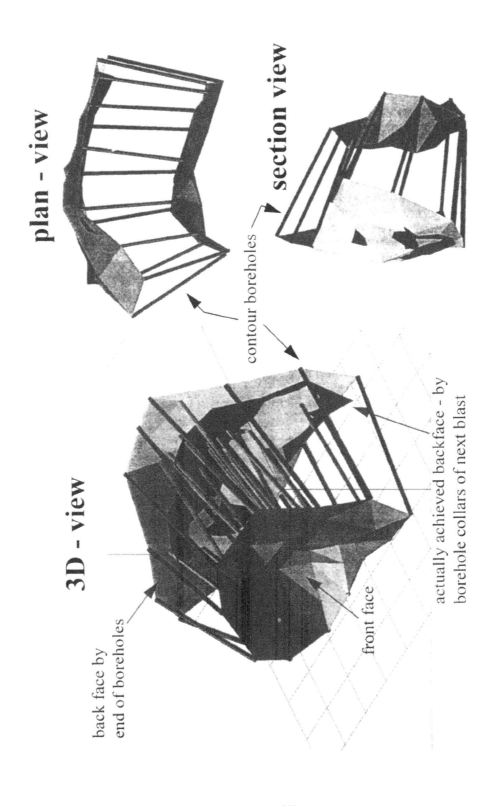

plan - view

section view

3D - view

contour boreholes

back face by
end of boreholes

actually achieved backface - by
borehole collars of next blast

front face

Figure 5: Example of viewing capabilities in 2D and 3D, showing front and end faces of the boreholes plus the front face of the next blast round

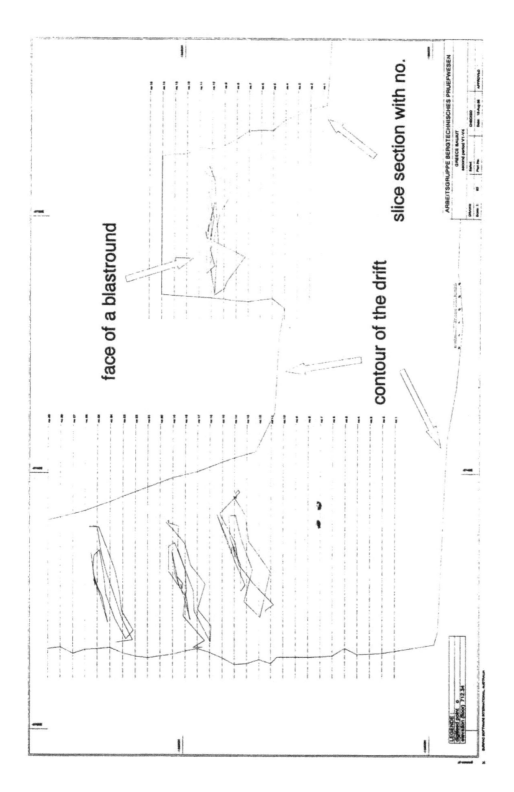

face of a blastround

contour of the drift

slice section with no.

Figure 6: Plan view of the drift faces and the position of the section slices for a specific test series

369

Rock Fragmentation by Blasting, Mohanty (ed.)© 1996 Taylor & Francis. ISBN 90 5410 824 X

Riprap production on the La Grande Complex, James Bay, Québec

Len Gagné
Ste-Thérèse, Qué., Canada

Zulfiquar Aziz & Sovanna Men
Société d'Energie de la Baie James, Montréal, Qué., Canada

ABSTRACT: The construction of the various structures on the phase 1 of the La Grande Complex, in northern Quebec, was done over a period of twelve years between May 1973 and December 1985. The project required the building of 215 embankment dams and dykes involving some 158 million cubic m of fill as well as the construction of three powerhouses with an installed capacity of 10282 MW. A decade after the impoundment of the reservoirs, damage to the riprap was observed on several structures, which necessitated maintenance and repairs. In all, 19 structures required various repairs involving the replacement of riprap with quantities varying from a few hundred cubic m to several thousand cubic m.

INTRODUCTION

In January 1992, Hydro-Quebec mandated the Société d'énergie de la Baie James (SEBJ) to review the criteria for determining the size of riprap and to evaluate different methods of repair from the techniques available.

An intense field campaign was undertaken in 1992 which was crucial to this study. The review of the historical data and the studies which followed the field investigation, allowed the intensity of the wave attack to be evaluated and the amount and methods of repair to be determined. It was found that 92% of the structures (excluding the freeboard dykes) had an acceptable to excellent performance. However, there were some structures where the upstream slope suffered local damage and needed local repairs, and then there were some structures that sustained generalized damage and required systematic repairs. (Levay et al 1994, and R. Arès et al 1995).

Technical specifications and bid documents were prepared, and the repair work was carried out, starting with the structures requiring the most urgent repairs, i.e., the:
- La Grande 4 Reservoir structures 1993
- La Grande 3 Reservoir structures 1994

Caniapiscau Reservoir structures 1994, 1995

The structures of the La Grande 2 and the Eastmain-Opinaca reservoirs, requiring repairs, are to be treated in 1996 and 1997 (Fig. 1).

SEBJ had estimated the riprap recovery rate that could be obtained from the various quarry sites (table 1). To ensure that all contractors bid on the same basis, and to minimize the element of risk and unnecessary claims, this riprap recovery rate was included in the contract documents.

GEOLOGICAL CONSIDERATIONS

The James Bay territory lies in the Canadian Shield, a glaciated peneplain developed on a precambrian basement of igneous and metamorphic rocks. The bedrock is generally massive, unaltered granite or granite gneiss varying in density from 2.6 to 2.85 gm/cc. Locally, mica schists and meta-sediments are also present in certain project areas. Usually, there are two sets of vertical, tight, conjugal joints. Near the surface, in the top 10 to 15 meters, a set of rebound joints are present throughout the Complex. The rebound joints are usually open and infilled with clay, sand or crushed rock fragments. The frequency and the spacing of

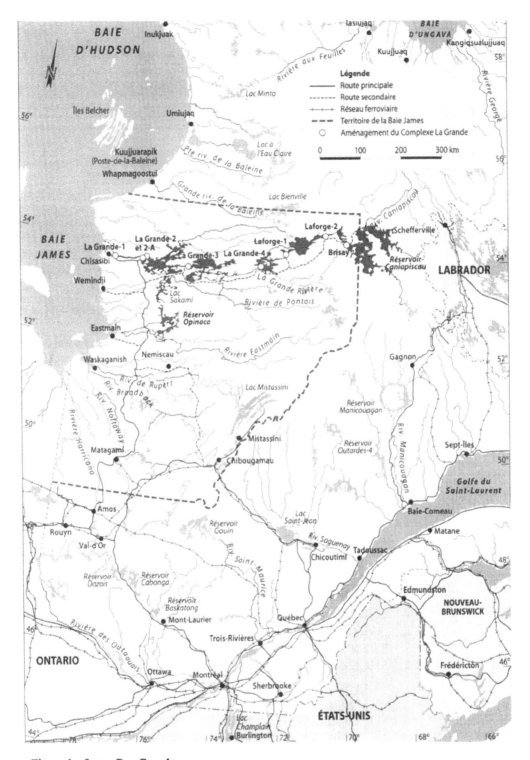

Figure 1. James Bay Complex

372

these joints is variable. In addition, minor faults and shear zones are present locally. The bedrock is usually covered with glacial deposits of variable thickness, and in the southwestern part of the project, the glacial deposits are generally covered with thick, sensitive marine clays.

In the riprap repair areas, an attempt was made to locate suitable quarry sites within a reasonable distance from the structures requiring repair. Suitable rock outcrops were located from the project geological maps and aerial photographs. In addition, partially exploited quarries used during the initial construction of the embankments were also inventoried. Site reconnaissance surveys were carried out by field geologists to assess the overburden thickness, the quality of the rock formation and the quantity of rock available. Authorization certificates were obtained for the exploitation of the selected quarry sites from the Environment Ministry of Quebec.

After having obtained the authorization from the Ministry, usually two to three standard NW-size diamond drill holes were drilled in strategic locations in the potential quarry site. The drill holes generally extended beyond the anticipated floor of the future quarry. The cores were carefully logged and the RQD index determined. RQD values of 80 to 100% were common. All the drill cores were photographed on color film. This information along with the borehole logs was provided to the potential bidders in the bid documents. In addition, during the bidding stage, a site visit to the designated quarry sites, including the inspection of the previously exploited quarries and the drill cores, was carried out.

Without much exception, the granite, granite gneiss and the massive schist of the James Bay territory form the most satisfactory rock types from a blasting and riprap durability point of view. It is to be noted that the riprap is subjected to repeated freeze-thaw cycles which are characteristic of this very cold region.

Due to the harshness of the climate, riprap production and placement can only take place when the reservoirs are free of ice, i.e., a period of roughly six months.

ENVIRONMENTAL CONSIDERATIONS

For environmental reasons, the overburden reject from the quarry was placed in a pre-selected area, away from water courses and drainage areas. The topsoil was removed and saved for reuse upon completion of the work, to replant the quarry area.

The reject from the quarry, as a general rule, was spread on the quarry floor on an on-going basis, except occasionally when it was transported to an adjacent dump area.

Riprap production involves several operations, including the removal of overburden from the quarry site, the drill and blast cycles, riprap size selection at the face, weighing of the trucks transporting the riprap on a weigh scale, the transportation and final placement of the riprap on the sloped embankments.

To begin the operation, the contractor removes the overburden from the site of the quarry to be exploited. Depending on local conditions, this operation is carried out on an on-going basis. Once the length of the quarry face is established, the overburden removal is maintained some 15 to 20 meters behind the existing quarry face.

FACE PREPARATION

In order to minimize the time required for the preparation of the face, the contractor establishes as long a face as possible. An important factor here is to initially blast as many rows of holes as possible, in order to rapidly establish a usable height of face for the riprap production. A typical length of face would be from 140 to 200 meters. In the interest of time and productivity, a usable face height is considered to have a minimum of 8 to 10 meters, while a typical production height can attain a maximum of 16 to 17 meters.

PRODUCTION HOLE DIAMETER

During the three (3) years that riprap has been produced for repairs in the James Bay area, hole diameters have varied from 3" to 3-1/2", but the 3-1/2" diameter hole, drilled with conventional hydraulic drills, has been generally used.

Initially, it was thought that a maximum height of face of 10 meters would be used, but it was quickly established that a higher face was more appropriate to maximize riprap production. To promote better hole verticality and alignment as well as more flexibility in hole loading, the 3-1/2" diameter hole became a standard in the riprap production.

RIPRAP SIZE

The size of the riprap to be produced depends on the design criteria and the size requirements at each particular site. The dimension of a typical stone is established by measuring the three largest principal dimensions, i.e., the length (a), the width (b) and the height (c); these dimensions are equivalent to those which could be found in a parallelepiped box containing the stone; the D dimension is obtained by using the following formula:

$$D = (a \times b \times c)^{1/3}$$

The dimension d_{min} of a rock is the smallest of the three principal dimensions (length, width, height). The dimension d_{max} is the largest of the three principal dimensions and must not exceed three times the dimension d_{min} with a maximum possible value of 2.10 meters in order to avoid extra long or extra flat rocks. Table 1 shows the range of riprap sizes and recovery requirements at some typical dam and dyke locations.

DRILL PATTERNS

Under normal blasting conditions, the hole burden is always equal to or smaller than the hole spacing. However, when blasting for riprap the reverse is true. The burden is drilled approximately 1.1 to 1.2 times larger than the hole spacing in order to yield a very coarse fragmentation. Typical blast patterns varied from 2.75 m x 2.4 m to 2.9 m x 2.6 m, depending of course on the riprap size required and the local geological conditions.

In order to maximize the riprap recovery from each blast, it was felt that the interaction between multiple-row blasting would be detrimental to riprap recovery, and hence the approach of single row blasting was adopted throughout. On a few occasions, two-row blasts were fired for schedule convenience and quarry face alignment.

The ideal blast for riprap production, is a single row of accurately spaced holes, drilled on an even burden on as long a face as possible, in order to permit several production crews to work simultaneously side-by-side.

SUBGRADE

The length of hole subgrade is critical in order to maintain the grade of the quarry floor. The subgrade length is also important in that it holds part of the relatively dense toe-charge below grade which helps to move the burden and allow the column of the hole to fall by gravity after blasting. Generally speaking, the subgrade length used was 1.2 meters.

Table 1. Blasting method, tonnage and % recovery

Location	Rock type	Blasting Method Used	Tonnes Req'd. '000	Sizes Required		% Recovery	
				Min.	Max.	Req'd.	Obtained
LG-3	Gr	1	428	0,7 m	2,0 m	30%	29,5%
LG-4	Gn/Sc	1	347	0,5 m	1,5 m	40%	33%
Caniapiscau	Gr	2	857	0,7 m	2,1 m	35%	40%
Gr = granite Gn = gneiss Sc = schist Method 1 = semi-continuous Method 2 = Airdek loading method.							

HEIGHT OF FACE

In order to maximize the recovery of usable riprap, the height of the face is an important parameter when blasting for riprap. As high a face as possible is generally sought, and of course this depends on the local geology and the practicality of the height with respect to safety, drilling penetration rates, accuracy of drilling, hole alignment, muckpile profile, ease of riprap selection at the face, etc.

As indicated earlier, the minimum usable face height is approximately 8.0 to 10.0 meters in order to start getting tangible riprap recovery results. The successful recovery of riprap in a given blast is contingent on essentially two factors, namely, the geological aspect of the rock formation being exploited and its inherent fractures, and of course the ratio of the drill hole column length to the total hole length. The higher the ratio the better. Face heights of up to 16 to 17 meters were successfully used.

BLASTING METHODS

Over a three (3) year period, two successful blasting methods were used for the production of riprap. The method selected depended on the contractor's preference. The main differences between the two methods was: a different length of toe load and in the approach to column loading.

METHOD NO.1 (a semi-continuous column loading method)

Toe loading

The length of the toe-load in this method was approximately 1.4 meters. Either a cast primer or 1/2 cartridge of a high explosive (65mm x 400mm) was used to initiate the AN/FO toe load.

Column loading

A typical 17.0 meter hole was loaded as follows:

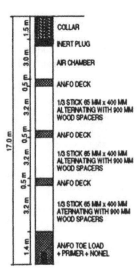

Figure 2. Method No.1

The 15.6 meter column was loaded with 3 distinctive group loads consisting of the following decoupled charges: { 1/3 stick of 65mm x 400mm high explosives, each alternating with a 900 mm wooden spacer } followed by an AN/FO deck of 0.5 m. These 3 groups were followed by a 3.0 meter air chamber (gap) and a 1.5 meter collar.

Powder factor

The explosives powder factor was approximately 0.20 kgs per cubic meter. The explosives distribution was 55% in the toe load and 45% in the column load.

METHOD NO.2 (Airdek blasting method)

The Airdek technique uses an air gap or gaps in the explosives column in place of conventional decoupled or continuous charges, in order to achieve the light loads needed for riprap production.

Toe loading

The length of the toe load in this method was approximately 2.7 meters. A 65mm x 400mm stick of high explosives was used to initiate the

AN/FO toe load. At times, in very hard rock formations, a small part of the AN/FO toe load was replaced with Nitropel (pelletized TNT) to increase the loading density in this area.

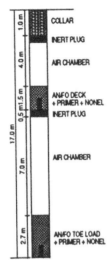

Figure 3. Method No.2

Column loading

Considering a 17.0 meter hole, the Airdek blasting technique consisted in using two and sometimes three airdeks or chambers in the borehole column. A typical loading would be as follows: after loading the toe, an airdek of 7.0 meters is used, followed by a 1.5 meter AN/FO deck primed with 1/2 stick of 65mm x 400mm high explosives. Above this, a 4.0 meter airdek is used, followed by a 1.0 meter collar. The airdeks are well sealed off from adjacent explosives charges by inserting a plastic hole plug at the proper location. The plastic plug is then topped off with 0.5 meters of inert material.

Each explosives deck is initiated by means of a NONEL detonator regardless of the loading method used. The single row blasts are hooked up by means of a primacord trunkline on surface. To promote a better shearing effect, all the holes are fired simultaneously.

Powder factor

The explosives powder factor used with the Airdek loading technique is similar to the semi-continuous method, i.e., approximately 0.20 kgs per cubic meter. However, the explosives distribution is quite different from method No.1, in that the toe load consumes 93% of the total charge and the column load only 7% of the total charge.

COMPARISON OF THE TWO LOADING METHODS

Both loading techniques yielded satisfactory results. However the Airdek loading technique is much simpler to use, requires fewer explosives products and takes less time to load.

MAGNITUDE OF BLASTS

The size and the time of the blasts were designed to accommodate the daily tonnage required to complete the work on time. Generally speaking, the number of holes per blast averaged about 40 holes. The daily tonnage of riprap required depended on the contract tonnage and the time allotted to complete all facets of the job. Daily tonnages varied between 3000 and 6000 tonnes, depending on the contract.

RIPRAP CALIBRATION

For visual calibration of the riprap at the face, a range of acceptable rock samples, selected according to size, was located in a specific area, adjacent to the quarry. This permitted the shovel operators to get acquainted with acceptable sizes. Quality control at the quarry face and the embankment sites was made regularly by quality control technicians.

RIPRAP RECOVERY

The constant monitoring of riprap production was based on the original ground surveys after the overburden removal and the final survey of the quarry floor. In order to evaluate the riprap recovery rate, all loaded trucks were weighed on a weigh scale. Based on the rock density, the percentage of rejects versus the percentage of recovery was calculated. The recovery rate at the various quarry sites are shown in table 1.

EQUIPMENT

Apart from the hydraulic drills used in the drilling operations, hydraulic backhoes were used for the removal of overburden, for riprap selection at the quarry face, as well as for levelling out the final gradient on the repaired embankments (Fig. 6).

Generally speaking, a dozer was used on the crest of the structure to be repaired, to push the dumped riprap parallel to the longitudinal axis of the embankment (Fig. 4 & Fig. 5).

Front-end loaders were used to displace the waste fines from the quarry face as well as to transport and place oversized rock in a special area for secondary fragmentation.

For secondary fragmentation, hydraulic rock breakers were used to break oversize rock

fragments (< 8.0 m³) down to acceptable size. For rock fragments greater than 8.0 m³, secondary blasting was used.

CONCLUSIONS

Sound blasting techniques, rock selection and good quality control are essential during construction in order to obtain acceptable size and durable riprap for embankment repair.

Inverted drill patterns and single row blasting using the loading methods described in this paper were conducive to the obtention of the relatively high riprap recovery rates at the various quarry sites, and this despite difficult geological conditions at times.

To adapt to these conditions, the contractor had to demonstrate flexibility in his operations, and be prepared to adjust drill patterns as well as explosives loading in order to maximize the production of acceptable size riprap.

A good quality control program must be in effect at all times, and include all aspects of drilling as well as the judicial use of explosives for both the toe and column loading.

Figure 4. Truck dumping riprap on embankment crest

Figure 5. Dozer pushing riprap off embankment crest

Figure 6. Backhoe levelling riprap to final gradient

Riprap selection at the face is a key operation and shovel operators must be trained to exercise good judgement as to size and careful handling of the riprap, keeping in mind that the stones are subjected to numerous handling operations before they end up on the embankments.

Although both loading methods have been successfully used for riprap production during the past three years in the James Bay area, the authors of this paper favor the Airdek blasting method because of its simplicity and slightly higher productivity.

The inclusion of a riprap recovery rate in the contract documents ensured that all contractors were able to bid on the same basis, and it was instrumental in the reduction of contractual risk and claims.

ACKNOWLEDGEMENTS

The authors are thankful to SEBJ for permission to publish this paper and to Mr. Jerry Levay, Director of Engineering and Environment at SEBJ for the review of this manuscript and for his many useful comments. The authors also thank Ms. Suzanne Vanier of SEBJ for her secretarial assistance.

REFERENCES

Levay, J., O. Caron, J.-P. Tournier & R. Arès 1994. Assessment of riprap design and performance on the La Grande Complex, James Bay, Quebec. Proc. 18th ICOLD, Durban, Q.68,R.25.

Arès R., J.-P. Tournier, L. Dionne & R. de Batz 1995. Design and construction of the berms for the riprap repairs on the La Grande Complex structures. Proc. 5th Uprating & refurbishing hydro powerplants conference, Nice, France, Volume 2.

Rock Fragmentation by Blasting, Mohanty (ed.) © 1996 Taylor & Francis. ISBN 90 5410 824 X

Determination of blasting criteria during the construction of an in-pit lime kiln

Paul Worsey – *Rock Mechanics and Explosives Research Center, University of Missouri-Rolla, Mo., USA*

Scott G.Giltner – *Econex Inc., Dixon, Ill., USA*

Terry Drechsler – *Chemical Lime Company, Ste. Genevieve, USA*

Ron Ecklecamp – *Geotechnology Inc., St. Louis, USA*

Ronnie Inman – *Tower Rock Stone, Ste. Genevieve, USA*

ABSTRACT: A blast monitoring program conducted at Chemical Lime Company's new lime calcining facility near Ste. Genevieve, Missouri USA is discussed. The purpose was to develop blasting criteria for the construction and operation of the lime plant within the quarry operated by Tower Rock Stone. Further, it was imperative to accommodate production requirements into the blasting criteria. The major concern was the effect of blasting on the curing of green concrete during the construction of the storage silos and kiln stacks. In addition to discussing the susceptibility of curing concrete to blast vibrations, the paper also details how this varies with time.

The following are addressed in detail: 1) effect of pre-splitting on ground vibrations; 2) formulation of scale distance models for blast vibrations in each bench; and 3) formulation of the effect of bench interactions at the kiln site when blasting out of bench horizon. A discussion of the instrumentation employed and the strategy used to achieve these is included. Results from the various instrumentation layouts are presented along with their use in establishing the blasting criteria.

An initial plan for coordinating the construction of the lime plant and quarry production blasting is presented. This incorporated the establishment of blasting criteria in relation to concrete pouring time and production blast location. Modifications to the blasting are also presented.

1 INTRODUCTION

An agreement between Chemical Lime Company (the largest lime producer in the United States) and Tower Rock Stone was made to build a lime plant at the Tower Rock site. The plant was to be located in the southeast side of the pit and designed to take advantage of the drop in elevation and proximity to rail and barge loading facilities. Advantages of this agreement included: the elimination of quarrying start up expenses, decreased construction costs, and increased sales potential for Tower Rock.

Due to the close proximity of the construction to current blasting operations, it was imperative to develop blasting criteria such that the construction and operation of the lime plant would be protected from damaging blast vibrations. Therefore, Geotechnology, consultants to Chemical Lime, contracted the Rock Mechanics & Explosives Research Center (RMERC) of the University of Missouri-Rolla (UMR) to provide expert personnel and equipment for blasting and vibration analysis.

The work entailed two objectives. First, determining whether the proposed presplitting around the site down to the bottom of bench 8 would protect the structures from blast vibrations. Second, the development of suitable blasting criteria prior to construction.

2 DESCRIPTION OF QUARRY & OPERATIONS

The Tower Rock quarry produces in excess of 4 million tons of limestone from the St. Louis and Salem formations. The quarry ranks as the largest in the state of Missouri and one of the largest in the United States. The rock produced is used for a variety of purposes including animal food supplements, agricultural lime, concrete aggregate and rip rap up to 5 tons and is shipped out by railroad or barge on the Mississippi River. Nine benches are currently blasted with the bottom 8 benches being in the Salem formation and the top bench in the St. Louis formation. The current pit is

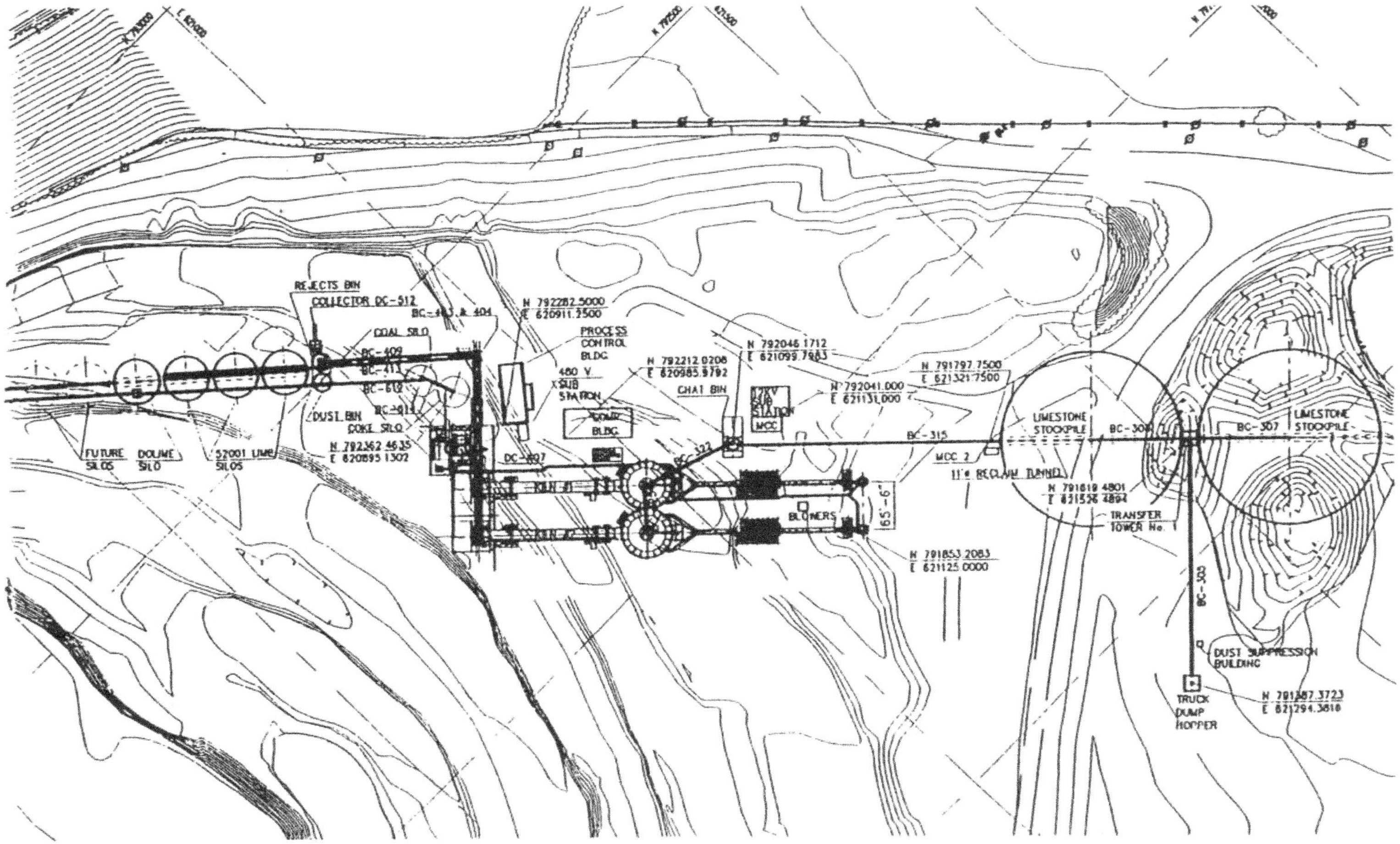

Figure 1. Plan Drawing of Lime Plant.

approximately 2000 ft by 2800 ft and 360 ft deep. The base of the pit is approximately 135 ft below the level of the Mississippi.

Bench heights vary from 14 ft for bench 5 to 104 ft for bench 8. Four drills are available for blasthole drilling: a Driltech D25Kd, a Tamrock CHA 1100 +20, and two Ingersoll Rand VR 140 airtracks. The D25Kd drills 6.5 inch blast holes on bench 8, the CHA 1100 drills 5 inch blastholes on benches 1A to 7 and the VR 140s drill 3.5 inch holes for the presplits and the overburden rock. All holes are vertical.

Standard hole loading practice is to use a bulk loaded 30/70 emulsion/ANFO blend in all dry holes and a 40/60 blend in wet holes. Priming is provided by a 1 lb. cast primer with a non-electric shock tube initiation system on the upper benches while electric detonators and a sequential timer are used on bench 8. Benches 1A - 2, and 5 - 7 are loaded as a single explosive column, while benches 3 and 8 are blasted using two and four decks respectively. Frequently benches 3 and 4 are blasted together at which time three decks are used. The decks are delayed at 25 ms intervals with the firing sequence bottom to top. Approximately 5 ft of 1/2 inch crushed stone is used to separate the decks. Presplitting is accomplished using continuous line presplit emulsion fired using detonating-cord in groups of 10 holes per delay (25 ms). Shooting is performed twice daily: during lunch break and at the end of the day shift.

Rock handling on benches 1A to 8 is accomplished with a Cat 5130 shovel, 3 Cat 992C and 3 Cat 998C front end loaders loading into 12 Cat and 12 Terrex 50-85 ton trucks.

3 DESCRIPTION OF LIME KILN PLANT

The 50 million dollar plant will produce a special lime called mag-enhanced lime, which is used for scrubbing in coal burning power plants. The planned starting production is 400,000 tons per year.

The plant illustrated in Figures 1 and 2 comprises 2 high tech rotary kilns fed by 2 limestone stockpiles and two 2,500 ton silos containing coal and coke. Exhaust gases from the kilns pass through two, seven cell baghouses prior to the two 225 ft stacks. The lime is conveyed to four (60 diameter by 108 ft high) 5,200 ton lime silos where it is blended and transported by conveyor to the barge/rail terminal. Other buildings on site include; electrical substations, bag houses, and the process control building. Plans exist for the future expansion of the plant including additional rotary kilns and silos.

Detailed specifications comprise 4,000 psi concrete for all floor slabs while all other structures utilize 3,000 psi minimum concrete (Watkins engineers).

Early in the planning stage, Chemical Lime identified the primary construction concerns as the kiln stacks and storage silos. Concern over the stack foundations, located on bench 1A, was expressed because of the high loads these foundations would be expected to carry due to the wind resistance and height of the steel stacks. Concern over the concrete "Lime Silos", located on bench 5, was expressed due to their size and important role in production.

Figure 2. Photograph of plant construction & production operations.

4 PLAN OF ACTION

The technical actions taken by RMERC at the site included the following:

1. Monitoring the effect of existing pre-splits on ground vibrations from production blasting.

2. Assessment of pre-split design and recommendations on pre-splits and other protection.

3. Monitoring production blasts in benches using multiple geophone arrays.

4. Formulation of scale distance models for predicting blast vibrations in each ledge.

5. Monitoring production blasts across ledges.

6. Formulation of effect of blasting out of horizon on key structures.

7. Liase with Chemical Lime and Tower Rock personnel to develop a blasting plan for the construction phase.

5 MONITORING STRATEGY

The blast monitoring program was conducted from 19 Sept. 1994 to 1 Dec. 1994. During this period, 32 individual blasts were monitored which included blasts on every bench except bench #6. Bench #6 was excluded as no production blasting occurred on this bench during the monitoring period. Approximately 200 seismograph readings had been taken by the end of the monitoring period. Seismic monitoring of blasts was undertaken using six White Industrial Seismology MiniSeis units. In order to ensure good coupling of the seismographs with the rock, bags containing approximately 20 pounds of finely crushed stone were placed on each seismograph. As it was necessary to obtain the maximum amount of useful data from each blast, the seismograph locations were changed for virtually every blast.

Three monitoring configurations were employed based on whether the subject of the monitoring was: 1) the effect of existing pre-splits on ground vibration, 2) characterizing the blast vibrations within each individual bench, or 3) characterizing the blast vibrations across the different benches. Survey stations were established to locate blasts and monitoring positions where direct tape measurements from the blast to each individual seismograph could not be made. Blast records were made for each of the blasts noting all pertinent information. Due to the differences in the elevations for the various benches, all surveyed points were located in three dimensions. By doing this it was possible to calculate the "direct path" or vector distance which was used in the determination of scaled distance.

5.1 Monitoring Effectiveness of Presplit for Vibration Control

The seismographs were arrayed so that two seismographs were placed 10 ft on the blast side of the presplit with two seismographs directly behind the first two and 10 ft on the opposite side of the presplit. This layout provided two sets of comparative data per blast. Pre-splits blasted for the construction on benches 1A and 1 were used for this purpose.

5.2 Monitoring Intra-bench Vibrations

During monitoring of in-bench blasting, 4 to 6 seismographs were laid out on the bench to be blasted. Whenever possible the seismographs were positioned between the blast and the construction area. Monitoring distances varied from 50 to 2000 ft

to give a range of values for scaled distance calculation.

5.3 Monitoring Inter-bench Vibrations

Once the initial intra-bench information was obtained, a seismograph was usually placed on bench 1A at the "stacks" while another was placed on bench 5 at the "silos". In order to acquire more detailed inter-bench vibration information on bench 5, seismographs were placed at fixed locations on bench 5 for selected blasts on the other benches.

6 MONITORING RESULTS:

6.1 Effectiveness of Presplits for Vibration Control

The data illustrated in Figures 3 & 4 indicate that presplits are ineffective in reducing blast vibrations. This was found to be the case for both shooting in and out of horizon. Shooting in horizon causes the presplit to close, thereby allowing the transmission of the vibration across. This is particularly true when the presplit is filled with either debris or water. When blasting out of horizon, it was found that the geometry of the blast relative to the presplit and the construction area was such that the direct path from the blast to the construction area would pass under the presplit. Due to the cost and time involved in pre-splitting to the bottom of bench 8, the plan to pre-split was dropped.

6.2 Characterization of In-bench Vibrations

The recorded data for each bench was processed and the scaled distance versus peak particle plotted. An example is given in Figure 5. The calculated scaled distances for each bench are given in Table I. (N.B.

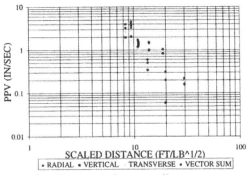

Figure 3. Vibrations before pre-split.

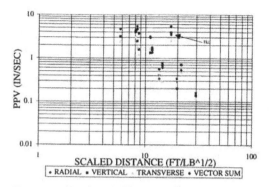

Figure 4. Vibrations behind pre-split.

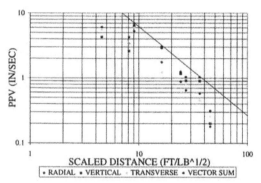

Figure 5. Scaled distance Vs. peak particle velocity graph for bench 3.

The scaled distances are calculated based on worst case vibration levels {Figure 5} and the quarries perspective of their blasting operations i.e. in some cases numerical delay overlap was ignored).

6.3 *Characterization of Cross-bench Vibrations*

A similar method for data processing was adopted. The calculated scaled distances for each bench when monitoring on benches 1A and 5 are given in Table II and
Table III respectively.

6.4 *General*

Seismograph readings indicate that the response of the rock to blasting varied from bench to bench. It can be seen in Table I that benches 1A and 5 attenuated the blast vibrations the most, while benches 2 and 7 transmitted the blast vibrations the best. However, readings taken on fill were much higher. It was therefore recommended, that all key concrete structures and foundations be located on solid rock rather than fill. This recommendation resulted in some construction design changes.

7 VIBRATION EFFECTS ON CURING CONCRETE / SETTING LIMITS

Geotechnology, Inc. reviewed numerous sources for information concerning effects of blast vibrations on curing concrete, particularly as can be related to the measured peak particle velocity. In general, published literature suggests that there is no effect of vibration on fresh concrete in the range of vibrations expected due to blasting. But, as a practical matter, allowable limits are employed. Vibration from blasting or other sources is known to affect the strength of concrete during curing. Allowable limits for both fresh concrete and concrete undergoing a 28 day curing process are often expressed as a function of peak particle velocity and/or charge weight per delay.

The allowable peak particle velocity selected for this project was developed from data provided by Oriard (1982) for structural concrete walls, structural slabs, etc. The allowable peak particle velocity is varied as a function of time measured from when the concrete was originally placed. The velocity is modified by multiplying the allowable velocity by a distance factor which ranges from 0.6 to 1.0; the lower value is for blasts at distances greater than 250 feet and was the value used for this project. The allowable maximum peak particle velocity within a time period was increased for this project as a straight line interpolation from the minimum value at the start of the time period to the minimum value for the next time period. A constant peak particle velocity is achieved at 10 days. For nonstructural fill and mass concrete, the allowable peak particle velocity values suggested by Oriard are 150 to 200 percent of those used for this project.

The peak particle velocities used are consistent with those reported by Gamble, et al (1985) up to approximately 5 days. However, after 5 days the allowable peak particle velocity suggested by them increases to a maximum value at 28 days of 10.16 centimeters per second. However, Hulshizer and Desai (1984) report an allowable peak particle velocity of 10.16 cm/s between 24 and 48 hours after the concrete is placed and 17.78 cm/s for concrete in place over 48 hours.

Table I
Calculated Scaled distances for shooting in horizon

Vibration Limits	1.4	0.3	0.53	0.8	0.98	1.2	1.7	2.1	2.6	3	6	ppv in/sec
Ledge #												
1A	16	35	27	23	20	18	15	14	13	12	8.2	Scaled
1	20	88	60	46	37	33	28	23	18	17	10	Distance
2	38	130	83	63	50	42	34	27	23	21	12	(ft/lb^2)
3	24	62	44	35	30	27	23	18	17	16	11	
5	16	57	35	27	22	18	14	12	9.4	8.2	4.5	
7	35	140	84	60	48	41	32	24	22	17	10	
8	31	74	53	44	37	35	28	24	22	19	13	

Table II
Calculated scaled distances for monitoring on bench 1A

Vibration Limits	1.4	0.3	0.53	0.8	0.98	1.2	1.7	2.1	2.6	3	6	ppv in/sec
All Ledges	26	85	55	42	34	30	23	18	16	14	8	(ft/lb^2)

Table III
Calculated scaled distances for monitoring on bench 5.

Vibration Limits	1.4	0.3	0.53	0.8	0.98	1.2	1.7	2.1	2.6	3	6	ppv in/sec
Ledge #												
1A	36	88	65	53	46	42	35	28	25	23	16	Scaled
1												Distance
2	27	60	44	37	33	30	25	22	20	17	14	(ft/lb^2)
3	26	63	45	37	32	28	24	21	18	16	12	
5	16	57	35	27	22	18	14	12	9.4	8.2	4.5	
7	24	82	54	40	33	28	23	17	15	13	7.5	
8	50	230	160	107	76	62	42	29	22	18	7.6	

Table IV
Scaled distances values Vs time for shooting in horizon

	Time After Initial Concrete Pour											
	0-4 hr	< 1	1.5	2	2.5	3	4	5	6	7	10	days
Vibration Limits	1.4	0.3	0.53	0.8	0.98	1.2	1.7	2.1	2.6	3	6	ppv in/sec
Ledge #												
1A	16	35	27	23	20	18	15	14	13	12	8.2	Scaled
1	20	88	60	46	37	33	28	23	18	17	10	Distance
2	38	130	83	63	50	42	34	27	23	21	12	(ft/lb^2)
3	24	62	44	35	30	27	23	18	17	16	11	
5	16	57	35	27	22	18	14	12	9.4	8.2	4.5	
7	35	140	84	60	48	41	32	24	22	17	10	
8	31	74	53	44	37	35	28	24	22	19	13	

8 DETERMINATION OF BLASTING PROGRAM DURING CONSTRUCTION

The seismic study was developed to consider two possible scenarios depending on the final vibration limits and the construction schedule. This approach was necessary as the Geotechnology's research on acceptable limits was not completed until half way through the monitoring period. In one scenario, the normal blast design is sufficient to meet the required production without exceeding the vibration limits. In the second scenario, the blast program would have to be modified in order to satisfy the production needs without exceeding the vibration restrictions

The upper limits of the scaled distance plots were used to derive individual scaled distance values for various curing times as defined by Figure 6 provided by Geotechnology. The Scaled Distance Value for each individual bench at various times after pouring is given in the accompanying Table IV.

The scaled distance equation was employed using normal blasting configuration and pounds per delay on each bench to calculate minimum distances for these typical blasts. Plots were then made for each bench on the quarry map showing the distance limits vs. days since last pour. An example for bench 5 is given in Figure 7.

In the event these distances were unworkable and the blast could not be scheduled when larger peak particle velocities and lower scaled distances were acceptable, then the pounds per delay for the blast had to be reduced.

At the pre-construction meeting in December, it was felt that although blasting schedules would have to be tailored to the concrete pouring schedule, the required production needs could be met without changing the blasting practice. This was based on the presumption of pours occurring 2 to 3 days apart.

However, when the construction contracts were let, the concrete contractor opted for jump forming simultaneously on multiple silos and bins, resulting in essentially continuous concrete pouring operations. This put excessive restrictions on rock production and made changes in blasting necessary. This unforeseen event resulted in concentrated effort and consultations, and the following blasting modifications.

9 MODIFICATIONS OF BLASTING

The blast logs, seismic records and observed loading practice were reviewed. From the seismograph traces it was apparent that for a number of benches,

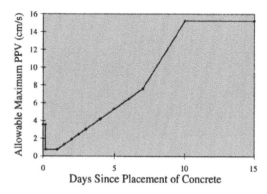

Figure 6. Vibration limits for curing concrete.

irregular vibration spikes occurred. It was deduced these were primarily due to delay overlaps and/or sympathetic detonation of decks, especially for bench 8. On other benches, involving a large number of holes, significant numbers of delay overlaps were found. After considerable telephone conversations with the powder company representative the following changes in loading and blasting practice were put into effect: 1) ensuring proper loading and blasting practices; 2) elimination of extra holes being tied into shots; 3) increased stemming in the collar and between decks; 4) incorporation of extra decks in some shots; 5) lifting the primers up a couple of feet on bench 8 to avoid sympathetic detonation; and 6) ensuring a clean face and properly cleared toe area existed prior to blasthole loading to avoid excessive burdens. In addition, in the vicinity of the plant large increases in stemming were used to eliminate flyrock.

10 MONITORING DURING CONSTRUCTION

Vibration monitoring during the construction of the plant was performed by Chemical Lime personnel using a Multi-seis V3550 and Evertlert BE2054 seismographs. The modifications to blasting worked extremely well and blast vibration levels were kept within tolerances, allowing production to run virtually uninterrupted. As the construction neared completion the critical structures were inspected and no visible damage was detected.

11 CONCLUSIONS/SUMMARY

1. The use of pre-splitting to reduce blast vibrations was ineffective and therefore it was

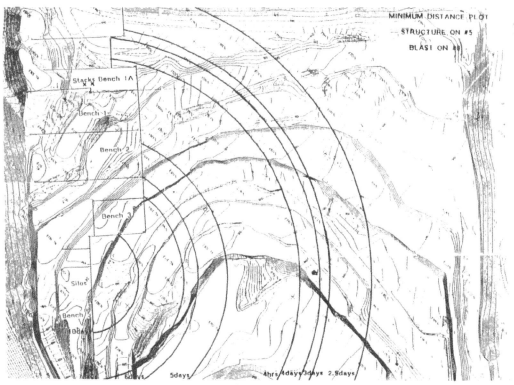

Figure 7. Distance limits after a pour on bench 5 for typical blasts on bench 8

possible to postpone pre-splitting of the final highwalls until production neared the final perimeter limits.

2. Effective vibration and blasting criteria was established to protect green concrete in critical structures from blast damage.

3. The construction of the lime kiln facility was completed with only the occasional delay of blasting operations.

4. Large scale quarry blasting and concrete pouring operations in close proximity can be compatible. However, modifications to the blast design may be necessary to reduce vibration levels.

12 REFERENCES

Gamble, D.L. & T.A.Simpson 1985. "Effects of blasting vibrations on uncured concrete foundations". *Proceedings of the 11th Conference on Explosives and Blasting Techniques*. International Society of Explosives Engineers.

Hulshizer, A.J & A.J. Desai June 1984. "Shock Vibration Effects on Freshly poured Concrete". *Journal of Construction Engineering and Management*. ASCE, Vol. 110, No. 2.

Oriard, L.L. 1982. "Blasting Effects and Their Control". *Handbook of Underground Mining Methods*. Society of Mining Engineers of AIME.

Watkins Engineers & Constructors 1995. "Technical Spec. No. 3046-TS-120294-01, Rev. 1". 9 pages.

Worsey, P.N., Giltner, S.G., Rupert, G. & H.S. Miller December, 1994. "Seismological Studies of Blasting and Design of protective Measures Involving Blast Vibrations for the Installation and Operation of proposed Lime Kiln Plant at Tower Rock.". Final Report of Investigations to Geotechnology Inc., 240 pages.

Rock Fragmentation by Blasting, Mohanty (ed.) © 1996 Taylor & Francis. ISBN 90 5410 824 X

Establishing a quantitative relation between post blast fragmentation and mine productivity: A case study

P.R. Michaud
ICI Explosives Canada, McMasterville, Que., Canada

J.Y. Blanchet
LAB Chrysotile Inc., Black Lake, Que., Canada

ABSTRACT: The physical characteristics of blasted rock such as volume, mass, and fragment size are fundamental variables affecting the economics of a mining operation and are in effect the basis for evaluating the quality of a blast. The object of this study was to quantify the effect of fragmentation on mine productivity. In order to be capable of achieving the latter, a relationship between fragmentation particle size and mine equipment productivity was required. Although the project entailed elements of drilling, blasting, loading and crushing, this paper focusses only on the hauling portion of the study. By utilizing a mine hauler equipped with a mass monitoring system and defining fragmentation by digital analysis of rock fragments carried from shovel to hauler, a linear relationship between the mass of material handled and a fragmentation index was established. The fragmentation index thus developed then becomes the criteria to designing the "optimum blast".

1 INTRODUCTION

The following analysis was conducted on the premise that a mining operation can achieve optimum production costs by controlling the size of rock fragments produced from a blast. To validate this assertion, a mine was chosen to conduct a data collection exercise focused on drilling, blasting, loading, hauling and crushing operations. For this presentation and as an example, only the hauling cycle is demonstrated. The study does not incorporate other productivity factors such as machine use and delay time, nor does it consider post-blast heave and swell as contributors to blast optimisation. Although information on drilling and blasting was collected, it is not the object of this study, nor is crushing. The data collection exercise compiled from one blast, comprises measurements of loading and hauling cycle times against mass and fragment passing size. Since the impact of the information and subsequent analysis is pertinent to the mining operation in which it was conducted, a brief coverage of the mine's installations and operating equipment follows.

LAB Chrysotile Inc. with head office in Thetford Mines, Quebec is a management contract company formed in 1986 to oversee the mining activities of three distinct asbestos mining companies then operating in the immediate region. Of these three, the

Black Lake operation is the focus of the study. The annual volume of rock excavated at this specific open pit operation is 14 600 000 tonnes of ore and waste. Stripping ratio is 2,1 to 1. The average bench height in the pit is 12 m and its present depth is at 330 m. The extreme pit dimensions are 1,7 km by 2,1 km. Production schedule is 24 hours a day, 6 days a week. The drilling is performed by two diesel-powered Bucyrus Erie 45R rotary drills, drilling 250 mm diameter holes. Hauling is managed by a fleet of 24 production haulers varying from 50 to 90 tonnes capacity. Loading equipment is comprised of four P&H shovels with 10 m^3 buckets and two 992C Caterpillar loaders with the same bucket capacity. Finally, the crusher is a 66" × 84" Allis Chalmers jaw crusher, set at an 18 cm opening.

2 COLLECTION OF DATA

2.1 Drilling and Blasting Parameters

Since the project's primary objective was to measure the sensitivity of productivity with respect to variations in fragmentation it was only necessary to measure these changes by discrete sampling within a unique blast. By doing so, geology and\or structural geology do not become variables that would otherwise be part of the equation. Having said this, one will still be confronted with the task of delivering required fragmentation by recommending the proper loading and blasting parameters!

The collection of the drill and blast parameters for the test blast #41-94-C-030 was done via the mine's blasting engineer on a tailored database application provided by the explosives supplier. Hole depths, collars, spacings, burdens etc. were physically measured and corrected on database if required. Input screens also provided information on blasting accessories and bulk explosives loading. Bulk explosives are specified in Table 1. The drill hole pattern and blast initiation patterns were documented. Rock samples were analyzed by the ICI laboratory and are shown in Table 2. Laser Profiling was not possible due to hardware limitations. A summary of drill and blast particulars can be found in Table 3. A high speed film of face velocities in granite and serpentine was taken at crest, mid-face and toe in each case and velocity of dtonation of three boreholes loaded with Magnafrac BR-9075 showed product velocities of 5.5 to 5.6 km.s^{-1}

2.2 Loading

The P&H 1900 shovel, fitted with a nominal 10 m^3 capacity dipper and identified by the number 3 was the only shovel monitored for the duration of the project. Monitoring during the day shift alone allowed for better control over operator turnover. In general, at the LAB operations, and more specifically at the B-70 level, queuing of haulers is virtually non existent. Prior to hauler spotting, the shovel operator has made provisions to load the hauler as quickly as possible. Therefore, once the shovel has dropped its first load in the hauler, loading is ideally continuous from bucket to bucket without the interruptions most often caused by hard toes, oversize or shovel positioning. In isolated cases where queuing is evidenced (highgrading conditions for example), the trucks are not loaded, let alone spotted, until the shovel operator has made provisions to load the truck without interruptions. Moreover, shovel operators are trained to fill their buckets by taking slices from toe to horizon in one smooth pass. In cases where the bucket's heaped capacity is not reached, a second pass is made prior to swing and drop. Bucket fill factors were determined by comparing the On Board Data Aquisition System (OBDAS™) dipper mass output with the calculated Society of Automotive Engineers (SAE) heaped dipper capacity.

Table 1. Bulk emulsion explosive properties

Emulsion	Magnafrac	Magnafrac
Brand	BR9075	B-965
Blend	30%Anfo	20%TNT
Density	1.25	1.35
RWS	86	92
RBS	128	142
VOD	5500	5000

Table 2. Mechanical rock properties

Rock type	Granite	Serpentine
S.G. (g cm^3)	2.56	2.5
P-wave (m s^{-1})	5350	5800
S-wave (m s^{-1})	3420	3780
Poisson ratio	0.15	0.13
Young's Modulus	69.1	83.4

Table 3. Design parameters for blast #41-94-C-030

Bench height (m)	12.0
Hole depth (m)	15.0
Spacing (m)	9.0
Burden (m)	8.5
Stemming (m)	5.0
Drill hole dia. (mm)	250
Number of holes	32
Number rows	3
Load per hole (kg)	656
Powder factor (kg\mt)	0.25

2.3 Hauling

The 100 ton (90.7 mt) DART 3100 hauler, identified by the number 724 was the only truck monitored during the complete data collection exercise. Day shift monitoring offered better control over operator changes. The automatic transmission controlled by an electric servo system provided consistent gear changes (independent of operator) during uphauling operations. Road conditions would however cause variations in gear shifts allowing for discrepancies in cycle times for a given load. The truck operator would at times elect to override the servo selection when lugging was apparent. The Truck body fill factors were determined by comparing the OBDAS™ truck mass output (total of shovel dipper loads per cycle) with the manufacturers specification for a heaped (2:1) load, which corresponds to the SAE standard.

3 MASS MONITORING

Monitoring of mass was made possible by equipping the DART 724 hauler with a commercially available mass-monitoring system supplied and fitted by Philippi-Hagenbuch, Inc. of Peoria, Illinois. The expandable OBDAS Truck Management System was installed with minimum options, mainly real time recording of shovel and hauler cycle times, including mass of each dipper load dropped in hauler. A printout of the OBDAS record for a typical load is shown in Table 4. The On Board Data Acquisition System is comprised of a compact 16 bit computer processor mounted in the hauler's cab. The CPU gathers information from four load sensors and produces a continuous load weight readout in real time. A schematic of the OBDAS system components is shown in Figure 1. The information gathered in the computer's memory was downloaded at the end of each shift and edited for future reference. The accuracy of this weighing system was claimed to be +/-2% of total

hauler payload. Calibration of the system is set in-shop prior to installation of sensors and did not require further in situ setting adjustments according to the manufacturer. It was found, however, that the system was not as simple as it was designed to be. Frequent hardware problems, notably the load sensors, required close monitoring on a day-to-day basis. An example of this was trying to compare OBDAS™ mass output with physical weighing of the same load on a public weight scale. Due to "Bottom Out" of one of the four sensors, only a total of 7 buckets out of 10 were registered, underweighing the load by 17 mt. Subsequently, more than half of the data coming out of the OBDAS™ had to be discarded with respect to individual hauls. Additionally, an intermittent leaking sensor caused irregularities in some of the loads that consisted of more than seven dipper loads. For this reason only full hauler loads comprising of five, six and seven dippers are considered whenever the analysis of the data collected is mass related. Please note that the supplier discontinued the manufacture of the OBDAS mass monitoring module shortly after this project ended.

Table 4. OBDAS print out from hauler

OPERATOR CHANGE 03/16/94 11:12:59
OLD OPERATOR: JOCELYN VACHON
NEW OPERATOR: PETE
TOTAL LOADS: 5
TOTAL T: 368.2
TOTAL TIME: 00 DAY 02 HRS 52 MIN 41 SEC
OPERATOR AVERAGE LOAD: 73.6
AVERAGE CYCLE MINUTES: 34.5
11:14 0.6 MTN NEUT 02 MIN 45 SEC
11:23 0.1 MTN FORW 08 MIN 55 SEC
11:23 0.2 MTN REVR 00 MIN 26 SEC
11:23 0.1 MTN FORW 00 MIN 27 SEC
11:24 0.1 MTN REVR 00 MIN 43 SEC
11:24 14.4 T BUCKET
11:25 14.7 T BUCKET 00 MIN 32 SEC
11:25 14.2 T BUCKET 00 MIN 56 SEC
11:26 15.0 T BUCKET 00 MIN 28 SEC
11:27 8.0 T BUCKET 00 MIN 56 SEC
11:27 83.2 MTN NEUT 03 MIN 06 SEC
AVERAGE BUCKET: 13.9 T
AVG BUCKET TIME 34 SEC
83.3 T TOTAL TIME: 02 MIN 50 SEC
11:45 77.2 MTN FORW 18 MIN 11 SEC
11:46 77.1 MTN REVR 00 MIN 31 SEC
11:47 77.3 MTN DUMP 36 MIN 03 SEC

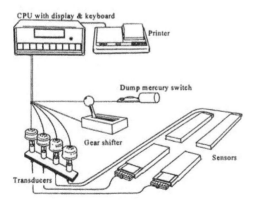

Figure 1. OBDAS™ system components

Figure 2. Digital photo capture of dipper loads

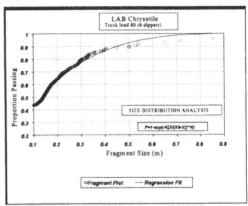

Figure 4. Appended fragmentation plot (6 dippers)

4 FRAGMENTATION ANALYSIS

In order to assess the degree of fragmentation proper to the mass of blasted rock being moved by the shovel and hauler, an approach previously adopted by Maerz et al. (1986), of using the pan of a hauler as a reference grid was used. Photographs (35 mm) were taken from above and behind the hauler being loaded (figure 2).

Figure 3. Digital fragmentation analysis

Care was taken so that the four corners of the truck pan were included in the picture as these points served as a scale reference defining a plane. Each dipper dumped in the truck was photographed and fragments were digitized on a PC monitor in a CAD environment. Rectification of digitized fragments was done via an Autolisp application which would simply compensate for perspective (refer to Figure 3). When photographed from an oblique angle, the perspective grid template defined by the four corners of the truck pan is not imaged as a rectangle but as a trapezoid. The digitized "rectified" image of fragments therefore represents a fragmentation profile which is proportioned against a rectified truck pan whose plane is perpendicular to the view angle and defined by a rectangle formed after rectification. Furthermore, since all the fragments digitized are not all in the same plane, let alone in the plane defined by the rectified grid template, further compensation for depth of individual fragments is required. Since the separation of digitized fragments with the normal plane used as the reference grid is not known it becomes practically impossible to determine absolute fragmentation. The fragmentation data processed in this report is therefore unique and relative to the methodology used. The output of the "rectified" fragmentation chunks from each set of pictures is in the form of an ASCII text file with the data on each individual dipper load appended one below the other. This file is then imported to a basic software package which outputs a distribution graph from which passing sizes can be extrapolated and later correlated to mass for individual dipper and hauler loads. Figure 4 shows the fragmentation plot of the six dippers in load #080.

5 ANALYSIS OF RESULTS

The data pertaining to shovel cycle times (252 observations) was analyzed in its entirety. The data pertaining to mass and fragmentation with respect to hauler activity required screening due to inconsistencies in the OBDAS™ output. OBDAS™ did not account dipper loads correctly in 70 out of 132 truck loads registered. Furthermore, invalid loads due to "Bottom Out" of a faulty sensor reduced the number of observations to analyze to 48 truck loads (294 dipper loads). The erratic behaviour of the faulty sensor caused concern over the final selection of the data that was retained for statistical analysis.

5.1 Fragmentation Analysis of Hauler Loads

By appending the digitized data obtained from individual dipper loads dropped in the hauler, a fragmentation plot representing percentage passing size against fragment size was drawn for each complete hauler load (Figure 4). The data was tabulated, to allow plotting of fragmentation against mass hauled.

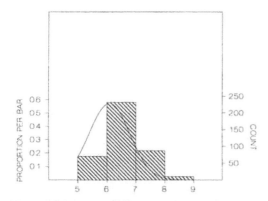

Figure 5. Histogram of dipper count per cycle

5.1.1 Analysis of hauler mass and fragmentation

Analysis of truck mass against fragmentation was done by considering only dipper loads that reached the nominal volumetric capacity of the shovel dipper. Truck loads that fall under this category are the five, six and seven dipper hauls. The models incorporating five and seven dipper loads, whether individually or together with the six dipper data, considerably affect the linear fit of the model. This is due to the fact that

the truck loads hauling five dippers are comprised mostly of fine to extra fine fragments almost soil-like in appearance and at times intermingled with large cap rock (dipper size !). The mathematical particle distribution functions of soil versus blasted rock are much too different from each other to allow any meaningful relationship within the range defining the fragmentation index. The fragmentation application used to plot the distribution in this case is not adequate in estimating passing size, as the areas of the chunks digitized are too dependant on operator interpretation and selected cut-off dimension. Moreover, when 7 dipper loads are considered, the lack of uniformity in fragment distribution per truck load measured against a quite narrow hauler mass spread, drastically affects the fit of the linear regression. Together with the intermittent malfunctioning of the OBDAS™ when more than 6 dippers were loaded on the hauler and the erratic effects on linearity brought about by inclusion of 5 and 7 dipper loads, the sample used for the regression model of truck mass against fragmentation passing size, consists of 31 truck loads by 6 dippers each, for which the number of photographs taken corresponds to actual OBDAS™ count. The Figure 5 histogram shows that 60% of the loads comprised 6 dippers. The regression model obtained for truck mass as a function of the fragmentation index is the following:

$$MASS\_TRK = 107.39 - 5.64*FRAG\_NDX \qquad (1)$$

The equation relates that the average mass of material (rock fragments) contained in the hauler pan is negatively bound by the value of the fragmentation

Result 1

DEP VAR:MASS_TRK	N	31	MULTIPLE R: 0.805	SQUARED MULTIPLE R:	0.648
ADJUSTED SQUARED MULTIPLE R:	636	STANDARD ERROR OF ESTIMATE:	3.282		

VARIABLE	COEFFICIENT	STD ERROR	STD COEF	TOLERANCE	T	P(2 TAIL)
CONSTANT	107.390	4.353	0.000		24.672	0.000
FRAG_NDX	-5.644	0.772	-0.805	1.000	-7.313	0.000

ANALYSIS OF VARIANCE

SOURCE	SUM-OF-SQUARES	DF	MEAN-SQUARE	F-RATIO	P
REGRESSION	575.977	1	575.977	53.481	0.000
RESIDUAL	312.320	29	10.770		

index, i.e. *an index developed by ICI to quantify relative fragmentation (mean particle size and spread of distribution).* In simpler terms, the smaller the fragments, the greater the mass hauled. A plot defined by equation (1) is shown in Figure 6. This relation is statistically significant in that the parameters associated with the measure of fragmentation is significantly different from zero (t_{29} = -7.313, p < 0.0001). More precisely, a drop of one unit in the

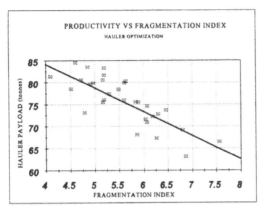

Figure 6. Hauler productivity plot

fragmentation index corresponds on average, according to the model, to an increase of 5.64 metric tons per haul. Moreover, a squared multiple R indicates that 65% of the variability observed in the average hauled mass is explained by the variable FRAG_NDX representing the fragmentation index .

Table 5. Predicted hauler mass for a given index

FRAG INDEX	MASS predicted	UPPER 95%	LOWER 95%
4.09	84.28	86.93	81.64
4.51	81.93	84.01	79.85
4.72	80.76	82.59	78.94
4.91	79.71	81.32	78.09
5.19	78.08	79.44	76.72
5.64	75.59	76.79	74.38
6.80	68.63	71.26	66.74

By retaining the simple regression model developed in Result 1, it is now possible to estimate the average mass of blasted rock fragments represented by 6 full dipper cycles in the truck haul, provided the appended fragmentation passing sizes are known. Additionally, a confidence interval can be built that gives us a specific idea of the variability of our estimate for the mean or the average hauler mass transported for each cycle. Typical calculated predicted values for the upper and lower 95% confidence intervals are shown in Table 5. For example, for a fragmentation measure of 5.19, the predicted value given by the model for the average mass carried by the hauler is 78.08 (107.39 - 5.644*5.19). The 95% confidence interval for such a

load is (76.72, 79.44). Indeed, with a 95% confidence interval, one can predict the mass of an average truck load containing a measured fragmentation of 5.19, to be situated between 76.72 and 79.44 metric tonnes.

5.2 Analysis of hauler fill factors and fragmentation

Hauler fill factors are calculated by relating the total weight of fragmented material loaded by the shovel per haul cycle (OBDAS™ output), to the SAE heaped rating of the hauler. By using an average rock density of 2.6 g cm^{-3} as was done with the individual dipper loads, the SAE body capacity of the hauler is 148 tonnes. Figure 7 shows the plot of the hauler fill factors against the fragmentation index for the 48 hauler loads previously analyzed. Result 2 shows on the following pagethe linear regression of the model comprising 6 dipper loads per haul (N = 31). Note the similarity of Result 2 below and Result 1 ,from Section 5.1: Fragmentation analysis of hauler loads. The difference lies in the application of the body capacity rating of 148 tonnes as a denominator to the right side

Result 2 (Fill factor regression for hauler loads comprising 6 dippers)

DEP VAR PAN_FF   N   31   MULTIPLE R 0.805 SQUARED MULTIPLE R   0.648
ADJUSTED SQUARED MULTIPLE R   636   STANDARD ERROR OF ESTIMATE 0.022

VARIABLE	COEFFICIENT	STD ERROR	STD COEF	TOLERANCE	T	P(2 TAIL)
CONSTANT	0.725	0.029	0.000		24.672	0.000
FRAG_NDX	-0.038	0.005	-0.805	1.00	-7.313	0.000

ANALYSIS OF VARIANCE

SOURCE	SUM-OF-SQUARES	DF	MEAN-SQUARE	F-RATIO	P
REGRESSION	0.026	1	0.026	53.481	0.000
RESIDUAL	0.014	29	0.000		

PEARSON CORRELATION MATRIX

	MASS_TRK	PAN_FF	FRAG_NDX
MASS_TRK	1.000		
PAN_FF	1.000	1.000	
FRAG_NDX	-0.805	-0.805	1.000

of the equation. The relation between pan fill factor and fragmentation can therefore be approximated by the following linear regression model:

$$PAN\_FF = 0.725 - 0.038*FRAG\_NDX \qquad (2)$$

The previous equation tells us that best pan fill factors can be achieved when fragment size is reduced. Pan fill factors for this exercise (N = 31) ranged from 0.43 for load #128 to 0.57 for load #031. The X50 and X80 passing sizes (cm) were respectively 24/40 and 5/20. A drop of 2.3 units in the fragmentation index represented by loads #31 and #128 is equivalent to an increase of 32% in the fill factor obtained in load #031 over load #128. Figure 8 illustrates visually the differences in fragmentation between the two loads. A squared multiple R of 0.648 indicates that 65% of the variability observed in the pan fill factor is explained by the fragmentation variable FRAG_NDX.

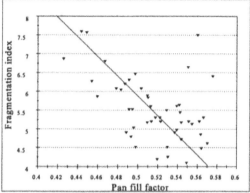

Figure 7. Fragmentation index against pan fill factor

LOAD #031 (above), Payload=84.5 mt, FragINDEX=4.60

LOAD #128 (below), Payload=63.1 mt, FragINDEX=6.90

Figure 8. Payload comparisson of hauls #31 and #128

6 DISCUSSION

The linear relationship of mass hauled with respect to fragmentation displayed in Figure 6 shows that there are significant gains in productivity that can be achieved by simply producing smaller fragmentation. For LAB Chrysotile, it is evident that smaller fragmentation will increase shovel and hauler cycle output. What is not clear is: can this finer fragmentation be achieved, and if so, will the drilling equipment at hand be suitable? Does additional capital investment and increases in drilling and blasting costs make it worthwhile? What fragmentation is required as feed at the primary crusher to optimize downstream milling operations? How will the complex structural geology affect the outcome? Is the predicted fragmentation from blast simulation programs consistent with the fragmentation obtained from the appended hauler loads? Homogeneous fragmentation is a term that does not exist at this mining operation as shown by the variety of fragmentation plots obtained even when taken within a specific hauler load. Post-blast fragmentation surveys of operations similar to LAB Chrysotile are necessary in order to locate actual and optimum productivity levels. Comparing powder factors from blast to blast would not produce measurable differences in productivity at LAB Chrysotile because of the complex geological formation. However given a "homogeneous rock formation", powder factors would indeed be the way to go. Limestone quarries are good candidates for studies involving productivity in relation to powder factors. Unless fragmentation analysis can be automated, projects such as LAB Chrysotile are practically impossible to evaluate in a time frame that is acceptable. Actual digitization of rock fragments from digital images, rectification and blocking out represents considerable unproductive time. PowerSieve™, a stand alone fragmentation analysis package developed by ICI Australia to quantify post-blast results, is presently being used to measure productivity variations attributable to fragmentation. The latter permits the analysis in quasi real time of digital photographs taken during the loading and hauling cycles.

7 CONCLUSION

A linear relation between mass of fragmented rock and the fragmentation index has been established. The significance of this relation is valid provided the passing sizes can actually be extrapolated from the observed fragmentation distribution function. The results of the study provide a graphical representation of productivity variations based on post-blast fragmentation. Typically, smaller fragmentation

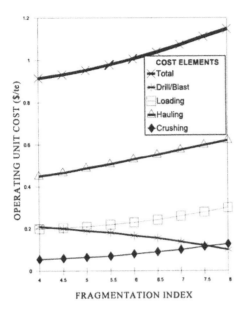

FRAGMENTATION INDEX

Figure 9. Mining costs and the fragmentation index

affects mine productivity by increasing tonnage of individual dipper and hauler cycles. Although the plot of hauler load against fragmentation index represented by Figure 6 is applicable to the LAB Chrysotile operation it can be used to demonstrate the implication of finer fragmentation on hauler productivity in any mine. An increase in the degree of fragmentation (finer) will forcibly increase the overall drilling and blasting costs, but will in turn have beneficial effects on the loading, hauling and crushing elements; the combined costs will define the "optimum blast" (see Figure 9). Capital investment required to achieve the required fragmentation must be considered, as it will affect the viability for change. Primary crushing is in many mining operations the controlling factor that dictates in absolute terms, the fragmentation required from a blast. If this is the case, we have a situation where loading and hauling productivity is controlled by the latter, regardless of any negative cost effects on the other elements of the mining cycle. However, the total mining cost which has been defined up to now by drilling, blasting, loading, hauling and crushing would require the addition of a sixth element in the equation: milling. By including milling, percent recovery becomes a major player in overall productivity costs and the determination of the "optimum blast".

8 ACKNOWLEDGEMENTS

The authors would like to extend their gratitude to LAB Chrysotile personnel, notably Michel Labbé, Production Mine Manager, André Duclos, Mine Superintendant, the Engineering and Maintenance Departments as well as the operators of heavy equipment that participated in this project. Also many thanks to the ICI technical team involved closely with the data collection and processing: Pierre Doucet, then mining research technologist, Samantha Phillips and Marie-Claude Turcotte both engineering mining students, respectively from Technical University of Nova Scotia and Ecole Polytechnique de Montreal. Finally thanks to Andrew duPlessis for providing the initial setup work required for the project.

REFERENCES

Hagenbuch, L.G., 1987. Mining truck dispatching: The basic factors involved and putting these factors into mathematical formulas (algorithms). *International Journal of Surface Mining*, pp. 105-129, January.

Lizotte, Y., and Doucet., 1992. Rock fragmentation assessment by digital photography analysis *CANMET MRL 92-116 (TR)* November 1992.

MacKenzie, A.S. 1966. Cost of explosives - do you evaluate it properly? *Mining Congress Journal*, May 1966: 32-41.

Maerz, N.H. , Franklin, J.A., Courson, D.L., and Rothenburg, L. 1987. Measurement of rock fragmentation by digital photonalysis.

Paley, N., 1992. Errors in sampling rock fragmentation from images. *Symposium on Sampling Practices in the Mineral Industry*, Mount Isa, Queensland, Australia, November 1992: 81-87.

Rock Fragmentation by Blasting, Mohanty (ed.)© 1996 Taylor & Francis. ISBN 90 5410 824 X

Successful long drift rounds by blasting to a large diameter uncharged hole

Stig Fjellborg
LKAB, Kiruna, Sweden

Mats Olsson
SveBeFo, Stockholm, Sweden

ABSTRACT:Over the years a large number of projects have been carried out in LKAB to improve the effectiveness and quality of tunneling and drifting.The project described in this paper involved the use of a large diameter hole as the opening cut for long (7.5m) round drifting and involved the Company's mines in both Kiruna and Malmberget. The paper includes a discussion of the (a) optimal hole diameter, (b) optimal blasting plan, (c) associated rock damage and (d) effect on scaling and reinforcement.

1. INTRODUCTION

The introduction of very large scale sublevel caving has lead to higher productivity and lower costs for LKAB's mines in Kiruna and Malmberget.Fig.1 shows the comparison of the mining scale to the 13-story administration building at the Kiruna mine.

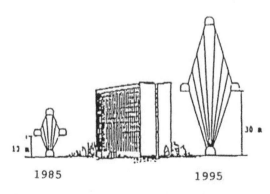

1985   1995

Fig.1 The change in sublevel caving
Geometry from 1985 to 1995.

Development is still the most expensive unit operation. Large scale means that the number of available development faces on any one level is very limited which means that the requirement for effective drift driving is increased. In order to increase the effectivity, the length of the round has been increased.In the period 1990-1991 as part of the Sofia project (Niklasson & Keisu, 1991),the technology and equipment for drilling and blasting rounds with a drilled depth of 7.7 m were developed. The advance per round was between 75% and 90% of the drilled length.The detailed results have been presented at the Frag-Blast Symposium in Vienna 1993 (Niklasson & Keisu, 1993).

The results were so positive that in 1992 it was decided to adopt these long rounds as standard in the Malmberget mine.After additional testing and improvement the average advance of these long rounds was 90% of that drilled. Measurments made with some of the rounds revealed problems associated with the opening cut.

In 1993 trials were conducted with various cut geometries based upon the use of 64mm holes. One of the cuts (Fig.2)provided advances up to 95% of the drilled length. However when a 300mm empty hole was substituted for the cut the advance per round increased to 100%.

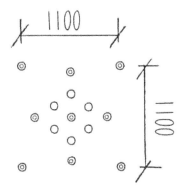

Fig.2 The cut based on the use of 64mm diameter holes.

Calculations showed that through cost savings associated with less (a) drilling, (b) explosive use and (c) face scaling and a better final drift contour, one could justify the costs of the pre-drilled large diameter hole.

In 1994 LKAB decided to purchase a special drill rig for drilling this large diameter cut hole.

A project was established around the use of this drilling machine to determine:

1. The optimal diameter of the large diameter hole.
2. The optimal blasting plan.
3. The associated blast damage zone.
4. The effect on scaling and reinforcement.

LKAB was in charge of the overall project while Kimit, Atlas Copco, Nitro Nobel, Kiruna Grus & Sten, Sandvik and Gruvprodukter took part in sub-projects. SveBeFo (the Swedish Rock Engineering Research Group) was responsible for assembling and reporting the results.

2. PROJECT GOALS

The main goals of the project were (a) to optimize the length of the rounds, maximize the advance and minimize the blast damage zone by using a pre-drilled large diameter cut hole and (b) to develop an optimized drilling and charging plan. Expressed in numerical terms the goal for the project was to achieve 99% advance in 90% of the

rounds. In addition the look-out of the contour holes should be limited to 20 cm outside the planned contour and the reinforcement costs should be reduced by 30%.

All together this should reduce the costs per meter by 1000 SEK (150 US$).

3. EQUIPMENT AND TEST AREA

The drilling rig (Fig.3) used for the large diameter cut hole is built on an AMV 26/35 chassis and propelled by a Deutz diesel engine. The drilling is done with a Wassara 6" diameter ITH water driven machine. To power the Wassara machine the rig is equipped with two water pumps with a total capacity of 285 l/min at 190 bars pressure. These pumps are driven by two 110 kw electrical motors. The magazine for the drill tubes contains 25 tubes, each with a length of two meters. Handling of these tubes is done using a crane which can be manoeuvred from the cab.

The drilling of the large diameter hole is done in two steps. First a hole of 165 mm diameter is drilled and this is then reamed to a diameter of 250 or 300mm.

The maximum hole length was determined to be 32m based upon the stated maximum hole deviation of 1% of the drilled length.

The time needed for drilling the 165mm diameter pilot hole in iron ore was 2.5-3.5 min/meter. The time for reaming was 6 min/meter for the 250mm and 9 min/meter for the 300mm large hole.

An Atlas Copco Rocket Boomer 353S (Fig 4) equipped with an automatic rod adding system (RAS) and Bever Control tunneling position system was used for drilling the long tunneling rounds. The boreholes had a diameter of 64mm which was determined by the drill rod coupling diameter. The average depth of the boreholes was 7.5m.

All the rounds were charged with Kimulux R which is a non-cap-sensitive, water resistant emulsion explosive manufactured by Kimit AB in Kiruna.

The main testing area was in the Norra Alliansen Orebody at the 790

Fig.3   The AMV Large hole drilling rig

Fig.4 The Atlas Copco Rocket Boomer
353 Long round drifting rig.

m level of the Malmberget mine.
Four 150 m (see Fig.5) long
drifts,to be driven primarily in
magnetite iron ore, were reserved
for the tests.

All of the tests concerning the
long rounds have been done in
Malmberget.The testing of scaling
and reinforcement was done in the
Kiruna mine.

4.THE TESTING PROGRAM

4.1 *Optimal diameter of the large
hole*

The purpose of this part of the
project was to compare the advance
between the standard long drift
rounds and rounds containing uncharged large-diameter cut holes of
250 and 300mm diameter.

The drill pattern used in both
cases was the standard pattern for
the long rounds shown in Fig 6.
When testing the large diameter
hole the center of the standard cut
was replaced by the large hole.

These tests includes 14 standard
rounds,7 rounds with the 250 mm
diameter hole and 5 rounds with a
300mm hole.

4.2 *The optimal blasting plan*

This part of the project included
tests with a drill pattern where
every borehole is located based
upon the expected rock removal produced by that hole (Fig 7).In that
way every charged hole is always
blasting towards a free face (Fig
8).

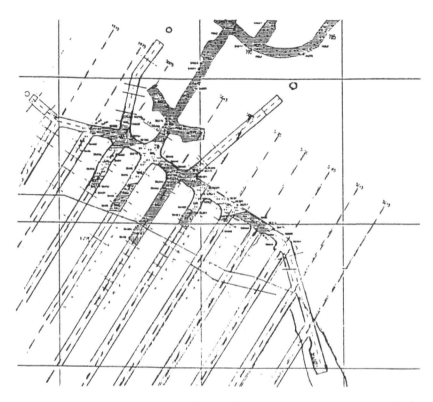

Fig.5 Plan view showing the Test area in the Norra Alliansen Orebody.
Scale 1:2000

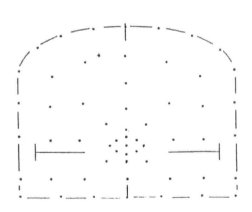

Fig.6 The standard drill pattern
and cut for long rounds.

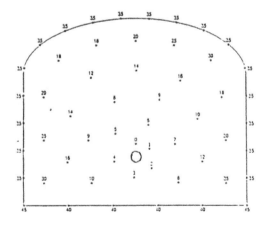

Fig.7 Hole positions and delay
numbers for the large
diameter cut hole tests.

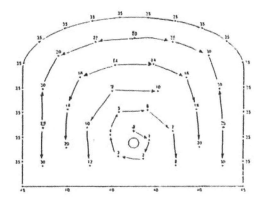

Fig.8 Successive relief of the
holes.

Tests were made using electronic
detonators,80g detonating cord, and
strings of Kimulux R charges in the
contour holes.

The tests with electronic detonators included tests of the two
systems developed by DNAG and Nitro
Nobel.

In the second part of the project
a total of 21 rounds with the 250mm
large diameter hole and 11 rounds
with the 300mm hole were blasted.

A short test series in which 51mm
boreholes were used in place of the
64mm holes was also made. The idea
of using smaller diameter boreholes
was to reduce the amount of explosive and as a consequence reduce
the damage zone.
Unfortunately the manufacture of
the drill steel was delayed so that
the tests have not been completed
at the time of writing.But testing
is continuing and hopefully the
results will be available for the
oral presentation.

4.3 *The associated blast damage
zone*

Two kinds of tests,borehole logging
and the slot technique were performed to determine the amount of
damage induced in the contour rock.
Borehole logging was applied in 4
diamond drilled holes. These holes
were drilled in two drifts at distances of 0.5 and 1.0 m from the
wall. Cracks were then observed
before and after drifting by a
borehole logging TV camera.

The tests using the slotting
technique were performed by SveBeFo
who have applied this technique
over many years to study blast
damage.

This method is described in detail in the paper "Crack length
from explosives in multiple hole
blasting" by M.Olsson & I.Bergqvist
presented at this conference.

Briefly, after blasting, slots
are excavated in the drift walls.
A special diamond saw is used to
make a series of vertical cuts 2m
long and 0.5m deep.The rock between
the cuts is then removed to create
a slot 2m x 0.5m x 0.5m. The remaining sawn walls can then be examined. Dye penetrants are sprayed on
the surface and photos are taken of
the crack patterns. The cracks from
the blasting can be clearly seen.
The technique was applied in four
selected rounds at LKAB.

4.4 *The effect on scaling and
reinforcement*

This part of the program involved
tests with different scaling and
rock reinforcement techniques.

All of these tests were made at
the Kirunavaara mine. Although the
length of the rounds was only 5m a
large diameter cut hole was used in
all of them.

A small number of rounds were
scaled with a new scaling method
called "Light mechanical scaling".
The equipment consists of a small
Montabert BRP 30 hydraulic hammer,
with a total weight of 85 kg attached to a crane arm and mounted on
a platform. The platform which is
also attached to a hydralic arm
(Fig 9) is carried by a 4-wheel
driven truck. The total reach is 7
meters horizontally and 10 meters
vertically. The idea of this method
is to replace the use of the old
scaling bar.The method has been
developed by LKAB in cooperation
with AMV to avoid scaling related
injuries.
Another method tested (Fig 10) was
scaling by the use of water under
moderately high pressure (100
bars).

In scaling with this method an
AMV 6400 shotcrete robot equipped
with a special nozzle to form the
water jet beam was used. To be able

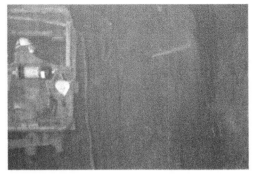

Fig.9 The AMV "Light Mechanical
     Scaling" Rig

Fig.10
          Water scaling of the walls.

to create this pressure a high
pressure water pump mounted on a
tank truck was attached to the
robot.
Rock reinforcement tests were also
made in combination with water-
scaling. Reinforcement in the form
of shotcrete was then applied
immediately after scaling, with the
cement being applied in a thin lay-

er to fill up the open cracks. Two
kinds of cement have been used,
the standard OPC cement for shotcrete and a new kind of fast curing
"super cement" called EMC.
A total of 6 rounds have been 100%
water scaled and 7 have been
partially water scaled (roof and
walls).

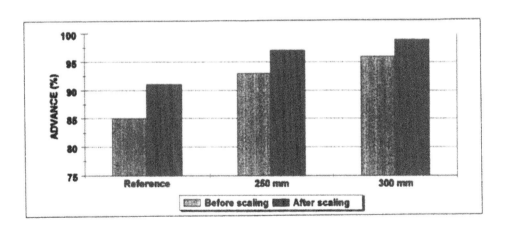

ROUNDS	ADVANCE			
	Before scaling (m)	Before scaling (%)	After scaling (m)	After scaling (%)
Reference	6.4	85	6.8	91
250 mm	7.0	93	7.3	97
300 mm	7.2	96	7.4	99

Fig.11 Results from the tests in
     Part 1.

5.RESULTS

5.1 *Advance*

The results from Part 1, as summarized in Figure 11, clearly show that the use of a large diameter cut hole increases the advance when using the standard drill pattern. Most remarkable is the advance for the 300mm holes which even before scaling has an average advance of 97%.It should also be noted that both of the large hole diameters have a better advance before scaling compared to the standard round after scaling.

No changes were made in the charging of the rounds with the large diameter cut hole compared to the conventional rounds. All boreholes with the exception of roof holes were charged with Kimulux R. In the roof holes a 40g detonating cord was used. This heavy blasting against the contour is revealed by the presence of half casts (Fig 12). In the standard rounds half casts only appeared in the roof. In the large diameter hole rounds half-casts were visible even in the walls.The appearance of half casts in walls increased with the diameter of the cut-hole.This evidently shows that the confinement of the boreholes goes down with increasing diameter of the large hole.

The amount of scaling was remarkably reduced both in time and volume for the rounds with the large cut hole. The average scaled rock volume was 3-4m³ for rounds with the large cut-hole and 9m³ for the standard rounds. The time reduction is harder to confirm since it depends upon the man doing the work but the time needed is approximately halved when using a large cut hole.

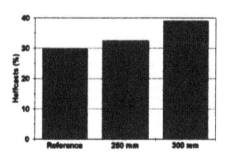

Fig.12 The presence of half casts in the contour holes.

5.2 *Hole deviation*

In all of the blasted rounds, hole deviation was measured by random sampling. The deviation varied from zero to a maximum deviation of 0.4m. The deviation is equally spread over the face but with the largest deviation observed in the contour holes. The average is 2.8% for the standard rounds while that for the large cut-hole rounds is 2.2%.

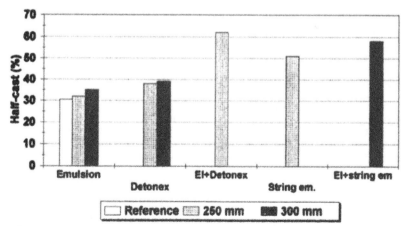

Fig.13 The presence of half casts in combination with different ignition systems and explosives.

5.3 Cautious contour blasting

In this part (Part 2) of the program the new drilling pattern was used for the rounds.

The appearance of half casts varies greatly with the type of explosive, the ignition system, and their different combinations (Fig 13).

Contour holes fully charged with emulsion and contour holes charged with 80g/m detonating cord (Detonex) when initiated by long period delay (LPD) caps have almost the same appearance of half casts. Standard LPD caps with a delay of 3500 ms were used in the whole contour. However when detonating cord was used in combination with electronic caps (EPD) the results were much better due to the instantaneous ignition.

Hence, the scatter in ignition times when using long period delays (LPD) has a great influence on the results and one must consider the initiation system and the contour explosive together. For example LPD caps of the same delay number used in combination with strings of Kimulux give a far better result than when used with detonating cord. The best results based upon half holes observations are obtained when using electronic caps (EPD) because of their instantaneous ignition.

When using the new drilling pattern the amount of scaling was almost the same as that for the standard drilling pattern.

The advance for the 250mm cut hole increased after scaling while for the 300mm hole it remained the same. The average advance for the large hole series had now increased to 99.5%. To estimate the improvement achieved in this project the advance of 20 standard rounds was measured.

The average advance was 94.1% after scaling. This result confirms that by using a large diameter cut hole and a modified drilling pattern the advance increased by about 5-8%. The scaling was reduced by 50%, and the final contour was better.

5.4 Blast damage

The results from TV logging are difficult to interpret but indicate, as expected an increase of cracks after blasting. A significant weakness of this method is that only a very small part of the rock (the periphery of the hole) could be examined.

The slotting method, which shows the result over a larger area, was more successful. The results showed very clearly different crack patterns with different explosive and initiation combinations. Contour blasting with a thin string of emulsion initiated with electronic caps showed no blast initiated cracks. Contour holes fully charged with emulsion ended up with crack lengths of at least 0.5m into the wall. The slotting method also revealed the thickness of the shotcrete which in this case far exceeded that planned.

5.5 Overbreak

The amount of overbreak was measured for every round. Since the amount of overbreak depends on a number of factors such as the accuracy of the reference laserbeam, the accuracy when drilling the contour holes, the scaling, and the rock mass structure it can not be used as a measure of the blasting quality.

The average overbreak at the beginning of the project was greater than 15% with peaks around 30%.

The average overbreak at the end of the project was about 12%.

5.6 Scaling and reinforcement

These tests were made in the Kiruna mine. The length of the rounds was 5m but a 250mm diameter cut hole was used in all rounds.

The tests with the "Light mechanical scaling" equipment indicated that this kind of scaling with a flexible and low-power machine is sufficient if the blasting has succeeded. Compared to ordinary scaling with heavier equipment, no time difference could be measured. Rounds with less successful blasting were also possible to scale satisfactorly but then of course the required time was much larger as compared to ordinary scaling machines.

The tests with high pressure water scaling gave unexpectedly good results when blasting had succeeded. The time needed for scaling was about 30 min for a 5m round. The water con sumption varied between 5m³ and 10m³ per round.

After scaling an examination was made to determine if reinforcement of any kind was needed. Walls and roof were then sprayed with a thin layer of cement to fill up open cracks and to secure the rock mass. About 2.5m³-3m³ of shotcrete has been used in every round. The time required for the whole operation is less than 2 hours. No deterioration of the roof when drilling the face in the following round could be measured.

However when blasting did not succeed as planned water scaling of the face was insufficient and ordinary scaling had to be done to get it approved.

Two kinds of cement were used after water scaling (a) the standard OPC shotcrete cement and (b) the new EMC cement. The EMC cement is manufactured by EMC Development AB company in Luleå and was originally intended for durable, water resistant concrete floors in factories and workshops. The most characteristic attributes of EMC cement are fast curing and durability.

The compressive strength graph (Fig 14) for EMC shows a very rapid strength development compared to standard OPC cement.

These results are based on 13 rounds reinforced with EMC and 12 rounds with OPC shotcrete.

6. DISCUSSION

This project which has been running from September 1994 through to November 1995 has included a total of 53 drift rounds involving the use of a large diameter hole.

By achieving an average advance of 99.5% and by halving the need for scaling when using a large diameter cut hole in combination with the modified drilling pattern, the main goals of the project have been fulfilled.

The results of the cautious contour blasting tests indicate that when using EPD-caps the type of explosives used have a minor influence on the results. The good results depend upon the instantaneous ignition. Detonating cord combined with LPD-caps yielded no improvement of the contour because of the scatter in ignition times. Scaling with light equipment or water jet is sufficient when using a pre-drilled large cut-hole.

EMC cement due to its fast curing time and high strength is a good alternative to standard OPC cement for reinforcement when rapid advance is needed.

LKAB has, based upon the project results, adapted the pre-drilled large diameter cut hole method in the Malmberget mine.

Tests with water jet scaling and the EMC-cement will be continued when suitable equipment which allows scaling and reinforcement application from the same vehicle has been developed.

7. REFERENCES

Niklasson, B., Keisu, M. 1993.
New techniques for tunneling and drifting. SveBeFo.
Niklasson, B., Keisu, M. 1993
Rock Fragmentation by Blasting.
Rotterdam: Balkema
Olsson, M., Fjellborg. S. 1996
Grovhål i centrum, Final report.
SveBeFo.

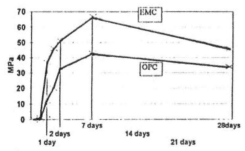

Fig.14 The compressive strength of shotcrete as a function of time.

Rock Fragmentation by Blasting, Mohanty (ed.)© 1996 Taylor & Francis. ISBN 90 5410 824 X

The impact of blast induced movement on grade dilution in Nevada's precious metal mines

S.L.Taylor, L.J.Gilbride, J.J.K.Daemen & P.Mousset-Jones
Department of Mining Engineering, Mackay School of Mines, University of Nevada-Reno, Nev., USA

ABSTRACT: A significant portion of the open pit precious metal mines in the western US are mining disseminated deposits with highly irregular mineralization. In some of these cases, migration of the ore zones caused by blast induced movement could possibly cause significant dilution when using standard ore control procedures.

The stated goal of this project is to develop a better understanding of the factors influencing how the material moves during a blast and to develop a simple method of accounting for the blast induced movement within the ore control procedures. This will reduce the dilution experienced by the mine at which the majority of the research for this project was conducted. Other mines in a similar situation may also benefit by adopting some variation of the modified ore control procedures.

This modified ore control procedure is not applicable to all open pit mines, or even all open pit precious metal mines with disseminated deposits. Those mines most likely to benefit from this project are precious metal mines with highly irregular mineralization and a tough host rock that requires a high powder factor to achieve adequate fragmentation.

1 PROJECT OVERVIEW

The project is being investigated along the following lines;

. general research into factors influencing the magnitude of the blasr induced movement, developing a method to measure the blast induced movement, how the blast material moves and factors affecting dilution levels (Zang, 1994); the blast induced movement using the Universal Distinct Element Code (UDEC). along with additional work on alternative method of measuring the blast induced movement Gilbride (1995); and modelling the effects of the blast induced movement on dilution and to develop a simple means of adjusting the ore control procedures at the subject mine (Taylor, 1995).

The main test site for this project was located at a surface mine in Nevada, which had been working to solve a serious dilution problem for several years. The initial study in 1992 resulted in modifications to the mine's blast design that helped to reduce the dilution by minimizing the magnitude of the blast induced movement. Unfortunately, this was not enough, as the dilution levels for gold were still regularly above 10%, and sometimes reached the 25% level for a given month. The overall dilution rates for gold and silver combined averaged at 5.3% over 34 months, with the rates slowly getting worse. The next logical step was to develop a means of accounting for the blast induced movement within the mine s ore control procedures. This paper mainly deals with the method of measuring the blast induced movement so that the ore control procedures can be adjusted to account for the blast induced movement.

2 INSTRUMENTATION AND EQUIPMENT

The only specialized equipment used for this project was a prototype version of a laser rangefinder. The IBEO Pulsar 500 (IBEO Systems, Sacramento, CA) was mounted on a Topcon theodolite and linked into a custom programmed datalogger. This system had three major advantages over a standard total station;

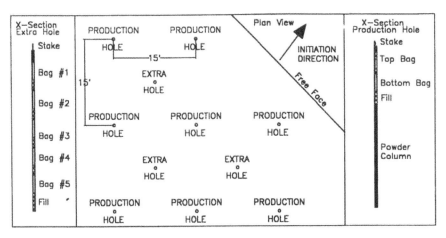

Figure 1. Configuration of Extra Holes and Production Holes

1) Safety, no rodman was required to be near the heavy equipment, or to be near potentially unstable working faces.

2) Accessibility, bags could easily be shot anywhere on the working face that had line of sight. A rodman would have trouble positioning a prism near the top of a 35 working face.

3) Speed, an average shot of an exposed bag (see research method) took between 20 and 30 seconds, thereby minimizing the projects impact on production.

The only drawback of this setup was the limited range where reliable distance measurements could be taken without a reflector. Side-by-side tests against a Topcon GTS-300 total station showed that the IBEO system was reliable to about 800 feet without a reflector when shooting a medium dark rhyolite. As a matter of policy, stations for face monitoring were always established within 300 feet of the working area and advanced if the distance became too great. A prism was also used with the system whenever the distance exceeded 400 feet, which was usually for the surface profiles.

Other equipment used for the project were as follows;

- Binoculars, a 35mm camera and a variety of hand tools
- A computer and a variety of software for cad, spreadsheets and statistical analysis
- 500 powder bags (3 feet long by 5 inch diameter) and 150 stakes (3 feet long)
- wire, three colors of paint, markers.

3 RESEARCH METHOD

The method used (marker bag method) to gather the basic information on the nature of blast induced movement is simple, reliable and fairly time consuming. Other methods are presently under study at the Mackay School of Mines, but none have yet proven to be feasible.

The marker bag method started with loading five color coded and labeled powder bags filled with stemming material into extra holes drilled between the regular production holes. The vertical distance from the collar to the bottom of each bag was measured, with the bottoms of the bags being nominally at 5, 10, 15, 20 and 25 feet below the collar. These bags had between one and five color coded stripes (three sets, top, middle and bottom) corresponding to the nominal depth of the bag and were labeled with the hole number in letters large enough to read through the theodolite scope (Figure 1). A color coded stake was then inserted into the hole so that only 3-4 inches showed above the stemming. A similar procedure was also used to place bags into the stemming column of some of the regular production holes. These bags were painted a solid color and labeled in large letters with the hole number.

The collars of all holes with marker bags were then surveyed and a surface profile of the blast was produced. With this information, the co-ordinates of the center of each bag could be estimated to within 4 to 6 inches. This level of accuracy was deemed to be adequate for the purposes of this study.

During the blast a 35mm camera with an autowinder was used to photograph the blast,

408

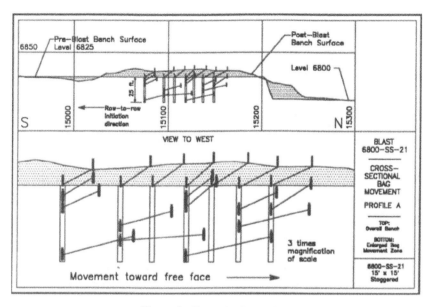

Figure 2. Example Cross- Section

which generally resulted in 8-10 frames taken at about 700-800 ms intervals. These photographs highlighted areas where the movement might have been altered by out of sequence holes, ejected stemming, etc. The photographs also made some of the surface movement easy to track, especially the free faces.

Directly after the blast a second surface profile was taken, with the locations of any exposed stakes or bags being surveyed. A station was then established on the pit floor or on a nearby bench, with good line of sight to the working face and in a safe area. These stations had to be kept within 300 feet of the initial working face and advanced as mining progressed to ensure accurate distance measurements without a prism.

The actual locating of the marker bags after the blast involved waiting until the loader uncovered a bag in the working face and then surveying in the location of that bag. With the marker bags covering an area of 10,000 to 20,000 square feet, in a muck pile averaging 35 feet in depth, this process took between 3 and 8 shifts.

4 ANALYSIS OF RESULTS

After resolving the survey data into pre and post blast locations for all of the recovered bags (34.5% overall recovery for 618 stakes and bags), the data collected was analyzed both graphically

and numerically. Plan and section (Figure 2) drawings were made, along with graphs of the horizontal and vertical movement components vs the original height in the bench. Figure 3 shows a graph of the horizontal movement data vs the original height in the bench for all five blasts monitored at the main study site. Note the discontinuity at the 15 foot level, which is about two feet above the nominal top of the powder column.

Statistical analysis of the horizontal and vertical magnitudes was accomplished by dividing the pre-blast bench height into two foot intervals (bottom interval was three feet), and using the descriptive statistics function included in the Microsoft Excel Statistics Add-on. Table 1 summarizes the results of the statistical analysis.

5 MOVEMENT PROFILE DEVELOPMENT

In order to be able to effectively model the ore dilution caused by the blast induced movement and to determine the optimal method for adjusting the ore control procedures, a two dimensional movement profile was developed. A three dimensional movement profile was not required since the third dimension would be parallel to the initiation line, i.e. perpendicular to the direction of movement caused by the blast. It had been observed that the movement in this third

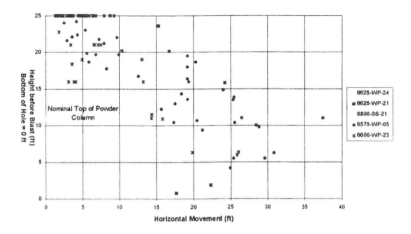

Figure 3. Horizontal Movement Summary

dimension is generally less than 3 feet, and is very erratic, probably being influenced mostly by the hole sequencing.

The profile was created by plotting the statistical means (of horizontal or vertical movement) for each two foot interval against the intervals'

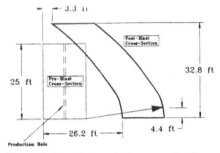

Figure 4. Blast Induced Movement Profile

original height in the bench (pre-blast). By then using curve fitting techniques to get a best fit line, two equations describing the horizontal and vertical movement were developed. By incrementing the pre-blast locations from (0,0) to (0,25), these equations resulted in the two dimensional movement profile shown in Figure 4.

6 MODIFICATION OF ORE CONTROL PROCEDURES

The ore control procedures in place at the mine site during this project were as follows;

· Sample drill cuttings using Thru-the-deck samplers
· Import assay data and survey data for each

Table 1. Descriptive Statistics

Descriptive Statistics for Vertical Movement Data

Statistic	Stakes	0 to 2	2 to 4	4 to 6	6 to 8	8 to 10	10 to 12	12 to 14	14 to 16	16 to 18	18 to 20	20 to 22	22 to 25
Mean	7.12	7.50	6.16	6.09	5.53	7.06	5.14	7.26	5.27	6.95	4.22	5.70	7.72
Standard Error	0.18	0.92	1.05	0.33	0.59	0.51	1.67	3.24	1.12	3.15	0.96	1.05	1.53
Median	7.02	7.50	6.00	5.97	5.52	6.34	4.92	7.01	4.66	4.05	3.66	5.06	6.58
Standard Dev.	1.59	1.30	2.96	1.80	2.37	2.45	4.08	7.26	4.62	5.46	2.55	3.15	4.34
Sample Variance	2.53	1.69	8.78	3.25	5.62	6.00	16.68	52.65	21.31	29.83	6.50	9.91	18.81
Range	6.51	1.84	8.47	6.76	8.79	10.03	12.61	17.67	17.71	9.70	7.90	9.14	14.16
Minimum	4.10	6.58	2.37	2.54	0.93	2.88	-0.28	-2.55	-1.85	3.55	0.92	1.71	4.06
Maximum	10.61	8.42	10.85	9.30	9.72	12.91	12.33	15.12	15.87	13.25	8.82	10.85	18.23
Count	78	2	8	30	16	23	6	5	17	3	7	9	8

Descriptive Statistics for Horizontal Movement Data

Statistic	Stakes	0 to 2	2 to 4	4 to 6	6 to 8	8 to 10	10 to 12	12 to 14	14 to 16	16 to 18	18 to 20	20 to 22	22 to 25
Mean	6.84	3.39	7.15	8.38	11.69	10.26	16.46	20.66	20.58	25.93	23.35	23.90	23.32
Standard Error	0.44	0.82	1.44	0.65	1.43	1.04	2.84	2.02	1.74	2.39	1.66	1.38	1.45
Median	5.97	3.39	5.41	7.39	11.88	9.63	16.41	19.21	19.96	27.75	21.80	24.04	22.90
Standard Dev.	3.86	1.17	4.33	3.55	5.73	4.98	6.97	4.52	7.16	4.15	4.39	4.15	4.10
Sample Variance	14.94	1.36	18.78	12.60	32.88	24.85	48.56	20.39	51.31	17.18	19.31	17.23	16.80
Range	14.99	1.65	13.44	13.85	16.72	16.73	16.28	9.79	32.39	7.67	12.28	12.45	13.08
Minimum	1.35	2.56	1.87	2.96	3.59	3.14	7.96	15.69	5.06	21.18	18.58	17.34	17.71
Maximum	16.34	4.21	15.31	16.81	20.30	19.86	24.24	25.48	37.45	28.85	30.86	29.79	30.79
Count	78	2	9	30	16	23	6	5	17	3	7	9	8

production hole into database
· Krige assay data using geostatistics
· Plot both the raw assays and kriged values on a
scale drawing
· Ore control engineer then draws the ore control
boundaries in by hand, then digitizes them into
the computer.
. A program then uses the digitized shapes and the
assay values to calculate tons and grade for the
blast
· A second program estimates the tons and grade
from the block model data which is also stored
in the database
· After the blast, the corner points for the ore
control boundaries are surveyed onto the muck
pile and the ore control flags are placed in
straight lines between them.

This method of ore control is based on the
pre-blast position of the ore, making no attempt to
account for the blast induced movement. Any
adjustments to this procedure had to be easy to
implement and not add significantly to the time
required to do the ore control. The solution to
this was to move the ore control boundaries in
such a way so as to reflect the blast induced
movement.

The next question was how far to move the two
dimensional ore control boundaries so as to
minimize the dilution in a three dimensional muck
pile. A series a profiles were taken of the
working faces and superimposed on the statistical
movement profile previously developed. An
advantage in this case was the fact that the blasts
are almost always mined in the opposite direction
that they were blasted (from the free face back,
not from the sides) and that the whole width of
the free face was advanced in a fairly uniform
manner. Figure 5 shows the average angle for the
working face superimposed over the blast
movement profile.

From this comparison and several other factors,
it was recommended to the mine that the ore
control procedures be adjusted as follows.

· Before surveying in the corner points, import
the digitized ore control boundaries into
Autocad.
· Move these ore control boundaries 10 feet in
the direction of the blast induced movement,
i.e.. towards the free face. Some further
adjustment of the ore control boundaries at the
front and back of the blast would be required
and were best left to the discretion of the ore

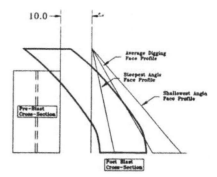

Figure 5. Recommended Adjustment
Distance

control engineer as how to adjust these in any
given situation. The magnitude of the blast
induced movement in these areas was different
than for the central portion of the blast, where
all of the marker bags were placed.

Use the modified ore control boundaries for
laying out the corner points.

These additional steps would be easy to implement
and add approximately 10 minutes to the 2-3
hours required to perform the ore control
procedures. Though far from a perfect solution,
these procedures would significantly reduce the
dilution due to blast induced movement.

7 ORE DILUTION MODELING

The ore dilution was modeled in order to estimate
the financial impact of the blast induced dilution.
The modeling used actual assay values and ore
control boundaries for three blasts used for the
blast movement monitoring. The initial intent was
to perform a three dimensional analysis. This
turned out to be very time consuming so a two
dimensional analysis was carried out. The third
dimension was used to determine the optimum
distance to move the drill hole polygons in order
to simulate the blast induced movement.

The first series of models (one for each of three
blasts) were done in order to check the accuracy
of the model against the original ore control
numbers from the mine. The drill hole polygons
were not moved, so the blast induced movement
was not accounted for in this series of models. A
comparison with the original ore control figures
from the three blasts resulted in the following;

0.08% difference in total tons

· 3.00% difference in contained oz of gold
· 2.60% difference in contained oz of silver

The second series of models (one for each of three blasts) took into account the blast induced movement. This would represent the actual material mined if the ore control boundaries were not moved. The basic procedure is outlined below;
· Divide the blast up into individual polygons representing the volume of rock associated with each drill hole assay. Autocad r12 was used for this.
· Move the drill hole polygons in order to simulate the blast induced movement
· Import the original ore control boundaries into the drawing one drill hole polygon at a time, find the area of that drill hole polygon that is inside an ore block as outlined by the ore control boundaries.
· Using a spread sheet, this area is entered into a cell, multiplied by an average depth and the tonnage factor and the gold and silver grades for that drill hole. This results in the assumed ounces mined for that drill hole.
· By summing the results for all of the drill hole polygons, the total recovered gold and silver ounces for that blast is estimated

These models will result in the maximum recovered ounces (minimum dilution) since they assume uniform movement over the whole blast and that the ore blocks are mined accurately. Also variations in powder factor and geology will influence the magnitude of the blast induced movement from blast to blast.

The results of comparing the first series and the second series of models are;
· First series 203,400 tons grading 0.0080 oz/ton Au and 1.41 oz/ton Ag mined
· Second series 204,000 tons grading 0.00·
Second series 204,000 tons grading 0.0079 oz/ton Au and 1.37 oz/ton Ag mined
· Lost values due to the blast induced movement were 20.2 oz Au and 8231 oz Ag
· This equates into 2.2% dilution for these three blasts, with the mine averaging 5.3% over 34 months
· This represents lost revenues (minimum) of US$47,000 for three blasts or approximately US$2.4 million per year. At 5.3%, the lost revenues would be approximately US$ 5.8 million per annum.

8 OBSERVATIONS ON BLAST INDUCED MOVEMENT

As a result of this project, the following conclusions were reached or confirmed about the nature of the blast induced movement at the study site and several other mines.

· The direction of blast induced movement is roughly parallel to the direction of initiation and towards the path of least resistance.
· A secondary free face does influence the direction of movement on a local scale, within approximately 40 to 60 ft of the secondary free face.
· At relatively far distances from the secondary free face, the deeper bags show a tendency to move in the initiation direction, while the shallow bags and especially the surface stakes tended to move towards the secondary free face.
· Blasting into a buffer can make the dilution problem worse depending on the type of material in the buffer. Blasting ore into ore or waste into waste does not cause any dilution except along the ore/waste boundaries.
· There is a significant increase in the horizontal movement below the top of the powder column. This is attributed to the fact that all of the explosives are located below this level and therefore, most of the energy produced by the explosives is applied to the material.
· The powder factor does influence the magnitude of the blast induced movement. This is offset by the variability of the rock properties with the blast foreman adjusting the powder factor based on the drilling rates for a particular pattern.
· The row to row delay timing does influence the magnitude of the horizontal blast induced movement. Shorter timing (35 ms) reduces the horizontal movement. It is thought that the shorter timing also causes more internal mixing of the material, possibly leading to an increase in dilution. The 35 ms timing definitely increases the vertical movement and causes high hills and deep valleys on the blast surface, a good indication of uneven internal movement.
· The size of an ore block can have a large impact on the dilution for that ore block. Generally the larger the ore block, the lower the dilution caused by the blast induced movement.
· The shape and orientation of the ore block can influence the dilution. Long and elongated

blocks perpendicular to the movement direction will experience high dilution, while similar blocks parallel to the movement direction will experience much lower dilution.

The location of the ore block within the blast can influence the dilution. Small ore blocks at the free face and the rear trough area are especially prone to high dilution.

9 CONCLUSIONS

It seems clear that in this case, blast induced movement has caused significant dilution and affected the profitability of the mine. Accounting for the blast induced movement within the ore control procedures should reduce the dilution levels experienced at the mine and increase revenues.

The marker bag method used is an easy, though labor intensive, way to measure the blast induced movement. Once a movement profile has been developed, it is a simple matter to adjust the ore control procedures in order to account for at least most of the blast induced movement, thus reducing the dilution caused by the blast induced movement and enhancing a mines viability.

This modification of the ore control procedures is not applicable or desirable to all open pit mines, or even all open pit precious metal mines with disseminated deposits. Those mines most likely to benefit from this project are precious metal mines with highly irregular mineralization and a tough host rock that requires a high powder factor to achieve adequate fragmentation.

REFERENCES

Gilbride, L., Zhang, S., Mousset-Jones, P., & Daemen, J. 1994. Blast Rock Movement and its Impact on ore Grade Control at the Rain Mine, Newmont Gold Co., *Proc. 3rd Int. Symp. on Mine Plg. and Eqpt. Selection*, pp. 713-719, Istanbul, Turkey, 18-20 Oct., A.A. Balkema, Rotterdam.

Gilbride, L., Zhang, S., Taylor, S., Mousset-Jones, P. & Daemen, J. 1994. Blast Induced Rock Movement Modeling for Nevada Gold Mines, *Proc. 12th Annual Workshop of the Generic Tech. Center on Mine Systems Design and Ground Control*, Alaska, Sept. 18-20.

Zhang, S., Hurley, J., Mousset-Jones, P., & Daemen, J. 1994. Blast Rock Movement and its Impact on Ore Grade Control at the Coeur Rochester Mine, *Proc. 20th Annual Conference on Explosives and Blasting Technique*, pp. 215-226, Jan 30- Feb 3, Austin, Texas.

Rock Fragmentation by Blasting, Mohanty (ed.) © 1996 Taylor & Francis. ISBN 90 5410 824 X

Rock and rock mass properties prediction based on drilling parameters for underground drift blasting

H.Wagner & P.Moser

Department of Mining Engineering and Mineral Economics, Montanuniversity Leoben, Austria

ABSTRACT: Monitoring of key drilling parameters can yield important information about rock properties and development blast design. As part of the EU-research project "Blasting Control" a development drill jumbo was instrumented to monitor drilling speed, percussion frequency, speed of rotation, torque and thrust. In total 40 development rounds of 3 m length each were monitored at the underground research site "VEBSTER" in Austria and a bauxite mine in Greece. In addition to the measurement of the drilling parameters tests were done on rock cores taken from the area of the cut to relate the drilling data to point load strength, uniaxial compressive strength, modulus of elasticity and seismic velocity. The first part of the paper gives an overview of the data that were recorded during the tests, the second part discusses the data analysis procedures, the third part deals with the correlation of drilling parameters and rock properties. In the final section of the paper the limitations of on-line drilling parameter monitoring (MWD) for the determination of rock and rockmass properties are discussed.

1. INTRODUCTION

In development work between 30 and 60 m of small diameter blast holes are drilled for every meter advance. It is argued that considerable practical benefit could be gained from the monitoring of critical drilling parameters as far as variations in rock strength, jointing and other parameters affecting blasting and tunnel stability are concerned.

As part of the broad based EU-research project "Blasting Control" the Department of Mining Engineering of the Montanuniversity Leoben in Austria was given the task to develop an on-line drilling monitoring system to predict rock and rock-mass parameters for blastability prediction. Other partners in the research project were the Ecole des Mines de Paris, the Politecnico Torino, Italy, the French explosives manufacturer Nitrobickford and the Greek mining company Bauxite Parnasse.

2 MEASUREMENT OF DRILL PERFORMANCE AND DETERMINATION OF ROCK PARAMETERS

A development jumbo provided by Böhler Pneumatic International was equipped with measuring devices for monitoring drilling speed, percussion frequency, drill rotation speed, torque and thrust during normal drilling operations. The measuring devices were located as close as practicable to the tip of the rockdrill. The electrical signals were collected in an on board data acquisition box where they were transformed in a RS 485 signal and transferred via cable to a PC which was housed in a suitable adapted underground room located about 100 m from the drill jumbo. The sampling rate was 2 Hz.

In total about one thousand 3 m long, 45 mm diameter blast holes, were monitored at underground research site "VEBSTER" of the Department of Mining Engineering in Austria and in a production section in a bauxite mine in Greece.

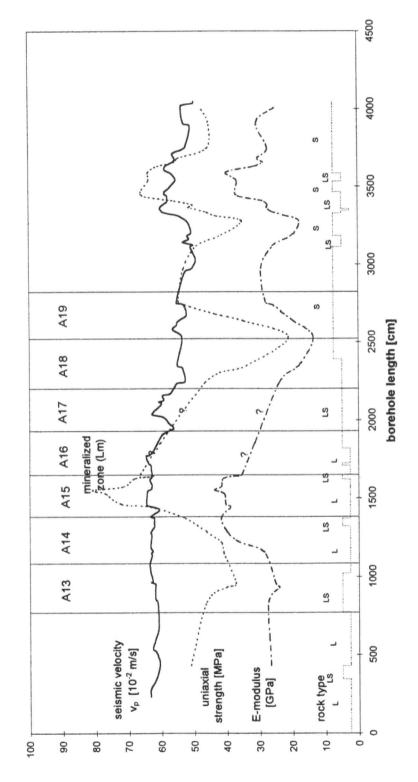

Figure 1: Rock and rock mass parameters determined on rock cores for blasts A13 to A 19

Some of the drilling parameters showed a variation of less than 5 % and were therefore virtually constant for all the holes drilled. The parameters in question were:
1. percussion frequency of the hammer at 50 Hz,
2. blow energy of the hammer at 170 J, and
3. drill rotation speed at 320 rpm.
Therefore these parameters were excluded from further analysis.

At the "VEBSTER" site the 76 mm diameter center hole of parallel cut of the blast was drilled with an exploration drill for the full length of test section of the development area. The resulting 45 m long continuous rock core with a diameter of 63 mm was carefully logged and the RQD value and rock jointing coefficients determined (Figure 1). The core was then used to determine the seismic wave velocity and to prepare samples for the radial and axial Point Load Tests as well as uniaxial compressive strength tests. Prior to testing the seismic wave velocity was determined for each sample. A feature of the development area at the "VEBSTER" test site was the jointed nature of the ground and the relative small angle between the major joint direction and the axis of the drift. This had adverse effects as far as the uniaxial compressive strength tests were concerned.

The drilling data for the 4 parallel cut holes surrounding the 76 mm diameter center cut hole were used to correlate drilling and rock parameters. In general the distance of each of the 4 parallel cut holes from the center hole did not exceed 200 mm. It can therefore be assumed that at any particular point the properties of the rock core were similar to the rock drilled by the drill jumbo.

In this paper only the results for 7 consecutive blasts in drift A of the underground research site "VEBSTER" in Eisenerz near Leoben are discussed. A specific feature of these blasts is that the rock conditions in the drift changed gradually from limestone into shale (minor limestone interbedded).

3 CORRELATION OF ROCK AND DRILLING PARAMETERS

3.1 Drilling data processing

Before the drilling data could be correlated with the rock data it was necessary to process them in order to:
1. remove the measuring error (negative drilling speed) at the beginning and end of every hole. This error was caused by the positioning and removal of the boom from the face;
2. eliminate the drilling data for the first 500 mm of every blast hole. The experience had shown that despite the barring of the face after each blast the first 500 mm of rock behind the face was still highly fractured and the drilling performance influenced by the fractures;
3. remove extreme variations in the drilling data which were caused by the low stiffness of the drilling system (movement of boom relative to rock face).

The drilling parameters which remained virtually constant throughout the test series, i.e. percussion frequency, blow energy of the hammer and drill rotation speed were excluded from further data processing.

Figure 3 shows the drilling data from Figure 2 after removal of the erroneous data. Even after excluding these values the variation of drilling data over short distances was very high. It is believed that these variations were due to the low stiffness of the drilling jumbo and not due to variations in rock properties. To minimize this effect on the correlation between drilling and rock parameters the drilling data were smoothed by calculating a moving average. All correlations between drilling data and rock parameters are based on a moving average.

3.2. Rock property determination

The following rock and structural parameters were determined from the rock core of the center cut hole:
1. Uniaxial compressive strength (UCS)

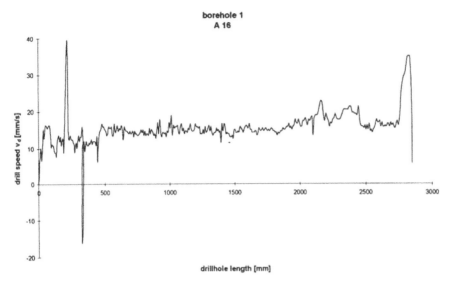

Figure 2: On-line monitored drilling data

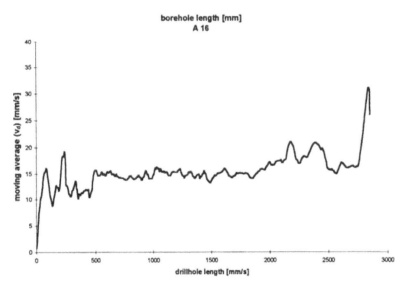

Figure 3:   Drilling data after removal of erroneous data and smoothed by calculating the
moving average

2. Modulus of Elasticity (E)
3. Point load index (radial) (PLI)
4. Compressional wave velocity (Vp)
5. Rock density (ρ)
6. Jointing coefficient
7. RQD value

A particular problem was the high degree of jointing of the rock mass in the area of concern. As a result only a limited number of core pieces were available to prepare test specimen for UCS tests. In many instances only PLI tests could be carried out. As indicated the compressional wave velocity was determined for all test specimen:

The results of the rock property tests are shown in Figure 1. The numbers A 13 to A 19 refer to the blast number.

3.3 Correlation of drilling and rock data

Before a correlation of the drilling and rock data could be attempted the exact position of the data points along the axis of the drift had to be determined. This proved to be difficult because the rock data were obtained from the 45 m long drill core which was drilled ahead of the drifting operations. Great care had been taken to account for core loss and the effects of several larger open joints which were intersected by the drill hole and the drift. After the drilling and rock data were properly aligned the drilling data were averaged over a distance of 100 mm at positions where rock samples had been taken for rock property testing. The average drilling speed and applied thrust were then correlated with the rock parameters (UCS, PLI, E and Vp).

4 RESULTS

It is a well known fact that up to a critical thrust value the drilling speed increases with the thrust applied on the drill steel. Figure 4 shows the relationship between drilling speed and thrust for the drilling jumbo used in the investigation. Also shown in Figure 4 is the relationship between applied thrust and resulting torque. The shaded area delineates the thrust range covered by the tests described in this paper. In the operating range the relationship between drilling speed and thrust is given by

$$V_d = 15 + 0,00057 \ T \qquad (mm/s) \ \textbf{(1)}$$

where T is the thrust in N.

According to (1) an increase in thrust from 10 kN to 12 kN corresponds to an increase in drilling speed of 1,14 mm/s or 5,5 %. It follows that variations in thrust in the range from 10 kN to 12 kN have a second order effect on the drilling speed and at this stage have been neglected in the analysis.

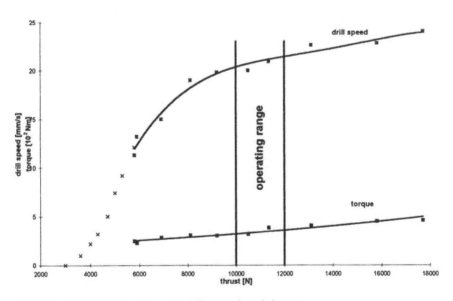

Figure 4: Relationship between drill speed and thrust

As expected the relationship between uniaxial compressive strength, UCS, and drilling speed, V_d, is rather poor. This has also been observed by other workers in the field (Schmidt 1972). In this particular investigation the situation was aggravated by the acute angle between the weaknesses in the rock mass and the axis of the drill core.

As a result many of the compression test specimen failed along planes of pre-existing weaknesses and not as a result of exceeding the compressive strength of the rock material. This also accounts for the large scatter in Figure 5.

Figure 6 shows the relationship between the point load index and the drilling speed.

A13 - A19 (borehole 1)

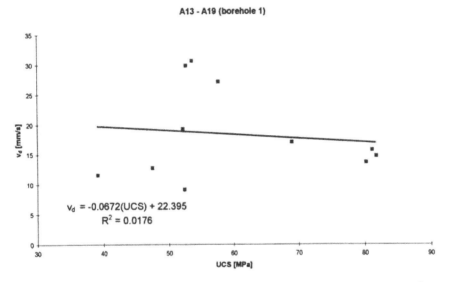

Figure 5: Relationship between drill speed and uniaxial compressive strength

A13 - A19 (borehole 1)

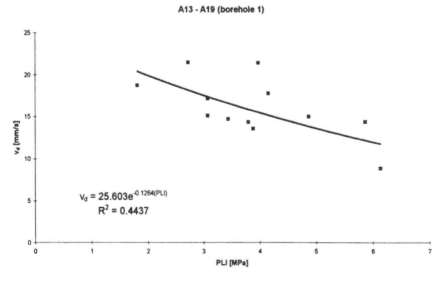

Figure 6: Relationship between drill speed and radial Point Load Index

Although there is still a considerable scatter a much more robust relationship is observed. Figure 7 shows the relationship between the modulus of elasticity, E, and the average drilling speed, V_d. A clearly defined decrease in drilling speed of rocks having higher elastic moduli is found. A similar trend has been established for the relationship between the seismic wave velocity and the average drilling speed, Figure 8. This is not surprising given the well established relationship between elastic rock properties and seismic wave propagation (Jaeger and Cook, 1979). Since V_p was determined for UCS and PLI samples

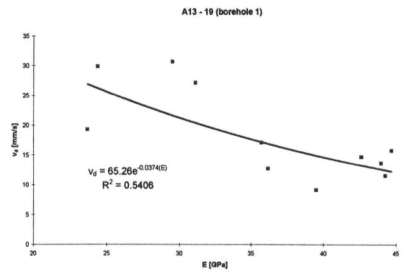

Figure 7: Relationship between drill speed and E-module

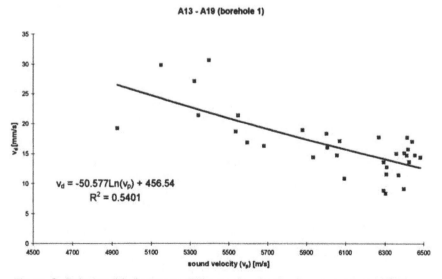

Figure 8: Relationship between drill speed and seismic wave propagation

the relationship shown in Figure 8 is based on a much larger sample population. According to Figure 7 and 8 the modulus of elasticity and the seismic wave velocity have a very pronounced effect on the drilling speed. Consequently variations in drilling speed offer a good potential to predict rock properties.

5. CONCLUSIONS

On-line measurement of drilling speed can provide valuable information about changing rock conditions. For the interpretation of the changes in drilling speed it is important that drilling parameters such as thrust, torque, speed of rotation and percussion frequency are monitored routinely. This information is required to ensure that the observed changes in drilling speed are not caused by changes in the operating parameters of the drilling system.

The stiffness of the drilling system is a critical factor as far as the interpretation of sudden changes in drilling speed is concerned. This aspect requires further investigation and is critical for the future development of "measuring while drilling" technology (MWD).

One of the original intentions was to locate structural discontinuities from the analysis of distinct anomalies in neighboring boreholes. In the case of the "VEBSTER" test site several major discontinuities have been intersected and it was possible to identify these in adjacent boreholes. However at a data acquisition rate of 2 Hz only joints with an opening in excess of 10 mm can be positively identified. In order to identify smaller joints the data acquisition rate would have to be increased markedly. Because of the stiffness effects on instantaneous drilling speed changes it is, however, doubtful whether it will be possible to positively identify smaller joints and structural weaknesses. At this stage drilling speed variations caused by a low system stiffness constitute to the practical limitations of the MWD approach. These could, at least in part, be overcome by a higher system stiffness and higher sampling rates.

There is no question about the ability of MDW systems to identify marked changes in rock properties. To do this effectively and reliably a considerable amount of data processing and data smoothing is necessary. There are no reasons why this could not be done on-line and in a mining environment.

The operation of an experimental MWD system has shown that commercially available pressure sensors to monitor thrust, torque and blow energies are reliable but care has to be taken that the sensors are capable of withstanding the exceptionally high short duration pressure pulses which do occur in rotary-percussion drilling. Once developed and properly tested no undue practical difficulties were encountered with the monitoring and data acquisition systems.

6 OUTLOOK

The results obtained to date indicate that MWD systems are a distinct possibility and can provide useful information not only about rock conditions but equally important on machine and operator performance. Significant improvements in operating performance and machine reliability can result from the introduction of such systems. Probably the most urgent need is for the development of an on-line blast hole positioning and orientation system to ensure that blast holes are correctly positioned and drilled not overburdened. In terms of the monitoring of the drilling speed there is an urgent need to develop a measuring system which measures the drilling speed with the rock face as reference point and not the boom of the drill jumbo. Only then will it be possible to obtain a true picture of the instantaneous drilling speed variations.

At this stage there is serious doubt whether it will be possible to identify and delineate individual small joints ahead of the development end. Zones of extensive jointing are more likely to be identified. This work is still in progress. The emphasis of the work should rather be directed towards the areas of drilling accuracy, determination of blasthole burden and identification of major structural changes and the routine monitoring of operational parameters.

7. SUMMARY

The work with an experimental MWD system at the "VEBSTER" test mine in Austria and a bauxite mine in Greece has shown that it is possible to monitor on-line the drilling speed of a drill jumbo as well as all important drilling parameters which can influence the drilling speed. The low stiffness of the drilling jumbo used for this investigation and the fact that the drilling speed was measured on the jumbo itself have made it necessary to introduce the algorithm of a moving average to eliminate the effects of sudden changes in drilling speed.

It was found that drilling speed relates well to the Point Load Index, the modulus of elasticity of the rock and the seismic wave velocity but not very well to the uniaxial compressive strength of the rock. Regional changes in rock conditions can be identified using MWD technology. Localized changes in rock conditions are more difficult to determine. At this stage the MWD technology does not allow to make predictions about the degree of jointing of the rock mass.

Acknowledgement:
The Austrian part of the project was funded by the Austrian Foundation for Research Promotion (FFF).

REFERENCES

Brennsteiner, E.: Investigation into the influence of parameters on percussion drilling with special reference to rock parameters. PhD Thesis. Department of Mining Engineering and Mineral Economics. Leoben, 1978. (In German).

Department of Mining Engineering and Mineral Economics: Rock and Rock mass parameter measurement. EU project Blasting Control. Research report. Leoben, 1995.

Jaeger, J.C. and Cook, N.G.W.: Fundamentals of rock mechanics. Chapman and Hall, London, 1979.

Schmidt, R.L.: Drillability Studies - Percussive Drillings in the Field. Report of Investigation 1684, United States, Department of the Interior, Bureau of Mines, 1972.

Spaun, G. u. Kurosch, T.: Untersuchungen zur Bohrbarkeit und Zähigkeit des Innsbrucker Quarzphylitts. Felsbau 12 (1994), 111-122.

Rock Fragmentation by Blasting, Mohanty (ed.) © 1996 Taylor & Francis. ISBN 90 5410 824 X

Applications of tracer blasting during stoping operations

S. P. Singh
School of Engineering, Laurentian University, Sudbury, Ont., Canada

Rodney Lamond
Royal Oak Mines Inc., Timmins, Ont., Canada

ABSTRACT: Tracer blasting is used in Canadian underground mines to minimize the overbreak in development headings. The miners feel much at ease with ANFO due to its safety and simplicity of loading. The lower cost and minimum overbreak makes it economically attractive to the mine management. This study has attempted to examine the application of tracer blasting during stoping operations. The field work was done at a Gold mine in Timmins, Ontario. The ANFO loaded holes in test rings were traced with a booster cord. The booster cord is different from a typical detonating cord and consists of a signal cord with bumps of higher core loads of PETN at equally spaced intervals.

There was a significant reduction in the explosive cost during ring blasting as compared to commonly used explosives in stoping operations. The peak particle velocity of ground vibrations was lowest in the case of tracer blasting as compared to the cartridged emulsion and ANFO. A reduction in the secondary breakage cost was observed during muck handling at different levels, which confirmed adequate fragmentation observed during visual examination of the muck pile.

1.0 INTRODUCTION

Tracer blasting is used in Canadian underground mines for overbreak control. This involves placing a detonating cord along the wall of a blasthole before charging the main column of ANFO. The detonating cord is taped to the detonator which is ideally kept in the centre, near the toe of a blasthole with the help of a spider (figure 1). The miners feel much at ease with ANFO due to its safety and simplicity of loading and the lower cost makes it economically attractive to the management.

During the last two decades tracer blasting has been mainly used in development blasting. This study is a follow-up of the previous study in which a mechanism of tracer blasting was proposed and the factors responsible for the inconsistency of the results were identified(Singh, 1996). The current work has attempted to examine the application of tracer blasting during Brow Retreat open stoping and sub-level caving operations with main focus on the later method.

1.1 Mechanism of tracer blasting

In a traced blasthole, the initiation of a detonator near the toe is followed by the simultaneous detonation of the detonating cord and the ANFO column in the longitudinal direction (figure 2). The VOD of detonating cord is much higher than the VOD of ANFO. Depending upon the location of the detonating cord, shortly after detonation, it causes side initiation of ANFO column before the detonation front from the detonator arrives there. Some or all of the following effects are produced during tracer blasting depending upon the explosive composition, core load and location of the detonating cord and blasthole diameter(Singh, 1996).

(i) Partial detonation and burning
(ii) Continuous side initiation
(iii) Partial decoupling
(iv) Desensitisation of ANFO
(v) Energy partitioning in favour of gas pressure
(vi) Reduction in ground vibrations
(vii) Reduction in total energy

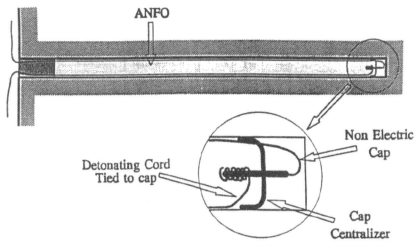

Figure 1. ANFO traced by a detonating cord in a blasthole.

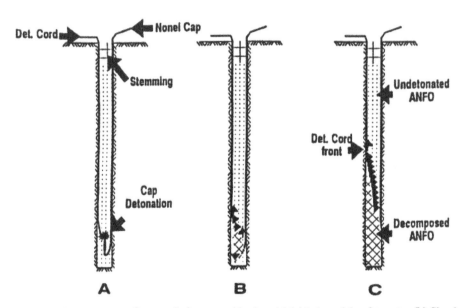

Figure 2. The sequence of events during tracer blasting: (a) Initiation of the detonator (b) Simultaneous detonation of the ANFO and the detonating cord (c) Side intiation of the ANFO column.

The role played by these effects in minimizing the blast damage has been depicted in figure 3.

2.0 FIELD WORK

This field work was done at a Gold mine in Timmins, Ontario. The mine is situated on the north limb of an overturned syncline, the axis of which dips steeply north at 70 degrees and strikes N70E. The south limb of this syncline has apparently been faulted off by the Destor-Porcupine fault which is located just south of the mine property. The property covers 2 miles (3.3 kms) of the Keewatin-Timaskaming contact where dominantly volcanic rocks of Pre-Cambrian age underlie sediments of Timiskaming age. All formations dip steeply north at 70-75 degrees and face south.

The ore body consists of polymictic conglomerate dipping at 68-70 degrees north and striking at Azimuth 072. The true width is 55ft (16.5

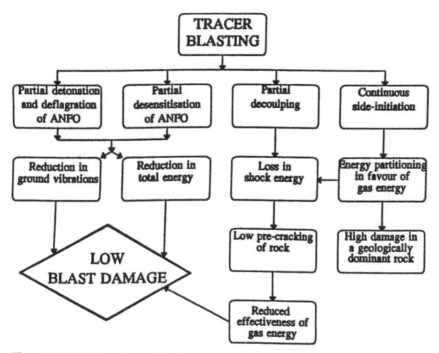

Figure 3. Mechanism of tracer blasting.

m) at the upper elevations, narrowing to 40 feet (12 m) at the lowest elevations. The hanging wall is unaltered, massive to weakly bedded greywacke of variable width, from nil to 20 feet (6 m). Otherwise, the hanging wall is chloritic basalt. The footwall usually contains a slate band of one to two feet (0.3-0.6 m) in width, immediately south of the conglomerate. South of the slate is interbedded, massive greywacke and slate bands.

2.1 Method of working

In the test area of the mine, Longitudinal sublevel caving method is used. Two parallel drifts 18ft (5.4 m) wide and 12ft (3.6 m) high were driven along the strike of the ore. The two sublevel drifts also called North and South drives were 15ft (4.5 m) apart. A slot raise was driven on the East end of the stope to retreat the sublevel cave towards the West end. Ring drilling from both North and South drives created transverse slices of ore which were blasted together during retreat. The vertical spacing between the sublevels was 85ft (25.5 m) with a maximum borehole length of 115ft (34.5 m). The diameter of the holes varied from 4.25 to 4.5 inches (108 to 114 mm) in different stopes, burden between the rings was 8ft (2.4 m) and the toe spacing was 10 to 11ft (3 to 3.3 m).

2.2 Blasting practice

During the last three years, different explosive products have been used at the test site. These include ANFO, cartridged emulsion and bulk emulsion. Tracer blasting, which is the focus of this paper has been predominantly used in the last year.

2.2.1 Loading of Traced Holes: The critical components of loading successfully in 4.5" (114 mm) uphole are the diameter of the hose and the loading hose centralizer. The loading is done by a low-stat 1.25" (32 mm) hose, which delivers a larger volume of ANFO at high velocity for better compaction of the prills. The centralizer is attached to the end of the loading hose to maintain a 90 degree angle between the velocity flow vector and the loading face. This minimizes the rebound or blow-back of the ANFO out of the hole. When the charging tank, loading hose and centralizer are in place, the loading process begins.

A small hole is made in the bottom of a locking plug through which the Booster cord (detonating cord) is threaded. The booster cord has a low strength signal cord of 14 grains per foot with a series of equally spaced "bumps" of PETN at 10ft (3 m) intervals. These bumps are 1.6ft (0.48 m) in

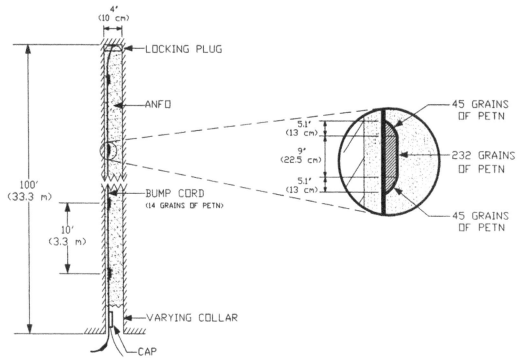

Figure 4. ANFO traced by the booster cord in a test hole.

length, the centre 9 inches (0.225 m) of each bump contains 232 grains of PETN and the remaining ends contain an additional 90 grains of PETN (Figure 4).

A knot is tied in the cord, large enough, to prevent it from pulling out of the locking plug. The locking plug is then placed over the end of the centralizer and inserted into the borehole. The hose is pushed up until it touches the toe of the hole. The hose is pulled back leaving the locking plug and the detonating cord at the toe of the hole. The loading of the hole with ANFO begins by keeping the hose at a stand-off distance of approximately 1.5ft (0.45 m). The hole is loaded leaving a predetermined collar length.

2.2.2 Tracing ANFO with booster cord: The booster cord is different from a typical detonating cord of uniform core load used in development headings. As stated earlier, the booster cord has a low strength signal cord of 14 grains per ft. with bumps of heavier core load at 10 ft (3 m) intervals. During pipe tests, it was found that the 14 grains per ft cord could not initiate ANFO. The next step was to examine the role played by the bumps in the booster cord. It was found during pipe tests that bumps do initiate the ANFO column. The bumps in the booster cord may cause (Figure 5):

(i) Some desensitization and incomplete reaction in the ANFO column depending upon the gap between the cord and the wall of the blasthole.
(ii) Radially propagating detonation of the ANFO column

The direction of radial detonation of ANFO is at an angle varying from 0 to 180 degrees to the direction of propagation of the detonation front in the booster cord. It appears that initial deflagration or burning of ANFO takes place which changes to radial detonation depending upon the available energy and number of nucleation points (Cook,1974; Singh,1993). Therefore the effect of the bump on the ANFO column may range from partial desensitization and burning to radial detonation of ANFO. The probability of dynamic desensitization decreases when the diameter of the hole exceeds 5 inches (127 mm) due to lower compacting effect. The signal cord may also cause partial desensitization and some decoupling effects for the later firing ANFO column.

3.0 RESULTS AND OBSERVATIONS

Tracing of ANFO with booster cord was introduced in stoping operations of the mine.

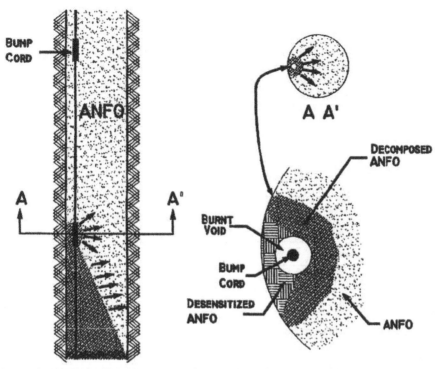

Figure 5. Radial initiation and desensitization of ANFO column by the bumps in a booster cord.

Though, it is different from typical tracer blasting technique commonly used in development or breasting headings, yet it was found to be effective both technically and economically. In order to examine its effectiveness, the results and observations during tracer blasting were compared with cartridged emulsion, bulk emulsion and standard ANFO. The comparison was made in terms of blast vibrations, fragmentation and explosive cost.

3.1 Ground vibrations

Ground vibrations were monitored for rings of holes blasted by cartridged emulsion, ANFO and tracer blasting. All the holes were blasted using short delays. The peak particle velocity for each ring was recorded which was normally generated by the earliest firing hole. The results have been provided in table 1. Six rings with cartridged emulsion, three with ANFO and eleven with tracer blasting were monitored. The average peak particle velocity was maximum for cartridged emulsion rings and minimum for tracer blasting. Table 1 shows a significant reduction in the ground vibrations for tracer blasting as compared to standard ANFO and cartridged emulsions. Firstly, the quantity of ANFO in a traced

hole is less than a normal ANFO hole. Secondly, some desensitisation and incomplete reaction in the ANFO column occurs during tracing of ANFO with booster cord. As the vibrations are directly influenced by the square root of the explosive quantity, the detonation of the reduced quantity produced less vibrations.

Ground vibrations is an indicator of the strain induced in the surrounding rock by the detonation of an explosive charge. Reduced ground vibrations suggested less fracturing of the surrounding rock and low dilution of ore.

3.2 Cost effectiveness of tracer blasting

Ring drilling from both the North and South drives divided the ore body into transverse slices which were blasted together during retreat. Drilling and blasting information for a typical ring blast has been given in figure 6.

Typical Length Of The Explosive Column Per Blast:
North Drive:
No. of holes	= 8
Total hole length	= 668ft
	(200.4 m)
Collar length	= 125ft
	(37.5 m)

Table 1. Ground vibrations for different explosives.

TYPE OF EXPLOSIVE	DISTANCE BETWEEN THE BLAST AND MONITORING POINT IN METERS	PEAK PARTICLE VELOCITY IN mm/sec
Cartridged Emulsion	376.3	1.36
"	377.9	1.87
"	383.7	1.70
"	381.6	1.81
"	375.9	1.18
"	379.8	1.75
Average Value	**380.8**	1.62
ANFO	407.1	1.49
"	401.3	1.60
"	395.4	1.38
Average Value	**401.3**	1.50
Traced ANFO	385.7	1.00
"	414.9	0.56
"	387.6	1.29
"	413.0	1.04
"	389.6	1.25
"	391.5	0.41
"	409.1	1.11
"	393.5	0.95
"	405.2	1.12
"	403.2	0.70
"	399.3	1.10
Average Value	**399.3**	0.96

Length of the explosive column   = 543ft
(162.9 m)

South Drive:
No. of holes   = 6
Total hole length   = 492ft
(147.6 m)
Collar length   = 105ft
(31.5 m)

Length of the explosive column   = 387ft
(116.1 m)

Total length of the explosive column
per blast   = 930ft
(279 m)

Explosive Cost Per Blast:
Cartridged Emulsion@$11/ft ($36.7/m)= $10,230

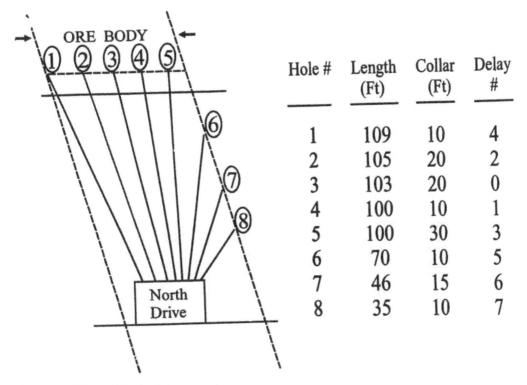

Hole #	Length (Ft)	Collar (Ft)	Delay #
1	109	10	4
2	105	20	2
3	103	20	0
4	100	10	1
5	100	30	3
6	70	10	5
7	46	15	6
8	35	10	7

Figure 6. Drilling and blasting information for a typical ring blast.

Bulk Emulsion@$6.65/ft ($22.2/m) = $6184.50
Tracer blasting@$1.73/ft ($5.77/m) = $1608.90

By using Emulsion explosives, the drill pattern can be expanded resulting in reduced drilling requirement. If Cost of drilling per ft is $7.10 ($ 23.7/m), to breakeven with tracer blasting cost, there should be drilling reduction of 35.8% in the case of bulk emulsion and 51.2% for cartridged emulsion considering only drilling and blasting costs. In order to make a more realistic comparison the cost of dilution, ventilation and secondary blasting etc. should be considered.

3.3 Rock fragmentation

As displayed in figure 7, tracer blasting produced good fragmentation of ore. A qualitative assessment of fragmentation was made by visual inspection. Significant reduction in the secondary breakage cost provided quantitative data suggesting an improvement in the fragmentation of the ore during tracer blasting (Table 2).

Secondary breakage is done at different stages of ore handling.

(i) At the top of the ore pass, big chunks are broken by Mechanical Rock Breaker.

(ii) Whenever needed, secondary blasting is done to clear the jammed ore passes.

(iii) Big chunks encountered by the scoop-trams at the draw points are moved to a convenient place and blasted.

The significant reduction in the secondary breakage cost at different levels was due to the combined effects of the introduction of tracer blasting and change in the mining method from slot and slash to sublevel caving.

The efficiency of an initiator is characterised by its detonation pressure and dimensions (Neil and Torrence,1989). When an ANFO column is radially initiated by the bumps in the booster cord, the run-up distance is much smaller than the end initiated

431

Table 2. Secondary blasting cost.

Type/Location of secondary breakage	Average annual cost with tracer blasting	Average annual cost prior to tracer blasting
Secondary blasting in draw points	$ 8,287	$ 26,196
Secondary blasting in ore passes	$ 1,138	$ 1,443
Rock breaker operations	$ 62,290	$ 172,155

Figure 7. Rock fragmentation in a tracer blasting stope.

charge. Due to the inadequate priming by the bump cord, the radially initiated ANFO charge may require considerable run-up to reach its steady state VOD. But the run-up distance is limited by the diameter of the blasthole. As a result of that the VOD of ANFO is well below the steady state velocity. This alters the partitioning of energy between shock and gas components. Gas energy increases at the expense of the shock energy for columns which detonate below their steady state value. In rocks with close spacing between bedding planes, partings and joints , and preconditioning by shock energy, the increase of gas energy results in good fragmentation, considerable movement and loose muck piles.

4.0 CONCLUSIONS

Tracing of ANFO with a booster cord was used during sublevel caving operations in a Gold mine. Though the practice was different from a typical tracer blasting technique used in development headings yet it was found to be effective both technically and economically. Ground vibrations during the blasting of rings were of lower magnitude for tracer blasting than in the case of normal ANFO and cartridged emulsion. This suggested that tracer blasting produces less overbreak and dilution. The cost of primary blasting by tracer blasting was lower than for other commonly used explosives. Visual inspection of the muck pile indicated good fragmentation, which was also supported by the reduction in secondary breakage cost at different muck handling stages. The method is particularly suitable for narrow ore bodies where the use of stronger and expensive explosive does not permit the expansion of the drill pattern.

ACKNOWLEDGEMENTS

Thanks to the management and bulk blasting crew of Royal Oak Mines Inc. for the support during this study and permission to publish the findings. Thanks to NSERC for the funding during fundamental research on tracer blasting.

REFERENCES

Bhushan V.; Konya C. and Lukovic, S.(1986)"Effects of detonating cord downline on explosive energy release", Proc. 2nd Mini-Symp. on Explosives and Blasting Research, Montville, Ohio, pp 41-55.

Cook M.A.(1974)"The science of industrial explosives" Graphic Service, Salt Lake City, Utah, 449p.

Neil I. and Torrence A.(1989)"The influence of primer size on explosive performance", Proc. Explosives in Mining Workshop, AusIMM, Melbourne, PP 329-341.

Singh S.P.(1993)" Investigation of blast damage mechanisms in underground mines (Phase 2)", Report to URIF, University Research Incentive Fund, Toronto, 44p.

Singh S.P. (1996)"Mechanism of tracer blasting", Geotechnical and Geological Engineering, 14, pp 41-50.

Posters

Rock Fragmentation by Blasting, Mohanty (ed.) © 1996 Taylor & Francis. ISBN 90 5410 824 X

On the fracture process in blasting: A numerical study by a discrete element method

Frederic Donze, Sophie-Adelaide Magnier & Jean-Claude Mareschal
GEOTOP, Université du Québec à Montréal, Qué., Canada

Jacques Bouchez
Laboratoire de Détection et Géophysique, Commissariat à l'Energie Atomique, France

A numerical model based on the discrete element method is used to investigate the importance of stress waves in the initiation and propagation of radial fractures during the dynamic loading phase of an explosion. An explosion occuring in a two-dimensional rock plate is simulated and the resulting fracturing process is detailed. First, during the pressure rise of the explosion source, a crushed zone is created in the vicinity of the explosion cavity. Then, because of the tensile tail of the propagating pressure wave, radial fractures propagate from the edge of the crushed zone with different average velocities. Higher crack velocities are observed for longer radial fractures. Tests with different explosive sources have shown that both the size of the crushed zone and the length of the radial fractures depend on their peak pressure and frequency content. Efficient sources, which generate long radial fractures with a small crushed zone, can be obtained with a low peak pressure provided that the frequency content is lowered. Finally, when the plate is subjected to uniaxial compression, the fractures are aligned along the main stress axis. In the light of these results, the method proposed here is greatly appropriate to study complex problems involving the creation and evolution of discontinuities.

Rock Fragmentation by Blasting, Mohanty (ed.)© 1996 Taylor & Francis. ISBN 90 5410 824 X

An investigation into the effects of some aspects of jointing and single decoupled blast holes on pre-splitting and boulder blasting

Syed M.Tariq
University of Engineering & Technology, Lahore, Pakistan

Paul N.Worsey
University of Missouri-Rolla, Mo., USA

ABSTRACT: Pre-splitting is a widely recognized perimeter control technique that is applied in many civil construction projects, open pit mining, and quarrying. The success of this technique depends largely on the nature of the geology, especially the presence and nature of discontinuities (joints). Many facets of jointing have been investigated in the past, such as the joint inclination, filling materials, frequency etc. This paper includes the results of the effects of varying joint separation, nature of joint surfaces on pre-splitting, and also the results obtained by blasting single decoupled blast holes and its application in pre-splitting and boulder blasting.

Specially prepared concrete blocks with 9.525 mm (3/8 inch) blast holes, loaded with 15-grain detonating cord were used in model testing. The research comprised of a) the modeling of both open and closed joints, b) a series of single hole testing conducted in approximately 30.48 x 22.86 x 15.24 cm (12 x 9 x 6 inch) concrete blocks. In single hole testing, spacers were used on one end of the block to create a free face, in addition tests were performed using unrestrained blocks to ascertain the resulting fracture pattern in boulder blasting mode.

It was found that precision ground joint surfaces did not reduce maximum blast hole spacing. The ground surfaces provide a relatively perfect match thereby acting as a continuous medium, and no significant loss in the transmission of energy takes place. In contrast, a simple machine-cut joint has undulations that cause a separation of the joint. This requires the blast holes to be brought closer to develop a successful pre-split plane. It was also found that a joint with a spacer thickness of .3048 mm (0.012 inch) acts like a free surface and almost the entire explosive energy is reflected back.

In single decoupled blast hole testing, it was observed that the cratering angle is greatest when the blast hole is near to the free face. During these cratering tests, it was noted that when a blast hole is placed close to the free face an unexpected split appeared away from the free face, all the way back to the abutted end, far in excess of what can be expected in pre-splitting. An extensive investigation was undertaken to find its possible application in pre-splitting and boulder blasting. This was also verified in tests without constraints, which illustrates its potential for application in boulder blasting.

Rock Fragmentation by Blasting, Mohanty (ed.) © 1996 Taylor & Francis. ISBN 90 5410 824 X

Predicting the envelope of damage resulting from the detonation of a confined charge

T. Michael LeBlanc
Inco Limited, Copper Cliff, Ont., Canada

Jason M. Ryan
Austin Powder Company

John H. Heilig
BLM Blastronics Canada Limited

ABSTRACT: Drill trajectory deviation is a recurring problem in vertical retreat stoping operations. As a result of this deviation, 60 kilogram (165 millimetre diameter) and 103 kilogram (203 millimetre diameter) explosive charges are routinely detonated beyond ore/waste contact zones. Consequently, there follows a reduction in finished stope wall stability and an increase in ore dilution. The added costs of dilution and unnecessary blasting are high. In an effort to reduce these costs, this project was initiated to explore the possibility of modeling the damaging effects of explosive charges. The final computer program is able to predict the dimensions of damaged rock zones resulting from the detonation of confined charges of various diameters and column lengths. The computer program calibrated with field collected data, uses a near field vibration equation to calculate predicted vibration levels generated from confined explosive charges. Predicted vibrations are then related to rock damage criterion to determine the dimensions of envelopes of damage surrounding the charge.

Experimental data indicated that the zone of freshly induced fracture could be delineated by using the 1,200 mm/s vibration contour, in the granodiorite rock where the experiments were conducted. Using this level, some 11,000 tonnes of rock were included in this region. The tensile failure criterion indicated this level with good accuracy, and the CANMET criterion for incipient damage levels generated the best results for the prediction of the onset of damage to the rock mass (320 mm/s). When envelope dimensions are determined for various charge diameters of the same explosive product, the relationship between charge configuration and induced damage levels can be presented graphically. Blast engineers will be able to utilize simple graphs relating to specific explosives and rock types for more efficient blast planning procedures. it is anticipated that knowledge of rock-failure envelope dimensions will allow instructions to be issued to blast loading personnel. Explosive charges may then be loaded to detonate at safe distances away from ore/waste contacts. This in turn reduce over-blasting and consequent over excavation of Vertical Crater Retreat (VCR) stopes which currently results in high percentages of ore dilution.

By using the computer program to predict the dimensions of damaged regions generated by the spherical charges used for VCR mining, an exponential rise in the dimensions of the damage ellipses is demonstrated. This may lead to more informed methods of blast hole layouts in future crater retreat stoping operations.

Rock Fragmentation by Blasting, Mohanty (ed.) © 1996 Taylor & Francis. ISBN 90 5410 824 X

The development of an inhibited explosive and safe and efficient blasting procedures to solve an explosive ground reactivity problem

Peter Bellairs
Dyno Wesfarmers Ltd, Singleton, Australia

Phillip Winen
Collinsville Coal, Australia

ABSTRACT: The   most   dangerous   of   all   mine   blasting   situations   arises   when   a   premature detonation/deflagration occurs as this event is triggered with no apparent warning. Death, serious injury, significant equipment loss or damage can all result from premature detonations. The most common cause for this phenomena is the explosive reacting with the rock via a strong exothermic reaction which results in a rapid rise in blasthole temperature, well in excess of 200°C, causing a detonation/deflagration. The hole does not need to be primed for a detonation to occur.

Explosive rock reactions are not isolated as is evidenced by the following Australian Mines which have experienced problems of this nature: Mt Whaleback, Mt Tom Price, Paraburdoo (Iron Ore); Mt Leyshon, Marvel Lock, Cosmo Howley (Gold Mines); Collinsville, Saraji (Coal Mines); and Mt Isa (base metal). Premature detonations/deflagrations usually result in lost production as the relevant mine personnel do not have the experience to safely and efficiently resolve these situations.

The strategy that has proved the most cost effective and efficient to solve these problems is the Team Approach with the team members being derived from the mine (Management - decision makers, blast crew - hands on, mine technical personnel - geologists, chemists, mining engineers and supervision) and the explosives supplier/manufacturer. Included but not active participants of the team are key personnel from the relevant regulatory authorities. They are fully informed of progress via reports distributed to them by the mine. The team approach gains ownership of the problem, commitment to solve the situation via an agreed common series of actions usually involving sampling, field and laboratory work. The conclusion of this work involves the formulation of an appropriate Safe and Efficient Procedure to blast the reactive rock types that is based on the results of the laboratory/field work, previous experience, accurate temperature logging and Australian Standards and Regulations. This procedure is then presented to the relevant regulatory authorities together with the results of the testwork before proceeding with blasting operations.

A case study is presented involving the Collinsville Coal Mine which has resulted in a decision to use an inhibited explosive to overcome an explosive ground reactivity problem that caused three blastholes to deflagerate towards the end of the first quarter of 1995.

Rock Fragmentation by Blasting, Mohanty (ed.) © 1996 Taylor & Francis. ISBN 90 5410 824 X

Shock wave propagation in a borehole

Kunihisa Katsuyama, Yuji Wada & Yuji Ogata
National Institute for Resources and Environment, Ibaraki, Japan

Fumihiko Sumiya, Kazuyoshi Kawami & Yoshikazu Hirosaki
NOF Corporation, Aichi, Japan

ABSTRACT: The shock wave which is called the precursor air shock wave(PAS) pre-compresses the explosives, and as a result of that failure of detonation occurs on occasion. This phenomenon is well known as the channel effect. The relation between the detonation wave and PAS was observed by high speed framing photography.

Some typical water in oil emulsion explosives were used. The explosives have the same matrix but different kinds of resin micro-balloons(RMBs) for sensitizer. By using different kinds of RMBs, a range of detonation velocities (300-4900 m/s) were achieved in the product. The higher the detonation velocity of explosives, the higher the PAS velocity became. When the decoupling coefficient was within the range of 1.2 ~ 2.7, the phenomenon of detonation failure was observed. The emulsion structure was destroyed after the detonation failure. The pre-compression speed of explosive was in the range of 12- ~ 170 m/s in these experiments.

Rock Fragmentation by Blasting, Mohanty (ed.) © 1996 Taylor & Francis. ISBN 90 5410 824 X

Blast designs for the excavation of an underground powerhouse cavern at Srisailam

A. Rajan Babu, R. Balachander & G. R. Adhikari
National Institute of Rock Mechanics, Karnataka, India

ABSTRACT: An excavation of a powerhouse cavern by drilling and blasting required designs for pilot headings, strip blasting, arch enlargement, slot blasting, horizontal and vertical benching. The details of the blast designs, field trials and analysis of the results are presented. Varying degrees of success of smooth blasting at different stages of the excavation are discussed. The influence of joints, delay timing and in-situ stress on blast results are also studied.

The work mainly dealt with an application of controlled blasting techniques to evolve suitable blast designs. A lot of care and proper supervision was required to control overbreak. The blast damage was minimised through the control of ground vibration and effective utilisation of the explosive energy. Vertical benching was found to be the most effective method for deepening of the cavern. The application of smooth blasting techniques effectively controlled overbreak and improved the stability of the excavation. Some contradictory results were obtained in the case of bottom priming. Bottom pilot heading complicated the excavation of the cavern and did not serve its purpose. A better understanding of the influence of joints, delay timing and in-situ stress was also obtained. Though good results were achieved, there was scope for further improvement provided suitable explosives for perimeter holes and proper delay detonators were available.

Rock Fragmentation by Blasting, Mohanty (ed.) © 1996 Taylor & Francis. ISBN 90 5410 824 X

Tectonic fabric and determination of drilling geometry for blasting

Branko Bozic & Stjepan Strelec
University of Zagreb, Geotechnical Faculty, Croatia

ABSTRACT: The paper deals with the relationship between the results of exploration in tectonic fabric of rock masses and results of geophysical exploration, and the blasting parameters which have optimal effects in open pits and quarries. A relationship between the burden, which is the distance between the charged borehole and the free rock surface, and the velocity of longitudinal waves in the rock mass is established, based on measured values for different limestone quarries. It is shown that the drilling geometry can also be determined from this relationship.

Relatively simple methods can produce sufficiently reliable parameters needed in quarry or open pit designs (optimal open face orientation, stabilities of working and end slopes) and in blasting designs aimed to achieve best results. These include, determination of explosive charge weights, fragment size distribution and forms of blasted rock material, seismic effects in surrounding areas. It is important to mention here that such simple explorations of rock mass undertaken before every blasting allow, even in highly anisotropic sediments, alterations of blasting conditions and their adaptation to local structural complexes. A relationship between the burden and the velocity of longitudinal waves in the rockmass is established based on measured values for different quarries.

Rock Fragmentation by Blasting, Mohanty (ed.) © 1996 Taylor & Francis. ISBN 90 5410 824 X

A superimposed TNT charge burst

V. N. Arkhipov, V. V. Valko & O. N. Ushakov
Central Institute of Physics and Technology, Defence Ministry of Russian Federation, Sergiev Posad City, Russia

ABSTRACT: Gas-dynamic processes occurring during an explosives charge burst on a ground massive surface are investigated with numerical methods. Geometric charge form and initial charge density are taken into consideration. The calculations are performed according to a numerical technique based on a completely conservative differential scheme in mixed Euler-Langrange variables. Physical processes in detonation products, soil and air are explored simultaneously. Formation and propagation laws of shock waves in ground and air, release processes of ground into aero-halfspace are analyzed. Gas-dynamic fields and pressure curves on a ground massive surface are obtained on the basis of the performed calculations.

Hodograph comparative analysis of explosives detonation scattering products, shock waves and pressure curves demonstrate good agreement of experimental data and numerical simulation results. However, the experimental and simulation results significantly differ from methodic dependence for "standard" hemispheric superimposed TNT charge. It shows that model problem formulation, simulation method and its program realization account satisfactorily for the field variables. The gas-dynamic parameter distribution required for mathematical simulation of succeeding stages of burst development (a burst cloud and seismoexplosive waves) are obtained as a result of numerical estimations of burst energy redistribution between air and ground as well as formation and propagation processes of shock waves in these media.

Rock Fragmentation by Blasting, Mohanty (ed.)© 1996 Taylor & Francis. ISBN 90 5410 824 X

Blasting techniques for reducing ground vibration at civilian structures near an open pit mine

Shantang Gao
China Society of Engineering Blasting, Beijing, People's Republic of China

The paper describes results from the in-situ blasting tests using varied blasting techniques for reducing blast-induced ground vibration under different blasting conditions.

From July 1992 to May 1994, the blasting tests were conducted in marble with a bench of 12 m height. The vertical blastholes, 250 mm in diameter and 13.5 m in length were drilled by YZ-35 type rotary drill, with 6 m spacing, 5 m burden, and 7 m toe burden. A slurry explosive of $1.09/cm^3$ density and detonation velocity of 4500-5000 m/s and nonelectric millisecond delay detonators of period #1 to #20 were used. The powder factor was 0.4-0.6 kg/m^3, with a 210-350 kg charge weight per hole. The explosive column in the hole was divided into two decks. The charges between the decks and between the adjacent holes were initiated by millisecond detonators initially at a delay period of 20-300 ms. The half period of blasting vibration wave, (i.e. 50 ms) was selected after observation of blasting vibration effects. The maximum charge per delay was 180-300 kg. The millisecond blasting was conducted by nonelectric detonators (#5 to #17) with the same delay interval of 50 ms. Presplitting technique was used near the open-pit slope. The initiation time of the presplit hole blasting before primary hole blasting was 100 ms. The direction of movement was changed from the original direction back from the residential structures to the direction parallel with it to minimize high vibration levels on the civil structures.

Also, from May 1992 to July 1993, the blasting tests were conducted in limestone with a bench height of 10 m. Seventyfive inclined holes 150 mm in diameter and 11.7 m in length, were drilled by YQ-150 type in-the-hole drill, with 5 m spacing, 4.5 m burden and 5-6 m toe burden. Comparative tests of single hole blasting and millisecond blasting between adjacent and initiation tests by zones were conducted with ANFO and millisecond nonelectric detonators.

The results from these tests are as follows: in marble at a distance 60-155 m from blasting center the peak vibration velocity was 1.10-0. 03 cm/s. At the civil structures 80-110 m from the blast the velocity was 0.62-0. 21 cm/s, which is much less than the safe vibration velocity of 2-3 cm/s for general brick structures, as specified by China National Standards Safety Code of Blasting. Blasting vibration velocity in front of the bast was 50% lower than the vibration velocity behind the blast. The vibration period for single hole per delay blasting in limestone was 64 ms. The delay interval was 25-50 ms. Proper selection of delay interval between the subzones can increase the basting vibration and significantly increase the vibration frequency. The vibration frequencies with initiation in two, four, six subzones were 22-88%, 85-168%, 128-208% higher than the frequency with single hole per delay blasting.

Rock Fragmentation by Blasting, Mohanty (ed.)© 1996 Taylor & Francis. ISBN 90 5410 824 X

An explosive pressure generator and its applications

Stanislav A. Novikov, Vladimir N. Khvorostin, Alexei V. Rodionov & Nikolai V. Brukanov
VNIIF, Sarov, Nizhny, Novgorod, Russia

ABSTRACT: Explosive Pressure Generators (EPG) are reusable powerful pulsed energy sources. In accordance with design of a tool being moved by EPG, it is able to make many various powerful operations. EPG utilizes energy of high explosive (HE) detonation in a chamber.

EPG contains an explosive chamber and a so-called "shock wave reducer", which saves the tool by lowering pressure peak in the shock wave front. When working as part of an installation EPG is connected with a working chamber, where high pressure occurs during the explosive process. The explosive techniques using EPG have the following advantages:

i) HE consumption is about two orders of magnitude lower
ii) there is no necessity to stop work because the explosion is confined in a closed volume
iii) environmentally safer
iv) it is possible to use EPG inside buildings or mines without having to interrupt work at adjacent sites.

Rock Fragmentation by Blasting, Mohanty (ed.) © 1996 Taylor & Francis. ISBN 90 5410 824 X

Explosion technology for extraction of gemstones

V.A. Borovickov & J.G. Dambayev
St.-Petersburg State Mining Institute, Technical University, Russia

ABSTRACT; Blasting leads to tectogenic defects in gem crystals, leading to significant loss in their value. Although the usual means of rock splitting by explosives destroys a significant part of these crystals, little research has been done to improve this situation. Experimental research on energy distribution in blasting has been carried out. Energy estimation in camouflet blasting has been obtained by the use of advanced impulse x-ray and electro-optical techniques.

Rock Fragmentation by Blasting, Mohanty (ed.) © 1996 Taylor & Francis. ISBN 90 5410 824 X

Fragmentation analyzing scale – A new tool for rock breakage assessment

P. Pal Roy & B. B. Dhar
Central Mining Research Institute, Dhanbad, India

ABSTRACT: From the angle between the strikes of major joints and bench face, a fragmentation analyzing scale has been developed and is extensively used in opencast mines to quantify the type of breakage to be produced in any bench blasting using conventional blasting practice.

This scale requires thorough geological mapping of exposed benches and their cluster analysis for the determination of dominant joints in each face. This is a useful tool in case of a newly developed mine where the face orientation may be chosen suitably so that the corresponding strike direction always makes angles between 25° and 65° with the dominant joints. If the mine is already developed then this scale would enable to predict the type of fragmentation to be produced in any bench blasting when combined with conventional blasting practice, and also to determine the parameters which may be changed or modified in order to get proper breakage.

Rock Fragmentation by Blasting, Mohanty (ed.) © 1996 Taylor & Francis. ISBN 90 5410 824 X

Realizing the potential of accurate and realistic fracture modelling in blasting

T. Kleine & P. La Pointe
Golder Associate Inc., Redmond, Wash., USA

B. Forsyth
Golder Associate Ltd, Burnaby, B.C., Canada

ABSTRACT: The mechanical and geometrical properties of rock and fractures control many aspects of mining. Rock, unlike artificial materials, is heterogeneous and anisotropic, which leads to complex mechanical responses. The unavoidable simplification that we are forced to make when describing rock, and in particular, the fractures in rock, is one of mining's and blasting's enduring problems. However, it is now possible to model rock discontinuities much more accurately and realistically through discrete fracture models. Discrete fracture models consist of deterministic and stochastic components. The deterministic component comprises those faults and joints that have been explicitly mapped or detected from boreholes, geophysics or underground mapping. However, the model may also include a portion of the rock mass that has not been directly sampled. This portion of the rock is represented by "stochastic fractures", which are fractures that have the same statistical distributions and geological correlations as the deterministic fractures. The new generation of sophisticated three-dimensional discrete fracture models can reproduce the true geological complexities of fractures, including their spatial distribution, size, orientation, roughness, aperture, and fluid flow or mechanical properties. This information alone has valuable application in support and damage calculations, and provides valuable input to a variety of analyses. This paper describes two potential applications of realistic fracture modelling: roof and wall support and blast damage control.

Rock Fragmentation by Blasting, Mohanty (ed.)© 1996 Taylor & Francis. ISBN 90 5410 824 X

Problem of crystal damage by blasting

A. L. Isakov & V. N. Beloborodov
Siberian Branch of the Russian Academy of Sciences, Institute of Mining, Novosibirsk, Russia

ABSTRACT: A model of behaviour of hard inclusions (crystals) located in dense medium during shock-wave action is presented. The concept being developed is as follows.

When an explosion-induced stress wave affects a crystal it begins to oscillate at its eignefrequencies. The crystal undergoes cyclic deformation as it oscillates leading ultimately to failure. The frequency spectrum of a real explosive pulse in rock is sufficiently broad to excite oscillations even in small crystals at very high frequencies of ultrasound range. The results of experimental and theoretical research are presented, with the following conclusions:

i)   under pulsed loads, crystals undergo oscillatory motion at resonance frequencies with clearly expressed sinusoidal cycles with a background component of the same form as the loading pulse;

ii)   the amplitude of the oscillations is determined not by the absolute amplitude of the explosive pulse but by the value of its spectral amplitudes at near-resonance frequencies;

iii)   lower damping of the pulse due to the surrounding medium results in higher damages to the crystal.

Under this concept the strength criterion of crystals is suggested. The breaking oscillatory strains for some crystals such as diamond and beryl are established. The possibility of estimation of crystal damage in rocks on the basis of explosive pulse spectrum is shown.

Rock Fragmentation by Blasting, Mohanty (ed.)© 1996 Taylor & Francis. ISBN 90 5410 824 X

Blast damage study in blasthole open stope mining

Qian Liu
CANMET Experimental Mine, Val d'Or, Que., Canada

Glen Ludwig
Mines Research, INCO Manitoba Division, Thompson, Man., Canada

ABSTRACT: Blast damage, as well as stress induced rock damage, were studied with instrumentations installed in the hanging-wall from an experimental drift. Blast induced damage, as measured by overbreaks and fracture creation in the hanging-wall was correlated with blast design and vibration levels. Consideration of rock mass quality and stress redistribution was emphasized in controlling blast induced damage.

It is concluded that evident blast damage was produced to the stope boundaries including the hangingwall during the mining of Block 19A. This damage produced operational problems such as mucking delays and secondary breaking. The rockmass damage measured on the hangingwall side of Block 21 was more stress induced in nature. It is recommended to take steps to reduce both blast induced and stress induced rock damages.

Rock Fragmentation by Blasting, Mohanty (ed.) © 1996 Taylor & Francis. ISBN 90 5410 824 X

The work performed by explosive

H. Honma
Hokkaido University, Japan

ABSTRACT: Cautious blasting requires a smaller and more accurate amount of explosive than usual. In such blasting, the gas expansion (the work performed by explosive) may well be regarded as the major source of rock breakage. Recently, the bubble energy test has been applied for indicating the explosive strength. However, there are some traditional methods to indicate the work performed by explosive, such as RWS by ballistic mortar and explosive force (f) by calculation. Simple and accurate characteristics are necessary for calculating the amount of explosive required for blasting. In addition to trial and error, theoretical understanding is also necessary for the design of the blasts. These traditional methods are re-examined in this paper. These experimental results show that the relationship between RWS obtained by ballistic mortar test and explosive force (f) by calculation can be written by the following equation.

RWS= { $(f_n/100 + a) / (f_s/100 + a)$ }. $\{(10/d_n+b)/(10/d_s+b)\}^{r-1}$

where, RWS : relative weight strength by ballistic mortar.

f: force per No. 6 cap 1 by calculation.

a: force per No.6 cap 1 by calculation.

b: additional initial volume by cap penetration in specimen.

d: density of specimen.

r: specific heat ratio of detonated gas.

Other results obtained from the experiments are as follows.

(1)   The specific heat ratio (γ) in the gas state is 4/3, which is obtained by a ballistic mortar test and as a matter of cause differs from the polytrop exponent (r) = 3, in solid explosives in detonation wave theory.

(2)   The explosive force (f) is useful and reliable for calculating the amount of explosive for blasting.

(3)   The slurry explosive is an exception because of the water ingredient.

Rock Fragmentation by Blasting, Mohanty (ed.)© 1996 Taylor & Francis. ISBN 90 5410 824 X

A method and optimization model for formulation design of multi-component mixed explosives

T.Jianjun & B.Wenjuan
The Industrial Explosive Research Department, Hunan, People's Republic of China

ABSTRACT: This paper presents a new method of formulation design for multi-component mixed explosives. Based on the theory of thermochemistry in detonation process, the mathematic model of formulation design is established by the optimization mathematics. Using the heat of detonation as target functions and taking the constraint conditions such as oxygen balance and costs etc into consideration, this model can provide an optimal solution quickly by means of personal computer. By use of this method, the calculation and the simulation tests for a series of constituent sets and the various constraint condition can be conducted easily. In this way, we determined the optimal formulation for the explosive to be developed, which can make the explosive have the maximum heat of detonation and lower costs.

The equation of explosive reaction for a specific type of multi-component mixed explosive is developed first. The mathematical model for formulation design is then derived according to the following steps:

1) The coefficients of all terms in the equation of explosive reaction are expressed by the explosive formulation $X = (x_1, x_2, \ldots, x_n)$.

2) Accoding to Hess law, the computational expression of the heat of detonation Q_v for this explosive is derived from the previous expression, i.e. $Q_v = f(x_1, x_2, \ldots, x_n)$.

3) The given values of constraint conditions are presented and expressed by equality or inequality.

Rock Fragmentation by Blasting, Mohanty (ed.) © 1996 Taylor & Francis. ISBN 90 5410 824 X

Explosive energy concept for drill pattern expansions field application: Productivity and cost benefits

S. R. Kate
IBP Co. Limited, Nagpur, India

ABSTRACT: The effectiveness of the explosives in a blast depends on the energy content and the velocity of detonation. Careful analysis of blast results involving a large number of blasting operations in hard and competent rocks leads to the following conclusions:

i)   'energy factor' seems to have better correlation with blasting performance than 'powder factor'.

ii)   'energy factor' for satisfactory fragmentation and blast results remains almost constant; the percentage increase in drilling pattern is almost the same as the percentage increase in borehole energy

iii)   better results are obtained with a staggered pattern, with lower burden and greater spacing.

iv)   a higher energy factor, for constant burden and spacing, usually results in higher machine (mucking) efficiency.

Rock Fragmentation by Blasting, Mohanty (ed.) © 1996 Taylor & Francis. ISBN 90 5410 824 X

Progress with computer aided blasting using (in-hole) programmable electronic delay detonators

C. Deacon
AECI Explosives Ltd, Modderfontein, South Africa

ABSTRACT: This paper presents the results of a series of production blasts on a number of South African quarries, designed to investigate the effects of a precision initiation system on fragmentation, vibration, airblast and muckpile shape.

In the high-tech world of blast analysis and research, it has become necessary to introduce a system that can overcome the inaccuracy and scatter of conventional systems. Computer Aided Blasting is the adoption of electronics and computer systems to achieve specific criteria and thus offer improved control over blasting, and so enhancing the effectiveness of operations.

The CAB system developed by Expert Explosives Pty Ltd., comprises a computer design software, the Blast Commander, Connection Testers that facilitate hook up procedures, safety and performance, the Blast Programmer that controls the blast, and the Ex-Ex 1000 electronic detonator. This system provides complete flexibility in timing control from 0 to 14999 ms, and the ability to program and change the delays of each detonator right up until the moment of initiation.

It has been proved through measurements on site that the precision of the CAB system has led to improved channelling of vibration frequencies and reductions in their amplitudes, important to today's environmentally aware quarries. Consistent fragmentation has lead to enhanced diggability and crusher throughputs, and the complete programmability and flexibility of the system results in the ability to control muckpile shapes.

Rock Fragmentation by Blasting, Mohanty (ed.) © 1996 Taylor & Francis. ISBN 90 5410 824 X

The DYNO Electric Detonator System – A versatile delay system

Johan Svärd & Magnus Wikström
Nitro Nobel AB, Gytorp, Sweden

ABSTRACT: Extensive tests within the last 4-5 years of field blasting with The DYNO Electric Detonator System, further described in this document, have shown that blasting results are improved when electronic detonators are used, in surface blasting as well as underground. The delay time can be chosen with an accuracy of 1 millisecond in the whole time range (1-6250 ms).

A summary of field blastings have shown that amongst others:

· the fragmentation is improved
· the ground vibrations are reduced
· the final contour is improved
· it is possible to blast larger rounds

These benefits can be achieved, in the same round, with the new electronic detonators with improved total economy. The use of single-hole blasts in obtaining a 'seismic signaature' serves as the key input information for optimizing delay pattern.

Rock Fragmentation by Blasting, Mohanty (ed.) © 1996 Taylor & Francis. ISBN 90 5410 824 X

Author index

Printed and bound by CPI Group (UK) Ltd, Croydon, CR0 4YY

23/10/2024

01777679-0017